International Conference on

SHALLOW OIL AND GAS RESOURCES

(1st: 1984: Norman, Okla.)

PROCEEDINGS OF THE FIRST INTERNATIONAL CONFERENCE

The United Nations Institute for Training and Research

Richard F. Meyer, Editor

UNIVERSITY OF TULSA-McFARLIN LIBRARY

Shallow Oil and Gas Resources

Proceedings of the First International Conference

Published by Gulf Publishing Company and UNITAR and its Rome Centre on Small Energy Resources.

Copyright © 1986 by UNITAR. All rights reserved. Printed in the United States of America. This book, or parts thereof, may not be reproduced in any form without permission of the publisher.

Library of Congress Cataloging-in-Publication Data
Main entry under title:
Shallow oil and gas resources.
Includes index.
1. Petroleum—Congresses. 2. Gas, Natural—Congresses. I. Meyer, Richard F. II. United Nations Institute for Training and Research.
TN863.S52   1986   553.2'82   85-17592

ISBN 0-87201-833-4

McFarlin Library
WITHDRAWN

SHALLOW OIL AND GAS RESOURCES

PROCEEDINGS OF THE FIRST INTERNATIONAL CONFERENCE

An international conference organized by the
United Nations Institute for Training and Research
and the
University of Oklahoma Energy Center

and co-sponsored by the

United Nations Department of Technical
Co-operation for Development

United Nations Development Program

Latin American Energy Organization

United States Department of Energy

United States Agency for International Development

Independent Petroleum Association of America

Peat Marwick International

25 July–3 August 1984
Norman, Oklahoma

Gulf Publishing Company
Book Division
Houston, London, Paris, Tokyo

Contents

Organizing Committee for The Conference xii

Acknowledgments xiii

Foreword, *Michel Doo Kingué* xv

Introduction, *Richard F. Meyer* xvii

Overview: The Outlook for Shallow Oil and Gas
Joseph Barnea xix

Section I. Exploration for Shallow Oil and Gas, 1

1. The History of World Shallow Oil and Gas
   Ruth Sheldon Knowles 3

2. Petroleum Source-Rock and Temperature-History Studies in Sparsely Tested Exploration Areas
   William E. Harrison 11

3. Time, Temperature, and Tectonics: Their Assessment as a Means of Evaluating Shallow Oil and Gas Potential in Frontier Areas
   John D. Pigott 35

4. Oil and Gas Seepages with Reference to Developing Countries
   Martin Forrer 45

5. Low-Cost Landform Analysis in Prospecting for
   Shallow Hydrocarbon Occurrences
   Richard L. Bowen 59

6. Satellite Remote Sensing in Exploration for Shallow
   Petroleum Resources
   John R. Everett 67

7. Satellite Remote Sensing—An Effective Tool for Oil
   and Gas Exploration
   John S. Carter 77

8. Use of Satellite Imagery in Discovery and Development
   of Shallow Oil and Gas Reserves
   J. H. Bartley and James R. Lucas 83

9. Seismic Reservoir Analysis
   John A. Ward 95

10. Petroleum Data: An Aid for Exploration Planning
    Jerlene A. Bright and Mary L. Fleming 99

11. The Microtechniques Method of Petroleum Exploration
    *Tony M. Preslar, Douglas Ball, and
    William P. Nash* 113

12. Successful Exploration on a Small Budget
    John H. Gray 125

Section II. Resource Occurrences, 131

13. Survey of World Shallow Oil and Gas
    Richard F. Meyer and Mary L. Fleming 133

14. Summary of Exploration History and Hydrocarbon
    Potential of Morocco
    A. El Morabet 173

15. Shallow Oil Fields of the Denver Basin, Colorado and
    Nebraska, U.S.A.
    J. F. deChadenedes 181

16. Petroleum Exploration Strategy for Zambia
    Chisengu Mdala 195

17. Note Concerning the Research and Development of Shallow Petroleum Fields in Senegal
    Babacar Faye 203

Section III. Shallow Oil Production and Technology, 205

18. Aspects and Specifics of Shallow Oil Production: Field Development Techniques, Equipment, and Economics of Production
    C. G. Rosaire, Jr., and Kenneth J. Schmitt 207

19. Producing Shallow Heavy Crudes Having High Water Content
    Charles D. Haynes 213

20. How to Predict Stability of Naturally Flowing Oil Wells
    Edward F. Blick 227

21. Enhanced Oil Recovery and Stimulation by High-Energy Gas Fracturing
    Dale A. Eastman 235

22. Progressive-Cavity Pump Applications
    J. C. Collum 243

23. The Economic Advantages of Microdrilling
    Josef Macsik, Tomas Dahl, and Lars Oldsberg 245

24. Subsidence and Oil Production from Shallow Reservoirs
    Erle C. Donaldson and Saeed N. Mogharabi 253

25. Control of Hazards in Oil Drilling and Well-Servicing Operations
    R. J. Murphy 265

26. Low-Cost Methods for Shallow Oil Exploration and Production
    Curtis L. Talbot 279

27. Exploration and Development of Shallow Oil Fields in Thailand
    Prakal Oudomugsorn 287

Section IV. Shallow Natural Gas Production and Technology, 295

28. Accurate Fluid Property Prediction in Natural Gas Resource Development and Custody Transfer
    Jeffrey L. Savidge, K. Hemanth Kumar, and Kenneth E. Starling 297

29. A Techno-Economic Model for Shallow Gas Reservoirs under Partial Water Drive
    Ashis K. Das and Djebbar Tiab 315

30. Natural Gas-Fired Co-generation Topping Plant
    Thomas J. George 337

31. Natural Gas—The Politically and Environmentally Benign Least-Cost Energy for Successful 21st Century Economics, the Energy Path to a Better World
    Robert A. Hefner III 345

32. Completion of Shallow "Barefoot" Wells in Appalachia
    Thomas J. George 367

33. The Natural Gas Resources Development of Kathmandu Valley
    G. S. Thapa 373

34. Prospects of Shallow Gas Resources in Bangladesh
    M. A. Ghafur 389

35. The Possibility of Exploring and Producing Shallow Oil and Gas Reserves for Rural Communities and Small-Scale Industries in Indonesia
    M. A. Warga Dalem and July Usman 405

Section V. Economics and Financing, 423

36. Drilling Statistics and Costs: An International Comparison by Target Depth
    Emil D. Attanasi 425

37. Attracting Capital for Oil and Gas Exploration in a Competitive Environment
    David Berberian and William J. Hibbitt 441

38. Problem of Financing of Investment in Exploration and Production of Petroleum in Third World Countries
    Gerald Jolin 455

39. Developing Country/Independent Oil Company: Considerations for Financing Exploration and Development of Shallow Oil and Gas Deposits
    Dorothy Mercer 459

40. The Independent Oil and Gas Company—An Essential for the Exploitation of Small, Shallow, or Marginal Hydrocarbon Deposits in Developing Countries
    G. C. L. Jones, J. H. Burney, and G. W. L. Cull 477

41. The Role of a Regional Bank in Financing Oil and Gas Exploration in Developing Countries
    J. Terry Aimone 485

42. An Emerging Alternative—A Trading Company Providing Energy Development Capability
    Douglas R. Courville 491

43. Risk: The Wildcatter's Partner
    Charles Robbins 499

44. OPIC Programs for the Energy Industry
    R. Douglas Greco 503

45. A New Strategy for Improving Exploration Incentives in Developing Countries
    Arthur E. Owen 509

Section VI. Legal and Institutional Considerations, 521

46. Shallow Oil and Gas Development: Proposals for its Encouragement
    Maxwell Bruce 523

47. A Producer's Perspective on Selected Issues in the Establishment of a Legal Regime to Encourage Marginal Petroleum and Natural Gas Development
    Sheila S. Hollis 533

48. A Natural Gas Energy System for Rural Communities in Developing Countries
    Christopher F. Blazek 543

49. Utilization of Small Shallow Heavy Crude and Tar Sands Resources
    Joseph Barnea and Ramon Omana 561

50. Gas Development Strategy and the Commercialization Process in International Exploration and Production Contracts
    Mohsen Shirazi 567

51. Recovery: The Saskatchewan Way
    Paul Schoenhals 583

52. The Management and Development of Oil and Gas Mineral Resources on the Osage Indian Reservation
    George E. Revard and Newell Barker 593

53. The Development of Small Fields: The Case of Guatemala
    Douglas Rosales Juarez 603

Section VII. Transfer of Technology, 619

54. The Use of Simulators in Operational Skills Training for Developing Countries
    L. Russell Records 621

55. Geological and Geophysical Technology Transfer
    Between U.S. Universities and Developing Countries
    John D. Pigott and John S. Wickham 639

56. Fossil Energy Assistance Activities of the United
    States Agency for International Development
    Charles Bliss and Pamela Baldwin 645

57. The United Nations Assistance Programme in
    Petroleum
    *United Nations Department of Technical
    Co-operation for Development* 651

Index 679

Organizing Committee for the Conference

Joseph Barnea, Senior Special Fellow, UNITAR
Michael Bloome, Research Officer, UNITAR
Charles Robbins, Consultant to UNITAR

Jay T. Edwards, Executive Director,
University of Oklahoma Energy Center

JoAnn McLin Carlson, Public Affairs Director,
University of Oklahoma Energy Center

Julia B. Chastain, Secretary,
University of Oklahoma Energy Center

Acknowledgments

The organizing committee wishes
to acknowledge with gratitude the support for the Conference
of the following organizations and individuals:

American Exchange Bank
Badger Oil Company
Belden and Blake Corporation
Berea Oil and Gas Corporation
DLB Energy Corporation
The Energy Resources Institute
Halliburton Services-International Operations
Kerr-McGee Corporation
Mr. James C. Leake, Muskogee, Oklahoma
Norman Chamber of Commerce
Oklahoma Center for Continuing Education
Oklahoma Geological Survey
Schlumberger Well Services
TOTCO
The Trail's Golf Club
Western Reserves Oil Company
Xerox Corporation

Foreword

As I said in the speech I made at the opening of the Conference, the proceedings of which are printed in this book, the Conference on Shallow Oil and Gas is the eighth in a series of international meetings which UNITAR has been organizing on energy issues since the mid-1970s. UNITAR entered this field of energy analysis through international conferences not only because energy development is vital to the world as a whole, to both developed and developing countries, but also because in the 1970s the view was widespread, and first expressed by the "Club of Rome," that the world was running out of energy and that we will have to reduce our energy consumption and, for developed countries, their development growth.

UNITAR did not agree with this view, and in its various international conferences, always organized with and for energy experts, assembled empirical data in order to change the outlook which was then prevailing.

From its first conference on "The Future Supply of Nature-Made Petroleum and Gas" held in July 1976, came the strong message that price increases exceeding cost increases expand the resources-base of energy resources, a tendency strongly supported by improvements in technology. This historical-analytical approach to energy resources has in time become the widely accepted view.

UNITAR has also organized conferences on specific energy resources, which because of improvements in prices and technology, have become attractive. Thus, UNITAR held in 1979 in Edmonton, Canada, the First International Conference on "The Future of Heavy Crude and Tar Sands," followed in 1982 by the Second International Conference on Heavy Crude and Tar Sands, held in Caracas, Venezuela. The Third International Conference on the same resources was held in California in 1985.

In 1982 UNITAR organized its First International Conference on Small Energy Resources, in Los Angeles, California. During that conference, it became obvious that small oil and gas resources have a large world-wide potential and that they were practically unknown outside North America. UNITAR therefore decided to organize the 1984 Conference on Shallow Oil and Gas Resources because shallow oil and gas are neglected resources in oil and gas exploration world-wide.

Companies, whether national or private, often do not measure the first thousand feet when they explore, as they assume that shallow resources might be insignificant. However, though they are usually small, shallow resources can, in certain cases, be quite significant, as evidenced by the well-known shallow Hugoton gas field, located in the United States of America, which has produced almost 28 trillion cubic feet of gas. It is one of the 10 largest gas fields ever found. Not only can shallow oil and gas resources be significant, they are also relatively easy to find, have low costs for exploration, and therefore are attractive as sources of oil and gas for local consumption in rural areas and small towns and industries.

The big success of the American independents, the pioneers in shallow oil and gas exploration and production, is a model which needs to be followed, especially in developing countries.

We therefore feel that developing countries should pay greater attention to ways and means of producing small energy resources for which investments are relatively cheap. These resources can help the countries concerned improve significantly the living conditions in their rural areas, thus increasing the conditions of comfort, the lack of which is one of the main causes of rural exodus. This concern of ours explains why UNITAR has established in Rome, the capital city of Italy, an International Centre on Small Energy Resources with the financial support of the Italian government.

We hope that the reports submitted to the 1984 UNITAR Conference on Shallow Oil and Gas, and the outcome of the discussions that took place during that conference, will contribute significantly to the development of energy resources all over the world.

Michel Doo Kingué
United Nations Under-Secretary-General
and Executive Director of UNITAR

Introduction

Richard F. Meyer

The First International Conference on Shallow Oil and Gas Resources was convened in Norman, Oklahoma from July 25-August 3, 1984 under the direction of the United Nations Institute for Training and Research and the University of Oklahoma Energy Center. This volume includes most of the reports presented during the course of the conference.

The premise of the meeting was that deposits of oil and gas not presently being exploited exist at shallow depths in many parts of the world and that these mostly small deposits can best be found and developed in a competitive environment by small, independent oil and gas operators. To accomplish this requires first that the small, shallow resources actually be present and second, that the financial incentives for development be provided by the host countries. The papers in the present volume examine various aspects of the premises and the requirements.

Following an overview prepared by Joseph Barnea as an introduction to the conference, the book is divided into 7 sections comprising a total of 57 chapters. The sections follow sequentially from exploration (Section I) to resource assessment (Section II), oil production (Section III), and gas production (Section IV). Section V then deals with the critical issue of shallow oil and gas economics and financing. It is critical because lenders are not prone to indulge in high-risk ventures and operators are equally prone to avoid unstable working environments. Nearly or quite equally critical is the subject of legal and institutional considerations dealt with in Section VI. This aspect of small and shallow resource development need not be intractable if certain basic conditions are met. Most importantly, the resource and its potential utilization must match; a small gas deposit, for example, may serve a village very well but be inappropriate as a source of supply for a distant city. This in turn makes the necessary incentives for the operator a sensitive subject. Lastly, Section VIII is concerned with a subject of vast importance to emerging industrial countries, the transfer of technology. Ultimately, for valid societal and economic reasons, every nation should and must have sufficient highly skilled manpower to either perform needed technical work or be able to fully evaluate work done under contract to others.

Regardless of cyclical aberrations, the world demand for oil and gas inevitably will rise over time, to the eventual point of resource exhaustion. In the process, some industrialized, oil- and gas-producing countries, faced with failing production capacity, will add to the intensity of demand. It therefore behooves each nation to take every possible step to enhance its hydrocarbon-production capabilities and if this is found wanting, to seek least-cost alternatives. It is for these reasons that conferences such as this one are of inestimable value.

Overview: The Outlook for Shallow Oil and Gas

Joseph Barnea

Seventy-five percent of the surface of this globe is covered by sedimentary rocks. But sedimentary rocks form only 5% of the earth's outer crust, which is about 10 miles thick. This relationship indicates that the bulk of all sedimentary areas are shallow and that thick sedimentary areas are comparatively rare. If all the 5% of sedimentary rocks would be evenly distributed over this globe the average depth of the sedimentary area would be about 3,000 feet. These few figures demonstrate two important effects: namely, (1) that many small countries and many areas in bigger countries will possess only shallow sedimentary rocks and (2) that we will miss many resources of oil and gas if we neglect the shallow sedimentary areas.

The Advantages of Shallow Oil and Gas

The most important advantages of shallow oil and gas resources are:

1. Low exploration costs as compared to deeper sedimentary basins.
2. Very much lower drilling costs as compared to deeper drilling, because drilling costs increase in a geometric progression with increase in depths.
3. Very much higher exploration success ratio for shallow resources, reaching, in certain areas of the U.S. with a fairly well known geology, 80%–90%, which means 8–9 exploration wells out of 10 are

successful.* It has to be added that shallow, low-cost oil or gas wells drilled by independents are regarded as successful even if the production amounts only to a few barrels of oil per day or a corresponding quantity of gas. The big companies with their very high overheads would probably regard such wells as uneconomic. More recently, many big firms pass on the small production discovered from shallow areas to independents, retaining only a royalty.

4. Shallow drilling often takes only a few days; it is possible to explore, drill, and complete a well within one year. As a result, shallow exploration and development is a faster way to oil and gas development. In favorable conditions this may allow return of the invested capital within 2–3 years, whereas deep sedimentary development or off-shore oil and gas exploration and development may take 5–10 years until production begins.

The Obstacles to Shallow Oil and Gas

In Europe and in developing countries, one obstacle is government ownership of all underground resources and the resulting legal structure, which requires long and detailed negotiations with the central government and which smaller producers find difficult. Furthermore, the present legislation in most countries provides for leases involving very large blocks, often extending over millions of acres with heavy front-end payment. In the developing countries, there is practically no distinction between shallow and deep areas. Thus a lease covers everything—shallow and deep—but the big companies are interested only in the discovery of large fields, to provide oil for export. In such a system the development of shallow, and often small resources, for rural and other local needs is obviously neglected. Furthermore, with the governments in those countries primarily interested in making money from big fields there is no interest in many countries in supporting shallow production, which may not be able to provide the government the large returns expected if large-

---

* In his paper "The Future Potential of the Appalachian Basin, With Special Reference to the Eastern Thrust Belt," Porter J. Brown, Chief, Geological Services, Columbia Gas Transmission Corporation, writes, "The number of wells drilled for oil and gas in the basin has steadily increased from 3,000 in 1971 to more than 10,000 in 1981. This represents more than 15 percent of all wells drilled in the United States. The reasons for the increase in drilling are the 89 percent success ratio and favorable price incentives." For further information please see "New Ideas New Methods New Developments" by Mathew Bender.

scale resources were discovered. This conference therefore should not only bring out the available technology for developing shallow oil and gas resources, and their attractiveness for local development, but it must also discuss the man-made, legal and institutional obstacles to the development of shallow resources which today exist in most countries of the world. The obstacles involve not only an inappropriate legal system but also a lack of local oil exploration companies, local service companies, local drilling companies, local pipelines, and so on, all parts of the infrastructure required for the development of shallow resources. I believe that in the United States, where independents drill more than half of all oil and gas wells, there is a model which should be followed in developing countries and in Europe. The obstacles which we have today in those countries are man-made and man can overcome them.

The Proper Management of Shallow and Small Resources

Small resources allow only a low overhead. Small companies are eminently fitted to manage shallow, and usually small resources, because their overhead is small, they are flexible, they can make decisions fast, and they can very often do their own drilling and make the completions, all the things that permit a profitable operation even with a small production. The big companies, with their many layers of committees for making a decision, cannot act fast. As a result their overhead is high; they become conservative. A geologist working for a big company with career prospects to consider, must be cautious in proposing exploration in an unknown area, because if the exploration should prove to be unsuccessful his career may be affected. As a result, such a geologist or committee of geologists will propose many different forms of seismic, geochemical, and other exploration methods, in order to minimize risks and distribute them over the whole organization. Surely the big organization is not the proper instrument for handling shallow oil and gas resources.

What is, then, the proper management of shallow oil and gas resources? Obviously the big companies or a big government cannot handle the small production from shallow wells. Decentralized development is, wherever possible, the most appropriate and efficient means to conduct shallow operations. Thus, a small gas development may allow a village or small town to have a gas supply without waiting for permits or government officials to provide their support and without waiting for budget approval by governments. If the local population, through a small company or cooperative, undertakes the small development itself, it will be faster and cheaper because decision-making will move from a centralized body to the local level. But it is not often realized how much cheaper decentralized development can be.

If a small gas field is found near a village and the gas is then brought in a cheap polyethylene pipeline to the village and if all the necessary investments are made to provide the village with the gas utilization equipment, then the total investment, as shown in the paper by Mr. Blazek, of the Institute of Gas Technology, Chicago, will amount to only $333,000. This estimate includes funds for a cable-tool drilling rig with all the accessory equipment, the transmission and distribution system for the gas, the gas-fired electric generator, other gas-utilizing equipment, and a water pump. The assumption underlying this estimate is that shallow gas is found about 40 kilometers or 25 miles from the village and that the gas will provide fuel for electricity generation for cooking and other purposes, especially water pumping, irrigation, and water supply for humans and cattle. The total investment for such a decentralized, small-scale system designed for a peak population of about 1,000 persons amounts, as mentioned, to only $333,000. Compare this with the cost of a high-tension electrical transmission system, which may cost as much for one mile. Investments in an electrical transmission line run between $150,000 and $500,000 per mile, depending on the voltage. Assuming that there is a power plant 25 miles from the village, the investment in the transmission line alone would amount to U.S. $3.65 million, or about 10 times the cost of the small gas system, and the electricity costs do not include the electricity distribution system in the village, the availability of a drilling rig, and the electricity utilizing equipment. This comparison demonstrates the great attractiveness of a local, decentralized system.

There is another paper before this conference, a paper on natural gas in the Kathmandu Valley in Nepal, a small shallow valley where gas was found. The paper contains not only a description of the geology and of the gas availability, but it also compares the cost of the gas supplied, with the cost of fuel wood which is now used and which results in deforestation. The paper demonstrates the attractiveness of such a gas supply system in spite of the fact that in this case the gas is in solution in underground water, technically known as hydropressured gas, which requires the gas to be separated from the water.

Decentralized development for shallow oil and gas is therefore a very important tool for the accelerated development of rural areas. From studies done by Resources for the Future and other research organizations we know that if electricity is provided to the villages life changes dramatically.

The Growing Recognition of the Advantages of Shallow and Small Resources

Over the last few years and in practically all areas of underground resources, there is a growing recognition of the advantages of developing shallow and small resources. Among the advantages is the fast pay-out of such resources as compared to the time requirements for the development of big resources. In 1984, with interest rates between 10%-15%, a big project requiring, say, 10 years to reach production, more than doubles its cost during that period whereas a small and shallow project can have its investment repaid in 2 or 3 years. We have noticed this situation in the development of heavy crude and tar sands in Canada, where even Exxon had now given up the development of a giant project in favor of smaller projects, which can be put into production in two years. This is now followed by Shell and most other companies. The same thing is noticed in the development of shale oil in the United States; in the mining industry, where the small gold and other mineral projects have increased sharply, and big projects are practically disappearing; in geothermal development; and with other commodities.

Another disadvantage of giant projects is the need to bring in many people to build roads, townships, schools, sewage systems, and other facilities. They are capital intensive and slow to bring to production. On the other hand, a small project does not require, in most cases, an addition to the existing infrastructure: it may not require new roads, new townships, etc. For developing countries this tendency is especially important because it can become a significant factor in speeding up development.

Many Different Views and Technologies

In this conference we have many different categories of experts and we have them from many countries. The papers are both academic, presenting new findings, and practical, describing experience and results. We therefore have a number of cases of differing views. Some of the papers have the view that for shallow oil and gas seismic is not necessary, other papers will argue that seismic is necessary. I believe that these differences of views are both interesting and important, because they allow experts to learn from other experts and to draw their own conclusions for the area where they will have to apply their skills. We will have some papers which describe new technologies, such as gas-pellet fracturing or drilling equipment. This demonstrates that the area of shallow oil and gas is not tied to a tradition-bound, old-fashioned technology, but is one in which the independents, the true pioneers of underground resources, continue with full vigor.

UNITAR hopes in the future to make a contribution to the monitoring of new developments through its new Centre on Small Energy Resources, which has just been established by UNITAR in Rome, Italy, with the support of the Italian government. This new Centre will have a membership open to private companies, universities, and everyone interested in small resources, and UNITAR hopes that in due course the Centre will make a contribution to the publicly-available information on small resources. The first resources to be dealt with in the Centre will be shallow oil and gas, which appear to us to be the most significant in the foreseeable future.

The interest in new technologies must not overlook the fact that exploration is still to some extent an art based on experience and that only those technologies which either reduce costs, reduce risks, or speed-up developments are really significant. Furthermore the reduction of risks and the speed-up of development must also be cost effective. Given these parameters, experience and tricks-of-the-trade possessed by many independents play a not-insignificant role. I know one independent who in principle buys only equipment which is discarded by the big companies, who always buy the latest equipment; thus his equipment cost is less than half that of the big companies. There are many other tricks, so that an experienced and successful independent is a good partner to have.

The Role of the Independents

Most of the independents in the United States are only involved in exploration and production. They usually sell their output to pipelines or to refineries. Thus most of the independents do not deal with the transportation, refining, and marketing of oil and gas.

If the independents should be invited to work in developing countries or in Europe, they will either have to find local partners to take over the transportation, refining, and marketing sectors, or local public bodies or the government will have to undertake to buy the production from the independents. Such activities outside the United States involve new types of commitments or new forms of cooperation, and some of these new issues will be discussed during the conference. Often the independents will be ready to take the risk of exploration in rural areas where the geology is not well known, provided they will be able to sell their production at an agreed price and provided that the financial and foreign exchange problems involved can be solved.

Financing of Shallow Oil and Gas

The independents are often called pioneers and promoters, because most of them have to raise the funds for their exploration. We will have papers which describe the fund-raising problems in the United States and the variety of systems which have been developed. However, more complicated and to some extent more important is the issue of finding financing for the development of shallow oil and gas in developing countries and how this can be done through independents. We will have a paper describing the type of insurance which the United States government can provide and similar insurance can be provided by other countries. But there are a number of other risks which cannot be covered completely by insurance. As a result, independents, when they intend to work in developing countries, will have also to exercise their double functions, not only as pioneers but also as promoters.

The Resources We Deal With

In this conference we are considering land-based shallow resources, which means resources which can be explored and developed at low cost; for this definition we have excluded off-shore resources, because the capital requirements are substantially higher. This includes shallow gas, the most widespread of the shallow resources, and in this category we have a number of different types of gas. We have associated gas found in pollution in conventional light oil and co-produced with the oil. We have natural gas not associated with light crude, which may be thermogenic or biogenic, produced by bacteria working on the organic material contained in the soil. We have gas which is in solution in undergound water, and we may have abiogenic and other types of gas. The variety of sources of gas we may find partially explains the wide availability of gas in sedimentary areas. Next we have conventional light crude, widely available in shallow areas but seldom in large fields, most of these having already been found.

Our shallow resources also include heavy crude, which is mostly found in shallow deposits. It is estimated that about 80% of the present production of heavy crude is derived from shallow reservoirs. Heavy crude can easily be found, and its exploration costs are low. However, production costs are much higher, and the technology is somewhat more complex. Nevertheless, heavy crude can be produced and upgraded to the level of a conventional crude at an estimated cost of between $6–$15 per barrel. Such crude would still be economic and attractive in rural areas and a paper on such resources is before the conference.

Our definition of shallow resources also includes tar sands, though the technology for developing small tar sands resources is still under development. But costs are going down, and wherever surface-mineable shallow resources of tar sands are found of sufficiently high grade, they may be produced at costs substantially lower than the present market price of conventional oil.

The Aims of the Conference

The purpose of this conference is first of all to draw the attention worldwide of the industry, of governments and of the public, to the attractive features and resource potential of shallow oil and gas. Secondly, we want to highlight the technical skill and initiative of the independents and hope that some of them will be able to work also in developing countries and in Europe on shallow resources. We also hope that in time independents will evolve locally, so that in less than one generation independents may be present in all countries with shallow oil and gas resources. Thirdly, we would like to bring out the attractive possibility that the shallow resources in developing countries can provide low-cost energy in rural areas, where today the lack of energy including the lack of electricity, prevents rural development and leads to the emigration from the rural areas to overcrowding in the cities. Finally we hope that the conference will draw attention to the legal and other obstacles which exist today in most countries to the development of shallow oil and gas. If the conference can fulfill these aims, it will make a significant contribution both to oil and gas supplies in the future, but more than that, it will contribute to a speed-up of economic development and to a revitalization of many rural areas.

Section I

Exploration for Shallow Oil and Gas

1

The History of World Shallow Oil and Gas

Ruth Sheldon Knowles

Abstract

The history of the search for oil and gas, and its development worldwide since the world's first commercial oil well in 1859 in the United States, has four distinct phases:

1. Drilling of shallow exploration wells in the areas of surface indications of oil and gas.
2. Application of evolving principles of surface and subsurface geology, indicating underground rock formations that might be favorable for the accumulation of oil and gas.
3. Use of geophysics to discover potential traps not indicated by surface and subsurface geology.
4. Search for oil and gas by deeper drilling in already productive areas.

From 1859 to the late 1930s the majority of the world's oil and gas discoveries were onshore and shallow, using a total depth of approximately 3,000 to 3,500 feet as the definition of "shallow." Many of these are still producing and remain among the world's greatest oil and gas fields. Although since the 1950s the exploration trend has been consistently toward deeper drilling and offshore exploration, shallow onshore drilling still remains relatively sparse worldwide. The world's total estimated area of possible oil-bearing sediments on the surface of the earth is about 15 million square miles and their volume is about 20 million cubic miles. Only a fraction of this has been explored, principally in the United States and whatever other countries have encouraged exploration by providing profit incentives, access to the land, and political stability. With these incentives, American independent operators historically have discovered three-fourths of America's new oil and gas. Exploration for shallow oil and gas today, with high energy prices, offers new horizons for private independent oilmen not only in America but also abroad, if other countries provide a sufficient profit incentive and political stability. Although there is still no method of direct oil and gas finding, and exploration is a high-risk operation, oil and gas will continue to be the principal source of world energy well into the 21st century. Oil and gas are more abundant in the earth than is generally thought, but they are found in abundance only if a great many people are looking for them at once in all sorts of unlikely places.

Man, from his earliest appearance on all the continents, marveled at everything he found in nature for his use. Among the most unusual substances were oil, asphalt, and gas, which he found in seeps, springs, lakes, and pits throughout the world. He called them by many names—slime, bitumen, pitch, asphaltum, tar—and from time immemorial found many uses for them—light, mortar, waterproofing, medicine, weapons, and religious rites.

In the Bible, following God's instructions on how to build an ark of gopher wood, Noah did "pitch it within and without with pitch." The Tower of Babel and the walls of Jericho and Babylon were made of bricks and "slime they had for mortar."

The "eternal fires" of Baku on the Caspian Sea—gas seeps ignited accidentally by lightning—inspired ancient religious fire-worshipping cults. Egyptian, Babylonian, Syrian, Assyrian, Persian, Greek, and Roman rulers fought for many centuries for control of the Dead Sea asphalt seeps.

In the Americas, thousands of oil, gas, and asphalt seeps and springs were used by the Indians for making medicine, caulking boats, waterproofing blankets, cementing beads, and creating decorative designs.

Mankind's progress has always been based on problems posed by his economic needs and his ingenuity at solving them. As civilization developed, he needed sources of energy to replace muscle and animal power in order to develop industries. It did not occur to him that oil and gas were a source for this. He relied on wind, water, and wood for fuel until the 17th century, when Europeans began to mine and use coal. As late as the 1860s, wood supplied three-fourths of the total fuel supply of the United States. However, the 19th century Industrial Revolution accelerated a greater demand for lubricants than vegetable oils and sperm oil from the ocean's dwindling number of whales could supply. Also, whale oil for illumination was scarce, and lamp oil from coal was costly.

The moment in time had arrived for the development and use of one of nature's greatest resources. Despite all the signposts nature had left for mankind for thousands of years throughout the world, no one had thought to explore beneath the seeps to see if oil could be found in quantities. The world's economic history and civilization's rapid progress would be changed dramatically by a small, shallow hole bored in the rocks in America's state of Pennsylvania in 1859.

A group of businessmen took a sample of oil from a Pennsylvania spring to Yale University for an analysis which confirmed it could be used as lamp oil. The group drilled a well to 69½ feet which found oil that could be pumped at the rate of 35 barrels a day. It sold for $20 a barrel, which sparked a rush to punch holes along Pennsylvania creek banks. Soon shallow wells were gushing as much as 4,000 barrels a day.

Europe was clamoring for cheap American lamp kerosene made from the new abundance of petroleum. John D. Rockefeller, an Ohio merchant, formed the Standard Oil Company and by 1880 owned 80 percent of America's refining capacity and 90 percent of its pipelines. His company emerged as the largest and richest manufacturing company in the world.

The Pennsylvania wells prompted others to drill near oil seeps. Between 1859 and 1900 big oil fields were found in 14 other states in America, most of them only a few hundred feet deep. Outside the United States, shallow oil was developed in Canada in 1860. Russia's first gusher was completed at a depth of 63 feet in 1866 near seeps in the Maykop region of the Black Sea. The shallow development of the first fields in the fabulous Baku "eternal fire" area on the Caspian Sea began in 1871. Shallow oil fields were found in Rumania in 1860 and in Poland in 1874. In South America shallow wells drilled near seeps started the oil industry in Peru in 1867 and in Venezuela in 1878. In the Far East, a successful five-barrels-a-day well drilled in Indonesia to 72 feet in 1885 near an oil spring launched an oil boom which was the cornerstone upon which one of the world's biggest international oil companies, Royal Dutch Shell, was built.

The oil well that changed the world and gave birth to the modern oil industry was Spindletop, an astounding southeast Texas gusher which on January 10, 1901, flowed 100,000 barrels a day from a depth of 1,020 feet. It was drilled on a low circular mound from which sulfurous gas seeped. Within three months, five other gushers were drilled, each equalling the first. Six wells in Texas were producing as much in one day as all the oil wells in the whole world.

Spindletop proved that there was an extraordinary abundance of oil in the earth. Inspired oil explorers were ready to go anywhere and everywhere to find more oil. Spindletop gave birth to some of America's greatest oil companies, broke Standard Oil Company's monopoly, and made the American independent oil and gas wildcatter one of the greatest forces in the development of world energy.

Spindletop marked the end of the Kerosene Age and the beginning of the Fuel Oil Age for factories, trains, and ships, and the launching of the Gasoline Age for automobiles and airplanes.

Not everyone wanted to find oil. A year after Spindletop's discovery, W. T. Waggoner, a rancher in North Texas, desperately needed water for his cattle. He drilled a series of wells as deep as 1,700 feet in search of it. He was infuriated when oil came up. "I wanted water," he said, "and they got me oil. I tell you I was mad, mad clean through. We needed water to drink ourselves and for our cattle to drink. I said, 'Damn the oil! I want water!' " When he discovered his problem was that he was sitting on top of a giant oil field he was slightly mollified, and named it Electra, after his daughter.

Oil men had been finding new pools for 54 years by following seeps, trends, hunches, divining rods, doodlebuggers, and spirit mediums. In 1913 the art of petroleum geology was born when Charles N. Gould, founder of the University of Oklahoma geology department and the Oklahoma Geological Survey, demonstrated that all Oklahoma's big oil pools lay under anticlines, places where rock strata arched, dipping or inclining in opposite directions from a ridge like the roof of a house. A rush to find anticlines began everywhere and more giant oil fields were discovered, most of them shallow.

Meanwhile, between 1900 and 1914 two more important new oil provinces were opened up. In Mexico small production at 575 feet was developed in 1901 on the Gulf Coast near one of many oil seeps. However, in 1910 the fabled "Golden Lane" was discovered with gushers of 100,000 or more barrels a day at depths of around 2,000 feet. The first indication of the enormous Middle East oil province came in 1908 when a big oil field was found at 1,180 feet in Persia (Iran), a land of many seeps. Iraq's first oil was found in 1919 at 2,200 feet.

The First World War, 1914–1918, was a war of fast ships, planes, and trucks. "The Allies floated to victory on a sea of oil," said Lord Curzon, British Secretary of State for Foreign Affairs. America supplied 90 percent of the great flow of oil. The country's oil resources were so drained that between 1920 and 1922 it became a net oil importer, drawing on the new fields American and British explorers had developed in Mexico. However, Congress quickly passed tax incentives and opened the public domain for leasing. Explorers redoubled their efforts. At the end of the 1920s oil production tripled. Forty-two giant oil fields (100 million barrels or more each) were found. In 10 years more giants were found than in the previous 20. Three-fourths of the discoveries were made by geologists.

The search for oil was becoming more scientific and the drilling, deeper. However, there were still big surprises. In 1923, 70,000 square miles of desert in West Texas was known as a petroleum graveyard, for every wildcat drilled was a dry hole. Two young Texans, with no experience or geology, drilled a well near Big Lake to 3,038 feet, and it flowed 100 barrels a day. They persuaded Mike Benedum, a famous Pittsburgh wildcatter, to drill eight wells to see if it was a field. Two had small, noncommercial oil shows and six were dry holes. Benedum decided to drill one more well. It roared in for 5,000 barrels a day, opening up the great Permian Basin. Within the next 30 years explorers would find almost 10 billion barrels of oil in its 70,000 square miles. There were more giant oil fields lying hidden under this wasteland than would be found in any other single area in the United States.

In 1927, a 70-year-old wildcatter, Dad Joiner, who had drilled mostly dry holes, leased some land in East Texas where most geologists scoffed at the idea of finding oil. By selling interests to farmers and townspeople Dad managed to raise enough money to drill. His equipment failed on two wells, but in May 1929, he started a third one. It was the onset of the Great Depression, and the well had become a community project and dream. When he announced he would drill the well in October 1930, 8,000 men, women and children came to watch. The miracle occurred. At 3,536 feet the well was a gusher and discovered the 6-billion-barrel East Texas oil field, America's biggest until Prudhoe Bay in Alaska 38 years later discovered 15 billion barrels.

The 1930s was another great oil-finding decade in America, primarily due to the success of geophysical prospecting pioneered and proven by the great geologist and father of petroleum geophysics, Everette Lee DeGolyer. Most of the shallow giant oil fields had been found by surface and subsurface geology. Petroleum seismograph revolutionized prospecting by locating unsuspected potential oil and gas structures at greater depths and to which no other clues existed. Also in the 1930s the great potential of Middle East reserves began to unfold further. Iran, of course, was being actively developed, as was Iraq. There were known oil seeps on the island of Bahrain, in the Arabian Gulf, and in Kuwait. In 1932 oil was discovered at 2,000 feet in Bahrain and in Kuwait in 1938 at 3,600 feet. These discoveries prompted exploration in Saudi Arabia, with no oil seeps, but oil was discovered in 1938 at 4,500 feet. The significance of the discovery would not be realized until after World War Two.

The Second World War was again "a matter of oil, bullets and beans," as Admiral Chester W. Nimitz put it. Almost 7 billion barrels of oil were consumed between December 1941, and August 1945 to meet the requirements of the United States and its Allies. Nearly 6 billion barrels of this enormous total came from the United States. Following the war America's major oil companies and 15,000 independent oilmen began a tremendous drive to replenish reserves and meet astonishing new consumption demands. Between 1945 and 1954 they doubled production. Also, technology was developed for offshore drilling and major companies began shifting to offshore exploration to look for big fields.

Natural gas was the glamor fuel after the war but it was placed under federal price controls in 1954 and exploration began to decline. However, in 1959, the centennial of the oil industry, America was the world's largest producing and exporting country.

In 1970 oil was placed under federal price controls in the United States and exploration was cut in half. The United States, with a potential estimated by the United States Geological Survey of having as much oil and gas yet to find onshore and offshore as had been found in 100 years, began a downhill slide. It began importing oil to the extent that the political Arab oil embargo in 1973 threw the country into a tailspin. By 1979 the United States was importing half its oil requirements. OPEC countries nationalized their oil resources, raised the price of world oil (in 1984 about $29 a barrel as contrasted to $4 a barrel for U.S. price-controlled oil in 1971). When the U.S. began to phase out natural gas price controls in 1978 and oil price controls were removed in 1981, the American oil industry exploded in record drilling of exploration wells. The steady decline in domestic production was halted and an upward trend began. New shallow oil and gas fields were developed all over the United States in response to increased prices and demand.

The American oil and gas industry still faces enormous problems in the development of its potential resources, for 50 percent of its best prospects for big new oil and gas fields lie under Federal lands onshore and offshore. Unfortunately, their exploration and development is under constant attack and delayed by environmentalist groups who have not yet realized the practicality and compatibility of multiple land use which, with new technology, can maintain the quality of the environment.

The American story is of vital importance to the less developed countries of the world. It illustrates on a grand scale and a practical one all the principles involved in encouraging and discouraging the development of natural resources.

Wallace Pratt, former chief geologist of Standard Oil Company of New Jersey (Exxon) was a philosopher-scientist who raised the profession of geology to an eminence and a dignity which it otherwise would not have attained and had a profound influence on the thinking of earth scientists. "It is the genius of a people that determines how much oil shall be reduced to possession," he said, "the presence of oil in the earth is not enough. Gold is where you find it, according to an old adage, but judging from the record of our experience, oil must be sought first of all in our own minds. Where oil really is then, in the final analysis, is in our own heads!"

Wallace Pratt believed that oil in the earth is far more abundant and far more widely distributed than is generally realized. But he was also firmly convinced that "the prime requisite to success in oil finding is freedom to explore and only slightly less imperative is freedom to develop and produce the oil once it is found." The freedom to explore produced what he called "the most precious heritage our experience as oil finders has bequeathed to us—our multitude of itinerant wildcatters. The entire American oil industry is but the lengthened shadow of the independent oilman whose form and substance are stamped indelibly on its whole structure." Historically, American independents have found and continue to find three-fourths of all America's new oil and gas.

A major reason why other countries have not had the intensive search for oil and gas as the United States has had is because of their lack of understanding of what Wallace Pratt said so well. The countries in the 1930s like Mexico, Argentina, and Bolivia, which expropriated foreign companies' holdings, delayed the development of their resources for many years. Nationalization of oil resources worldwide has only been ameliorated by the introduction of production-sharing contracts innovated by Indonesia.

In recent times, whenever a country has unilaterally tried to take too much of the profit incentive away from foreign contractors, major and independent oil companies have packed up their rigs and gone home. Basically, in any country of the world, price controls and lack of profit incentive mean decreased energy exploration and development. Furthermore, developed and less-developed countries alike must keep in mind that the consensus of all government and industry surveys is that oil and gas will continue to supply the majority of the world's energy needs well into the 21st century.

The 1980s are seeing increased exploration deep-drilling onshore and offshore for both oil and gas, but worldwide the potential of shallow oil and gas discoveries is still great. According to Lewis Weeks, the world's greatest authority on sedimentary basins having potential for oil and gas deposits, possible oil-bearing sediments on the land surface of the earth total about 15 million square miles in area and 20 million cubic miles in volume. Only a small fraction of this potential has yet been explored.

Oil and gas explorers, major and independent, are eager to explore wherever conditions are favorable—potential resources, profitability, and political stability. There is still no way to find oil and gas directly and the odds remain that out of every nine exploration wells, eight of them will be dry holes. It has been estimated that more money has been lost in looking for oil and gas than has been made by finding them. But the great gamblers, large and small, are still with us.

Energy is the key to creating wealth not just for the few but for the many. As the American experience has demonstrated, oil and gas are found in abundance only if a great many people are looking for them at once in all sorts of unlikely places.

2
Petroleum Source-Rock and Temperature-History Studies in Sparsely Tested Exploration Areas

William E. Harrison

Introduction

In order for commercial deposits of petroleum to exist in a given area, a relatively small number of geologic requirements must be satisfied. Modern basin-evaluation programs seek to answer the same geologic questions that confronted explorationists 20 years ago. Although the tools that are currently available for exploration programs are more sophisticated than ever before, studies related to potential traps, reservoirs, seals, source rocks, and temperature history remain the critical components of reconnaissance-level investigations.

Early in the exploration history of a basin, emphasis usually is concentrated on evaluating geologic structures. If these efforts are successful, the established production constitutes *a priori* evidence that the requirements for commercial petroleum deposits have been satisfied. Should preliminary exploration programs fail to establish production, large multinational oil companies often will redirect their efforts in the hope of discovering giant accumulations elsewhere. Under these conditions, questions pertaining to the lack of production are often left to later exploration efforts.

The reasons for such unsuccessful programs can be varied. In some prospect areas, the lack of suitable facies and/or diagenetic processes has precluded lithologies of adequate reservoir quality. If specific geologic structures originated relatively late in the history of a prospect area, then the resulting traps can be charged only with the hydrocarbons formed late in the generation sequence. Under these circumstances, liquid petroleum will not be available to fill traps, and natural gas will be the dominant type of hydrocarbon mixture. Temperature history also plays an important role in the type of hydrocarbons one may anticipate in a given region. Most liquid petroleum exists at reservoir temperatures of 300°F or less. Natural gas is more likely to occur at higher reservoir temperatures. Under some geologic conditions, structures that appear to be favorable for oil or gas accumulation do not have adequate seals. Petroleum introduced into such structures migrates up through multipay zones and is eventually lost to the structures, rendering them noncommercial.

The type of organic matter contained in potential source rocks also controls the type of petroleum that may occur in a specific region. Some source rocks yield oil at one level of thermal maturity and gas at a higher level, whereas others are capable of yielding only gas regardless of maturity level.

It is in prospect areas where relatively few data are available to the exploration geologist that petroleum-source-rock and thermal-maturation studies make the greatest contributions. Preliminary exploration efforts in frontier areas are usually dictated by economic considerations. Because of difficulties in transporting natural gas in regions that do not have adequate gathering and distribution systems and ready markets, gas-prone prospect areas are less likely to attract as much attention as oil-prone prospect areas. The results of source-rock and maturity studies provide the explorationist with information on whether a specific prospect area is likely to yield oil, gas, or a mixture of the two. Equally as important, such studies also can be used to avoid comprehensive (and usually expensive) programs in areas where there is a low probability of discovering significant reserves.

Petroleum Source Rocks

Almost all earth scientists subscribe to the organic origin of petroleum. The inorganic or abiogenic theories were summarized in a well-written and thought-provoking paper by Porfir'ev (1974) and were treated recently by Gold and Soter (1980). The inorganic reactions that yield hydrocarbons are well known and testify to the fact that all hydrocarbons need not have their origin in sedimentary organic matter. It remains to be demonstrated that commercial quantities of petroleum have resulted from such reactions, however.

Assuming that most of the world's petroleum originates in fine-grained sedimentary rocks, the next questions that confront explorationists concern parameters that can be used in evaluating specific basins or prospect areas. Three basic questions must be addressed when evaluating source-rock characteristics and temperature history:

1. How much organic carbon do the potential source rocks contain?
2. Is the organic matter of the oil- or gas-generating type?
3. What is the maturity level (temperature history) of the source rocks?

These parameters are discussed thoroughly in Tissot and Welte (1978) and Hunt (1979) and require only brief comment here.

Petroleum geochemists agree that some minimum level of organic carbon is required in order for fine-grained sedimentary rocks to serve as petroleum source rocks. The generally accepted limits are 0.5 and 0.3 percent organic carbon, respectively, for shale and carbonate source rocks. These are essentially minimum criteria, and many well-documented cases show known source rocks to contain more than 2.0 percent organic carbon. Considerably more is known about organic-rich shales as sources of petroleum than is known about other types of sedimentary rocks. Some recent studies (Malek-Aslani, 1979; McKirdy and Kantsler, 1980; Tissot, 1981), however, focus on the source-rock potential of carbonate, evaporite, and lacustrine deposits.

Although organic carbon is an essential component of petroleum source rocks, the nature of the carbonaceous material is also important. Sedimentary organic matter derived from higher plant material, such as lignin and cellulose, has a relatively high oxygen content and a low atomic H/C ratio. Such material is unlikely to be a source of liquid petroleum but may be a source of gas. By comparison, organic matter derived from marine or fresh-water algae has chemical characteristics that favor the generation of oil. This is largely because of the similarities between a class of biochemicals called lipids (which constitute a major part of algal biomass) and the hydrocarbons that exist in petroleum mixtures. Thus, sedimentary organic matter derived from the woody components of higher plants typically will yield gas, and material that owes its origin to algal lipids may be oil sources.

The generation of petroleum from sedimentary organic matter requires that the latter be reactive under the influence of temperature. An oil-prone kerogen at a low maturity level may contain a few hundred parts per million soluble or extractable organic matter (SOM or EOM), much of which will be hydrocarbons. At progressively higher levels of maturity the extractable hydrocarbon content increases to some maximum level and then becomes relatively low. Figure 1 shows this relation for kerogen in the Uinta Basin. The interval from 3,800 to 5,800 m corresponds to the zone in which maximum oil generation occurs. At depths below 5,800 m the extractable hydrocarbon values are quite low, and oil is thermally cracked to gas. Pyrolysis techniques also can be used to assess source-rock potential (Espitalié et al., 1977). A practical example involving the relation between pyrolysis data, visual kerogen descriptions, and petroleum potential is found in Harrison and LaPorte (1977).

The use of extractable hydrocarbon content to rank or evaluate petroleum source rocks was proposed by Philippi (1957). Baker (1962), in his study of mid-continent organic-rich shales, normalized extractable hydrocarbon yields to organic-carbon content (Figure 2). A 5,000 ppm (parts per million) extractable hydrocarbon yield might be normal for a sample containing 20–30 percent organic carbon, but it would have completely different implications if the rock contains only 0.2 percent carbon. Figure 3 is a variation of this relation, which is used to evaluate petroleum-source-rock potential.

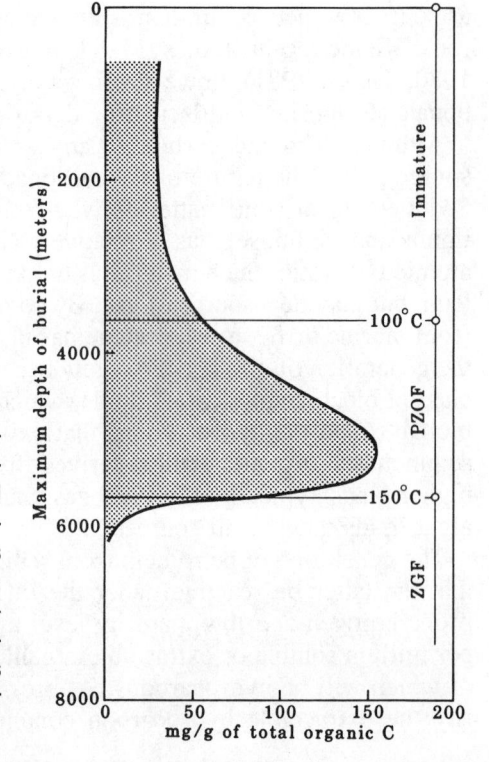

Figure 1. Relation between maturity level (depth) and C_{15+} extractable hydrocarbon content for Type I (oil-prone) kerogen in the Uinta Basin. Maximum hydrocarbon values occur at about 5,100–5,500 m, which corresponds to about 135°–145°C. At greater depth (higher maturity levels), liquid hydrocarbons are thermally cracked to gas. (After Tissot and Welte, 1978.)

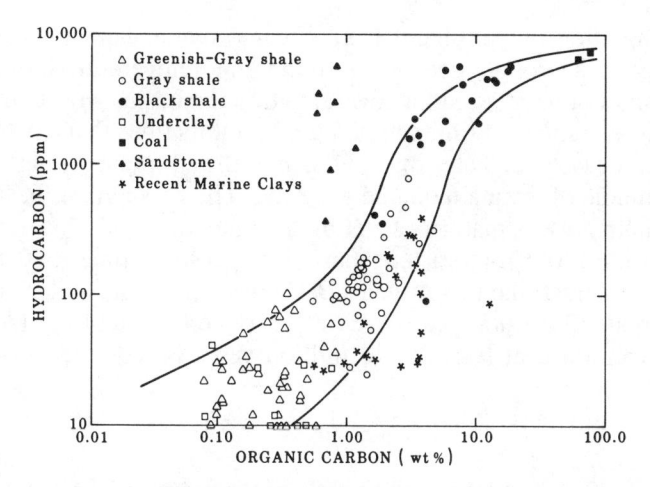

Figure 2. Organic-carbon and extractable-hydrocarbon content for various Cherokee Group (Kansas and Oklahoma) and modern marine clay samples. Compare with Figure 3. (After Baker, 1972.)

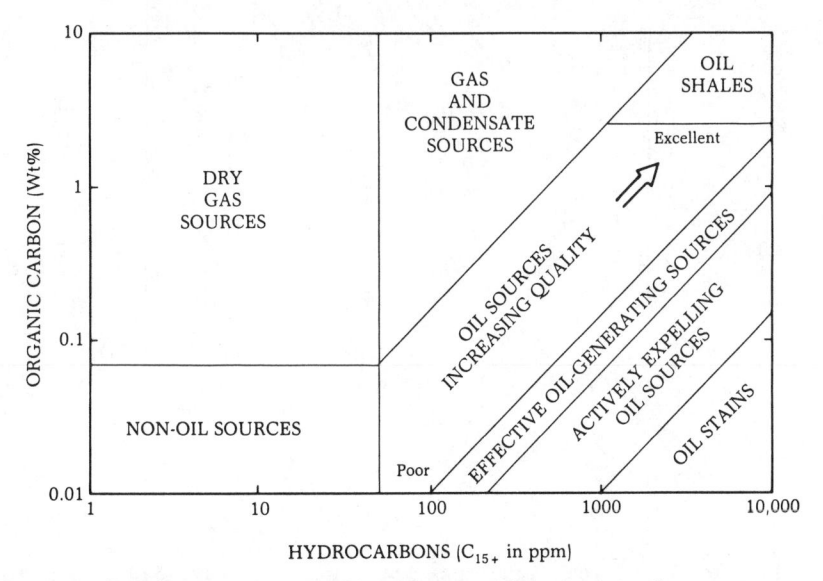

Figure 3. Relation between organic-carbon and C_{15+} hydrocarbon content and variation in petroleum-source-rock quality. High-quality petroleum source rocks have higher carbon and hydrocarbon concentrations than do poor source rocks. Samples with high hydrocarbon levels and low carbon values usually indicate nonindigenous (migrated) petroleum. (After Geochem Laboratories Source Rock Evaluation Reference Manual, undated.)

16 Shallow Oil and Gas Resources

The generation of petroleum from organic matter contained in fine-grained rocks is strongly dependent on the time-temperature interaction. Figure 4 shows this relation for several petroleum basins worldwide. Notice the large number of time-temperature combinations that exist for the main oil and gas zone. The Amazon Basin in Brazil (point A, Figure 4) is near the middle of the main oil and gas zone. The rocks in this basin have attained their present maturity level by a combination of a low temperature (approx. 140°F) and a long period of geologic time (360 million years). By contrast, the Los Angeles Basin has reached a similar maturity level in about 12 million years at 240°F. Notice that at 140°F (Amazon Basin temperature) at least 38–40 million years is required in order to

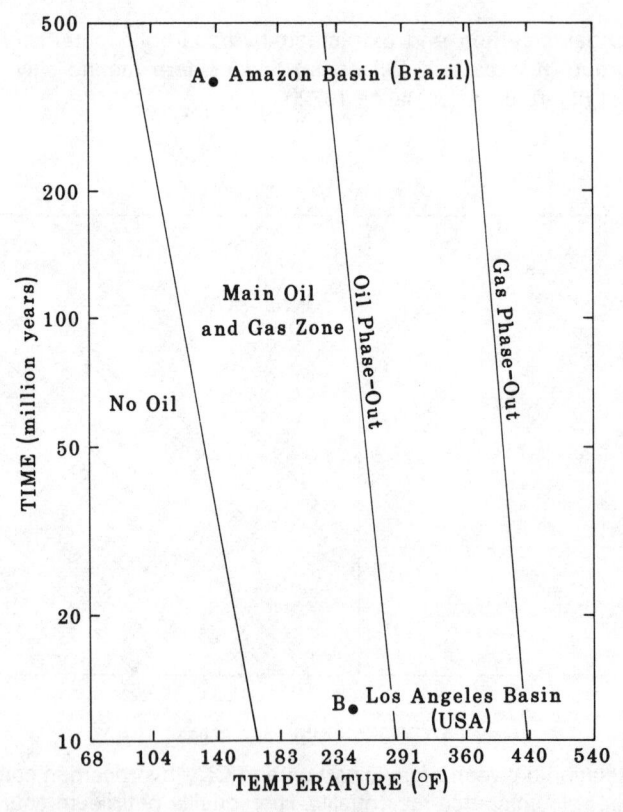

Figure 4. Time-temperature interaction for several petroliferous basins worldwide. Geologically younger basins require higher temperature levels to achieve suitable maturity levels than do older basins. See text for details. (After Connan, 1974.)

arrive at the maturity level necessary for significant oil and gas accumulations. Geologically young basins with high geothermal gradients (e.g., Indonesia and Sumatra) can have excellent oil and gas production at shallow depths. Such high gradients can be responsible for maturation conditions suitable for oil generation at depths of 2,000–3,000 ft. Waples (1980) discusses the TTI (Time-Temperature Index) scheme developed by Lopatin (1976) and demonstrates the use of this parameter in hypothetical examples. The time-temperature interaction is incorporated in the LOM (Level of Organic Metamorphism) scale developed by Hood et al. (1975).

How are source-rock and maturity studies used in frontier exploration areas? Although the presence of petroleum source rocks and an adequate thermal history are basic requirements in order for basins to be petroliferous, these studies also can be used in evaluating specific prospect areas. Data from the Douala Basin in Cameroon (Africa) can be used to illustrate how source-rock and maturity studies can be applied to exploration programs. The data shown in Table 1 are taken from Albrecht et al.

Table 1
Quantitative Data on Samples from Douala Basin, Cameroon, Africa

Sample Number	Depth (m)	C_{org} (wt.%)	Bitumen Ratio (mg EOM/g C_{org})	R_o	H/C (atomic)	O/C (atomic)
102-1	775	0.87	--	0.47	0.99	0.196
102-2	910	1.49	50.9	0.51	0.90	0.185
103-3	1,250	1.40	--	--	0.80	0.161
103-8	1,447	1.38	49.4	--	--	--
103-11	1,564	1.36	23.8	0.73	0.76	0.103
103-14	1,668	1.59	38.6	--	0.76	0.094
103-14K	1,688	--	--	0.79	--	--
103-17	1,788	1.65	55.0	0.82	0.73	0.092
103-23	1,945	1.67	88.4	0.72	0.81	0.061
103-29	2,041	1.32	100.3	0.85	0.71	0.085
103-32	2,280	1.29	80.6	1.05	0.65	0.075
103-35C	2,377	1.19	64.3	1.29	--	--
103-36	2,488	1.23	98.4	1.53	0.57	0.064
104-2C	2,740	--	--	1.80	--	--
104-38	2,746	1.06	28.3	1.87	0.49	0.058
104-3C	2,855	--	--	1.90	--	--
104-6	3,160	1.26	7.3	--	0.50	0.054
104-7C	3,290	--	--	2.37	--	--
104-10K	3,585	--	--	2.81	0.45	0.039
104-10	3,597	1.22	7.6	--	--	--
104-11C	3,700	--	--	3.00	--	--
104-13C	3,955	--	--	3.26	--	--
104-14	4,018	--	18.1	3.57	0.43	0.023

After Albrecht et al., 1976; Durand and Espitalié, 1976.

(1976) and Durand and Espitalié (1976). Information similar to that in Table 1 covering types of samples (outcrop, composite cuttings, sidewall, and conventional core) can be obtained from commercial service organizations. With a minimum of planning, effort, and expense, this type of data can be gained for virtually any well.

The Upper Cretaceous Logbaba series of the Douala Basin is somewhat unusual with respect to source-rock characteristics. The kerogen contained in these shales has atomic H/C and O/C ratios that correspond to the Type III kerogen of Tissot and Welte (1978). This type of kerogen usually is considered to have poor oil-generating capabilities; in the Douala Basin, however, it has apparently yielded "significant amounts of oil" (Durand and Espitalié, 1976). The input of terrestrial plant debris to a marine environment was relatively constant throughout deposition of the Logbaba series, resulting in kerogen of uniform chemical characteristics. And finally, the geochemistry of these Cretaceous shales has been well documented in the literature (Albrecht et al., 1976; Durand and Espitalié, 1976).

The data in Table 1 provide information that can be used in evaluating petroleum potential. For example, every organic-carbon value in Table 1 exceeds the minimum required for source rocks (0.5 percent, by weight). Thus, the first requirement, a sufficient quantity of organic carbon, has been satisfied. The question of the type of hydrocarbons that might be anticipated from a particular type of kerogen can be partially resolved by plotting elemental ratios on a Van Krevelen diagram (Tissot and Welter, 1978). This diagram is shown in Figure 5. Type III kerogen has limited potential for oil generation when compared to Types I and II. Notice that significant oil generation from Type III kerogen is initiated at atomic H/C and O/C ratios of approximately 0.8 and 0.10, respectively. Figure 6 shows the Logbaba kerogen samples plotted on a Van Krevelen diagram. With the exception of a few samples, the atomic ratios plot in the zones of oil and gas formation (compare Figures 5 and 6). Although Type III kerogen is considered to be more of a gas source than an oil source, it appears that the chemical characteristics and maturity level of the Logbaba series source rocks are adequate for generation of liquid petroleum.

Having established that Logbaba series kerogen is capable of generating oil, the final item to be considered concerns maturity level. This matter is especially important in terms of economic assessment, because maturity parameters can be used to discriminate between oil-prone prospects and gas-prone prospects. Figure 7 shows two parameters, both of which are related to temperature history, plotted against depth. Figure 7A demonstrates the relation of EOM as a function of depth. The EOM/C_{org} ratio varies from about 25 to 50 down to a depth of 1500–1600 m and

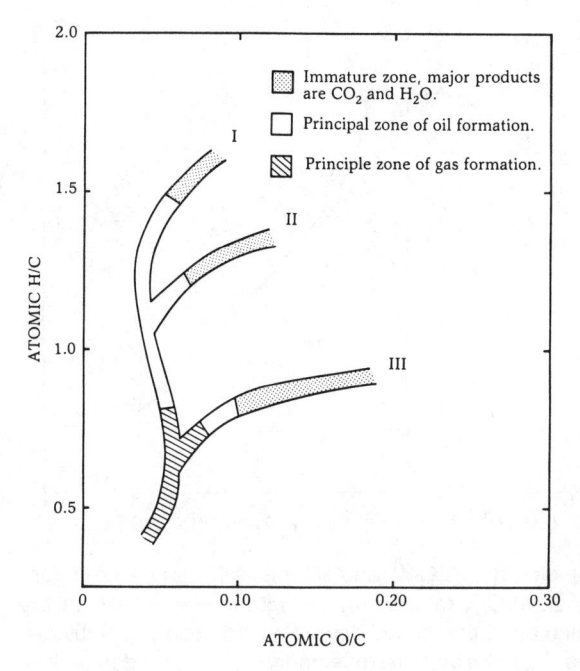

Figure 5. Van Krevelen diagram showing atomic H/C and O/C characteristics for various types of kerogen. Notice the relative capabilities for oil generation in kerogen Types I, II, and III. (After Tissot et al., 1974.)

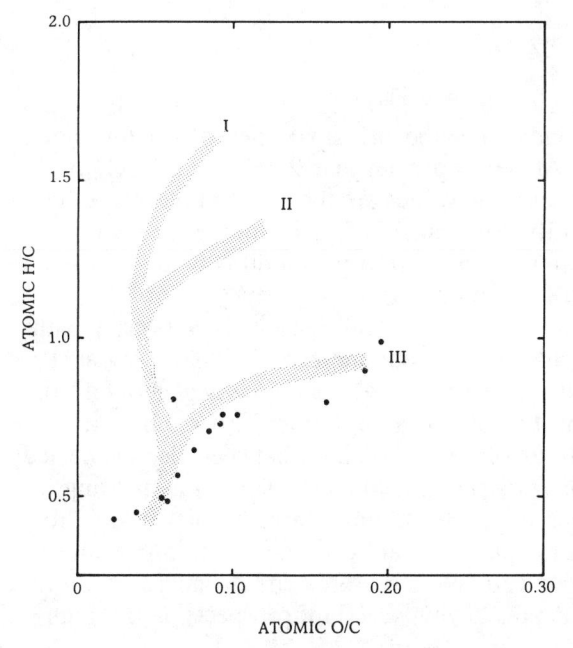

Figure 6. Van Krevelen diagram for Douala Basin samples. This kerogen is unusual in that visual and chemical determinations indicate Type III (gas-prone) source material, but apparently it has generated liquid petroleum. (Data from Albrecht et al., 1976; Durand and Espitalié, 1976.)

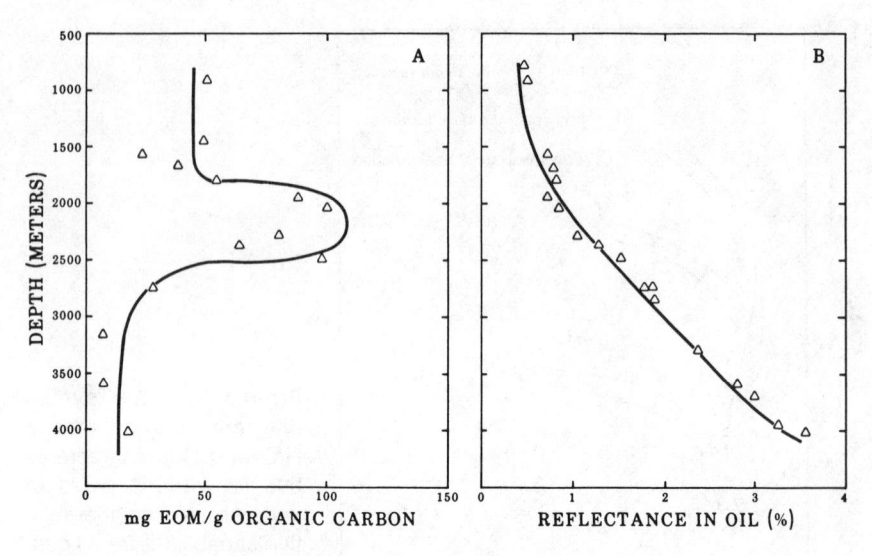

Figure 7. Extractable-organic-matter (EOM) and vitrinite-reflectance data for Douala Basin samples. The EOM/C_{org} ratio increases with increasing depth to about 2,100–2,400 m, at which point the values decrease and become relatively constant at 15–20 mg EOM/g C_{org}. The vitrinite-reflectance values that define the zone of maximum EOM/C_{org} yields are approximately 0.6 and 1.5. (Data from Albrecht et al., 1976; Durand and Espitalié, 1976.)

indicates thermal immaturity. In the 1800–2500 m interval, the ratio ranges from 55 to 100, indicating the principal zone of oil formation (Tissot and Welte, 1978). At depths greater than 2500 m, EOM/C_{org} ratio values are less than 20. These low values are the result of thermal cracking and demonstrate the difference in EOM yields in the oil-generating interval (1800–2500 m) and the interval in which oil is being destroyed and converted to gas (2500 m and deeper).

The visual "best fit" line in Figure 7A defines an evolutionary pathway that is related to petroleum potential and economic assessment. Petroleum that exists at depths of less than 1800 m (and was generated from Logbaba source rocks) must result from migration. This requires (a) porous and permeable beds to serve as conduits between deeply buried source rocks and traps or (b) structural movement, such as faulting, to bring mature source rocks in juxtaposition with reservoir rocks. Prospects in the 1800–2500 m depth interval are in the most important oil-generating zone, and major geologic structures and distances of migration are not required. At depths beyond 2500 m, prospects have a higher probability of containing gas than oil. Thus, the position of a given prospect on the evolutionary curve of Figure 7A has an important bearing on whether oil, gas, or some mixture of the two should be anticipated.

A commonly used parameter to predict whether a given prospect area will be gas-prone or oil-prone is vitrinite reflectance. Vitrinite is a coal maceral that exhibits the unique characteristic of reflecting plane-polarized light as a function of temperature history. The use of this techinque as a maturity parameter for shales has become routine in basin-evaluation studies and is treated comprehensively in publications such as those by Hood and Castaño (1974) and Dow (1977). There is general agreement that significant oil generation is initiated at R_o (reflectance in oil) values of 0.5 to 0.6 percent, continuing to R_o values of 1.3 to 1.4 percent. Gas should be expected at R_o values greater than about 1.4 to 1.5 percent. Projection of vitrinite-reflectance data (Figure 7B) to the EOM curve (Figure 7A) reveals that the principal zone of oil formation from Logbaba source rocks corresponds to R_o values of approximately 0.6 to 1.5 percent.

In the Douala Basin, the necessary geologic conditions for "significant" oil production have been satisfied. The fact that the Logbaba organic-rich shales have been so thoroughly examined makes this basin one of the best documented models for source-rock and maturity studies. A brief look at a few additional areas will serve to illustrate further the application of petroleum-source-rock and maturity studies in exploration programs.

Selected Case Studies

Lower Cook Inlet, Alaska

The data shown in Figures 8, 9, and 10 were taken from a report entitled "Hydrocarbon Source Facies Analysis, C.O.S.T. Lower Cook Inlet #1 Well, Cook Inlet, Alaska," which is available for public inspection at the U.S. Geological Survey office in Anchorage. The section penetrated by the subject well can be divided into three zones on the basis of gross lithology. The interval between 1,355 and 4,900 ft consists of glacial till(?), siltstone, sandstone, and minor coal seams. From 4,900 to 7,450 ft the section consists of sandy siltstones interbedded with gray shale. Between 7,450 and 12,375 ft, silica-cemented silty sandstones are interbedded with light- to dark-gray shale. The study was designed to investigate the quantity, type, and maturity level of organic matter contained in the sedimentary strata penetrated by the C.O.S.T. well.

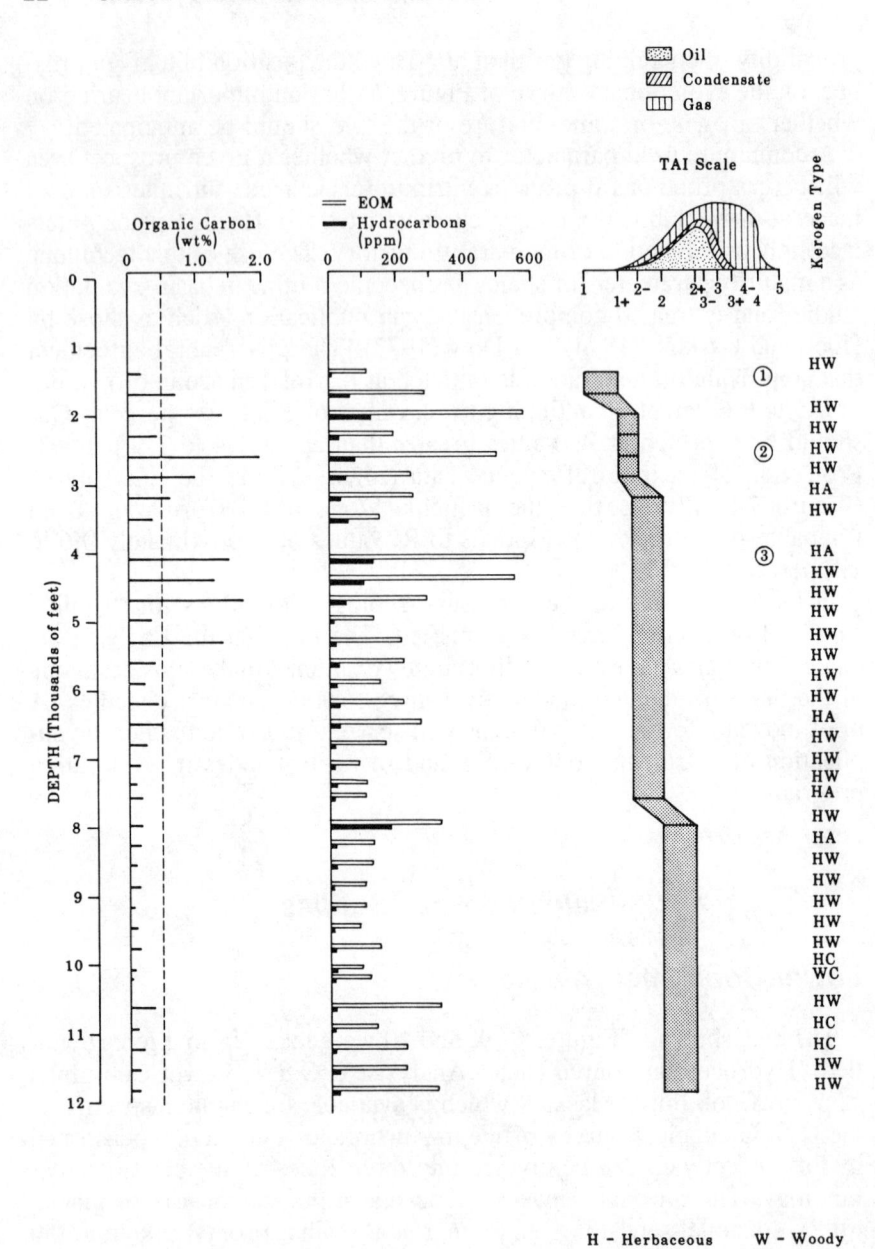

Figure 8. Organic-carbon, extract-data (extractable organic matter and hydrocarbons), and TAI (Thermal Alteration Index) determinations and visual kerogen descriptions for the C.O.S.T. Lower Cook Inlet 1 well. (U.S. Geological Survey Open File data.)

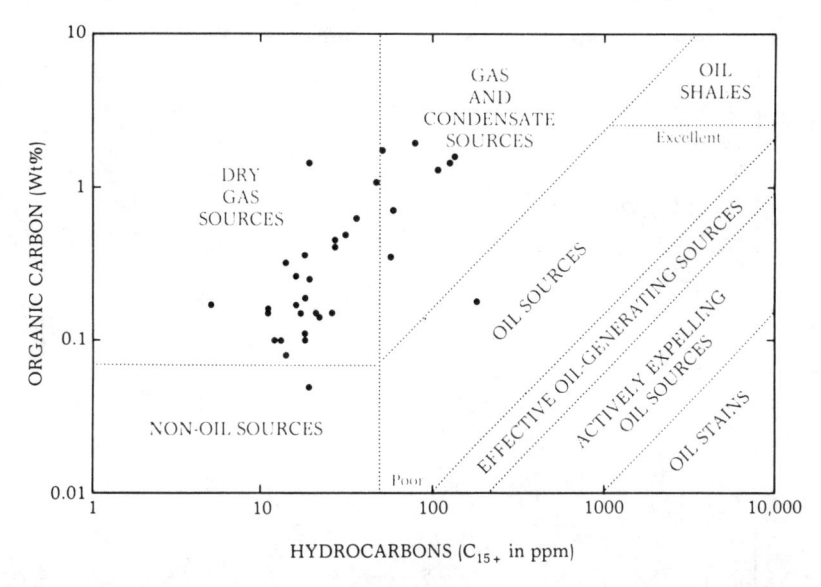

Source-Rock And Temperature-History Studies 23

Figure 9. Extractable C_{15+} hydrocarbon and organic-carbon data for the C.O.S.T. Lower Cook Inlet 1 well. Most of the samples fall in the dry-gas region of the source-rock diagram (see Figure 3) and are unlikely to be prolific oil sources. (U.S. Geological Survey Open File data.)

Figure 8 shows organic carbon, EOM, TAI (Thermal Alteration Index), and kerogen type plotted as a function of depth. Ten samples contain at least 0.5 percent organic carbon, and all these occur at depths between 1,000 and 6,500 ft. In general, the EOM yields are adequate for marginal-quality oil source rocks. The kerogen described from these samples is primarily of the herbaceous-woody type, although a few samples contained coaly and amorphous organic matter. TAI values indicate that optimum maturity conditions exist from about 8,000 to 12,000 ft. Figure 9 shows organic-carbon and extractable hydrocarbon data plotted on the conventional source-rock diagram. Most of the samples plot in the dry-gas source region and thus are not likely to serve as oil sources. The relatively high carbon and hydrocarbon yields in the 1,000–5,000-ft range can serve as oil sources under more favorable maturity conditions. If so, an obvious possibility would be to seek prospects where these rocks are buried more deeply. The Van Krevelen diagram (Figure 10) shows three kerogen samples from depths of 1,480–1,510, 2,560–2,590, and 4,060–4,090-ft. These three kerogens have elemental ratios that plot in

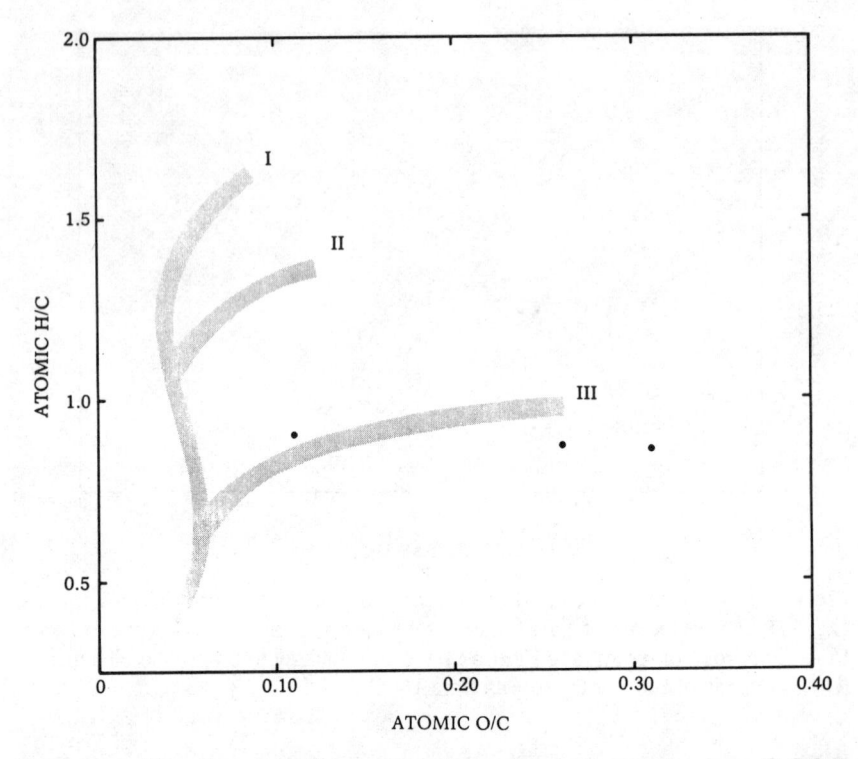

Figure 10. Van Krevelen diagram for three kerogen samples from C.O.S.T. Lower Cook Inlet 1 well. All three samples have atomic H/C and O/C ratios indicative of Type III kerogen and agree with visual kerogen descriptions (see Figure 8). (U.S. Geological Survey Open File data.)

the Type III (humic or lignitic) region on the Van Krevelen diagram. Type III kerogen is gas-prone and has limited potential as a source of liquid petroleum. Visual kerogen descriptions (Figure 8) for samples from 1,000 to 5,000 ft are similar to those from deeper, more mature zones. These samples still do not have good oil-source characteristics, although this is due in part to the lack of adequate quantities of organic matter. Thus, on the basis of source-rock and maturity parameters, Lower Cook Inlet appears to be essentially a gas-prone region. Because the kerogen is Type III (herbaceous-woody), significant quantities of oil are unlikely to be generated, even under more favorable maturity conditions.

Officer Basin, South Australia

McKirdy and Kantsler (1980) investigated the petroleum-source-rock potential of the Cambrian Observatory Hill sequence in South Australia. This study is somewhat unusual in that the potential source rocks are carbonates and evaporites and because the organic matter is devoid of contributions from higher plants.

The Observatory Hill and equivalent strata consist primarily of carbonates and silty shales. One of the wells in the study penetrated more than 100 m of bedded salt. Data taken from McKirdy and Kantsler (1980) are shown in Figures 11 and 12. Organic carbon, EOM, and extractable hydrocarbon concentrations for five wells in the Officer Basin are shown in Figure 11. Because the generally accepted criteria for fine-grained clastic source rocks may not apply to carbonate and evaporites, the extractable hydrocarbon/organic-carbon ratio was employed as a means to help assess source-rock potential. Samples from the Byilaoora, Wilkinson, and Wallira West wells were judged to be marginal to good oil source rocks. Oil-rock correlation studies suggest that the oil encountered in the Byilaoora well had its source in the "organic-rich dolomites and calcareous argillites" that occur in the Observatory Hill sequence.

The Van Krevelen diagram for demineralized kerogen samples from Observatory Hill beds in the Byilaoora, Wilkinson, and Wallira West wells is shown in Figure 12. With one exception (the sample from 1,032 ft in the Wilkinson well), all kerogen samples analyzed are either Type I or Type II and have excellent oil-generating potential. In addition, all but one of the samples are of a maturity level favorable for liquid hydrocarbons. Based on these results, it appears that the quantity, type, and maturity levels of the organic matter contained in the Observatory Hill beds indicate that the eastern part of the Officer Basin is oil-prone.

Takutu Basin, Guyana

The Takutu graben is a relatively unexplored geologic feature that lies along the Brazil-Guyana border in South America. Crawford and Szelewski (1984) published the results of preliminary exploration efforts in this basin and suggested that "likely additional oil finds will be made in this graben." This optimism is based on a combination of source-rock and temperature-history data and anticipation of encountering reservoir-quality lithologies in untested areas.

Studies of source-rock quality and maturity level were made on samples from the Home et al. 1 Lethem and the Home 1 Karanambo wells; the unpublished data from these wells are shown in Figure 13. Although hand-picked samples from the Lethem well appear to contain enough or-

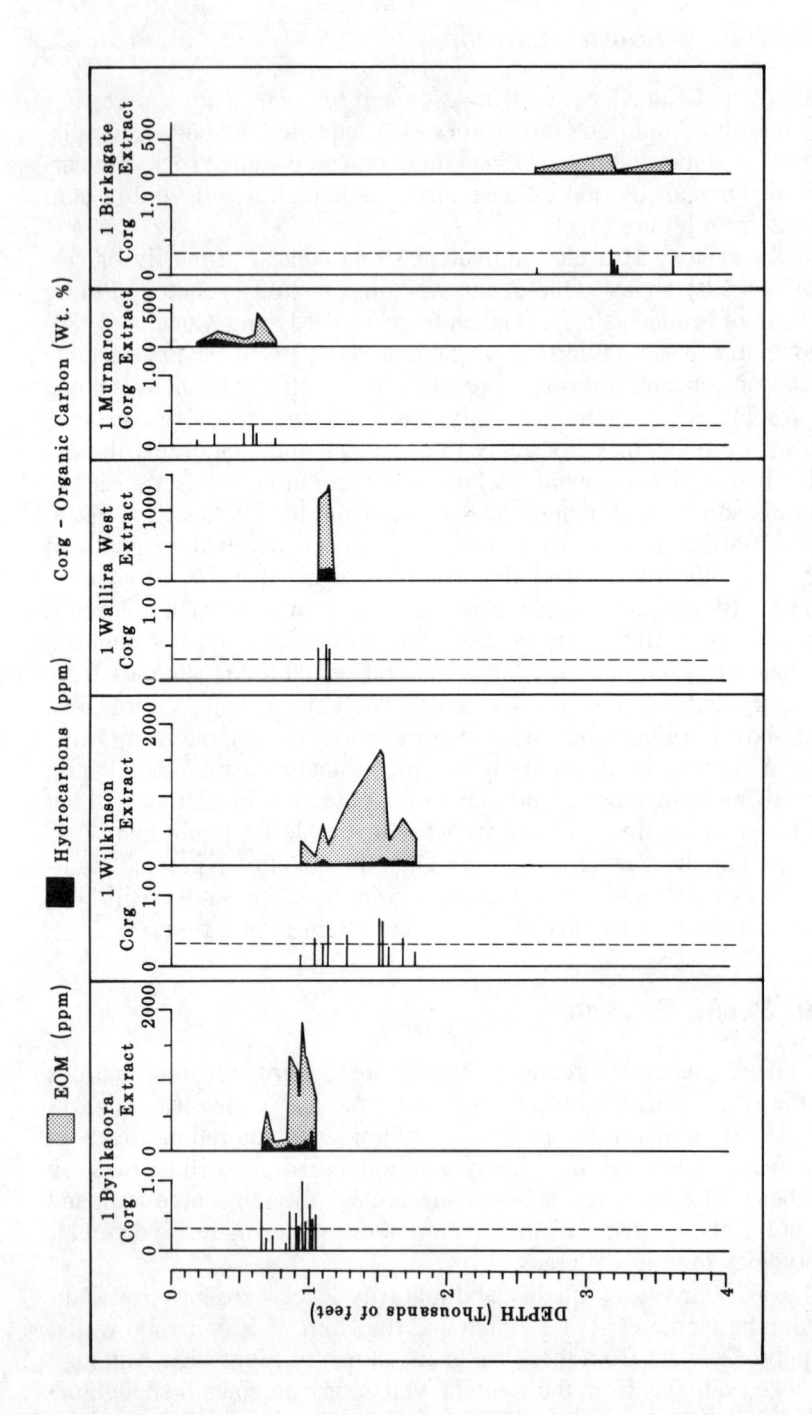

Figure 11. Organic carbon, EOM, and hydrocarbon data for samples from four wells in the Officer Basin, South Australia. (Data from McKirdy and Kantsler, 1980.)

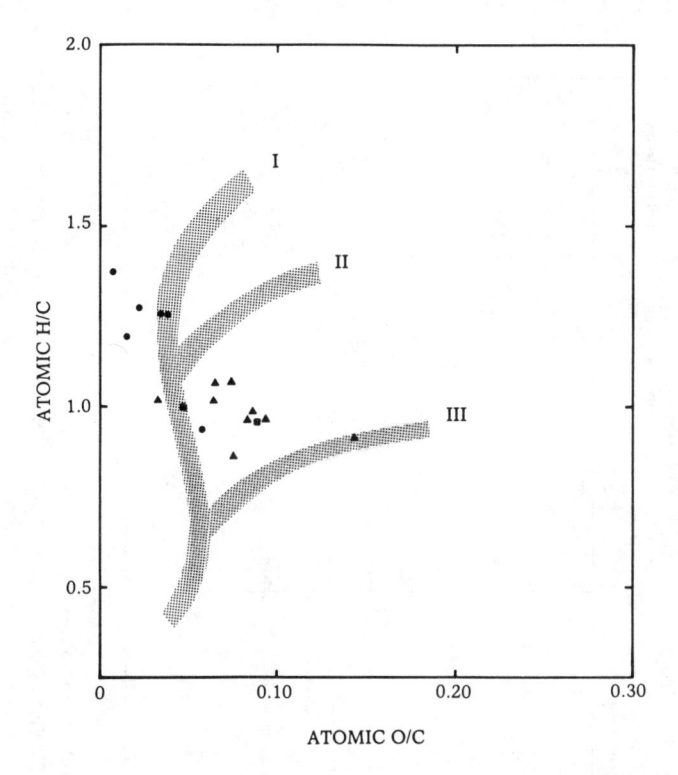

Figure 12. Van Krevelen diagram for kerogen samples from Byilaoora (circles), Wilkinson (triangles), and Wallira West (squares) wells in the Officer Basin, Australia. Atomic H/C and O/C ratios indicate kerogen Types I and II and maturity levels favorable for oil generation. (Data from McKirdy and Kantsler, 1980.)

ganic carbon to be considered potential source rocks, visual kerogen assessment indicates that the type of organic matter is not suitable for oil generation. The Hydrogen Index, a parameter obtained by pyrolysis, is a measure of the hydrocarbon-generating potential of sedimentary organic matter. Compare the Hydrogen Index values for samples from the Lethem well with those plotted in Figure 14. In general, the good oil-prone source rocks in Figure 14 have HI (Hydrogen Index) values greater than 300 mg HC/g C_{org}. Only a single sample (at 5,775 ft) from the Lethem well has an HI value over 100. Thus, the pyrolysis data support the visual kerogen data, and both indicate poor oil source rocks. The ma-

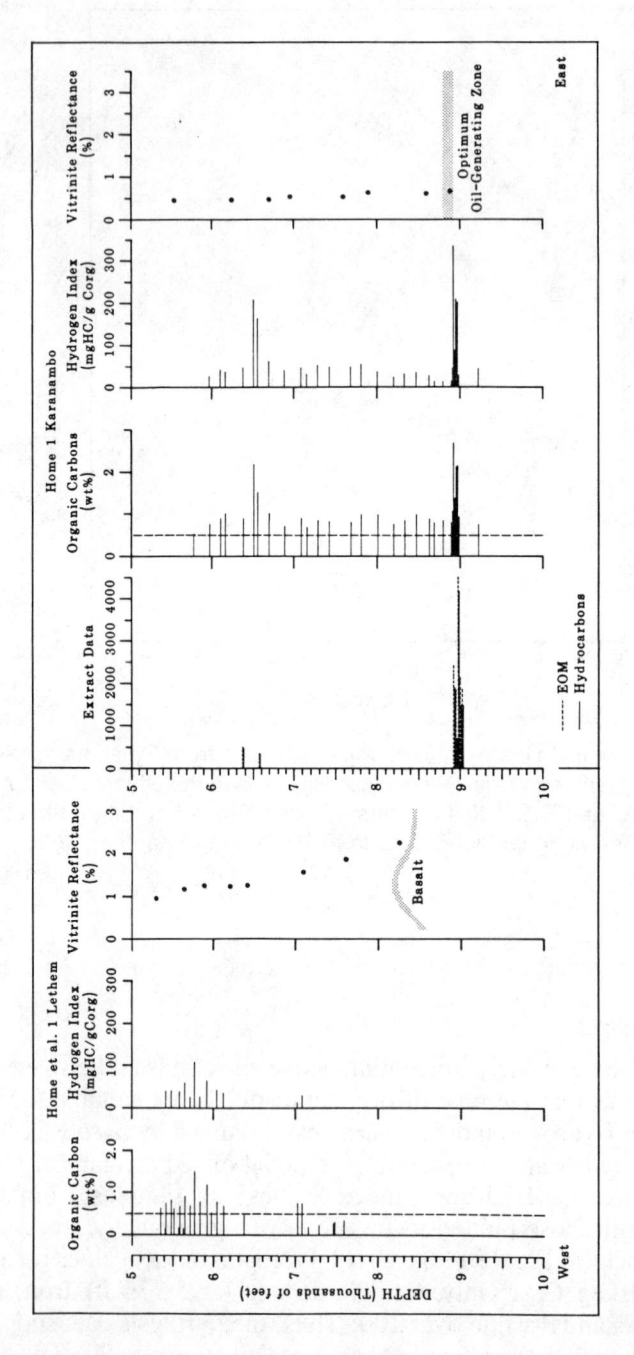

Figure 13. Organic-carbon, Hydrogen Index, and vitrinite-reflectance data for the Home et al. 1 Lethem and Home 1 Karanambo wells in the Takutu Basin, Guyana. The data suggest poor oil source potential for rocks penetrated by the Lethem well. The interval between about 8,918 and 8,960 ft in the Karanambo well has good oil source potential and is at a maturity level that is optimum for oil generation. Compare Hydrogen Index data with Figure 14. (Data furnished by Home Oil.)

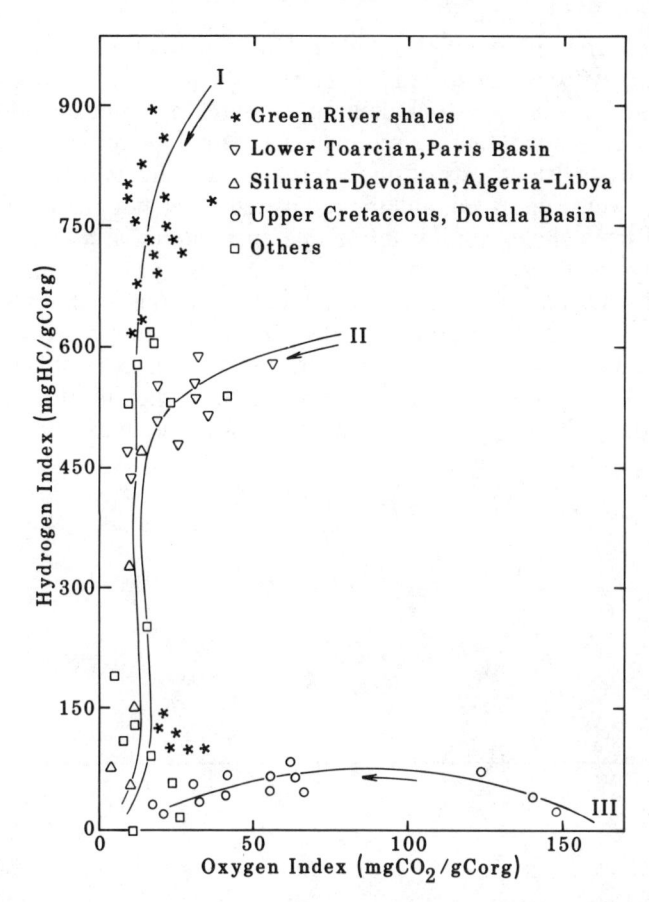

Figure 14. The use of pyrolysis data to evaluate petroleum source rocks. The Hydrogen Index is the ratio of hydrocarbons generated by pyrolysis to organic-carbon content and approximates the atomic H/C ratio of the Van Krevelen diagram. The Oxygen Index is the ratio of generated carbon dioxide and organic-carbon content and approximates the O/C atomic ratio. These data are comparable to the Van Krevelen diagram.

turity levels are interpreted as being "late mature" down to about 6,250 ft and "post mature" at greater depths. The sequence penetrated by the Lethem well is unlikely to be oil-prone owing to the quantity, type, and maturity level of the sedimentary organic matter.

Source-rock quality and maturity levels are quite different in the Karanambo well, approximately 25 miles northeast of the Home et al. 1 Lethem. The shale that was cored between about 8,918 and 8,960 ft has good oil-generating potential, as indicated by extract data and organic-carbon and Hydrogen Index values. The extract and organic-carbon data, when plotted on a source-rock diagram (Figure 15), indicate good-quality oil and gas-condensate source material. Small quantities of oil were recovered in drillstem tests of this interval. The extracts from the shales and the oil were compared, and the recovered oil is thought to have been generated from shales similar to the one encountered at about 8,925 ft.

Maturity levels, as shown by spore-carbonization and vitrinite-reflectance data, indicate optimum oil-generating conditions at about 8,860 ft. Crawford and Szelweski (1984) predict more favorable reservoir facies east of the Karanambo well. If this interpretation is correct, and if the source-rock quality and maturity level are similar, then most of the components for additional oil accumulations exist in the unexplored eastern part of the Takutu Basin.

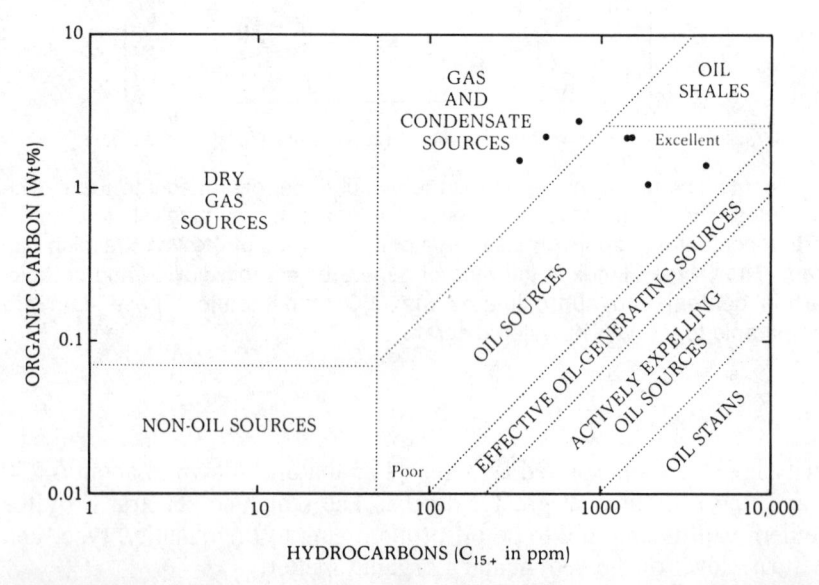

Figure 15. Extractable C_{15+} hydrocarbon and organic-carbon data for the Home 1 Karanambo well in the Takutu Basin, Guyana. These data indicate good to excellent oil source potential. (Data furnished by Home Oil.)

Fushan Depression, Eastern China

A single-well source-rock and temperature-history study from Eastern China will serve as the final case study from the literature. Benshan et al. (1983) studied the W-2 well in the eastern Gulf of Tonkin in an effort to gain a better understanding of the source-rock potential and temperature history of the Oligocene organic-rich shales of lacustrine origin. The geochemical and maturity data for well W-2 are shown in Figure 16. The

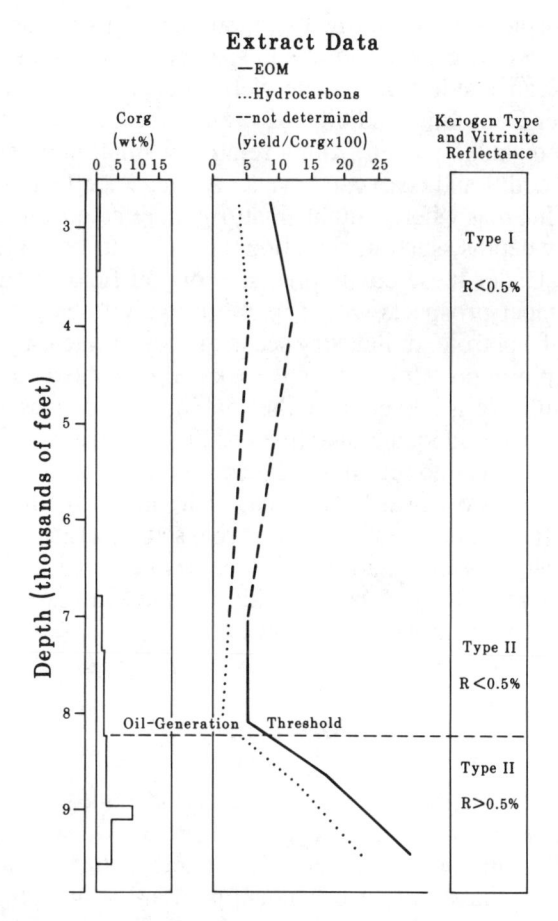

Figure 16. Organic-carbon and extract data, and vitrinite-reflectance characteristics of the W-2 well in the Gulf of Tonkin. Kerogen type and maturity level indicate good oil source potential beginning at about 8,200 ft. (Redrawn from Benshan et al., 1983.)

quantity, type, and maturity level of organic matter contained in strata penetrated by the W-2 well are optimum for oil generation at about 8,200 ft. If the conditions that exist at the W-2 well persist to areas where adequate traps and reservoirs occur, one might reasonably expect to find oil accumulations.

Summary

Petroleum-source-rock and maturity studies can provide valuable information for exploration programs in sparsely tested areas. This type of information can result from single-well studies, such as the C.O.S.T. well in Lower Cook Inlet and the W-2 well in the Gulf of Tonkin. In such cases, the knowledge gained from a relatively small number of analyses (and modest additional cost) can have a pronounced effect on exploration philosophy. In areas where suitable maturity parameters are known to exist at shallow depths, such information can be quite valuable in exploration efforts. Under these conditions, shallow oil fields would be anticipated and deeper prospects would be more likely to be gas-prone.

Although the petroleum industry seeks to concentrate on positive items related to exploration efforts, source-rock and maturity studies that have negative results (e.g., excessive thermal history or absence of fine-grained rocks with adequate quantities of organic matter) often are considered to be of economic significance. Exploration budgets are frequently adjusted downward for areas that lack any of the required components (including adequate source rocks and maturity level) for significant petroleum accumulations. Companies involved in the search for giant oil accumulations usually prefer to make investments in areas where no known negative parameters exist.

References

Albrecht, P., Vandenbroucke, M., and Mandengue, M., 1976, Geochemical studies on the organic matter from the Douala Basin (Cameroon)—I. Evolution of the extractable organic matter and the formation of petroleum: Geochim. et Cosmochim. Acta, v. 40, pp. 791–799.

Baker, D. R., 1962, Organic geochemistry of Cherokee Group in southeastern Kansas and northeastern Oklahoma: Am. Assoc. Petroleum Geologists Bull., v. 46, pp. 1621–1642.

Benshan, Wang, et al., 1983, A preliminary organic geochemical study of the Fushan Depression, a Tertiary basin of eastern China, *in* Bjoroy (Ed.) Advances in organic geochemistry, 1981: New York, John Wiley & Sons, pp. 108-113.

Connan, Jacques, 1974, Time-temperature relation in oil genesis: Am. Assoc. Petroleum Geologists Bull., v. 58, pp. 2516-2521.

Crawford, F. D., and Szelewski, C. E., 1984, Geology and exploration in the Takutu graben of Guyana: Oil and Gas Jour., v. 82, no. 10, March 5, pp. 122-129.

Durand, B., and Espitalié, J., 1976, Geochemical studies on the organic matter from the Douala Basin (Cameroon)—II. Evolution of kerogen: Geochim. et Cosmochim. Acta, v. 40, pp. 801-808.

Dow, W. G., 1977, Kerogen studies and geological interpretations, (Hitchon, Ed.) Application of geochemistry to the search for crude oil and natural gas: Jour. Geochem. Explor. Spec. Issue, pp. 79-100.

Espitalié, J., et al., 1977, Méthode rapide de caractérisatin des roches méres de leur potentiel pétrolier et de leur degré d'évolution: Rev. Inst. Fr. Pétr., v. 32, pp. 23-42.

Geochem Laboratories, (undated), Source rock evaluation reference manual.

Geochem Laboratories, 1977, Hydrocarbon source facies analysis, C.O.S.T. Lower Cook Inlet #1 well, Cook Inlet, Alaska.

Gold, Thomas, and Soter, Steven, 1980, The deep-earth-gas hypothesis: Scientific American, v. 242, no. 6, pp. 154-161.

Harrison, W. E., and LaPorte, J. L., 1977, Shipboard organic geochemistry, *in* (Montadert and Roberts, Eds.) Initial reports of the Deep Sea Drilling Project, Vol. XLVIII: U.S. Govt. Printing Office, pp. 959-964.

Hood, A., and Castaño, J. R., 1974, Organic metamorphism: its relationship to petroleum generation and application to studies of authigenic minerals: United Nations ESCAP CCOP, Tech. Bull. 8, pp. 85-118.

Hood, A., Gutjahn, C. C. M., and Heacock, R. L., 1975, Organic metamorphism and the generation of petroleum: Am. Assoc. Petroleum Geologist Bull., v. 59, pp. 986-996.

Hunt, J. M., 1979, Petroleum geochemistry and geology: San Francisco, W. H. Freeman and Co., 617 p.

Lopatin, N. V., 1979, Temperature and geologic time as factors in coalification: Akad. Nauk SSSR Izv. Ser. Geol., no. 3, pp. 95-106.

Magoon, L. B., and Claypool, G. E., 1981, Petroleum geology of Cook Inlet Basin—an exploration model: Am. Assoc. Petroleum Geologists Bull., v. 65, pp. 1045-1061.

Malek-Aslani, M., 1979, Environmental and diagenetic controls of carbonate and evaporite source rocks: Am. Assoc. Petroleum Geologists Bull., v. 63, pp. 489-490.

McKirdy, D. M., and Kantsler, A. J., 1980, Oil geochemistry and potential source rocks of the Officer Basin, South Australia: Australian Petrol. Explor. Assoc. Journal, v. 20, Part I., pp. 68-86.

Philippi, G. T., 1957, Identification of oil source beds by chemical means: Internat. Geol. Cong., 20th, Mexico City, Sec. 3, pp. 25-28.

Porfir'ev, V. B., 1974, Inorganic origin of petroleum: Am. Assoc. Petroleum Geologists Bull., v. 58, pp. 3-33.

Tissot, B. P., 1981, Generation of petroleum in carbonate rocks and shales of marine or lacustrine facies and its geochemical significance, *in* (Mason, Ed.). Petroleum Geology in China: PennWell, pp. 71-82.

Tissot, B., et al., 1974, Influence of native and diagenesis of organic matter in formation of petroleum: Am. Assoc. Petroleum Geologists Bull., vol. 58, pp. 499-506.

Tissot, B. P., and Welte, D. H., 1978, Petroleum Formation and Occurrence: Berlin, Springer-Verlag, 538 p.

Waples, D. W., 1980, Time and temperature in petroleum formation: application of Lopatin's method to petroleum exploration: Am. Assoc. Petroleum Geologists Bull., v. 64, pp. 916-926.

3
Time, Temperature, and Tectonics: Their Assessment as a Means of Evaluating Shallow Oil and Gas Potential in Frontier Areas

John D. Pigott

Introduction

Prior to a substantial economic commitment for exploration of shallow oil and gas in a frontier area, preliminary evaluation of a region's geochemical potential, i.e. source-rock maturity, should be made. Time, temperature, and tectonics are controlling parameters of maturity, and their assessment helps select areas as candidates for more intensive geological-geophysical investigations.

Time and Temperature: Kinetics of Maturation

The maturation of amorphous kerogen into hydrocarbons can be described as a first approximation by Model 1.

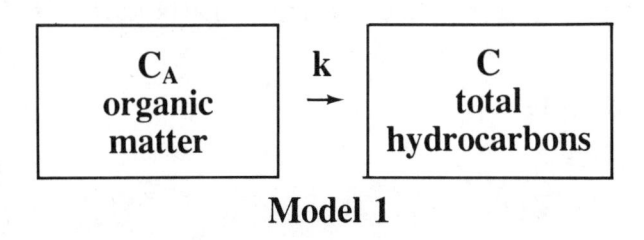

$$\boxed{\begin{array}{c} C_A \\ \text{organic} \\ \text{matter} \end{array}} \xrightarrow{k} \boxed{\begin{array}{c} C \\ \text{total} \\ \text{hydrocarbons} \end{array}}$$

Model 1

Assuming first-order reaction kinetics (Tissot, 1969), one has:

$$\frac{dC_A}{dt} = -k_1 \, C_A \tag{1}$$

where C_A represents the concentration in moles of carbon as unreacted kerogen.

However, there is generally a competing reaction, that of produced oil also generating gas. A simplified model assuming consecutive first-order reaction kinetics is:

$$\boxed{\begin{array}{c} C_A \\ \text{organic} \\ \text{matter} \end{array}} \xrightarrow{k_1} \boxed{\begin{array}{c} C_B \\ \text{oil} \end{array}} \xrightarrow{k_2} \boxed{\begin{array}{c} C_C \\ \text{gas} \end{array}}$$

Model 2

Then for box C_B we may write:

$$\frac{dC_B}{dt} = -K_1 \, C_A - K_2 \, C_B \tag{2}$$

and for box C_C we have:

$$\frac{dC_C}{dt} = k_2 \, C_B \tag{3}$$

where C_B and C_C represent concentrations of moles of carbon as oil and gas, respectively.

Assuming conservation of mass, that is $d\Sigma C/dt = 0$, and $C_{B_{t=0}} = 0$, and $C_{C_{t=0}} = 0$ (negligible biogenic gas), integration of Equations 1, 2, and 3 yields:

$$C_A = C_{A,t=0} \, [e^{-k_1 t}] \tag{4}$$

$$C_A = \frac{k_1 \, C_{A,t=0}}{k_2 - k_1} \, [e^{-k_1 t} - e^{-k_2 t}] \tag{5}$$

$$C_C = C_{A,t=0}\left[1 - \frac{k_2 e^{-k_1 t} - k_2 e^{-k_2 t}}{k_2 - k_1}\right] \tag{6}$$

Therefore, knowing the value of k would allow one to quantify the production of oil and gas through time (in gallons/acre-feet, etc.). Now, k is also a function of temperature, and can be described by the Arrhenius equation (Connan, 1974):

$$k = A\, e^{\frac{-E_A}{RT}} \tag{7}$$

where A = the frequency factor
       E = the activation energy
       R = the gas constant

However, owing to our present inability to agree upon a consistent A or E in natural and experimental data (Snowden, 1979), one is unable to supply accurate or consistent values of k to the preceding equations. Although calculations are presently made as to the number of gallons/acre-feet, the actual validity of such numbers is open to question. The k's may be forthcoming with additional research, but in the meanwhile we may choose an additional, less theoretical methodology. The natural data are amenable to being statistically analyzed by brute, empirical force (i.e., the best-fit line provides the most accurate model). From a compilation of threshold temperatures or the "oil ceiling" and sediment ages within cratonic sedimentary basins from different regions of the world (Table 1), the statistically best derived relation is:

$$T = 164.4 - 19.39 \ln t, \quad r = -0.89, \quad df = 16 \tag{8}$$

where T = temperature in °C
       t = time (sediment age) in 10^6 years

These data and the best fit line are illustrated in Figure 1. Similarly, the necessary time for maturation of organic matter to the threshold at a particular temperature is statistically best described as:

$$\begin{aligned} t &= 1910\, e^{-0.0408T} \\ r &= -0.89 \\ df &= 16 \end{aligned} \tag{9}$$

The data and best-fit lines are illustrated in Figure 2. An indication of the error in the predictive abilities of Equation 8 is shown in the plot of the

Table 1
World Basin Data*

Basin	Formation Age 10^6 years	Threshold Temp. °C
Douala (Cameroons)	70	65
Los Angeles	12	115
Ventura	12	127
Paris	180	60
Aquitaine (France)	112	90
Aquitaine (France)	135	72
Camargue (France)	38	106
El Aaiun	105	85
Sulu Sea	12	102
Offshore Taranaki	70	80
Amazon (Brazil)	359	62
Onshore Taranaki	32	95
Dongyang	35	93
Qianjiang	35	90
Songliao (1)	110	70
Songliao (2)	100	65
Songliao (3)	90	63
Liaohe Sag	50	81

* Compiled from Connan (1974) and Yu (1983).

residuals in Figure 3. Note that 90% of the natural observations fall within ±15°C of predicted relationship. Knowing, then, a region's geothermal gradient enables the determination of the necessary depth for the "oil ceiling," or D_{oc}.

$$D_{oc} = \frac{T - T_s}{dT/dZ} \tag{10}$$

where T is calculated from Equation 8, T_s is mean surface temperature, and dT/dZ is the geothermal gradient in °C/100 m.

As 150°C is generally considered the temperature at which oil is degraded thermally to gas and condensate (Pusey, 1973), the depth at which this temperature occurs is correspondingly termed the "oil floor," or D_{of}, and may be calculated from the following relationship:

$$D_{of} = \frac{150 - T_s}{dT/dZ} \tag{11}$$

Time, Temperature and Tectonics 39

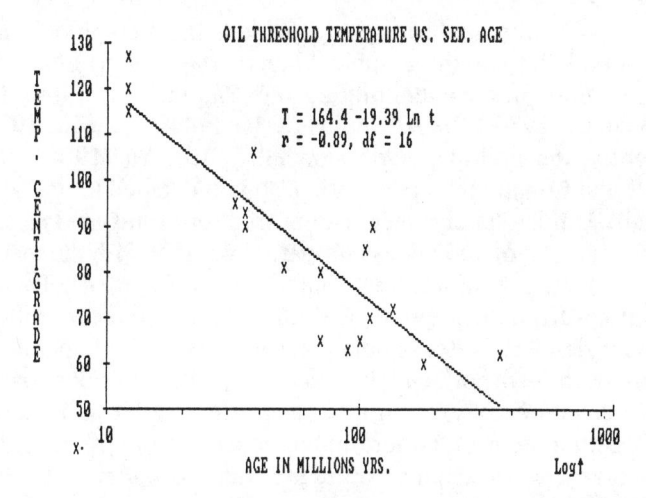

Figure 1. Oil threshold temperature plotted as a function of the logarithm of sediment age. Line represents best fit of world basin data (N = 18).

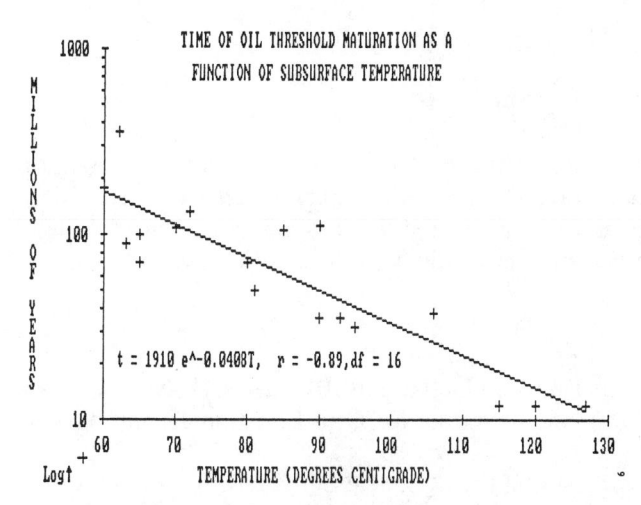

Figure 2. Logarithm of time necessary for maturation to the oil threshold plotted as a function of temperature. Line represents best fit of world basin data (N = 18).

Conventionally, the nonproprietary methods most commonly used for maturation modeling are the graphical Hood et al. (1975) LOM (level of organic metamorphism) technique, the Waples-calibrated Lopatin method (Waples, 1980), the Arrehnius-derived plot (Connan, 1974), and more recently, the multiple regression method of Yu (1984). Both the Hood et al. and Lopatin methods work best with detailed temperature and age analysis. Unfortunately, such data in frontier basins are typically unavailable. Lopatin's method does compare favorably with the method derived here (Equation 8) as a fit to the natural data (Figure 4). The Connan formula suffers from the previously discussed present uncertainties in activation energies. Yu offers more recent time and temperature data which have been incorporated here but includes an additional parameter, depth. The equations offered in this paper are simpler and instead rely upon age and temperature (where in application the depth would be implicitly involved, because temperature is a function of depth, the geothermal gradient).

Temperature as a Function of Time and Tectonics

Geothermal gradients can vary both in time and space. In contrast to the preceding cratonic crustal basins used for the maturation models, the heat flow of oceanic crust is not steady state and decays as the power function of time (Royden et al. 1980). Their empirical formula is:

$$Q(t) = 11.3 \, t^{-1/2} \, \mu cal \, cm^{-2} \, sec^{-1} \tag{12}$$

where Q is heat flow in HFU

Using this relationship, one may derive the change through time of the geothermal gradient of a sedimentary basin floored by oceanic crust, assuming it undergoes continuous, passive, thermal decay. Taking Fourier's law of heat conduction in one dimension:

$$Q = -k/10 \, dT/dZ \tag{13}$$

where k is thermal conductivity in 10^{-3} cal cm^{-1} sec^{-1}°C^{-1}, and dT/dZ is in °C/100 m, we substitute for Q in Equation 12 and have:

$$dT/dZ = 10/k \, (11.3 \, t^{-1/2} \, \mu cal \, cm^{-2} \, sec^{-1}) \tag{14}$$

Therefore, geothermal gradients of such basins tend to decrease, or cool, through time. Such decreases in temperature must be accounted for in

Time, Temperature and Tectonics 41

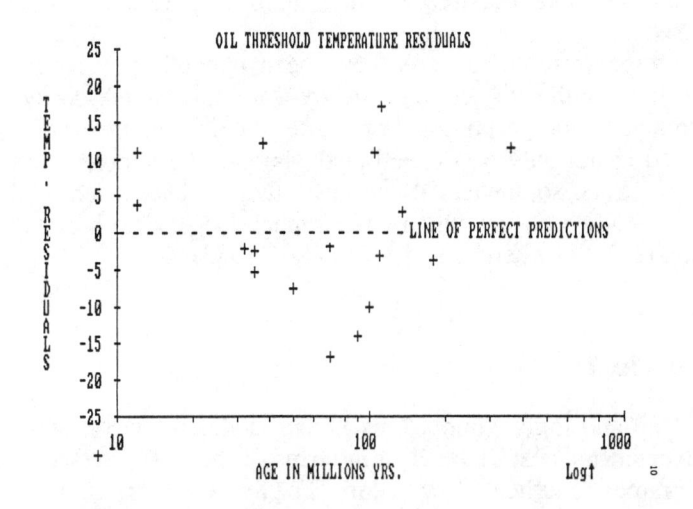

Figure 3. Oil threshold temperature residuals (difference between predicted and observed values) calculated from the relationship illustrated in Figure 2.

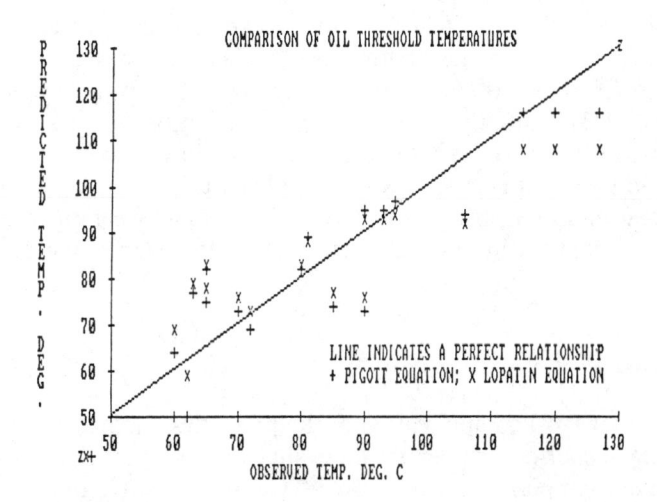

Figure 4. Oil threshold temperatures predicted by Lopatin method and method proposed here compared to the natural data.

order to correctly apply the maturation models. As the reactions of Model 2 are nonreversible, the maximum temperature encountered may be used.

If temperatures of an area have been irregularly perturbed through time, then neither the cratonic steady-state nor passive-decay models of basin maturation pertain, and only after the thermal history has been delineated (from outcrop or well analysis) can the appropriate models be applied. Such sophisticated computer models have been applied to intensely-studied basins which have a much larger data base than frontier basins (e.g., Mackenzie and McKenzie, 1983).

Example 1

O'Neil and Pigott (unpub. ms.) have modeled the St. Ann's Basin on the Northern Coast of Jamaica in terms of the post-Cretaceous geothermal gradient and heat flow history. For 88×10^6 yrs shaley sandstones with a conductivity of 4 the application of Equation 14 yields a geothermal gradient of 3.0 °C/100m and with Equation 13 a heat flow of 1.20 HFU. Supporting the essential correctness of the model, a proximal deep sea heat flow measurement is 1.18 HFU (Jessop et al. 1975).

Spatially, geothermal gradients also vary as a function of tectonic setting. For the classic convergent tectonic area of the Indonesian region of Southeast Asia, the geothermal gradient of the trench averages 3.2 °C/100m (N = 5), the fore-arc basin averages 2.5 °C/100m, the volcanic-arc is greater than 20 °C/100m, the back-arc basin averages 4.7 °C/100m (N = 48), and the continental (Malaysian) massif averages 2.5 °C/100m (N = 46) (data calculated and compiled from Rutherford and Qurishi, 1981). Although there is some basin-to-basin variation in these averages, in Indonesia as is the case for most convergent margin settings intrabasin variation is much less and the back-arc basins are significantly the hottest and the most petroliferous. Early classification of a frontier basin's tectonic setting aids its thermal typing, even if the geothermal gradient is not initially known.

Example 2

As another worked application of the preceding methods, consider the following scenario: an unexplored basin in a region of the world where the surface temperature averages 28°C. Tectonically, the basin fits a back-arc setting where 100 km away one well yielded a gradient of 6.6 °C/100m. If drilling costs dictate a shallow well of 1 km depth maximum to reach mature oil source rocks, what age should the sediments be? As the temperature at 1 km would be 94°C (28°C + 66°C), direct application of Equation 9 suggests the age of the rocks should be greater than 41 million years (Eocene or older).

Example 3

Within the same basin, where would one anticipate thermally degraded gas source rocks to occur? Using Equation 11, one calculates thermally degraded gas source rocks to occur at about 1.8 km depth (the oil floor).

Conclusions

The hydrocarbon maturity of a basin can be described in terms of its age, thermal history, and tectonic setting. As a first approximation to the geochemical modeling of a frontier area, the derived equations afford a rapid and significant means for just such a task. With the derived information, one may then systematically make the initial economic evaluations, design an appropriate field geological/geophysical acquisition program, and test the basin for oil and gas.

References

Connan, J., 1974, Time-temperature relation in oil genesis: A.A.P.G. Bulletin, v. 58, pp. 2516–2521.

Hood, A., Gutjahr, C. C. M., and Heacock, R. L., 1975, Organic metamorphism and the generation of petroleum: A.A.P.G. Bulletin, v. 59, pp. 986–996.

Jessop, A. M., Hobart, M. A., and Sclater, J. G., 1975, The World Heat Flow Data Collection: Geothermal Series Number 5, Ottawa, Canada, 125 pp.

Mackenzie, A. S., and McKenzie, D., 1983, Isomerization and aromatization of hydrocarbons in sedimentary basins formed by extension: Geology Magazine, v. 120, pp. 417–528.

Pusey, W. C. III, 1973, Paleotemperatures in the Gulf Coast using the ESR-kerogen method: Gulf Coast Assoc. Geol. Socs. Trans., v. 23, pp. 195–202.

Royden, L., Sclater, S. G., and von Herzen, R. P., 1980, Continental margin subsidence and heat flow: important parameters in the formation of petroleum hydrocarbons: A.A.P.G. Bulletin, v. 64, pp. 173–187.

Rutherford, K. J., and Khalig Qureshi, M., 1981, Geothermal Gradient Map of Southeast Asia: S.A.P.E.X. and I.P.A., 51 pp.

Snowden, L. R., 1979, Errors in extrapolation of experimental kinetic parameters to organic geochemical systems: A.A.P.G. Bulletin, v. 63, pp. 1128-1138.

Tissot, B., 1969, Premieres donnees sur les mecanisms et la cinetique de la formation du petrole dans le sediments: simulation d'un schema reactionnel sur ordinateur: Inst. Francais Petrole Rev., v. 24, pp. 470-501.

Waples, D. H., 1980, Time and temperature in petroleum formation: Application of Lopatin's method to petroleum exploration: A.A.P.G. Bulletin, v. 64, pp. 916-926.

Yu, Zhijun, 1983, New method of oil prediction: A.A.P.G. Bulletin, v. 67, pp. 2053-2056.

4
Oil and Gas Seepages with Reference to Developing Countries

Martin Forrer

Introduction

Present direct exploitation of seeps is of small importance, and not likely to increase in the near term. Seeps can be guides to buried deposits, but the connection is often complex, and each case requires specific evaluation.

Following a short general review of seeps, this paper will address two aspects of oil and gas seepages:

1. The value of seeps for direct exploitation.
2. The value of seeps as guides to (shallow) buried hydrocarbon deposits.

Link (1952) defines a seep as follows: "A seep is . . . a place on the present earth's surface where liquid and gaseous hydrocarbons are reaching, or have formerly reached, the surface and can still be seen."

We add that a seep can be seen "by the naked eye," or discovered by very simple tests, and we include shallow pits or (water) wells as "at the surface."

We exclude surface hydrocarbon occurrences of such minute or dispersed character that they only can be detected by sophisticated (geochemical) equipment.

Oil seeps can be of asphaltic or paraffinic oils. The former are generally marked by brown to black oil, stains, or tarry residues; the latter are light brown to colorless. They may be "dead" seeps, i.e., oil no longer flows at the surface, or "live." Live, active seeps are generally mixed with gas or salt water. In area extent, seeps can range from pinpoint exudations on a rock face to many square kilometers of an oil- or tar-covered surface. Flow rates can vary from a slow ooze to several thousand liters a day.

Gas seeps are either pure gas, or accompanied by oil and/or saltwater. They range from pinpoint vents to the spectacular "eternal fires" of the Middle East.

Some seeps are easily detected by untrained people; others are only recognized by trained observers. Table 1 lists some of the characteristics of seeps.

Seeps occur:

- Geographically, all over the world; there are few countries or basins without seeps.
- Under any kind of past or present climatic condition.
- At all elevations, from below sea level to Andean plateaus.
- In any type of rock.
- In rocks of all geological ages.
- Onshore and offshore.
- In basins with or without known oil or gas fields.

Seeps are more likely to occur:

- In geologically young rocks (Cretaceous to Recent).
- In sedimentary and, secondarily, volcanic rocks.
- In earthquake belts, i.e., in areas of recent tectonic activity.
- In oil- or gas-producing areas.
- Around basin margins where hydrocarbon-bearing rocks crop out and are breached.

Seeps are more likely to be seen in low, wet areas.

The Economic Value of Seeps

Direct Surface Exploitation

In prehistoric and later times, some seeps were directly exploited (the export of bitumen from Hit in Mesopotamia, and from the Mare As-

Table 1
Recognition of Oil and Gas Seeps

Clues	Oil	Gas
Direct		
Visual	• Oil springs (dark brown to colorless). • Oil lakes or puddles. • Oil drips from rocks. • Rocks or soil are stained brown or black. • Asphalt or tar residue and mounds, often mixed with earth or vegetation (in dry climates, many paraffinic oils evaporate without a trace). • Oil floats on water in puddles, creeks or lakes and forms iridescent films (these can also be formed by iron, but break into fragments when disturbed). • Mud volcanoes (if not solfataric). • Positive ultraviolet or chloroform test.	• Gas bubbles in water or oil. • (Eternal) fires (could be marsh gas or coal gas). • Mud volcanoes and mud springs.
Smell	• Oil smell on outcrop or in broken rock fragments.	• Gas smell (esp. the "rotten egg" smell caused by poisonous H_2S).
Noise	• "Plopping" noise from mud volcanoes or mud springs.	• Gas vents give off a hissing sound. • Gas can explode. • "Plopping" mud in mud volcanoes and springs.
Indirect		
Visual	• Tar lumps floating on water or washing up on beach (could be pollution). • Stunted vegetation, bare patches in forest. • Yellowish stain from sulphur; whitish/many-hued stain from bacteria. • Wild or domestic animals coated with oil or tar. • Burnt clays (porcellanites). • Ozarkite, manjak veins, etc. • Saltwater springs and licks (may be seep-connected).	• Stunted vegetation. • Yellowish, whitish or varicolor stain (see oil). • Burnt patches in forest (could be from lighting, etc.). • Salt springs or salt licks (sometimes). • Preferred watering places for animals ("carbonated" H_2O). • Accumulations of dead animals, especially in low spots due to poisoning from H_2S gas.
Other	• Native place names, eg. manjak, naphtun, etc.	• Native place names (see oil).

phaltites, the Dead Sea). None of this bitumen was used as a significant source of heat or other energy. Seeps were also exploited for the medicinal qualities of the oil, e.g., in Europe, or by New World Indians. The quantities produced were insignificant, and so was the technology applied to this production. Mostly, the oil was allowed to accumulate in hollows and pits, or on water, whence it could be drawn off periodically. Shallow wells were dug which sometimes improved production, e.g., in China, Burma, Mesopotamia, Germany, Poland, and many other areas. Pitch lakes were exploited manually with some mechanical aids, or tar sands were spread directly for road material. Only recently has seep exploitation used sophisticated technology (Table 2).

Direct seep exploitation has declined since the 19th and early 20th century (except in Athabasca in Canada) and today, the exploitation of seeps as an energy source is minimal in a world context. The Athabasca Tar sands in Canada—probably the largest seep in the world and located in an industrialized country—are only fractionally exploited. Large pitch lakes in Venezuela and Trinidad are no longer producing significantly. Over the next 10-15 years, direct seep exploitation may rise to between 500,000 and 1,000,000 barrels per day, but this would still be only about 1% of the 85-million barrels oil and gas equivalent per day produced worldwide during the first two months of 1984.

The reason for this low production is mainly three-fold: firstly, seeps rarely yield enough oil or gas to make exploitation economical (Table 2); secondly, the oil is often of poor quality (tar) and intimately associated with its host rock, thus requiring high-cost mining and upgrading technology; and thirdly, seeps are often located far from consumers. Gas seeps are difficult to capture, and gas is expensive to bring to market. Cheaper energy sources will be available for some time. The picture may eventually change: known "dead" seeps like Athabasca represent truly gigantic potential "reserves" (UNITAR, 1981), but the change will not occur soon. Present oil and gas supplies will have to become scarce so that prices rise very significantly, or technology needs a breakthrough which dramatically lowers the cost of winning and upgrading. Even then, there will be years of delay between inception and completion of huge projects like Athabasca.

Regarding Table 2, reliable figures on quantities produced by oil seeps are rare, and production from gas seeps is even less well documented. The "eternal fires" of the Middle East, Iran, and the Baku area must have yielded large amounts of gas over long periods, but there are no good estimates. In some areas, e.g., Illinois, in the United States, Barbados, and Guyana, houses and small industries have been heated or powered, but there are no production statistics.

Table 2
Some Examples of Direct Production From Oil and Gas Seeps

Location	Production	Remarks	Reference
Natural flow production			
Worldwide range	1 to many 100s liters per day		Various
Tar Springs, California, USA	2–4 liters/day		Blumer, 1922
San Timoteo, Venezuela	50 liters/day		"
Canyon City, Colorado, USA	75 liters/day		"
Sulphur Mt., California, USA	150 liters/day		"
Coal Oil Point, California, USA	est. 8,000 to 140,000 liters/day	Offshore	Wilson, 1973 Mikolaj, 1972
Santa Monica, California, USA	320–3,200 liters/day	Offshore	Mikolaj, 1972
Oil Creek, Penna., USA	10–20 liters/day		
Ahwaz, Iran	80–100 liters/day		Coomber, 1938
Maidan-i-Naphtun, Iran	3,000 liters/day		Cunningham-Craig, 1920
Estimate of worldwide marine seepage	640,000 to 16-million liters/day "Best" estimate is 1.6-million liters/day		Wilson, 1973
Umiat, Alaska, USA	800 liters/day		Johnson, 1971
Shelikoff Strait Alaska, USA	100 liters/day	Offshore	Johnson, 1971

Table 2 continued

Location	Production	Remarks	Reference
La Brea Pitch Lake, Trinidad	100–150 tons/day	Estimated possible influx	Author
Seep production from hand-dug pits or shallow wells			
Ghana	800–1,200 liters/day	Wells	Khan, 1970
Barbados	700 BOPD and 1 MM cu ft gas per day	± 20 wells	Bull AAPG Annual Review, 1983
Dominican Republic	"several hundred BOPD"	Shallow wells Information not verifiable, not sustained	Vaughan, 1921
Guyana	140,000 cu ft gas per year	Water well	Kugler 1942
Burma	32,000 to 64,000 liters/day	± 500 pits	Tainsh 1950
Dagah Shabell Somali Republic	10 liters/day	Pit	Beeby-Thompson 1918
Production from seeps using moderate to sophisticated technology			
Athabasca, Canada	160,000 to 170,000 BOPD	Sophisticated high cost mining, upgrading and transportation	

Oil and Gas Seepages

Pitch Lake, La Brea, Trinidad	11.3 million tons from 1893–1980	Mostly manual mining, but mechanized and relatively sophisticated upgrading and shipping 1963 figure	Various
Selenizza, Albania	11,000 tons per year	Production ceased	Various
Val de Travers, Switzerland	15,000 to 50,000 tons/year	Production ceased	Various
Teca and Cocorna, Colombia	±5,000 BOPD	Production from 50 wells on or near seep (±100 BOPD on pump) and using steam stimulation	*The Texaco Star*, 1984
Austria	<100,000 Bbls cumulative	No production now	
France	<300,000 Bbls cumulative	No production now	
West Germany	300,000 Bbls cumulative	No production now	
Thailand (Mae Fang)	±100,000 Bbls cumulative; individual wells avg. 10 BOPD appr. End 1958 est. 350–500 BOPD from 15 wells	Government made major efforts to develop this seep area between 1930–1960	Various
Japan (Kanto)	Approx. 300– million m^3 gas p.a.	In shallow wells	Various

The question of how much reserve is represented by a seep is a very complex one (see the next section). A "dead" seep like Athabasca in Canada allows estimation of oil or tar in place (for other examples, see UNITAR 1981, 1984; Chilingarian and Yen, 1978) but it is impossible even there to say in a meaningful way how much of the 100 billion tons constitutes reserves, the quantity economically exploitable with present-day technology. It becomes even more complex to try and relate a "live" active seep to its buried source and to estimate those reserves.

Seeps as Guides to Shallow Oil and Gas Deposits

Early drilling near seeps soon led to oil or gas discoveries underground, and to a general recognition that seeps were often in some way indicative or buried exploitable deposits. Seeps thus became the earliest, and one of the most powerful tools for the discovery of oil and gas fields. Most, if not all, of the world's classic producing areas owe their discovery and early development to the leads provided by seeps. However, drilling near to or on seeps did not always lead to success, and conversely, there was some success where no seeps were known.

The following general relationships emerged:

1. Seeps occur frequently in sedimentary basins with oil and gas fields.
2. Less frequently, there are seeps in a basin without fields.
3. Very frequently, basins without seeps have no oil or gas fields.
4. Very rarely, a basin will have oil or gas fields, but no seeps.

Seeps, or their absence, are clearly not infallible indications of the presence or absence of buried oil or gas fields. While a total absence of seeps in a basin is always quite unfavorable, the opposite, namely the presence of seeps can have several meanings. Seeps, of course, at least indicate that the basin contains the right elements, the right mix, to generate hydrocarbons; this does not necessarily mean enough hydrocarbons—or the right traps—to form exploitable hydrocarbon deposits. Nor do seeps always directly indicate where fields are located.

Regarding the amounts of oil and gas generated: many basins contain enough oil or gas to form small seeps, but not enough for commercial fields. Or else hydrocarbons may be plentiful, but so dispersed throughout the rocks that they will form seeps but not commercial fields. Also, today's seep may be the relic of a deposit long destroyed by leakage and breaching. Indeed, seeps have been considered a bad sign by many investigators, since they may have depleted a natural reservoir. However, many seeps occur in areas where giant fields are still being found.

Among the largest active seeps known today are the California offshore seeps at Coal Point, etc. These seeps have been estimated to flow as much as 150 tons of oil per day into the ocean. If the seeps had been active for the last 10,000 years, which is entirely possible, they could have dissipated some 550 million tons. However, fields with 700 million tons and more reserves are still being found or produced in the areas nearby.

Regarding seeps as a direct indication of buried oil and gas, early drilling showed that only occasionally do buried deposits directly underlie or adjoin a seep. Oil and gas do not always move vertically to the surface from their buried source rocks. Where this is the case, drilling a well on the seep can indeed be successful. Deposits, however, need a cap to retain the oil and gas in the reservoir rock, and those hydrocarbons which manage to escape do so often by circuitous routes (faults, fractures, lateral porosity, and permeable channels), which are only indirectly, and often no longer even indirectly, connected to the source. Oil and gas may migrate many kilometers underground before emerging as a seep.

The common association of seeps with production has led to serious efforts in all areas of the world with known seeps, and which are politically and technologically accessible, to find the potential buried wealth. In many areas this resulted in commercial discoveries. In others, even considerable investment in the search has not turned up any fields and the possibility that deposits have been missed must always be considered.

Even in those areas where production was found, the question remains whether some shallow deposits were overlooked or neglected on the way to more promising deep resources. During the energy crisis of the mid-1970's, many operators, even in old heavily drilled areas, like the United States, re-evaluated the shallow horizons, often successfully developing new reserves.

Ultimately, the significance of any seep needs to be determined by a trained observer. He can often trace the cause and origin of a seep with fair confidence, especially where ample supporting data are available. This is known as seep analysis. One of the best analyses of seeps, their geological meaning, and relation to underground deposits was published by Link (1952) (but see also Blumer (1922), Cunningham-Craig (1920), and Beeby-Thompson (1934)). From a study of many seeps it becomes clear that no general rule can be laid down other than that each case needs to be evaluated on its own geological aspects.

Seep analysis attempts to trace the oil and gas in a seep back to its source, and thus to determine whether the seep is the expression of a useful deposit. Seep analysis can often be helpful, and at least a preliminary effort should always be made to see whether further analysis is desirable and likely to succeed.

Occasionally, investigation of the seep and its geological setting will immediately indicate a likely cause. It can often determine whether there might be a field or whether the seep merely originates from very diffuse hydrocarbons in breached shaly rocks, etc. More often, the relationship is more obscure, and the analysis requires application of sophisticated methods and means. The exact location of the source may remain a mystery, even in areas with a great deal of relevant information.

Each seep analysis has its own special questions and answers; here we can only indicate in a general way some of the more common problems and approaches. Specific examples would not have application in other areas, but one example of seep analysis is given by Reitsema and others (1978). In general, the expert must:

1. Determine whether the seep is genuine, or merely pollution, a misleading indication, or even due to fraudulent action. This generally requires field investigation and laboratory analysis.
2. Try to accurately locate the seep(s); this is often difficult under heavy vegetation, in the offshore, or in seeps covering very large areas.
3. Attempt to tie the seep to a specific rock or geological unit. Again, this may be difficult if the seep issues from extensive and thick soil or mud cover. Geological and geophysical information can help and so can modern geochemical methods, which permit "fingerprinting" the oil. This may lead to correlation with a source or host rock even where no direct tie can be seen. Geochemistry may also help to determine whether the seep owes its origin to a single source, or several; and it can show whether a gas seep is likely to be associated with oil or is only dry gas.
4. Try to determine the local geological conditions which give rise to the seep; does it issue from a porous and permeable carrier bed, from fractures, or along a fault? Is it connected with volcanic intrusives?
5. Tie this information into a broad geological framework which reconstructs the migration path taken by the oil and gas from the source to the present seep locality. This means establishing the underground configuration of a reservoir bed or tracing a fault to its intersection with a possible reservoir containing oil or gas. In trying to retrace the migration path, the expert will need as much subsurface geological and geophysical information as he can get. Unfortunately, the most critical information concerns the shallow "window" between the surface of the earth and, say, 500–1,000 m depths, and this zone is often very incompletely known. Explorers have long treated this shallow zone as a "quantité négligeable," sav-

ing themselves the expense of sampling or logging it in wells. Exploration geophysics (seismic) for underground deposits is purposely biased in favor of bringing out the deep information at the expense of the shallow data, and unless special methods and techniques are applied, the shallow zone remains blanked out.
6. Finally, using all data, he must try to determine the optimum location for drilling into the buried hydrocarbon deposit.

Recommendations

Any country with known seeps but no production should make seep analysis a part of its energy exploration program. Countries, or basins, with known seeps but no production which might benefit from such efforts include Costa Rica, Dominican Republic, Equatorial Guinea, Ghana, Guyana, Jordan, Kenya, Malagasy Republic, Nepal, Panama, Portugal, Somali Republic, Suriname, Tanzania, Uganda, and Yemen.

Any country with or without seeps but with established production may still find it useful to go through the same seep analysis and shallow-section review. Examples of old productive areas where shallow deposits were indicated by seeps but were either overlooked or ignored are quite numerous and include even countries with long production histories, like the United States. One may reasonably say that *any* producing province needs to be reexamined for shallow production potential. Better documentation of shallow geological and geophysical information should be required from oil and gas operators as part of their exploration and/or producing agreements.

References

Abraham, Herbert, 1960, Asphalts and Allied Substances, Sixth Ed., Volume 1: Van Nostrand Co., Inc., New York & Princeton.

Beeby-Thompson, A., 1918, Report on the Dagah Shabell Oil Field (Brit. Somaliland): Cairo, Government Printer.

Beeby-Thompson, A., 1934, The economic value of surface petroleum manifestations: World Petr. Congr. Proc., v. 1, London.

Beeby-Thompson, A., 1950, Oil field exploration and development: The Technical Press, London.

Blumer, E., 1922, Die Erdoellagerstaetten: verlag Ferd. Enke, Stuttgart.

Chilingarian, G. V. and Yen, T. F., 1978, Bitumens, asphalts & tar sands: Elsevier, New York & Amsterdam.

Clapp, F. G., 1927, Fundamental criteria for oil occurrence: Amer. Assoc. Petr. Geol. Bull., v. 11, pp. 683–703.

Coomber, S. E., 1938, Surface indications of oil: Science of Petroleum, v. 1, pp. 291–293, (A. E. Dunstan et al., Eds.).

Cunningham-Craig, E. H., 1920, Oil finding: Edward Arnold & Co., London.

Dalloni, M., 1922, La géologie du pétrole en Algérie: Alger.

deGolyer, E., 1940, Direct indications of the occurrence of oil and gas: Elements of the Petr. Industry, AIMME, New York, pp. 21–25.

Demaison, G. J., 1979, Tar sands and supergiant fields: Amer. Assoc. Petr. Geol. Bull., v. 61, pp. 1950–61.

Denser, W. G., et al., 1973, Gas in Lake Kivu: Science, v. 181, pp. 51–54.

Forbes, R. J., 1936, Bitumen and petroleum in antiquity: E. J. Brill, Leiden.

Harvey, G. R., et al., 1979, Observation of a sub-surface oil-rich layer in the open ocean: Science, v. 205, pp. 999–1001.

Illing, V. C., 1938, The significance of surface indications of oil: Science of Petroleum, v. 1, pp. 294–296 (A. E. Dunstan et al., Eds.).

Johnson, T. C., 1971, Natural oil seeps in or near the marine environment: U.S. Dept. of Commerce NTIS AD-723.310.

Khan, Mohsin H., 1970, Cretaceous and Tertiary rocks of Ghana, etc.: Ghana Geol. Survey Bull., no. 40, Accra.

Kugler, H. G., and MacKenzie, S. C., 1942, Report on exploration for oil in British Guiana: Brit. Guiana Government Geologist Bull., no. 20, Georgetown.

Landes, K. K., 1959, Petroleum geology: John Wiley and Sons, New York.

Landes, K. K., 1973, Mother Nature as an oil polluter: Amer. Assoc. Petr. Geol. Bull., v. 57-4, pp. 637–641.

Levorsen, A. I., 1967, "Geology of petroleum, second ed.: W. H. Freeman and Co., San Francisco.

Link, W., 1952, Significance of oil and gas seeps in world oil exploration: Amer. Assoc. Petr. Geol. Bull., v. 36-8, pp. 1505–40.

Mikolaj, P. G., Allen, A. A., and Schlueter, R. S., 1972, Investigation of the nature, extent and fate of natural oil seepage off Southern California: Paper no. OTC-1549, v. 1, pp. 365–378, AIMME Offshore Technol. Conf., Dallas.

Phizackerley, P. H., and Scott, L. O., 1978, Major tar sand deposits of the world: Bitumens, asphalts and tar sands, (G. V. Chilingarian and T. F. Yen, Eds.) Dev. in Petr. Sci., Elsevier.

Reed, W. E., and Kaplan, I. R., 1977, The chemistry of marine petroleum seeps: Jour. Geochem. Explor., v. 7, pp. 255-293.

Reitsema, R. H., Lindberg, F. A., and Kaltenback, A. J., 1978, Light hydrocarbons in Gulf of Mexico water: Sources and relation to structural highs: Jour. of Geochem. Explor., no. 10, pp. 139-151.

Rice, D. D., and Claypool, G. E., 1981, Generation, accumulation and resource potential of biogenic gas: Amer. Assoc. Petr. Geol. Bull., v. 65-1, pp. 5-25.

Tainsh, H. R., 1950, Tertiary geology and oil fields of Burma: Amer. Assoc. Petr. Geol. Bull., v. 34.

UNITAR, 1981, The future of heavy crude and tar sands: McGraw-Hill, New York.

UNITAR-UNDP, 1984, UN agencies tally heavy oil, bitumen reserves: Oil and Gas Journal, May 14, pp. 66-67.

Vaughan, T. W., et al., 1921, A geological reconnaissance of the Dominican Republic: Geol. Svy., Dom. Rep., Memoirs, v. 1.

Wayland, E. J., 1926, Petroleum in Uganda: Memoir #1, Geol. Svy. of Uganda.

Wilson, R. D., 1973, Estimate of annual input of petroleum to the marine environment, etc.: NTIS v. 1, pp. 59-96, U.S. Dept. of Commerce.

Wilson, R. D., et al., 1974, Natural marine oil seepage: Science, v. 184, pp. 857-865.

Woodring, W. P., et al., 1924, Geology of the Rep. of Haiti: Geol. Svy. Dept., Port-au-Prince.

Zuber, St., 1934, On some kinds of outcrops of oil-bearing strata and their prospecting: World Petr. Congr., Proc., v. 1, London.

5

Low-Cost Landform Analysis in Prospecting for Shallow Hydrocarbon Occurrences

Richard L. Bowen

Abstract

Landform analysis, while widely used as a tool in petroleum search during the last century and the early half of this century, became rarely utilized in the 1950s. Now, however, with the extensive world wide mapping programs on lands and continental shelves, generally available air photographic coverage, and inexpensive access to satellite imagery, this tool merits re-evaluation and inclusion in exploration programs. Because landforms and drainage networks largely develop in response to near-surface geologic conditions, regional reviews of their character and implications for successful petroleum discovery by experienced landform analysts should constitute a low-cost initial step in the search for shallow hydrocarbon occurrences.

Introduction

Before shallow oil and gas resources can be developed, they must first be discovered. The discussion which follows aims at emphasizing landform analysis as an initial program in assisting in the reduction of costs of petroleum exploration activities as well as in improving the odds towards discovering exploitable hydrocarbons. This tool of petroleum search was widely applied in the early and middle portion of this century, but has fallen into disuse in recent decades. However, recent developments in atmospheric and satellite sensing and photography, combined with more refined national mapping programs onshore and hydrographic surveys offshore, have made landform analysis a low-cost and useful early step in the search for petroleum.

Historical Background

Discovery by Direct Observation

Soon after the discoveries of oil fields in the eastern portions of the United States and Canada in the latter half of the 19th century, recognition emerged that oil fields frequently and regularly occurred in anticlines of that part of North America. Domestically, the idea of drilling into anticlines to find oil was rapidly applied in Oklahoma, Colorado, and, with particular success, in California. In these states, the arid to semi-arid climate made anticlines far more visible than in the humid east of North America. At the same time, concessions were being taken up and recognizable anticlines drilled (with many a discovery) in Burma, Indonesia, Iran, and Venezuela, as well as other overseas regions.

Often, the areas in which the drilling of anticlines resulted in discoveries were also areas in which tar deposits, gas seeps, oily springs and similar surface manifestations of buried hydrocarbons occurred with considerable frequency. Drilling in areas of known seeps has generally been successful worldwide.

The next step in petroleum search based on observable surface features of the landscape was that of drilling around domes, features with circular outlines in contrast to the typically elongated, elliptical shapes of anticlinal oil fields. In the U.S. Gulf Coast, such features typically formed over salt domes. Bloom (1978, p. 49) notes that only five Gulf Coast salt domes have positive relief, although many others are characterized by annular drainage patterns. While quite a few shallow hydrocarbon reservoirs have been discovered in the deformed strata above these salt intrusions, producible traps occur with somewhat greater frequency around the sides of these domes, and others like them, in Europe, Africa, and elsewhere.

An anomaly discovered through geologic mapping led to the discovery of the shallow (about 1,000 m) Jackson, Mississippi, gas field. This domal structure formed over a buried, extinct volcano rather than over a salt dome.

From the earliest days of flight, but more particularly in the time of the slow, low-flying aircraft of the 1920s and 1930s, airborne observation of the land surface revealed structures difficult to see on the ground. Rather frequently, drilling such structures resulted in discoveries of shallow producible reservoirs. Stegner (1971, pp. 36-44) describes how from 1934, in the first years of the Aramco concession in Saudi Arabia and prior to the initial discoveries, air photography analysis and overflights were extensively used in the exploration program to study landforms and reveal indications of geologic structures.

Inferential Analysis

Through the years of successful exploration involving analysis of the drainage patterns and land surfaces associated with success in petroleum discovery, whether from on-the-ground observation or from airborne platforms, a data base was constructed. The principal impact of this collection of data was the demonstration that, far more pervasively than earlier imagined, surface drainage and landforms reflect subsurface conditions and structures. Moreover, shallow petroleum reservoirs have a high frequency of association with anomalous landforms and drainage features. Howard (1967) published a particularly extensive collection and codification of examples of normal and anomalous drainage patterns.

Meanwhile, national programs of photogeologic analysis and mapping, utilizing air photography of the 1940s and 1950s, were undertaken. Two large-scale country programs—Brazil and Australia—resulted in maps on scales of 1: 250,000 and 1: 500,000. These maps involved the principles of landform analysis applied to photogeologic mapping and were supplemented by only a small amount of on-the-ground geologic control. Such maps, of course, had great utility in planning future programs of more detailed studies.

Impact of the Oil Crisis

Several factors from the late 1950s through the early 1970s together reduced the utilization of landform analysis in petroleum search. The long-term flat price for petroleum until the early 1970s caused exploration efforts, domestically and internationally, to be directed toward potential large-discovery, large-payout ("bonanza") exploration plays. These largely involved campaigns in frontier areas, such as the eastern foothills and plains bordering the Andes Mountains, the interior, rather than coastal, portions of Africa or the offshore area of the Arctic Ocean, the North Sea, and other continental shelves. The geographical expansion of exploration search was accompanied by progressively deeper drilling with correspondingly greater costs.

Just as the frontier regions were being tested, petroleum prices jumped in late 1973. This price incentive to wildcat exploration was accompanied by a qualitative and quantitative improvement and change in geophysical technology, which made possible the deep-drilling of small but geophysically defined structures, which with high prices for product could become economically attractive.

These shifts of exploration practice largely bypassed landform analysis as a tool, although as early as 1952 standard exploration practice for offshore drilling in California involved the geomorphologic analysis of bathymetric and hydrographic charts in order to determine whether shelf warping or faulting appeared indicated.

Why Landform Analysis Now?

The thrust of petroleum exploration in the past decade was mostly focused on potential large-payout targets, both domestically and internationally. Highly sophisticated drilling and geophysical technology became available and provided support for such searches. Now, however, it has become important to consider that often significant production and profits can come from finding and developing shallow reservoirs. To maximize such production and profits, it becomes imperative to minimize the costs of successful exploration. Landform analysis, or geomorphologic evaluation, utilizing information largely unavailable 20 years ago, when the practice largely disappeared from exploration programs, can help significantly to reduce exploration costs.

Applicable New Information

Airborne

In contrast with conditions prevalent 20 or 30 years ago, almost all land areas of the world now have been photographed at low to high altitudes in surveys used for national planning, forestry studies, agricultural programs, geographic and geodetic surveys, and the like. Often such photography is available in several different kinds of black and white processing as well as true-color and false-color processing. Thus, a data bank exists already for most parts of the world and awaits only the interpretation of a landform analyst, or exploration geomorphologist, to define areas of potential interest for shallow petroleum.

Satellite

With the ready availability of satellite imagery world wide, any exploration program involving areas of more than a few hundred square kilometers should begin with the geomorphologic evaluation of all displays available from satellite studies. Such displays are available in several spectral bands, including infra-red, along with many types of processing in black-grey-white, true color, and false color. At ground coverage of a thousand square kilometers and up for each print, this provides a remarkably inexpensive first look at the landforms of the area concerned and how they reflect geologic conditions in the subsurface. Good examples of the use of remote sensing imagery appear in papers by Bentz and Gutman, Miller, and Worthing in the First Pecora Symposium on Remote Sensing (Woll and Fischer (Eds.), 1977). Halbouty (1980) has shown that features associated with giant oil fields are recognizable on satellite imagery.

Marine

The satellite photography programs have also provided useful images of the shallow sea floor, particularly in areas of clear water. Moreover, commerce of the last three decades has required the extensive preparation of bathymetric and hydrographic maps over nearly all the waters of the world which are less than 200 m in depth. While the detail of such maps and charts varies considerably, they generally are readily available and inexpensive, thus constituting a highly useful form of data for geomorphologic analysis of shelf regions. One example is a recent evaluation of the continental shelf of Portugal (Vanney and Mougenot, 1981). In addition to charts of these types from national hydrographic services, other information on shelf morphology is available from published or open-file scientific studies in oceanography or marine geology.

Mapping

Both through the efforts of national mapping programs and those of international agencies, large portions of the world's lands have now been mapped planimetrically or topographically or both at scales of 1: 25,000 to 1: 250,000. Such maps are particularly useful to the landform analyst for the details of drainage systems available from them. In an in-progress study of salt-dome stability beneath 12,000 sq km of southeastern Mississippi, the author, using such maps, has found that contrasts in drainage nets over known salt domes reveals whether such domes have experienced significant deformation in geologically recent times. Berger and

Aghassy (1980) utilized similar procedures in evaluating a select group of shallow salt domes throughout the U.S. Gulf Coast, as potential sites for nuclear waste isolation.

In recapitulation, then, photography, imagery, and mapping from shipboard, atmospheric investigations, and satellite observations now constitute a great information bank of excellent quality. This collection of extremely useful data, available at minimal expense, should provide the basic raw material for geomorphic analysis in planning exploration programs for shallow hydrocarbon accumulations.

Landform Analysis Potential

A landform analyst has a particular place in exploration for shallow hydrocarbon concentrations in the initial stage of the exploration program. This is time when the most promising portions of the region under evaluation are being defined and selected for detailed and intensive investigation.

Precisely because the satellite imagery, atmospheric photography, land maps, and marine charts already exist for most of the world's land and offshore areas, and are inexpensively available and accessible, the costs for an initial stage of an exploration program of this type are low. They involve only the cost of materials and the time required for the landform analyst to evaluate them and prepare his report and recommendations.

Moreover, because exploration programs here considered are exactly those that involve shallow hydrocarbon concentrations, the expertise of the landform analyst becomes especially applicable. It is the near-surface folding, faulting, fracturing, and lithologic variability of the rocks which control the occurrence of shallow petroleum accumulations; these are the factors that most affect the development and shaping of surface landforms and drainage patterns.

The landform analyst's studies result in the most accurate evaluations of shallow structure, rock character, and possible sites of hydrocarbon accumulation in arid and semi-arid regions, for under these climates relations between subsurface and surface phenomena are rather straightforward. However, in humid regions of extensive and dense vegetative cover, an evaluation by experienced landform analysts becomes particularly important. For, in such conditions of little or no rock outcrop, analysis of the topography and drainage patterns is often the only procedure short of expensive geophysical study and drilling that can indicate potentially anomalous geologic conditions.

Conclusion

In the search for shallow hydrocarbon accumulations, the potential of landform analysis, utilizing presently available mapping and imagery to discriminate between more and less favorable portions of the region under consideration at a minimal cost, make such an analysis an eminently desirable early step in exploration programs. Moreover, discovery of shallow production often carries a hoped-for bonus—a large portion of shallow hydrocarbon discoveries often prove later, on deeper drilling, to also have producible deeper reservoirs.

References

Berger, Z., and Aghassy, J., 1980, Geomorphic manifestations of salt dome stability, in Applied Geomorphology, (R. G. Craig and J. L. Craft, Eds.): London, Allen and Unwin, pp. 72–84.

Bloom, Arthur L., 1978, Geomorphology: Prentice-Hall, Inc., Englewood Cliffs, New Jersey, 510 pp.

Halbouty, Michel T., 1980, Geologic significance of LANDSAT data for 15 giant oil and gas fields: *Amer. Assoc, Petroleum Geologists Bull.*, v. 64, pp. 8–36.

Howard, Arthur David, 1967, Drainage analysis in geologic interpretation: A summation: *Amer. Assoc. Petroleum Geologists Bull.*, v. 51, pp. 2246–2259.

Stegner, Wallace, 1971, *Discovery! The Search for Arabian Oil:* Middle East Export Press, Inc., Beirut, Lebanon, 190 pp.

Vanney, Jean-Rene, and Mougenot, Denis, 1981, La Plate-Forme Continentale du Portugal et les Provinces Adjacentes: Analyse Geomorphologique, *Serv. Geol. Portugal, Mem. 28*, 86 pp. + 41 plates.

Woll, P. W., and Fischer, W. A., Eds., 1977, Proceedings of the First Annual William T. Pecora Memorial Symposium, October 1975, Sioux Falls, South Dakota, *U.S. Geol. Surv. Prof. Paper 1015*, 370 pp.

6
Satellite Remote Sensing in Exploration for Shallow Petroleum Resources

John R. Everett

Abstract

The concepts of regional geology provided the geologist by use of satellite remote sensing data and the likelihood that surface manifestations are more likely to mark shallow rather than deep hydrocarbon accumulations makes data from space-borne sensors valuable tools in the search for shallow oil and gas, particularly in the less-explored areas of the world. The synoptic nature of these data and their relatively low cost make them appropriate to the search for smaller accumulations and useful for guiding and focusing more expensive conventional exploration tools and techniques. This is especially true in remote areas of difficult access.

The role of remote sensing in an exploration program can range from simple logistical support, such as planning access routes and seismic surveys, to regional exploration and definition of plays and, in some instances, to aid in selection of drill sites. Even though remote sensing is a relatively new addition to the petroleum explorationist's arsenal of tools, there are many examples of its practical contribution to exploration programs. Regional analysis of the Western Overthrust Belt of the U.S. suggests that transition zones where regional strike-slip faults become thrusts are the most attractive exploration targets in the geologic province. Many potentially productive structures are clearly visible on Landsat imagery. In the U.S. most of these "obvious" structures have been tested, but in less explored parts of this world this is not the case. Under specific conditions some of the newer sensors can detect the results of hydrocarbon microseepage, thus greatly aiding exploration.

To date, satellite remote sensing has made major contributions to petroleum exploration by reducing costs and increasing success. As our understanding of these tools increase through experience and as sensors with improved spatial and spectral resolution begin operation, this contribution to the search for shallow oil and gas will increase.

Introduction

Perhaps the place where satellite remote sensing makes its greatest, or at least most obvious, contribution to the quest for hydrocarbons is in the search for shallow petroleum accumulations. This comes about as a conjunction of the perceptual powers of regional geological patterns provided the geologist using high-quality data from satellite systems and the probability that shallow accumulations are more likely to have surface manifestations than those lying at greater depths. In well-explored parts of the world, such as the United States, most of the "obvious" (those with strong surface expression) shallow accumulations have been discovered. However, in less explored parts of this world the possibility exists for identification of many potential, as yet untested, shallow hydrocarbon traps.

The synoptic view, on the order of thousands of square kilometers, of these systems and the relatively low cost of acquiring, processing, and interpreting the data make these tools particularly attractive for the pursuit of shallow accumulations. Except in unusual circumstances satellite data will not by themselves lead to the discovery of oil or gas. Usually the results of this interpretation must be combined with information from conventional petroleum exploration tools in order to define viable prospects and select drill sites. However, information from satellite remote-sensing systems can guide and focus more conventional exploration approaches, resulting in substantial savings of time, money, and effort. Reducing the amount of seismic data needed to locate and define a prospect by only 10 kilometers will more than pay for the cost of using most types of satellite remote-sensing data.

Examples of Remote Sensing in Exploration

The contribution of satellite remote sensing has ranged from simple logistic support to, in a few instances, selection of drill sites. The planimetric quality of the data and the abundance of terrain detail make several types of data, e.g., Landsat and shuttle imagery radar (SIR), useful for planning routes of access, location of base camps, and the orientation and location of geophysical survey grids based on the structural grain of the area.

Applications that are more directly associated with exploration include establishment of a regional structural framework, location of prospective structural traps, and identification of areas of possible hydrocarbon microseepage.

In all instances a great deal of information not discernable from the remotely-sensed data must be brought to bear in order to produce an exploration success. These types of information include the nature of the geologic section (age, lithology, source rocks, reservoir rocks, seals), nature of the structure beneath an unconformable cover, and tectonic and depositional history of the area. Simply stated, the application of remotely sensed data to petroleum exploration in a vacuum of supporting geologic information and experience does not work well in most instances. The more information and experience that is brought to this interpretation the more valuable it is and the greater the likelihood of exploration success.

Possibly the best documented example of this integrated use of remote sensing in a successful petroleum exploration campaign is Trend Explorations effort in Irian Jaya, Indonesia (Foster and Soepaijadi, 1974; Jordan, 1974; Soepaijadi, 1974; Vincelette and Anwar, 1974; Vincelette and Soepaijidi, 1974; and Warters and Najoau, 1974). This project involved locating pinnacle reefs in the Salawati Basin, using regional geologic knowledge and existing, limited subsurface and seismic data, guided by the ability to recognize the geomorphic expression of the buried reefs on black-and-white, color, color infra-red, and low-illumination-angle imagery. The result was a series of discoveries of giant proportions.

Wyoming Overthrust

A widely used and very powerful role for satellite imagery data is as a basis for looking at entire geologic provinces in order to focus exploration efforts. An obvious advantage of the data in this role is that a single image can span a mountain range and a few images can cover an entire exploration province. The Wyoming Overthrust is an example. The sawtooth line on the map (Figure 1) shows the position of the Western Overthrust. We shall focus our attention on the eastern bulge in Wyoming, which is covered by the Landsat Multispectral Scanner (MSS) mosaic (Figure 2A).

The regional geology (Figure 2B) is dominated by north-trending thrust faults. These faults are obvious in the topography as long linear ridges in the MSS mosaic. Bear Lake, Fossil Basin, and the Uinta Mountains are good points of reference.

The structure (Figure 2B) emerges as a pattern of left-lateral, west-northwest-trending, strike-slip faults that swing to the right into the north-trending thrusts; these in turn swing to the southwest and become right-lateral strike-slip faults. Figure 3 diagramatically portrays this pattern. During the Laramide orogeny (about 60 ma) North America telescoped in an east-west sense, producing this pattern of structure. The distribution of oil and gas, shaded gray in Figure 2B, reveals another inter-

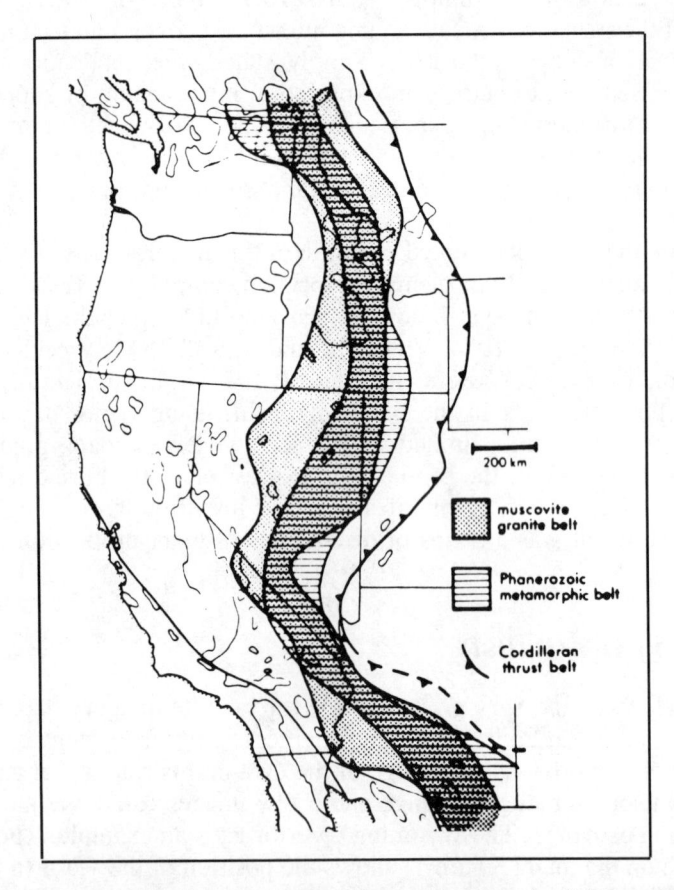

Figure 1. Map of the western United States showing the Western Overthrust Belt (sawtooth line). The easternmost bulge in the belt is in Wyoming and is the focus of Figure 2.

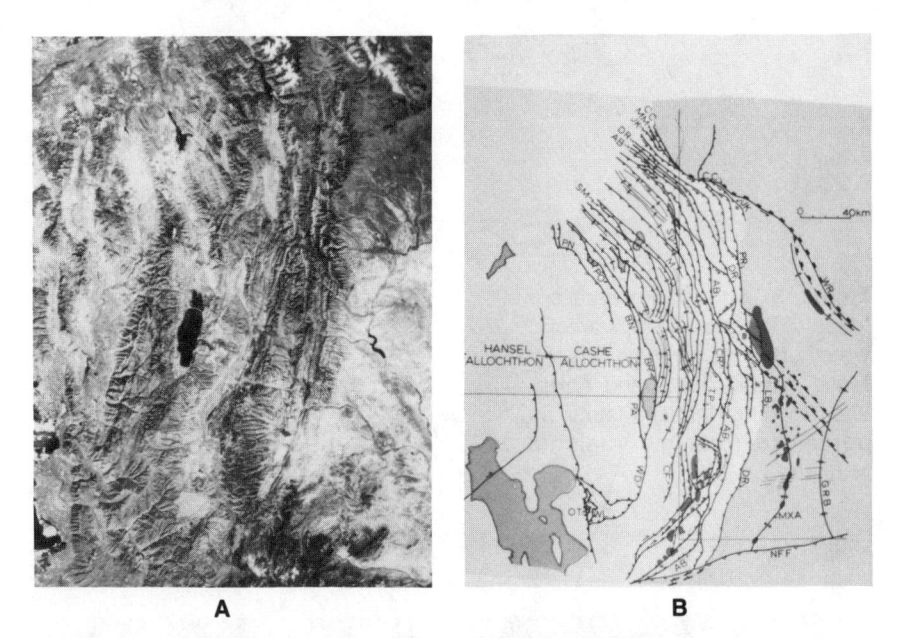

Satellite Remote Sensing for Shallow Petroleum Resources   71

Figure 2. (A) Landsat mosaic of Wyoming portion of the Western Overthrust Belt. Bear Lake lies near the center of the mosaic; Fossil Basin is the plateau area east of Bear Lake. (B) Generalized structure of the area of the mosaic compiled to and interpreted from the mosaic. Black areas indicate current hydrocarbon production (notice production tends to lie in transition zones from thrusts to strike-slip faults). GRB = axis of Green River Basin, MXA = Moxa Arch, LB = LaBarge Thrust, DR = Darby Thrust, PR = Prospect Thrust, AB = Absaroka Thrust. The mosaic (A) is at a slightly larger scale than this map.

esting pattern. All of the known accumulations occur in the transition zones where thrusts swing into strike-slip faults.

Figure 4 shows the south end of the Fossil Basin. The major accumulations lie within the zone of strike-slip/thrust transition. The numbered boxes delineate structures that are analogous to traps at these nearby fields.

In Wyoming, careful interpretation of a dozen Landsat images coupled with geologic insight allowed the interpreters to quickly proceed from considerations of the regional structure to local areas, where one can focus more expensive exploration tools and effort.

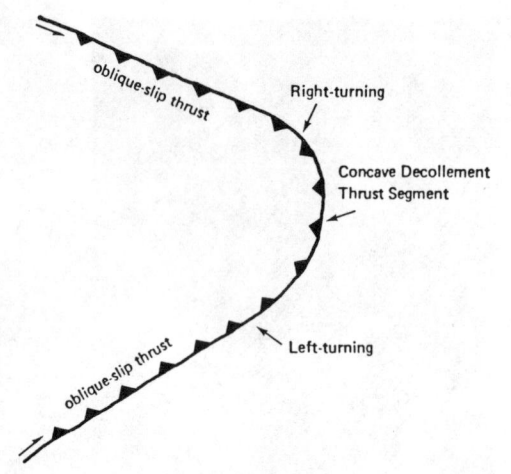

Figure 3. Schematic representation of the relationships among west-northwest trending left-slip faults, thrust faults, and northeast trending right-slip faults, as perceived from the regional structure.

Figure 4. (A) An enlarged portion of the mosaic in Figure 2A showing Bear Lake with Fossil Basin to the east. (B) A close up of a portion of the Landsat image that comprises the southeastern portion of mosaic of Figure 4 with an overlay that shows interpreted structure, the location of existing oil and gas production (stippled), and areas of exploration interest (numbered boxes).

Bold Surface Features

Anomalous surface features mark the location of many shallow oil and gas accumulations. These features include obvious anticlines, anomalous topography, unusual drainage texture or patterns, and in some instances vegetation communities that contrast with the surrounding vegetation. The spatial and spectral resolutions, synoptic view, and the resulting uniform lighting conditions over large areas and the regional perspective inherent in many satellite imagery systems make many of these features obvious. In addition the large areas covered by each frame of imagery make it possible to examine known producing areas and quickly search for analogues. In well-explored areas such as the United States most of the obvious features have been drilled. However, even in some mature exploration areas there are attractive features that are as yet untested. In less thoroughly explored or remote areas this potential for new discoveries is even greater.

The Rocks Springs Uplift, a large, highly productive anticline lies in the central portion of the Green River Basin. The Church Buttes oil field, a less obvious feature with anomalous drainage texture, lies in the southwestern part of the uplift. In less-explored parts of the world equally obvious anomalies remain untested. Satellite imagery offers an efficient, inexpensive tool for rapidly locating features of this type.

Paradox Basin, Utah

Where the proper conditions exist, microseepage of hydrocarbons from oil and gas reservoirs alters the surface rocks in ways that change their spectral and weathering characteristics. The Paradox Basin of Utah is one such area. Vegetation is sparse and the rocks are well exposed, making it an excellent area for studying such effects.

Two outcrops, both in Wingate Sandstone, are less than five kilometers apart. One is the usual red color and appearance of the formation, the other is bleached. The bleached outcrop overlies Lisbon oil field. This type of change is common in areas of red sandstones. The hydrocarbons and associated hydrogen sulfide leaking from the underlying reservoir reduce the relatively insoluble red ferric iron to ferrous iron, which is more soluble and is removed. Thematic Mapper (TM) data is excellent for detecting these color changes because of the sensitivity of the short wavelength bands to iron absorption. As in the other examples discussed, other geologic data must be considered in order to precisely select a drilling location. This is because other factors may produce similar bleaching or the escaping hydrocarbons may not have moved vertically from the reservoir.

Raton Basin, Colorado

In the Raton Basin of southern Colorado, one of the problems in exploration is avoiding igneous dikes when drilling a well. Figure 5A is a Landsat-4 Thematic Mapper (TM) land 1,4,5 composite showing the Sangre de Cristo Range on the west, covered with snow. The Raton Basin lies east of the range. The Spanish Peaks, with their spectacular set of dikes, are near the center of the picture, and the area of exploration activity surrounds the peaks to the north, east, and south. Figure 5B is a portion of the interpretative map made from the TM image. Mapping was done at a scale of 1:48,000, which is a common oil industry map scale in this country. Using TM imagery we found that we mapped slightly more dikes than were mapped in the field and that the interpreted dikes tended to be longer than those on the map. We did miss a few dikes that were mapped in the field, but found that we mapped more than twice as many dikes from the TM as we mapped from black-and-white stereo air photos. This apparent improvement of TM over air photos may be a function of color, low sun angle, and perhaps the spectral characteristics of TM. This example suggests that TM data can assist the explorationist at the well-location level of detail.

Conclusion

Data from existing satellite sensing systems (MSS, TM, Shuttle Imaging Radar) can contribute to the discovery of shallow oil and gas accumulations. This is true in the relatively well-explored parts of the world, but the contribution will be greatest where exploration is less intense. The data are particularly valuable in remote areas or areas of difficult access. Insights from the data allow explorationists to concentrate or focus more expensive conventional geologic and geophysical tools on areas with the highest probability of success. This conservation effort reduces costs, saves time, and makes it feasible to search for smaller accumulations.

Future satellite systems will have higher spectral and spatial resolution, broader spectral range, and potential for acquiring stereo imagery. Each of these improvements will increase the value of satellite remote-sensing data in the search for shallow oil and gas accumulations.

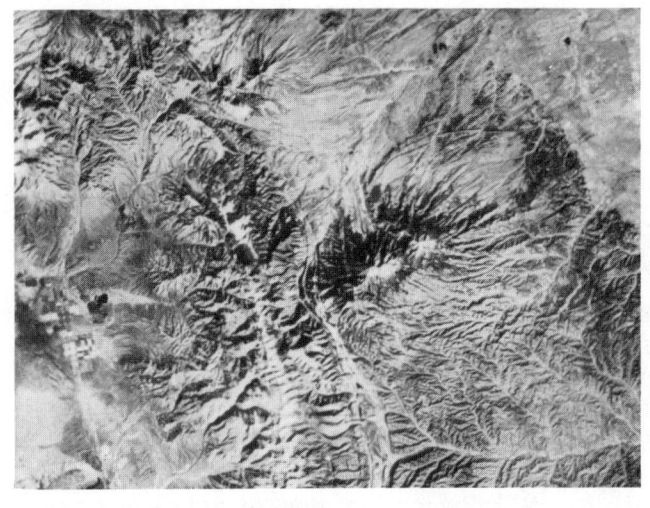

Satellite Remote Sensing for Shallow Petroleum Resources 75

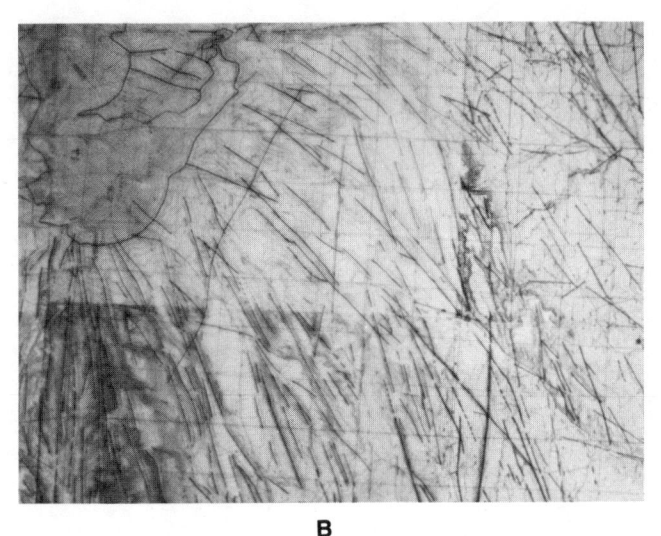

Figure 5. (A) Landsat TM (bands 1, 4, 5,) image of the Sangre de Cristo Range, Spanish Peaks (near the center of the frame), and Raton Basin (area to north, east, and south of Spanish Peaks), Colorado. Note dikes radiating from the Spanish Peaks. (B) To avoid drilling the igneous dikes this map was prepared from the TM image. The interpretative map shows about the same number of dikes as detailed field maps and more than twice as many dikes as can be interpreted from air photos.

References

Foster, N. H., 1974, Geomorphic expression of pinnacle reefs in Salawati Basin, Irian Jaya, Indonesia: Amer. Assoc. Petrol. Geol., Annual Meetings Abstracts, v.1, pp. 35-36.

Jordan, T., 1974, Reef exploration in Irian Jaya, Indonesia: Amer. Assoc. Petrol. Geol., Annual Meetings Abstracts. V.1, p. 51.

Soeparjadi, R. A., 1974, Changing outlooks of Irian Jaya: Amer. Assoc, Petrol. Geol., Annual Meetings Abstracts, v.1, p. 84.

Vincelette, R. R., 1974, Geology of productive oil fields, Salawati Basin, Irian Jaya, Indonesia: Amer. Assoc. Petrol. Geol., Annual Meetings Abstracts, v.1, pp. 92-93.

Vincelette, R. R., 1974, Geology of Salawati Basin, Irian Jaya, Indonesia: Amer. Assoc. Petrol. Geol., Annual Meetings Abstracts, v.1, p. 93.

Warters, H. R., 1974, Regional relations of reef development in Salawati Basin, Irian Jaya, Indonesia, as compared to other reef terrains in world: Amer. Assoc. Petrol. Geol., Annual Meetings Abstracts v.1, p. 95.

7
Satellite Remote Sensing— An Effective Tool for Oil and Gas Exploration

John S. Carter

Introduction

Satellite data contain extractable information relative to three major aspects of oil and gas exploration:

1. Buried structure: geomorphology, drainage, etc.
2. Effects of tectonic and post-tectonic activity: lineaments.
3. Hydrocarbon micro-seepage: tonal or spectral anomalies.

It should be noted, however, that a great deal of caution must be exercised in properly extracting this information, especially with respect to the latter.

Three examples of Landsat-generated "wildcat" oil and gas prospects, supported by subsurface, surface, and geophysically-acquired geochemical information illustrate that multi-spectral scanner satellite data can be used to locate minor low-relief buried structures, minor faulting, and related fracture systems and in some instances, geochemically altered surface rocks and soils resulting from hydrocarbon micro-seepage. Once this information has been obtained from the satellite data, it is then necessary to follow up with detailed subsurface geology, if available, seismology, other geophysical tools, when economically prudent, and geochemistry, in order to develop a drillable-prospect model.

In the illustrations to follow, the initial investigation, and subsequent prospect development, were promulgated by the use of computer-processed Landsat Multi-Spectral Scanner (MSS) data. The threshold question: "Is there any apparent need for further geologic investigation?" was answered affirmatively. Thereby, the prospect development process was put into motion. The following is a brief outline of the procedures which resulted in the geologic models for the prospects:

1. Detailed remote sensing and geomorphic study.
   - Computer-processed Landsat Multi-Spectral Scanner data.
   - USGS topographic maps (scales 1:250,000 and 1:24,000).
   - County soil conservation maps.
   - Texas geologic maps.
2. Local production evaluation.
3. Acquisition of geochemical data.
   - Computer-processed aeromagnetic detection of diagenetic magnetite in near-surface rocks.
4. Extensive subsurface study.
   - Structural mapping.
   - Cross-section of structural markers and producing horizons.

Remotely sensed multi-spectral scanner satellite data provide a viable, cost-effective reconnaissance tool for oil and gas exploration, when properly weighted and integrated into prudent geologic models.

Spreen Ranch Prospect

Location

The Spreen Ranch Prospect is located in southeast Runnels County, Texas, (Figure 1) approximately 10 miles southeast of Ballinger. Thin limestone and shales of the Permian Talpa and Grape Creek Groups crop out, forming gently rolling hills and valleys in and around the Spreen Ranch Prospect area.

Geology

Regionally, the Prospect is situated on the Eastern Shelf of the Permian Basin, where regional dip is approximately 40 feet per mile. Runnels County was first established as part of this oil and gas province in 1927, and at present yields approximately 2 million barrels of oil per year. The Spreen Ranch Prospect has 16 potential producing zones present in surrounding wells, most of which have had shows or tested gas or oil, while 12 of the 16 formations have established commercial production within the area covered by the Prospect maps (Table 1).

Satellite Remote Sensing—An Effective Tool 79

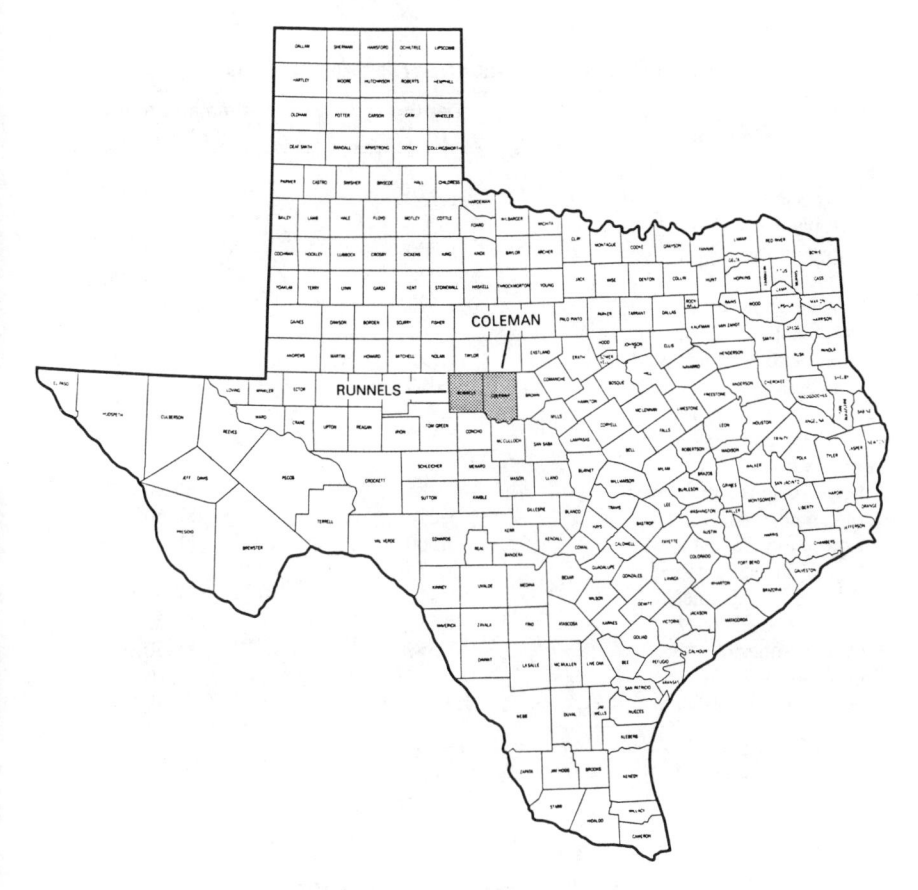

Figure 1. Index map of Texas.

Conclusion

The Spreen Ranch Prospect is well located to test one of the more significant structures in southeast Runnels County, Texas. The type of structural development present on this prospect will afford favorable conditions for both stratigraphic and structural traps in the sandstone and limestone beds proven to be productive in the area. Subsurface, remote sensing geomorphic, and geophysical information strongly suggest the presence of such a structure with a high probability of extensive fracturing in the deeper beds. Additionally, geochemical information of the area reinforces the probability of economic hydrocarbon accumulation.

Table 1
Producing Formations of Spreen Ranch Area

Formation	Depth (ft)	Estimated production (bbl oil/well)
Permian-Wolfcamp		
Cook Sandstone*	1,750	35,000-50,000
Pennsylvanian-Cisco		
King Sandstone*	1,900	30,000-60,000
Serratt Sandstone*†	2,040	40,000-60,000
Gunsite Limestone*	2,130	Unknown
Pennsylvanian-Canyon		
Home Creek Limestone*	2,425	20,000-35,000
Winchell Limestone (Reef)*	2,480	Unknown
Palo Pinto Limestone*	2,900	25,000-50,000
Dog Bend Limestone*†	3,100	20,000-40,000
Pennsylvanian-Strawn		
Upper Capps Limestone*	3,185	20,000-50,000
Fry Sandstone*†	3,270	35,000-60,000
Lower Fry Sandstone	3,360	35,000-60,000
Goen Limestone (Reef)	3,375	50,000-300,000
Jennings Sandstone*†	3,450	45,000-60,000
Gardner Sandstone*	3,530	20,000-50,000
Caddo Limestone*	3,657	20,000-50,000
Cambro-Ordovician		
Ellenburger Dolomite†	4,150-4,200	35,000-100,000

* Productive formations in Spreen Ranch Prospect area.
† Primary producing formations on Spreen Ranch Prospect.

South Rowena Prospect

Location

The South Rowena Prospect is located in southwest Runnels County, Texas (Figure 1) approximately 3½ miles south of Rowena. Farming is the primary use of the Permian (Clearfork) surface overlying the prospect area.

Geology

Regionally, the prospect is situated on the Eastern Shelf of the Permian Basin, where regional dip is approximately 40 feet per mile. Runnels County was first established as part of this oil and gas province in 1927, and at present yields approximately 2 million barrels of oil per year. The primary objectives in the South Rowena Prospect are five of the Strawn

Age carbonates: the Upper Capps Limestone, Lower Capps Limestone, Goen "Reef" Limestone, Jennings Limestone, and Gardner Limestone. These formations, with the exception of Goen, consist primarily of bedded limestone units of fairly uniform thickness ranging from approximately 40 to 100 feet throughout the area.

The Rowena Field, 3 1/2 miles northeast of the South Rowena Prospect has produced in excess of 4 million barrels of oil from the Upper Capps, Lower Capps, and Jennings Limestones, while the Gardner Limestone has yielded in excess of 750,000 MCF of gas. Presently the average cumulative production per well is approximately 300,000 barrels of oil, with ultimate recovery estimated to be 6 million barrels from 560 proven acres.

Conclusion

The South Rowena Prospect is well located to test one of the more significant structures in southwest Runnels County, Texas. The type of structural development present on this prospect will afford favorable conditions for both stratigraphic and structural traps in the sandstone and limestone beds proven to be productive in the area. Subsurface, remote sensing geomorphic, and geophysical information strongly suggest the presence of such a structure, with a high probability of extensive fracturing in the deeper beds. Additionally, geochemical information of the area reinforces the probability of economic hydrocarbon accumulation.

Bush Ellenburger Prospect

Location

The Bush Prospect is located in northwest Coleman County, Texas (Figure 1) 1 1/2 miles northeast of Goldsboro and approximately 6 miles north-northwest of Novice, Texas. Shales and thin limestones of the Permian Lueders and Talpa Formation crop out along a fairly irregular surface carved primarily by Jim Ned Creek and its local tributaries.

Geology

Regionally, the Bush Prospect is situated on the Eastern Shelf of the Permian Basin, where regional dip is to the west at approximately 40 feet per mile. This maturely drilled oil and gas province contains numerous pay zones from the Wolfcamp, Cisco, Canyon, Strawn, and Ordovician series. Oil and gas entrapment is both stratigraphic and structural; the majority of the traps seem to be a combination of carbonate development and sandstone pinch-outs in the Strawn and younger reservoir rocks.

The primary objective of this prospect is the Ordovician Ellenburger formation, a dense cherty dolomite with very little primary porosity and permeability in the upper units when devoid of fracturing. Thus, it is largely dependent upon post-depositional fracture systems for migration, accumulation, and subsequent production of oil and gas.

The most significant tool for detecting the minor faults and fracture zones has been the MSS satellite data. The major fracture zones associated with the Hrubetz Ellenburger Field and other recently discovered Ellenburger production may be clearly delineated with this remotely sensed satellite data. Subsequent drilling and development of the Hrubetz Field and surrounding area have confirmed the presence of fracturing, associated minor faulting, nosing, structural ridges, and in some cases low-relief closure against regional dip. These structural features are now mappable on the top of the Ellenburger and in many cases are reflected throughout the Strawn, Canyon, Cisco and Wolfcamp stratigraphic marker beds, and into the competent Permian surface rocks. It is important to note that the Ellenburger is an erosional surface and as such, is not structurally dependent for the entrapment of oil and gas. The migration and subsequent stratigraphic accumulation is, however, dependent upon fracturing in the dense, cherty, dolomitic reservoir rock. Therefore, the key to the accumulation and subsequent production from the Ellenburger in this province is the areal extent of the fracture system. The Bush Ellenburger prospect is on trend with the recently discovered Hrubetz Ellenburger Field.

In the subsurface, the prospect is situated on a broad west-southwest plunging asymetrical nose. This structural feature is also apparent in the remote sensing data and is transected by two major northeast trending lineaments, characteristic of lateral or strike-slip faulting, positioned along the east and west sides of the Bush lease. These fracture zones are further reinforced by two northwest trending lineaments that bound the prospect on the north and south, thus substantially increasing the probability of fractured porosity and permeability in the Ellenburger formation as well as in other potentially productive carbonate and sandstone reservoirs in the prospect area.

Conclusion

This prospect is well situated to test a significant, probably highly fractured, proven structure.

8
Use of Satellite Imagery in Discovery and Development of Shallow Oil and Gas Reserves

J. H. Bartley and James R. Lucas

Introduction

The examination of the earth's surface through satellite and aircraft imagery (termed remote sensing) has been proven to be an effective and relatively inexpensive method for the initial identification of oil and gas prospects. Satellite images have become a particularly useful tool for prospecting because

1. Regional geologic structural trends can be identified on these images that are virtually undetectable on the ground.
2. The ability to view the earth's surface in different parts of the visible and infrared spectrum have allowed earth scientists to identify changes in the reflectance over hydrocarbon reservoirs that have been correlated with microseepage from the oil and gas accumulations below.

In this paper, both structural and alteration patterns are analyzed and interpreted from Landsat satellite images acquired over four diverse shallow oil and gas producing areas in the United States. These areas include:

1. The Williston Basin of North Dakota.
2. The Powder River Basin of Wyoming.
3. The Illinois Basin of Illinois.
4. The Permian Basin of West Texas.

Landsat Overview

The first Earth Resources Technology Satellite (ERTS-1) was launched in 1972. In 1975 the second ERTS satellite was launched and, shortly afterwards, was renamed Landsat-2 with ERTS-1 being retroactively renamed Landsat-1. Subsequent Landsat satellites have been launched in 1978, 1982, and 1984; all have carried a Multispectral Scanner (MSS) to produce images of the earth's surface at 570 miles (917 km) altitude for Landsats-1, -2, and -3, and 435 miles (705 km) altitude for Landsats-4 and -5. Images from the Multispectral Scanner are recorded in four parts of the spectrum. Two spectral-band passes are in the visible (green light—0.5μ to 0.6μ and red light 0.6μ to 0.7μ) and two are in the near-infrared (0.7μ to 0.8μ and 0.8μ to 1.1μ).

The original digital image data are received at the Goddard Space Flight Center in Greenbelt, Maryland and later transmitted to the EROS Data Center in Sioux Falls, South Dakota to be processed into photographic products. False-color infrared images are produced in the photographic laboratory by using the green, red, and infrared spectral bands with a tri-color printing technique.

The resolving capability of the MSS depends on an object's reflectance and its contrast with surrounding background. Generally, the Multispectral Scanner can differentiate objects that are more than 80 meters apart. This resolution allows for the location of geographical as well as geological patterns on an image. Each Landsat scene covers approximately 13,000 square miles (34,000 sq km), often giving a regional perspective that is valuable for the understanding of basin tectonics.

Remote Sensing Methodology

The utilization of Landsat imagery in the search for hydrocarbons is broken down into two distinct operations. The first operation involves the analysis of image patterns such as linear features or anomalous color variations that may relate to oil and gas accumulations. The second operation compares the extracted image patterns with known production and related subsurface data. The prediction of oil and gas accumulation is an interpretation based on an understanding, guided by the surface expression delineated from Landsat imagery, of the geology of the study area.

Lineaments

Faults, fractures, and joints are either breaks or weaknesses in the earth's crust. Faults may serve as barriers to the migration of hydrocarbons forming structural traps. Fracturing or jointing may form secondary porosity important to the accumulation and recovery of hydrocarbons. Such geologic phenomena generally have a surface expression in the form of subtle linear patterns termed "lineaments." A lineament is an anomalously straight pattern resulting from drainage, landforms, or cover patterns (or a combination of all three). Often associated with lineaments are curving line patterns which have been termed "curvilineaments" by some.

Many linear patterns on the earth's surface are man-made and are termed cultural lineaments. These patterns are of no use to the earth scientist in the search for hydrocarbons. Therefore, care must be taken to map only linear patterns with geological significance.

The philosophical foundation for lineament mapping is based on the concept of continuous upward propagation of geologic structures to the surface. Scientists have attributed this upward extension of fracture zones of deep-seated origin to earth tides. Others have thought that the continuous seismic activity of the earth's crust is the mechanism that forms these linear patterns. The significance of lineaments in a geologic interpretation depends on the geologic model used to predict oil and gas accumulation.

Surface Alteration Above Hydrocarbon Accumulations

The escape (microseepage) of such gases as helium and carbon dioxide and light hydrocarbons can be measured over oil and gas accumulations by various surface geochemical sampling techniques. This seepage of gases causes surface alteration of rocks, soils, and vegetative cover that can be detected on false-color imagery. Image patterns over alteration areas take the form of mottled or honeycomb-like patterns that are located immediately above the subsurface hydrocarbon accumulation. The degree of mottling is interpreted to be directly proportional to the bottom-hole pressure of the reservoir. Experience seems to indicate that the more the bottom-hole pressure exceeds hydrostatic pressure, the brighter will be the alteration pattern on the Landsat false-color images, regardless of the depth of the hydrocarbon accumulation.

Significant alteration areas are usually identified by their association with circular or curvilinear patterns. Some of these patterns are believed to have a structural basis that is a result of differential compaction. Others are believed to be associated with alteration from microseepage. Curvilinear patterns are commonly circular in nature, but in some instances they do not make a complete circle because they may be disrupted by associated linear patterns.

Application of the Remote Sensing Methodology

To demonstrate the surface expression of hydrocarbon accumulations as seen on false-color Landsat imagery, four diverse shallow oil- and gas-producing areas were chosen for analysis. The study areas were located within

1. The Williston Basin of North Dakota.
2. The Powder River Basin of Wyoming.
3. The Illinois Basin of Illinois.
4. The Permian Basin of West Texas.

False-color MSS Landsat imagery was acquired over each of the study areas and was analyzed at a scale of 1:250,000. This is the same scale as the 1 degree by 2 degree U.S. Geological Survey topographic maps, making it easy to match map locations with image patterns.

Detailed pattern analyses were accomplished to identify the existing linear, curvilinear, and alteration patterns by overlaying clear film on the Landsat image and drafting the desired patterns on the overlay with a marking pen. Portions of these pattern-analysis studies are shown in the following discussion, but only patterns associated with current production are reproduced. Patterns outside of established production are not discussed because several sites are being considered for potential prospect development.

Williston Basin

Within the Williston Basin significant shallow producing fields occupy the eastern flank. The study area chosen for remote sensing analysis is located principally in Bottineau County, North Dakota (Figure 1). The oil fields in this county produce from depths of 3,100 to 4,200 ft, primarily from the basal Triassic (Spearfish) sandstones and Mississippian limestones (Charles of the Madison Group). They have average cumulative recoveries of 20,000 to 250,000 barrels of oil per well.

The Newburg field, discovered in 1955, which produces from the Spearfish and Charles Formations, is typical (Figure 2). The reservoir wedges out updip on the east flank of the basin (Figure 3). Such traps are

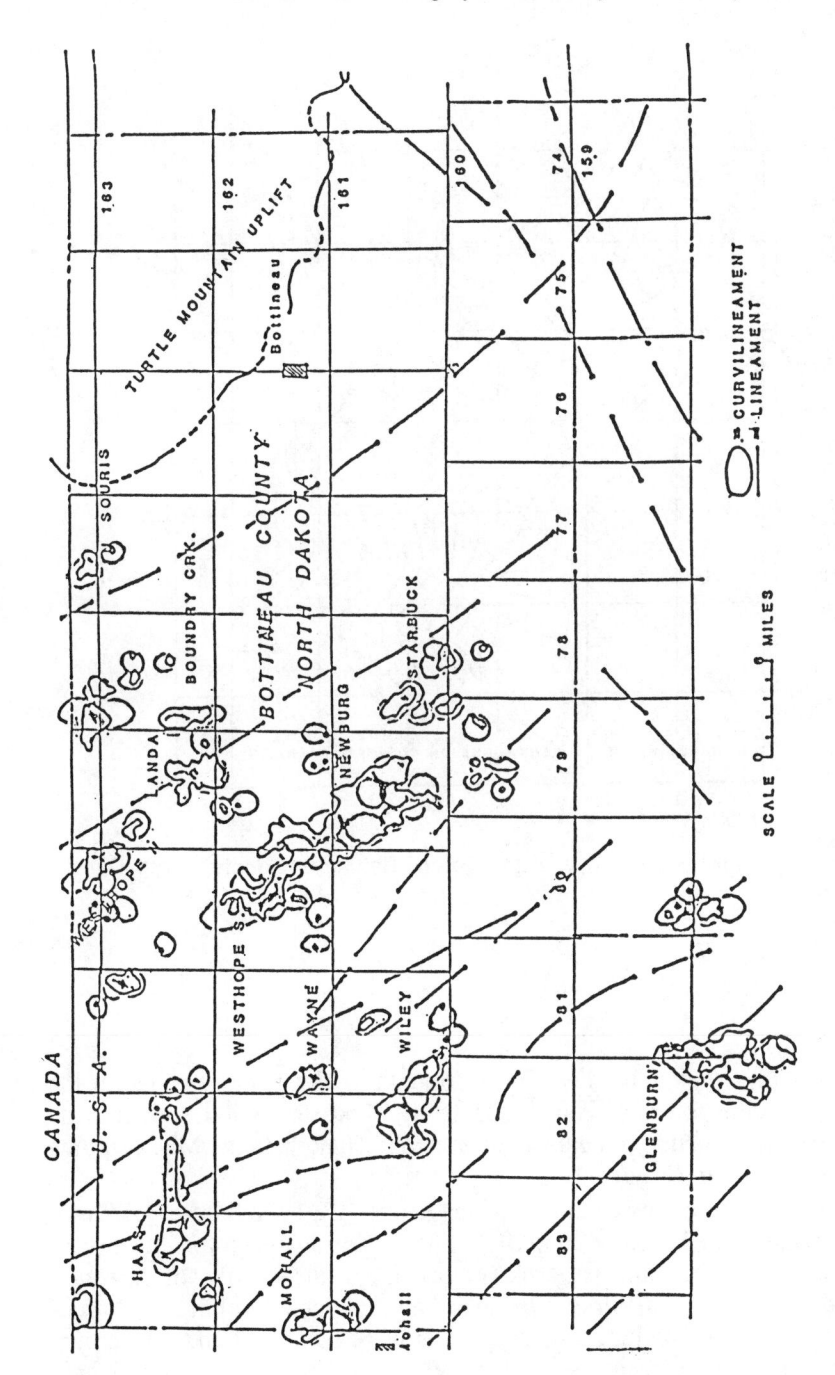

Use of Satellite Imagery in Discovery and Development   87

Figure 1. Shallow oil fields, showing lineaments, east flank of Williston Basin, Bottineau County, North Dakota, U.S.A.

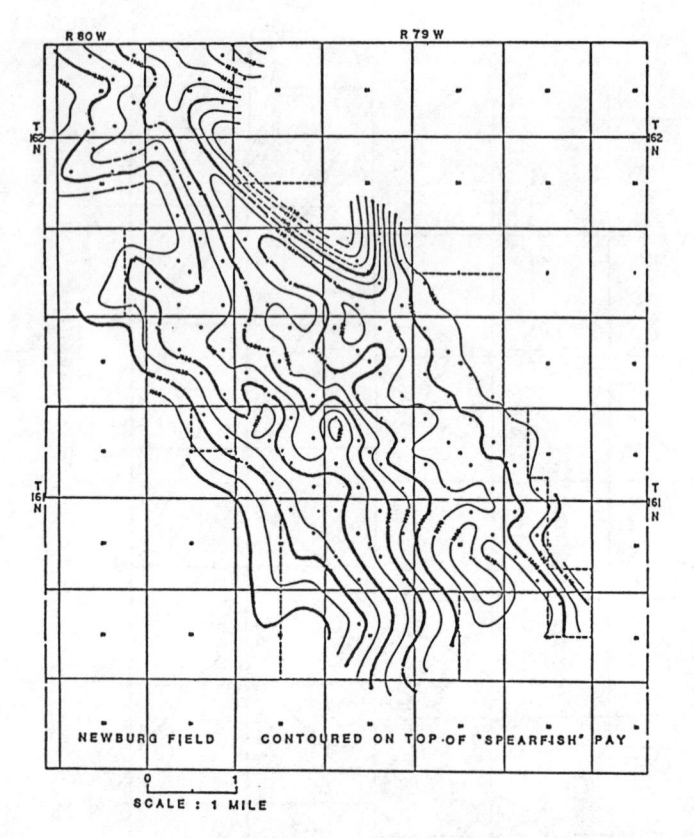

Figure 2. Newburg Field, Williston Basin, Bottineau County, North Dakota, U.S.A.

a result of a regional uplift (Turtle Mountain) along the east side of the Williston Basin which ended during the Triassic period, causing the Spearfish Formation to be deposited unconformably on the Mississippian Madison Group (Figure 3).

For the remote sensing analysis, an October 12, 1972 false-color Landsat image (Path 35, Row 26; #1081-17052) was analyzed for linear, curvilinear, and alteration patterns (Figure 1). In the vicinity of currently producing fields, curvilinear patterns were mapped. Within these circular delineations subtle, mottled patterns were identified and interpreted to be alteration patterns associated with microseepage. The high degree of correlation with these image patterns and current production has

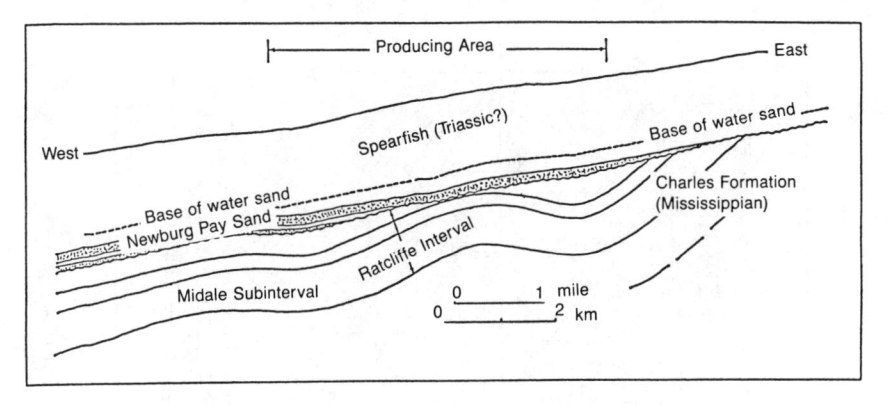

Figure 3. West-east cross-section of Newburg Field, Bottineau County, North Dakota, U.S.A.

formed a strategy for the identification of potential exploration prospects in this area.

Powder River Basin

The Powder River Basin occupies the northeast quarter of Wyoming, extending approximately 210 miles north-south and 125 miles east-west (Figure 4). The remote sensing study area covers the eastern flank of the basin in Crook and Weston counties, Wyoming. Oil fields in this area produce from depths ranging from 400 ft on the east side of the basin to 2,200 ft on the west side. The Newcastle (Muddy) and Dakota sands of Lower Cretaceous age are the primary horizons of hydrocarbon accumulation. The reservoirs wedge out updip, where the producing sands occur as lenses of irregular shape surrounded by shale.

The largest shallow producing fields in the study area are the Osage and Mush Creek fields in Weston County (Figure 5). The Osage Field was discovered in 1919 while searching for water and has recovered approximately 28 million barrels of oil. The Mush Creek Field was discovered in 1943 by surface geology and has recovered approximately 15 million barrels of oil.

The remote sensing analysis utilized a May 2, 1976 false-color Landsat scene (Path 37, Row 29; #5379-16360 EDIES). Extensive lineament patterns were mapped in conjunction with production. The characteristic mottled patterns were also identified within curvilinear patterns associated with production.

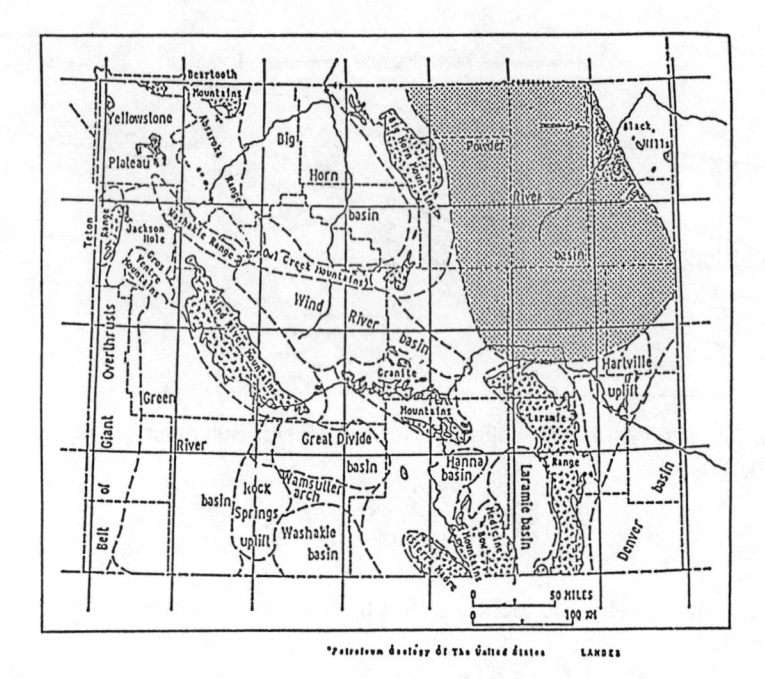

Figure 4. Wyoming tectonic map with area of interest along east flank of Powder River Basin, Wyoming, U.S.A. (From Landes, 1970).

Illinois Basin

The Illinois portion of the Illinois Basin features shallow producing horizons ranging in depth from 1,000 to 3,600 ft. Primary production is from Mississippian and Pennsylvanian sediments, mostly sandstones, located in stratigraphic traps. Numerous Silurian (Niagaran) reefs, however, are developed along the rim of the basin on the northwest and northeast. There are approximately 22 pinnacle-reef fields in the Illinois portion of the basin. Most of these fields have been very prolific, producing a cumulative average of 78,000 barrels of oil per well from the Silurian reefs, with as much additional production from the Devonian, Mississippian, and Pennsylvanian horizons above. An excellent example of this type of production is the Marine Field in Madison County, Illinois, which has recovered in excess of 157,000 barrels of oil per well (75 wells at an average depth of 1,770 feet).

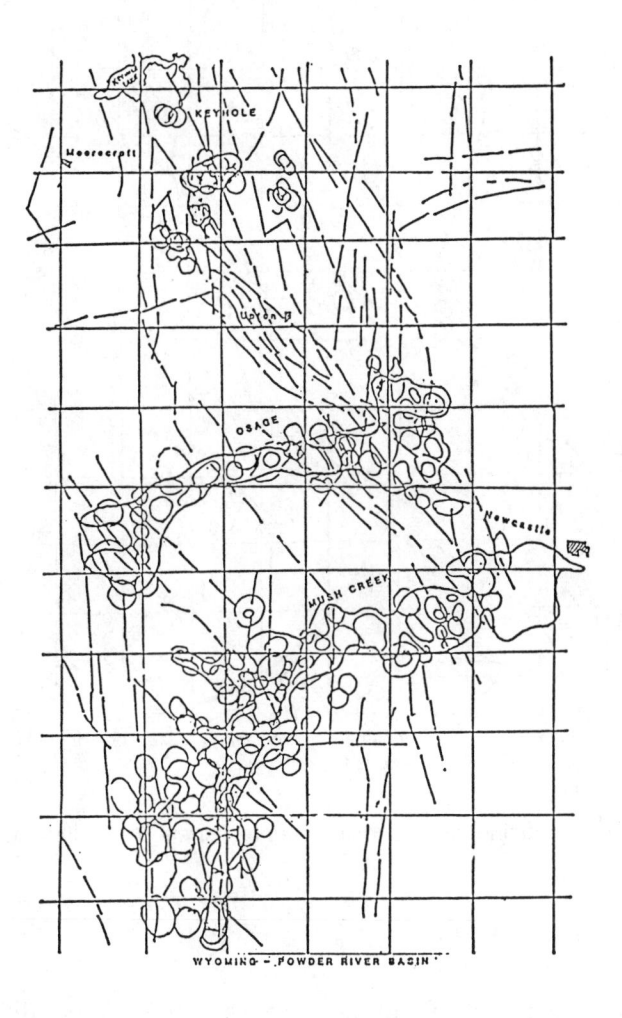

Use of Satellite Imagery in Discovery and Development 91

◯ = CURVILINEAMENT AND — — = LINEAMENT OF PRODUCING AREAS
AS DETERMINED BY REMOTE RENSING ANALYSIS OF SATELLITE IMAGE
(PATH 37 – ROW 20)   SCALE: 0 ⎯⎯ 4   4 MILES

Figure 5. Producing fields and lineaments, Powder River Basin, Wyoming, U.S.A.

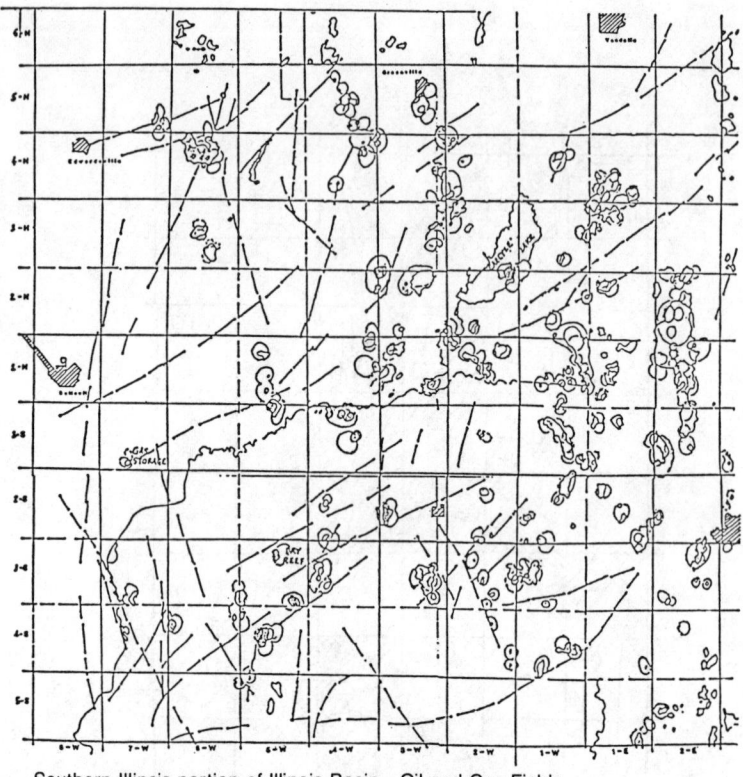

Southern Illinois portion of Illinois Basin—Oil and Gas Fields
◯ = Curvilineament and — — = lineament of producing areas as determined by remote sensing analysis of satellite image (Path 25–Row 33)
Scale: 4 Miles

Figure 6. Producing fields and lineaments, Illinois Basin, Illinois, U.S.A.

Satellite imagery analysis of the southern portion of the Illinois Basin was applied in the same manner as for the other study areas (Figure 6). A computer-enhanced Landsat false-color image (July 10, 1976; Path 25, Row 33; #2535-15523) from the Environmental Research Institute of Michigan (ERIM) was the primary image data source. Lineaments in the basin were found to trend northeast-southwest, at right angles to the direction of regional dip. These lineaments do not indicate vertical displacement in the subsurface, but do identify sharp dip in the subsurface. Curvilinear patterns were also found with characteristic alteration over producing areas.

Permian Basin

The Permian Basin of western Texas and southeast New Mexico in 1983 produced 22 percent of U.S. oil and gas. Production ranges in depth from 75 ft to 25,000 ft from Paleozoic rocks in many types of traps.

The area chosen for detailed study was the Howard-Glasscock Field, which is located on the Eastern Shelf of the Permian Basin. This field (Figure 7) in Howard and Glasscock Counties produces from depths of 1,250 to 3,800 ft. Entrapment is over an anticlinal structure approximately 15 miles long, trending generally east to west. The field was discovered in 1925 and has produced in excess of 300 million barrels of oil. Production is principally from the limestones and dolomites from the Yates, Seven Rivers, Queen, Grayburg-San Andres, and Glorieta-Clearfork zones of the Upper Permian.

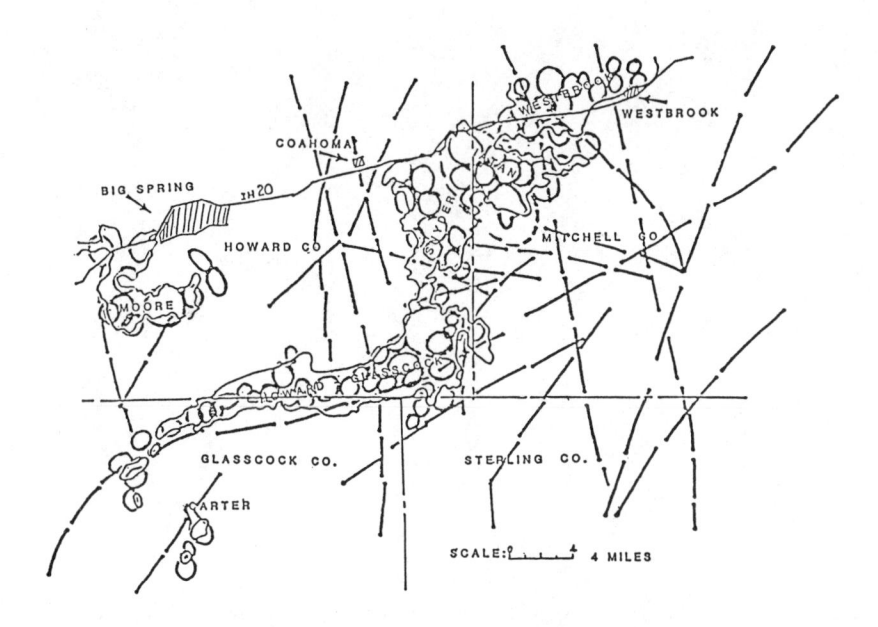

Figure 7. Howard-Glasscock Field and lineaments, Permian Basin, Texas, U.S.A.

A computer-enhanced Landsat image from ERIM (March 7, 1977; Path 31, Row 38; #2775-16203) was used to map the desired surface patterns (Figure 7). The lineaments transversing the study area are believed to be fractures that provide shallow reservoir porosity. Curvilinear and alteration patterns are prevalent over the producing areas.

Conclusions

Two fundamental conclusions may be drawn from this Landsat investigation:

1. Landsat data may be used as an inexpensive guide to the discovery of new prospects. These data may also be used effectively to extend known producing areas.
2. Landsat data allow more expensive exploration methods to be sharply focused on the most promising targets.

9
Seismic Reservoir Analysis

John A. Ward

Seismic data have classically been utilized to resolve structural and stratigraphic problems in an exploration environment. In some complex structural settings, like Gulf Coast salt domes, or in well-defined stratigraphic anomalies like Michigan basin pinnacle reefs, seismic information has provided criteria for selection of development locations. Modern seismic data processing and interpretation techniques allow geophysics to move into a new field—the solution of complex reservoir problems. An ideal complex geologic situation to examine the effect of seismic data on reservoir analysis is that of interfingering sandstone bodies. This geologic environment is typified by ambiguous log correlations which do not resolve reservoir affinities during the early stages of drilling. Often, seismic data provide the key link to the proper lateral relationships of sandstone lenses in the subsurface. The case histories selected in this review contrast the approach to seismic reservoir analysis for onshore exploration for high-velocity sandstones with offshore development in low-velocity, gas-filled sands.

Exploration exercises utilizing geophysical data usually rely heavily on the integrated borehole sonic log to provide both a correlation tool and an indicator of time-depth relationships. The two examples discussed are virtually independent of the sonic log. In the onshore example, electric logs are common, whereas sonic logs are scarce. In the offshore case acoustic logs were run, but hole rugosity and invasion make these measurements a poor indicator of gas saturation in the reservoir. In neither area can detailed study be done based on sonic log control. An alternative approach available is to integrate geologic and geophysical data. Instead of converting the borehole sonic log data to the time domain, the seismic information can be transformed into the depth domain. In depth format the seismic data can be compared directly with borehole logs, sample information, and well-evaluation data.

Techniques for generating a depth display of seismically-derived sonic logs (Seislogs) with a depth vertical axis have been described by Lindseth (1979). In the process each stacked trace is inverted from reflection coefficients to velocity on a 2-millisecond sample-interval basis. The inverted trace containing the high-frequency data from the seismic is then corrected for density and modulated with a low-frequency component extracted from velocity calculations, also derived from the seismic data. The resulting Seislog can then be compared directly to the well derived sonic log.

A Seislog can be treated on a trace-to-trace basis much like the interpretation of a set of sonic logs on a log-to-log basis. Therefore, the Seislog display is three-dimensional, with space, depth, and transit time presented simultaneously. Each trace has its own scaling and iso-transit-time values have been joined from trace to trace. The contouring of these equal-transit-time (velocity) values is completed by machine. The last step involves color-coding each contoured interval, assigning generally darker yellow-brown for higher velocity and restricting the greens to shales and using pink-orange for very low-velocity, gas-filled sand.

In the onshore example, data were acquired across a producing field in which gas is stratigraphically trapped in a Tertiary sandstone. The sand unit is high velocity, cemented with calcite, approximately 30 ft thick, and is encased in low-velocity shale at a drilling depth of 2,600 ft. A geologic cross-section parallel with the seismic line indicates that there is sandstone in the same stratigraphic position as the producing unit, updip of the field units. The log data also suggest that a slightly deeper sandstone, at 3100 ft, is a sheet sandstone.

The Seislog was produced at the same scale as the electric log cross-section at 1 in.=100 ft. The input data to inversion processing was a 12-fold seismic line acquired in 1978 which had a dynamite source with 220 ft shots and 110 ft groups. At a point corresponding to the updip limit of the field there is a subtle phase change in a reflector at the top of the producing sand. The geologic significance of this anomaly is not known.

On the color display at 1 in.=100 ft, the electric log units are easily recognized, with sands corresponding to the orange-yellow units and the lower velocity shales in the green colors. The high velocity Seislog sand units are arranged in a distinctive en-echelon pattern. Apparently, sands were deposited in a prograding sequence, with younger units downdip of older beds. This clearly explains the presence of a sand comparable to the productive unit updip of the field limits. The lower sand is likewise indicated to be lenticular rather than sheet.

The very close spacing of the Seislog traces in this type of environment allows interpretation of abrupt facies changes which cannot be determined by log correlation alone. This facilitates interpretation of the productive history of the area.

The offshore example is considerably deeper (7000 ft) but the reservoir is also thicker, with gas-filled sands exceeding 100 ft in thickness. The conventional seismic data in the area are 24-fold, using Air Gun as the energy source. On the seismic sections gas is indicated by a conspicuous "bright spot" anomaly. The "bright spot" is extremely complicated, consisting of one or more "cycles." Two wells that evaluated the zone tested multiple-pay sands, with one well encountering three productive intervals, the other with one definite gas sand and one possible gas sand. With only two logs it is difficult to determine the lateral correlation of individual sands. Similarly, the total pay volume was ambiguous since the overall "bright spot" anomaly could represent one or more productive zones.

Seislog sections were processed in a grid through the field at a vertical scale of 1 in. = 100 ft, to correspond with the borehole logs. The gas-filled sands are represented by low-velocity anomalies in the Seislog that are colored pink-orange. On the Seislog sections there are multiple low-velocity lenses within the overall anomaly. At the tie well these lenses correspond to the individual sands in the wells. When these are traced laterally along the lines, the correlation of sands between the wells can be discerned.

Individual sand lenses exhibit very abrupt terminations and variations in thickness on the sections. On the depth data it is possible to measure the thickness of each lens. When placed in map view, the geometry of the sand bodies indicates the model of deposition. This model provides an overall theme that can be integrated with paleontologic data, petrographic data, and core analysis. This geophysical study provided both an understanding of the geology of the area and the net pay distributions.

The examples illustrate the applicability of geophysical analysis in the depth domain to the solution of complex reservoir analysis. Principal problems to be resolved by the technique are lateral correlation of beds and measurement of reservoir thickness between subsurface data points.

10
Petroleum Data: An Aid for Exploration Planning

Jerlene A. Bright and Mary L. Fleming

The University of Oklahoma has assumed a leadership role in a world of very sophisticated computerized energy resource information, keeping pace with fast-changing technology in both computer hardware and software. The wealth of experience that the University has gained in compiling its many data files in the United States and Canada is directly applicable to other countries. It role in this area began in 1969 when the U.S. Department of the Interior contracted with the University to build a Petroleum Data System (PDS). This activity is now funded by subscriptions from the U.S. Department of Energy and from major oil and gas producing companies in the United States. In this paper we will examine data structure and applications.

A network of databases containing similar information to that in PDS should be of equal value in the development and management of any country's hydrocarbon resources. PDS now contains ten databases with information gathered from various publicly available sources.

TOTL is a master file of data on more than 108,000 fields and reservoirs from the 32 oil and gas producing states.

The LPR5 and LPRX databases contain lease bidding information, including the names, bidders, and amounts of all bids received on Federal Outer Continental Shelf (OCS) leases. These files also contain annual production, royalty, and rental amounts for OCS leases.

The SECR file contains secondary recovery project information for the states of Texas, Louisiana, Kansas, and Illinois. Data consist of location information, name of operator and lease, lithology, porosity, permeability, viscosity, acreage, number of wells, pressure, and recovery information. The file also includes production data and information on the types of injection and fluids injected, by project.

The BRIN file is a database of approximately 77,000 oilfield water-sample analyses. Each sample consists of a single analysis taken at a specific well location. The data can be very useful for enhanced oil recovery, geochemical exploration, petroleum exploration and production, and environmental impact studies.

The COIL file contains over 8,000 crude oil analyses for the United States and 600 analyses from wells located outside the United States. These analyses, donated to the Federal Government by industry, computerized, and placed in PDS, include information on location, field name, formation, age, depth, lithology, sulphur content, nitrogen, pour point, carbon residue, distillation residue, and percent of crude. They also include volume percent of crude oil, viscosity, cloud point, and specific gravity.

The GANL file consists of approximately 11,000 gas analyses. Each record consists of a single-well analysis or, in some cases, an analysis from a gas pipeline. Data elements reported include the location, the name and age of the producing formation, the date sampled, the depth, thickness, and wellhead pressure of the reservoir, and the specific gravity, heating value, and molecular percent of the gas.

The GC02 file contains approximately 630 analyses of gases from 27 states in which the percent of carbon dioxide is greater than five percent.

The CHST and CNDN files are field and pool production and injection information, as well as reserve information for the years 1962-1981, for the Province of Alberta, Canada.

In Chapter 13 in this volume by Richard F. Meyer and Mary Fleming, the TOTL datafile was used extensively to analyze shallow oil and gas fields. There are many other obvious applications of the data in PDS. Clients are finding many ways to use the information for different types of studies.

In exploration project evaluations, companies use PDS to provide input data relating to the economics of drilling wildcat wells based on analogy with existing fields.

Data from the files may be used to provide preliminary information for property evaluation before specific data from leases are available.

In crude oil supply studies, PDS can provide historical production data, and reserve data, where available, as a basis for projections of future supply.

Oil and gas field production from PDS may be used to provide field production capacities for pipeline feasibility studies.

Providing information on past performance of various basins enables explorationists to compare technology and exploration logic from basin to basin. PDS can be used to provide a statistical basis for deriving such items as chance of success for a drilling venture or field size distribution. Performance data on existing fields are used for NGL plant feasibility studies and for projecting future potential.

PDS is being used to provide data on reservoirs which are being screened as potential secondary and enhanced oil recovery candidates. In many cases PDS has been used to find reservoir candidates. A description of such an application appeared in the June 4, 1984 issue of the *Oil and Gas Journal*.

Crude oil prices have awakened interest in heavy oil reserves. PDS has been used to help provide estimates of these reserves and to locate occurrences of heavy oil reservoirs.

Figure 1 depicts the 1982 annual crude oil production for the state of Oklahoma, with each spike representing county production. Another example of the way these data can be displayed is in a pie chart showing production percentages from major basins in Oklahoma (Figure 2). Figure 3 presents a nonassociated gas production for 1982, by county.

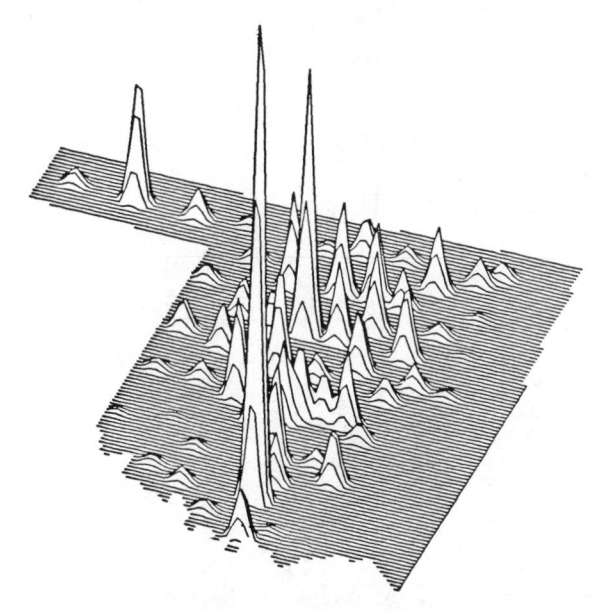

Figure 1. Annual crude oil production for the state of Oklahoma, U.S.A., 1982.

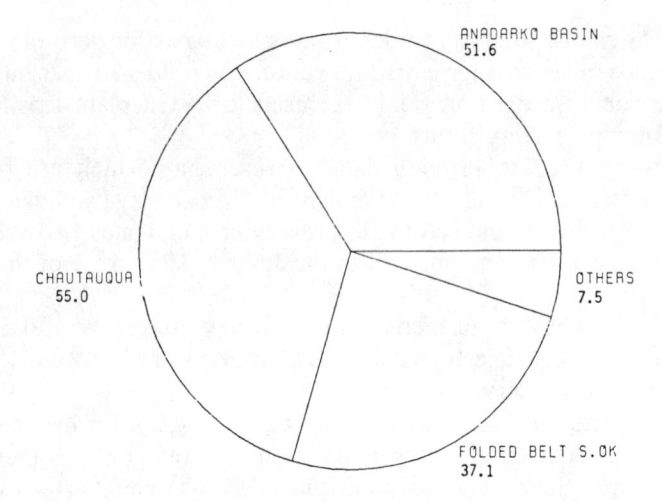

Figure 2. Crude oil production by basin (thousand barrels) for the state of Oklahoma, U.S.A., 1982.

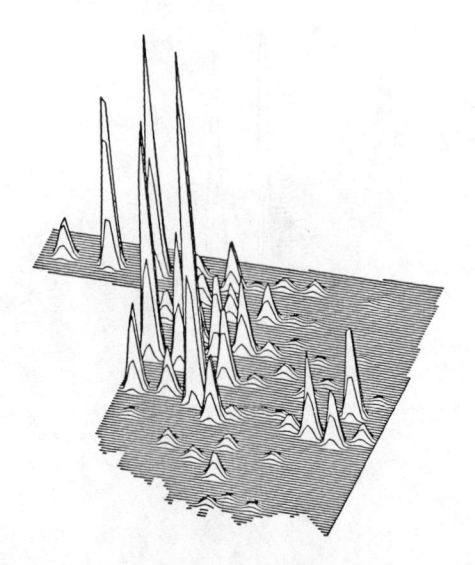

Figure 3. Nonassociated gas production for the state of Oklahoma, U.S.A., 1982.

Data from the TOTL file can be used to examine production in a particular area. For example, a detailed look at production in the Texas Gulf Coast Basin (Figure 4) can be obtained. Further detail is available from Figure 5, a graph of 1983 annual crude oil production by depth of perforation. Figure 6 shows that in Duval County, Texas, of the 152 reservoirs producing in 1983, the largest volume of production is from those reservoirs shallower than 5,000 feet. A similar retrieval (Figure 7) for nonassociated gas shows the opposite, as might be expected. However, the amount of 1983 annual gas production and cumulative gas for this county from shallow fields is significant. Additional information on the Texas Gulf Coast is presented in Figure 8, which shows cumulative oil production by year of discovery. This graph indicates the maturity of the production. Figure 9 shows a count of the number of oil producing reservoirs, by year of discovery, in the Gulf Coast Basin.

The BRIN file contains 77,000 oil field water samples for the United States with 9,500 of those samples reported for the state of Oklahoma.

(Text continued on page 108.)

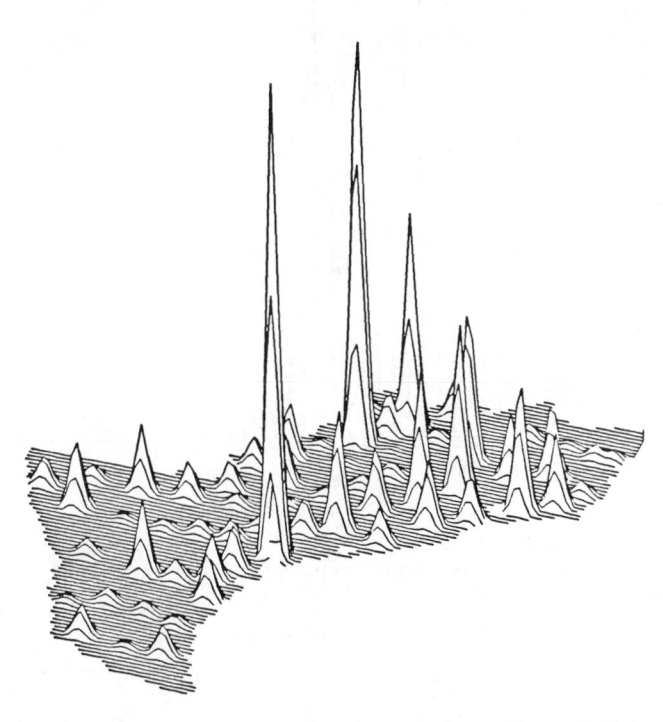

Figure 4. Annual oil production for the Texas Gulf Coast Basin, U.S.A., 1983.

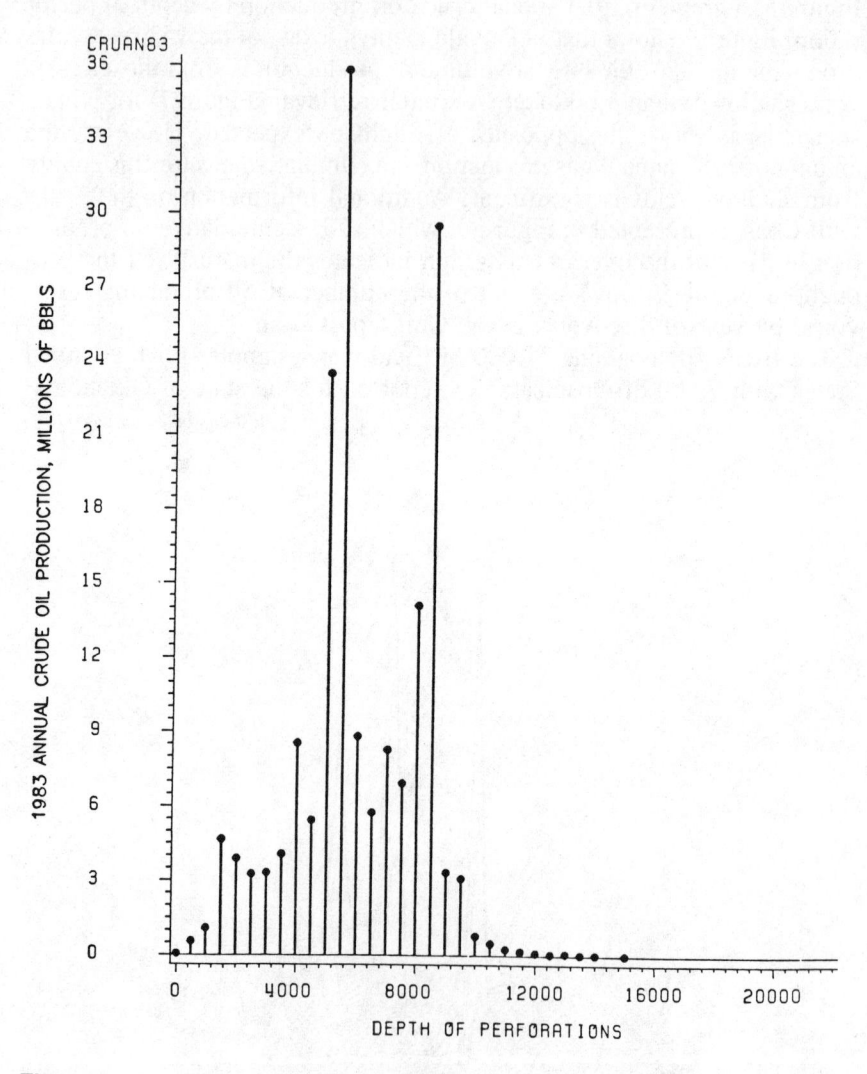

Figure 5. Annual crude oil production by depth of perforations for the Texas Gulf Coast Basin, U.S.A., 1983.

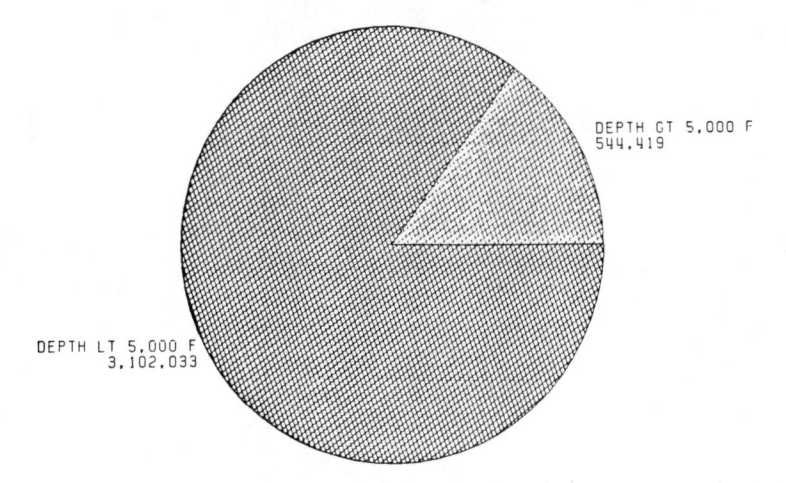

[Pie chart: DEPTH GT 5,000 F 544,419; DEPTH LT 5,000 F 3,102,033]

Figure 6. Annual crude oil production by depth (in barrels), Duval County, Texas, U.S.A., 1983.

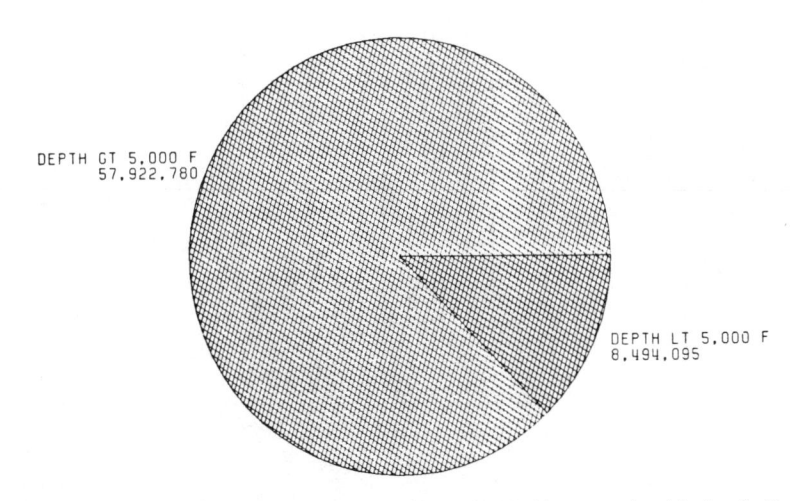

[Pie chart: DEPTH GT 5,000 F 57,922,780; DEPTH LT 5,000 F 8,494,095]

Figure 7. Nonassociated gas production by depth (thousand cubic feet), Duval County, Texas, U.S.A., 1983.

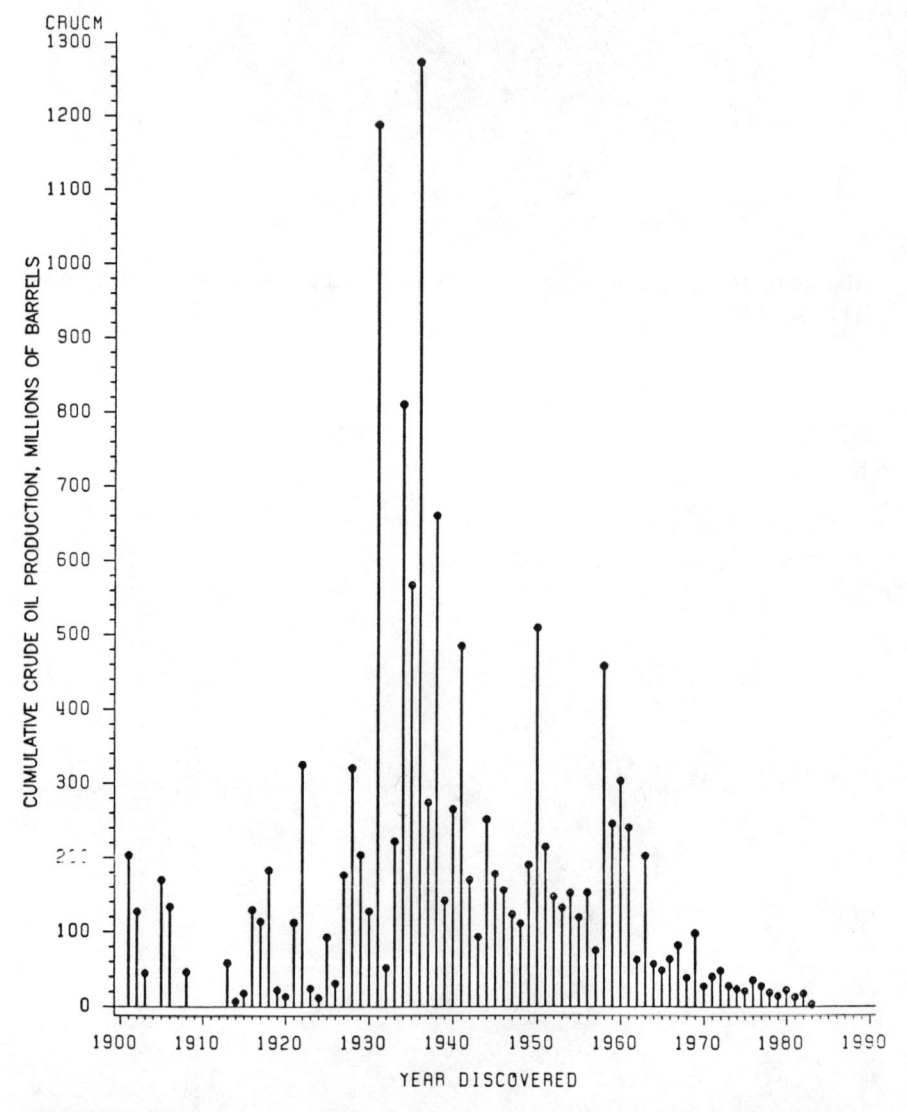

Figure 8. Cumulative crude oil production by year discovered, for the Texas Gulf Coast Basin, U.S.A.

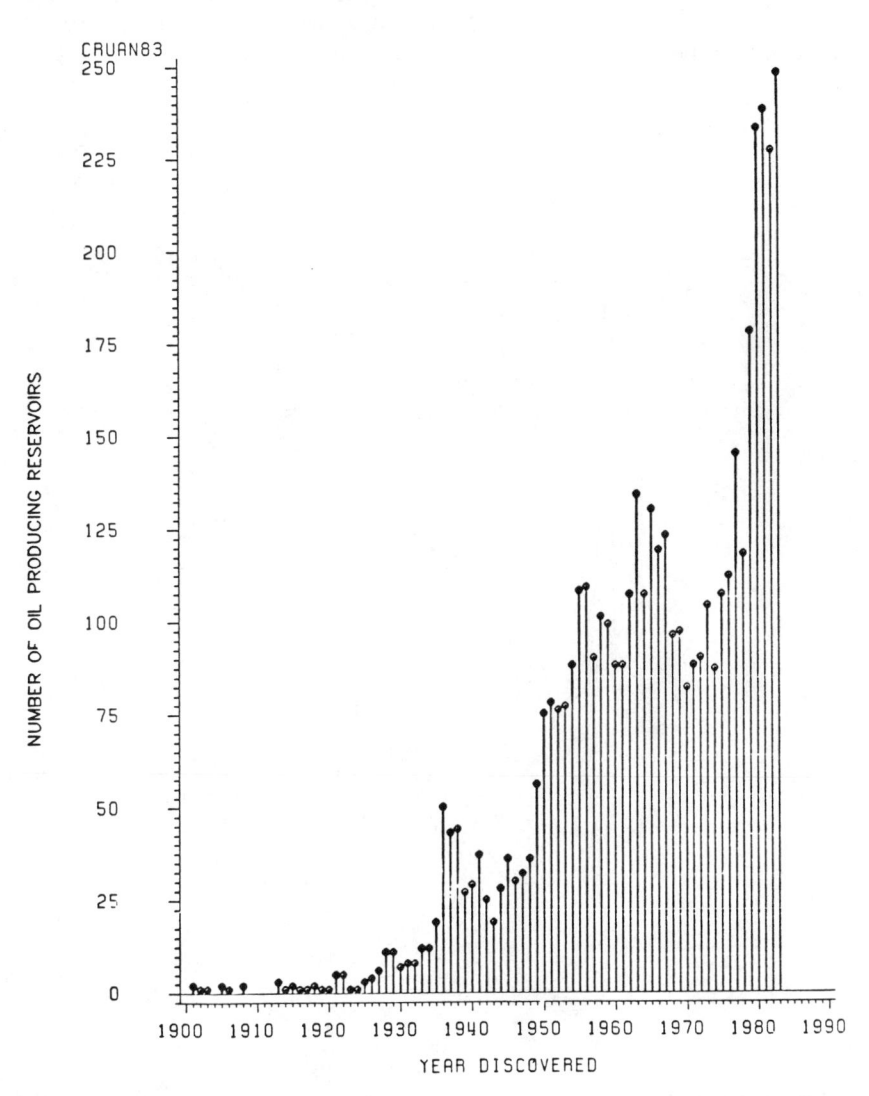

Figure 9. Number of oil-producing reservoir by year discovered, for the Texas Gulf Coast Basin, U.S.A.

108 *Shallow Oil and Gas Resources*

Figure 10 shows, by county, sulphate concentrations for the state of Oklahoma. Data included in the BRIN file allow the user to get an overview of variations in elemental concentrations.

AAPG, the Exploratory Well File of American Association of Petroleum Geologists' Committee on Statistics of Drilling, is also available through ISP. This file includes data on exploratory wells drilled between 1954 and 1982. Basic well data pertaining to wells drilled in the United States is reported by means of an individual well ticket initiated by respondents to the American Petroleum Institute. The University of Okla-

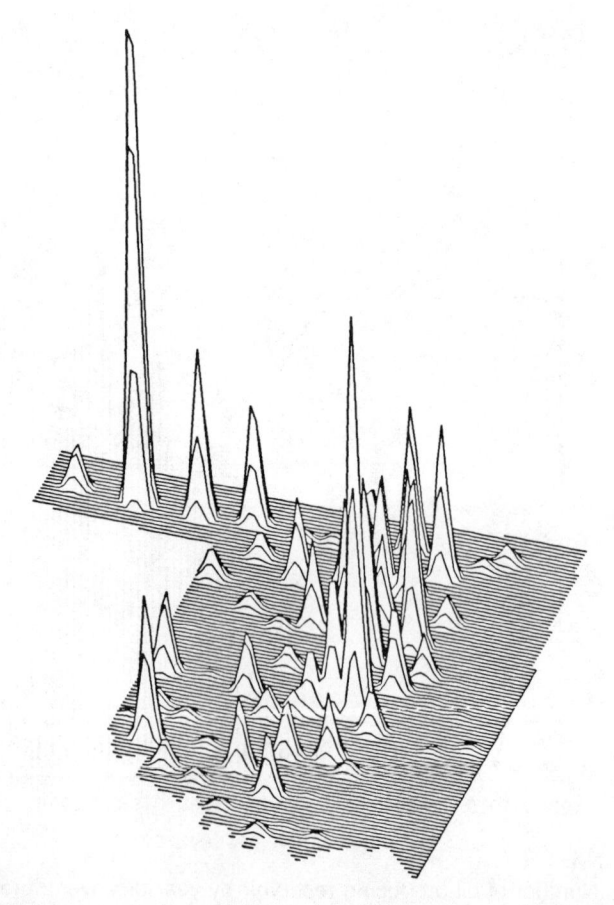

Figure 10. Sulphate concentrations form brine analysis in Oklahoma, U.S.A.

homa is the repository for the data. The AAPG Exploratory Well File includes information on the total number of wells drilled, the hydrocarbon involved, and whether the well is successful.

Figure 11 shows the total number of exploratory wells drilled for the state of Oklahoma in 1982. At present (July 1984), there is increasing interest in shallower-pool tests. Of all the exploratory well categories, shallower-pool tests are the most successful.

A particularly useful information label in the Exploratory Well File is the estimated ultimate yield of successful exploratory wells. Analysis of the average yield for gas completions for the state of Oklahoma by well depth in 1982 gives the results shown in Figure 12. The shallow wells drilled to less than 1,000 ft are probably producing nonassociated gas. The middle interval, up to 16,000 ft, is probably associated gas, as it is primarily in the liquid envelope. The deeper wells have a higher proportion of nonassociated gas, with the very deepest being all nonassociated gas. These data reflect a bias toward higher yields because low-yield, noneconomic wells would be shut in and thus not included in this analysis. Figure 13 depicts the number of successful gas wells drilled in Oklahoma for the years 1962–1982.

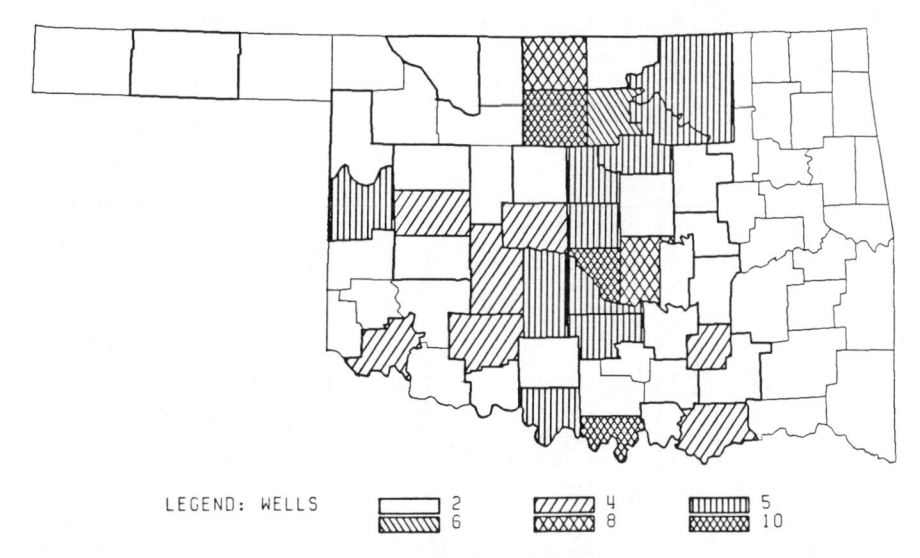

Figure 11. AAPG exploratory well data by number of oil wells drilled, 1982.

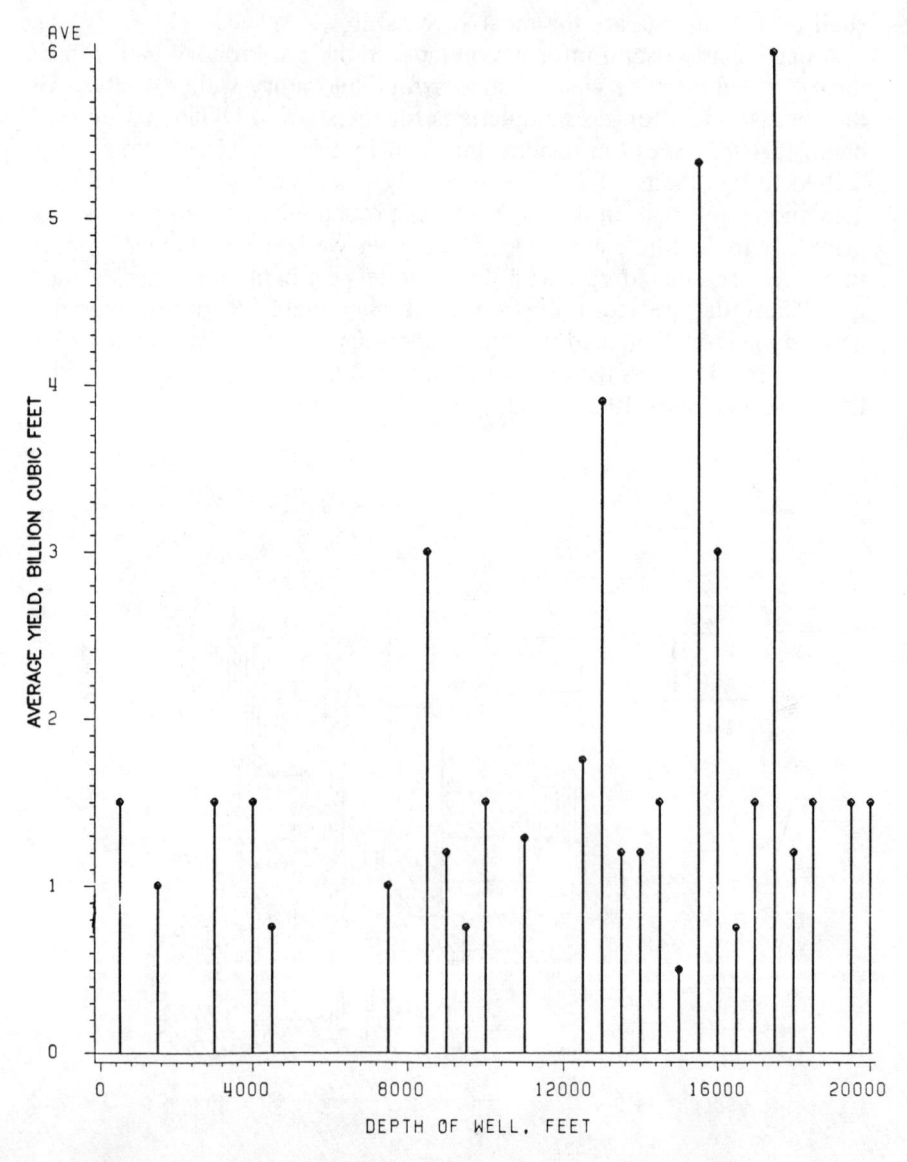

Figure 12. Average yield for gas completion by depth, 1982.

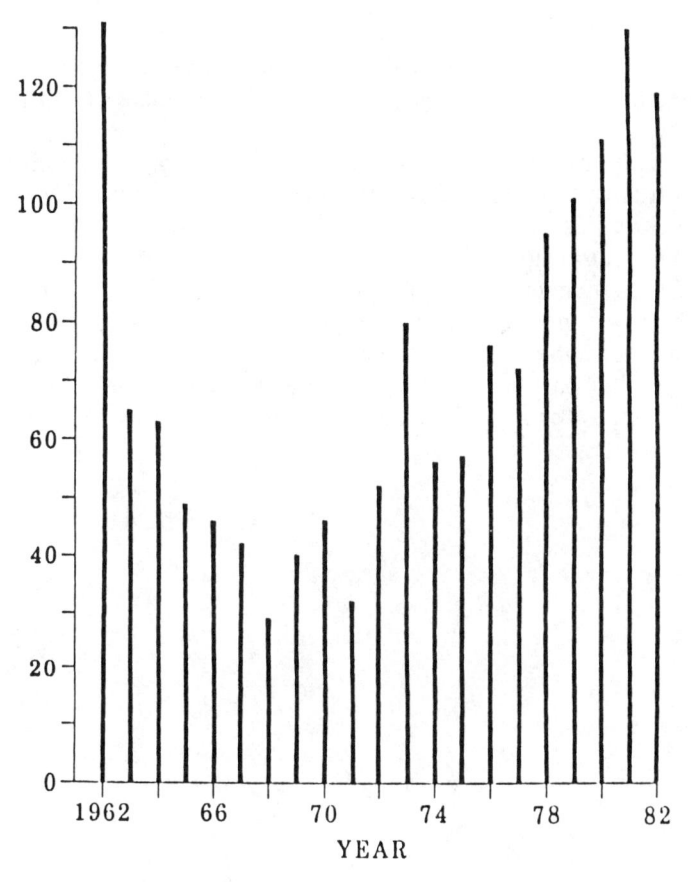

Figure 13. Number of successful gas wells drilled.

In conclusion, the Energy Resources Institute, working in conjunction with the Oklahoma Geological Survey, plays a major role in furnishing needed information for both the public and the private sector. It is our goal to increase and refine the data which are publicly available in order to meet the needs of those in decision-making roles. Because of the increasing amounts of information available, the job of designing, developing, and maintaining usable computerized data resources remains a very creative and challenging one.

Economic conditions facing the oil industry in 1982 changed the exploration focus of companies of all sizes. As a result of the decline in revenue, many small oil and gas companies went out of business, and lending institutions failed because loans were not fully secured. Mergers of major companies continue to create havoc in the traditional exploration departments.

There is no question that, as a result of this economic situation, exploration will have to make use of all available tools, including such information sources as PDS.

The Petroleum Data System is used actively in many of the large U.S. oil and gas companies to provide basic data for strategic planning relating to the sharing of market production and new venture costs. In combination with other databases from commercial and nonprofit organizations, PDS plays a key part in providing basic input data to models that evaluate geological areas for new field and reservoir discoveries.

11
The Microtechniques Method of Petroleum Exploration

Tony M. Preslar, Douglas Ball, and William P. Nash

Abstract

Since geology became important in finding oil, research institutions, private companies, and individuals have tried exploring with a variety of inexpensive geophysical and geochemical methods with varying success. Considering subsurface and seismic studies conventional, we label the inexpensive methods "unconventional" herein. They have found petroleum in both stratigraphic and structural traps, as documented in the literature.

The Microtechniques method uses hand-held instruments to display or record geophysical information, and the other hand-carried equipment to sample the soil or soil gases. This method can be used almost everywhere people can walk. Survey costs, including costs of data and sample gathering, analysis and data reduction, are very low compared to seismic costs.

A motor van contains and transports the crew and surveying equipment, though the field stations are a few yards or farther from the van. With no need for a seismic-wave source or vehicles off the road, unconventional exploration disturbs the land, the users of the land, about as much as land surveying.

The Microtechniques survey team considers costs, local geology, and exploration history to determine with which techniques to start the survey. The first field work, reconnaissance, enables the definition of the areas and techniques for later detailing, if any. Thanks to modern field, laboratory, and office equipment, the wealth of field information evolves quickly into useful maps, sections, and other graphic and numeric displays. Several illustrations show this.

Microtechniques composite anomalies, coupled with known favorable geology, will locate petroleum in many, perhaps most areas. Thus Microtechniques exploration can find high-quality drilling targets at low cost.

Finding Petroleum

Oil and gas, lumped here into petroleum, are found in the rocks of the earth, usually with water or brine. Petroleum is not found in underground lakes filling vast caverns, nor in streams flowing in subterranean channels. It is found in minute pores or voids of sandstones and limestones. It does not collect in the troughs and low places of such rocks—it accumulates in traps in their higher parts, held there by water that fills the pore spaces of the rest of the reservoir rock.

The three essentials for a commercial accumulation of petroleum are source beds, reservoir beds, and traps, and these are the things for which a geologist looks. If there has been no source, there will be no petroleum—so he looks for beds that seem to have contained enough organic material to have furnished some petroleum. If there is no porous or reservoir bed, the oil and gas will still be disseminated through the source beds, so he looks for a rock stratum porous enough to hold fluids, pervious (permeable) enough to permit movement and accumulation. If there is no trap, the oil will be distributed through the reservoir bed in quantities too small to be useful, so he looks for a place where the folding or the termination or interruption of the reservoir bed may have caused a concentration of the oil.

The source beds will probably be shale or limestone, which, when it was deposited as mud or ooze on the sea floor millions of years ago, contained an abundance of animal or vegetable remains or both. The reservoir bed will probably be a sandstone but may be porous limestone or dolomite. The structure, or trap, will probably be an arch, or upfold, in the strata, formed by the folding or settling of the earth's crust in the course of some crustal movement, or a place where the continuity of the reservoir pore space is interrupted or terminated.

There is nothing mysterious about all this—geologists have observed what is happening on the earth's surface today, and have projected their observations back through the eons of geologic time (Ball et al., 1965). The mystery is that we can never see petroleum or the rocks which surround it deep in the earth and therefore must use surface clues, measuring the effects deep petroleum accumulations cause at or very near the surface.

Conventional Exploration

The drill finds oil directly, but the first drill-hole into an oil field usually must be supplemented by a second, perhaps a third, before we know whether or not the oil field will pay the cost of finding it. All other exploration finds less-direct evidences of oil or evidences of a petroleum trap. With such evidence, where a proper combination of source bed, cap rock, and reservoir seems to exist we can designate, with reason, a place to drill a test hole. This is making a location.

Oil seekers record every bit of information possible from drilled wells, both dry holes and petroleum producers. The data from several wells, or many wells, can be combined in a number of way to help interpret rock structure, stratigraphy, and fluid contents. Such studies are the mainstays of subsurface geology, and subsurface geology probably finds more petroleum than any other exploration tool.

Where well-control is scanty, conventional exploration is based mostly, sometimes exclusively, on seismic work. The seismograph maps rock structure very well in most sedimentary areas. Where everything is just right, the seismograph even detects petroleum. Where seismograph surveys can be run between wells or dry holes, the seismograph is a useful adjunct to subsurface studies.

Though seismic mapping is getting better every year, it is also becoming more expensive, and it mainly maps rock structure and not petroleum. Exploring with a modern common-depth-point seismograph on land costs at least $4,000 a line mile, and gains two to six miles of information a day. In remote areas, costs can be $30,000 a line mile and productivity is lower.

Other Petroleum Finding Tools—Inexpensive Geophysics and Geochemistry

Local structure, meaning variations in the shapes and conformations of near-surface rocks, forms some of the traps geologists seek, so gravity meters and magnetometers, which measure local variations, have been used to find petroleum fields. So have other instruments that measure the earth's gamma-ray emission, and so have analytical measurements of numerous gases in the near-surface soils. The common characteristic of all these geophysical and geochemical methods is that they cost little per survey. The industry view is that none of them is a very reliable oil finder. The seismograph has found more oil by finding "structure" than all of them put together.

An important characteristic of the geochemical techniques is that they measure petroleum traces that have leaked from petroleum reservoirs, or the effects these traces have had on near-surface soil-chemistry, or soil-air chemistry. Gamma-ray surveying, or radiometric surveying, is closely related to the geochemical techniques. The literature reports finding petroleum by each of these inexpensive techniques, particularly in recent years, since sampling and analytical methods for most surveys have improved considerably.

Microtechniques Surveying

Reconnaissance and Detail Surveying

Microtechniques surveying uses several of the high-technology, inexpensive geophysical and geochemical techniques, combined with subsurface and other geological studies. It generates up to 25 separate anomalies and shows them on contoured maps.

Microtechniques reconnaissance measures four-channel gamma-ray emissions and magnetic intensity, and samples soil and soil-air at each survey station. Areas where the resultant anomalies coincide usually justify further work, which includes detailing with the same techniques more closely spaced, and probably with additional techniques.

The detail surveys further determine the characteristics of the composite anomaly. Computers process all the field data, and evaluate enormous amounts of data in a very short time. The over-all geologic analysis enables choosing the most-favorable drilling locations.

Costs

Microtechniques surveying costs should average less than $250 a line mile, and the crew will cover about 10 miles on the average day. A single crew should be able to survey four to seven prospects in a year.

Effectiveness

As an oil finding method, Microtechniques exploration has no track record, but the individual techniques have furnished favorable, even glowing, reports in the literature.

The Microtechniques strengths are the combinations of techniques that prove themselves in a given sedimentary basin, and the flexibility that derives from the empirical approach as a survey is started. Any technique that yields no usable signal or yields misleading results can be abandoned and other techniques substituted.

Site Selection

Under the best of circumstances, drilling for oil is expensive; all other exploration is directed toward finding petroleum with the fewest drill holes. This principle directs the petroleum seeker to inexpensive exploration so as to limit the areas that later, more expensive exploration must cover.

Literature Search

Whether the objective is finding petroleum efficiently or evaluating an area for its petroleum possibilities, applying all known pertinent information will render an exploration program the most effective. Accordingly, searching the literature and the records of such institutions as geological surveys and universities and commercial enterprises for information on the rocks and fluids in them usually turns out to be a worthwhile endeavor. The student can usually make some geological interpretation from what he finds and use the new-found knowledge to direct further exploration with greater efficiency.

Regional Geophysics, Geochemistry, and Photogeology

In many areas, regional geophysics and/or geochemistry can help select sites for further local study. Airborne magnetics, airborne radiometrics, and the several forms of aerial photography and radar scanning are good preliminary tools. The cross-country variations of magnetics and radiometrics, whereby the instrument is carried in an automobile or truck, can cover tens to hundreds of line-miles a day, and help select sites of a few square miles.

Hydrogeochemical studies may help select sites for further local exploration. Table 1 outlines the site selection considerations.

We use hydrogeochemistry here to mean the study of concentrations of iodine, chlorine-to-sulphate, and other ratios reported to be associated with petroleum in places.

The United States Geological Survey and other agencies monitor the water chemistry of potable-water aquifers in the United States. Data-sets are available to the public and some may be adequate for hydrogeochemical studies.

Wide-area-magnetic and wide-area-radiometric maps show geologic structure under favorable conditions. These also permit mapping such tectonic features as sedimentary basin outlines and uplifts that separate the basins.

**Table 1
Site Selection**

I. Literature search
   A. Geologic interpretation
II. Regional geophysics and geochemistry
   A. Moving vehicle surveying
      1. Magnetics
      2. Radiometrics
      3. Photogeology and radar scanning
   B. Hydrogeochemistry

Pirson and Pirson (1976) have written on magnetoelectrics and how magnetic data can be converted to magnetoelectric data for use in petroleum exploration.

An Exploration Program

A petroleum exploration system should be a combination of the exploration tools and procedures that are the most compatible, inexpensive, rapid, and effective. We outline such a program on Table 2. Note that the reconnaissance tools are also used for detailing, which includes more closely spaced lines or stations with the reconnaissance tools and the additions of other tools and procedures. The tools and procedures in Table 2 include structure-finders and oil-trace detectors and are more comprehensive than practice will encourage in most areas. The first reconnaissance days will limit the procedures that prove effective in a given area; thereby they will simplify further field work and lower the cost of the finished survey. The techniques retained will be those that provide the greatest accuracy in delimiting the prospect area and choosing drill sites.

In most surveys, the soil samples will be collected at each station, but only the samples from areas found favorable by the other procedures will be analyzed. Microtechniques intends to conduct most, but not all, of the soil-air analyses in the Microtechniques laboratory. Using outside laboratories for some analyses assures automatic quality control to the overall procedure.

No one ever knows everything about a geological problem; answering specific questions about a project will probably call for trial-and-error combinations of procedures not used throughout the project generally.

Table 2
Microtechniques Geophysical and Geochemical Methods

GEOPHYSICS

R & D*	Radiometrics	R & D	Magnetoelectrics
R & D	Microradiometrics	D	Self-Potential
R & D	Magnetics	D	Microgravity
R & D	Micromagnetics		

GEOCHEMISTRY

SOIL-AIR INERT GASES

R & D	Helium	D	Krypton
D	Argon	D	Xenon
D	Neon	D	Radon

SOIL-AIR HYDROCARBONS

D	Methane	D	Pentane
D	Ethane	D	Hexane
D	Propane	D	Heptane
D	Butane		

SOIL-AIR ISOTOPES

D	Carbon 12 & 13	D	All Soil-Air Inerts

SOIL-AIR OTHER GASES

D	Hydrogen Sulfide	D	Iodine
D	Carbon Dioxide	D	Bromine

SOIL ANALYSES

D	Delta C Carbonates	D	Hexane
D	Methane	D	Heptane
D	Ethane	D	Carbon 12 & 13
D	Propane	D	Iodine
D	Butane	D	Bromine
D	Pentane		

---

* D - Probably used only to add detail.
  R & D - Probably used both in reconnaissance and detailing.

Self-potential and microgravity determinations and some parts of soil geochemistry are new techniques, but the literature reports oil fields they have found.

How Microtechniques Exploration Works

In the oldest application, magnetic geophysics is used to map local variations in the earth's magnetic field. The new "micromagnetics" and all the other Microtechniques exploratory methods are used to map constituents leaked from subsurface accumulations or the changes these leaked-hydrocarbons have made on near-surface soils and on the air in the soil pore spaces.

Geochemists believe all petroleum fields leak, and that the leakage is vertical or nearly so, except where brine or water is flowing through the reservoir or through a higher aquifer (the hydrodynamic effect). Solid-earth-tide pumping, chemical-potential drive mechanisms, and the variance in pressure and temperature with depth aid the upward migration of hydrocarbons and associated trace elements about petroleum fields. The results are various dynamic and static, chemical, radiometric, electrical, and magnetic surface anomalies which reflect the presence of the petroleum accumulation and even reflect characteristics of the trap in places. The Microtechniques precept is to detect and map the significant anomalies and thereby to find the petroleum.

Pandely et al. (1974) and other workers report compelling evidence supporting upward migration of hydrocarbons through shale, limestone, and dolomite caprocks.

The work of companies and research institutions shows two basic types of anomaly patterns, apical and halo, associated with petroleum accumulations. Under apical theory, leaked hydrocarbons concentrate over the bulk of a petroleum accumulation, giving an apex to the geochemical anomaly. A halo anomaly, with a geochemical low over the bulk of the petroleum accumulation, results from the accumulation blocking upward movement so the leaked hydrocarbons concentrate at the boundaries.

Such theoretical considerations help explain the physical processes by which the anomalies are generated and can lead to understanding the composite characteristics of multi-technique anomalies and give them greatest usefulness in finding petroleum.

Some oil fields have given strong apical anomalies, others strong halo anomalies. Various combinations are known; the pattern depends on the degree of leakage or blockage and the extent of each. A host of companies have contributed information on what to look for to determine the type of leakage contributing to composite anomalies. Sufficient here is acceptance that leakage exists and the resultant direct and indirect geochemical effects can be reliable indicators of petroleum when considered in combination.

Radiometrics

Almost every part of the earth emits gamma rays, and for 40 years people have observed lows in gamma radiation over petroleum fields. With many of these are high-gamma halos around the edges. The most accepted view of the cause is that leaked hydrocarbons above a petroleum field induce chemical-reducing zones above the petroleum, and the reducing zones block some of the natural radiation.

Helping the primary blockage may be the adsorption of radioactive materials by the petroleum hydrocarbons and the blockage and diversion of upward-bound formation fluids by the hydrocarbon accumulation.

Thermal buoyancy—the rising of formation fluids because they are warmer than their superjacent counterparts—explains the rising of dissolved solids, including radioactive solids. That a petroleum accumulation can turn aside rising solutions and concentrate them at the edges of the petroleum is easy to visualize, too; hence, the radioactive halos around many petroleum fields.

The major radioactive elements that concentrate near the surface include uranium, thorium, and potassium 40, and radioactive daughter products. All these are gamma-ray emitters with long half-lives. They provide "permanent" gamma ray sources in the near-surface soils.

Weart (1981) and others reported a 79.7 percent success ratio in more than 70 Sunmark Exploration Company radiometric surveys in the United States. They conclude: "Our experience to date strongly implies that the proper utilization of this type survey in an integrated exploration program can be effective in the reconnaissance phase for developing leads and prospects, and can provide supplementary evidence in localizing and evaluating prospects."

Micromagnetics and Magnetoelectrics

Microtechniques surveying can be used to map two other data-sets derived from magnetic field measurements. These are micromagnetics and magnetoelectrics. Micromagnetics is used to map the near-surface reducing environment, in effect. This changes high-valence iron compounds to those of lower valence, such as magnetite, the most magnetic mineral, and the magnetometer maps the resultant anomaly.

Donovan et al. (1979) and Roberts (1981) reported such micromagnetic anomalies over the Cement oil field in Oklahoma and several other oil fields.

Magnetoelectrics also respond to the reducing environment over an oil field. With the surrounding oxidizing environment this amounts to a reduction-oxidation (redox) cell that induces electrotelluric currents which flow from the oxidized region to the reduced region. The redox cell, continually charged by the upward seepage of formation waters and hydrocarbons, maintains the disequilibrium and consequent current flow as long as the petroleum is present.

The electrotelluric currents induce magnetic fields, and these perturb the force lines of the earth magnetic field. By measuring the perturbation Microtechniques permits mapping of the effects of electrotelluric currents as one of the magnetic-effect oil-finding techniques.

Soil-Air Analyses and Soil Analyses

Several elemental and compound gases concentrate in the near-surface soil over petroleum fields, including the rare or inert gases. Helium is the best known for its association with petroleum. Microtechniques surveys use soil-air sampling and helium analyses as a standard, or primary, exploration technique.

Helium, produced by radioactive decay in crustal rocks, is flushed into hydrocarbon traps by rising formation fluids, as described above, and becomes part of the natural gas in the petroleum. Most of it stays with the hydrocarbons in the gas, but enough leaks to be detected in the air in the near-surface soil over the petroleum. Helium is chemically inert, is not produced biogenically, and is very mobile because of its small atomic size and atomic weight.

Iodine, another elemental but not inert gas, is a good surface indicator of deeper petroleum. Other anomalous gas concentrations and other forms of soil and soil-air geochemistry may become important in Microtechniques exploration, but they will be less used and are of lesser importance in any Microtechniques program we anticipate now.

Conclusions

From our studies, and the voluminous literature, we believe Microtechniques composite anomalies, coupled with known favorable geology, will locate petroleum in many, perhaps most, areas. Thus Microtechniques exploration can find high-quality drilling targets at low cost.

References

Ball, Douglas, 1965, This fascinating oil business: Bobbs Merrill Company, Inc., pp. 22-23.

Donovan, T. J., et al., 1979, Aeromagnetic detection of diagenetic magnetite over oil fields: American Association of Petroleum Geologists Bulletin, v. 63, no. 2, pp. 245-248.

Pandely, G. N., et al., 1974, Diffusion of fluids through porous media with implications in petroleum geology: American Association of Petroleum Geologists Bulletin, February, pp. 291-303.

Roberts, A. A., 1981, Unconventional methods in exploration for petroleum and natural gas—Symposium II, (B. M. Gottlieb, Ed.): Dallas, Texas, Southern Methodist University Press, 257 pp.

Weart, R. C., 1981, Unconventional methods in exploration for petroleum and natural gas—Symposium II, (B. M. Gottlieb Ed.): Dallas, Texas, Southern Methodist University Press, 257 pp.

12

Successful Exploration on a Small Budget

John H. Gray

The focus of this paper will be on an exploration technique based on many years of experience operating as a small company without access to sophisticated data. This technique has resulted in many discoveries, extensions, and new pays in old areas.

Operating on a small budget, it is necessary to reduce the reasons for commercial accumulations of oil and gas to a lowest common denominator. When I rate a prospect, these are the conditions which I apply. Having confidence that these conditions exist, any prospect then can be graded. I will also discuss some of the prospects I have drilled.

- *Condition No. 1: Porosity.* For a reservoir to exist, porosity—primary, secondary, fractured, reefing, leached, etc—is essential. It has been my experience that if paleostructure exists, then porosity very likely exists.
- *Condition No. 2: Structure on a time marker below the producing interval.* Sometimes the structure above a particular target horizon does not extend to deeper beds. An excellent example of this is in South Arkansas, where there is both Cretaceous and Jurassic production. Many of the shallow Cretaceous producing fields show no relationship to the deeper Jurassic (Smackover Lime) fields. Correspondingly, many of the deeper Smackover Lime fields are not reflected in the upper Cretaceous horizons. There is a regional unconformity on the top of the Jurassic Cotton Valley. There is an additional unconformity at the base of the Upper Cretaceous. Efforts by this author to use the synchronous-high technique, by making use of markers in the interval stretching

across both unconformities, have proven to be of little value. The main focus of this paper is not towards structural mapping; therefore condition number two should be utilized only as a supplement to condition number three, paleostructure, for in many cases there is not enough well control to make a structural map.
- *Condition No. 3: Paleostructure.* A paleostructure has also been called a residual high or synchronous high. This concept is the main focus of this paper, first on technique and then on actual field examples. It is necessary to have some well control, but through experience I believe that if enough acreage is available, any area can be evaluated with only a minimum amount of drilling. The term paleostructure implies that a structural anomaly was being formed simultaneous with deposition.

The general concept is that if distance $T^1 > T^2$ vertical movement is indicated in the lower bed prior to the deposition of the upper marker. This then is the preferential habitat of the oil. This concept tends to support the early accumulation of hydrocarbons. It can also be observed that when a paleo low, or residual low, such as near a salt dome, is mapped it is in proximity to a residual high. Nonproductive salt domes that I have mapped, for example, in the North Louisiana Salt Basin indicate uplift long after deposition. However, one of the productive domes, the Minden Dome, shows periods of movement simultaneous with deposition.

An ideal interval to isopach is about 366 m (1,200 ft). Seldom are two markers available that approach the ideal interval. Sometimes in a closer interval the contrasts between the thick and thin areas are so small that they are difficult to map. An example of a thin interval is the Buckner Formation of South Arkansas, which at times may be less than 30 m (100 ft).

If an interval is too thick it is possible that one or more unconformities may be crossed; other factors then interfere with the timing necessary for the synchronous high concept.

In some cases an unconformity has to be crossed in order to obtain reliable time markers. In South Arkansas there are many stratigraphic changes above the Jurassic Smackover Limestone in the Buckner, Cotton Valley, and Travis Peak formations. The first reliable time marker is the lower Cretaceous Sligo Formation. Examples of mapping the Sligo-Smackover interval are the Shuler and Days Creek fields.

The technique is very effective for determining where not to drill and when to stop drilling. The Eocene Wilcox of Louisiana and Mississippi is an area of multipay, lenticular sands and cyclic deposition. Synchronous highs are developed by draping over sand buildups. Once the drill has penetrated the cyclic deposition series, there is no reason to drill any deeper below the synchronous high. Carbonate reefing can also act in the same manner. In the Canyon Reef trend of Scurry County, Texas, the synchronous high is related to draping over the reef. The formations below the reef do not show the movement developed by the draping and compaction of shales over the reef. The breaching of the sealing mechanism explains why some paleostructures are dry.

This technique also may be applied to production from growth faults. Trap loss can result if the prospective reservoir is opposite a porous zone across the fault. The porous zone should be traced up dip to the point where migration is interrupted, for a trap may exist there containing hydrocarbons.

In many cases, such as lenticular sands or pinnacle reefs, where there was little chance for long-distance migration, the technique works best because the hydrocarbons remain trapped in the original reservoir.

To develop the technique, start with the available control, picking as many different points in one well as possible that correlate with other wells in the area. Then determine which interval shows the maximum amount of thickening and thinning; it is not necessary that it be a previously identified and named marker. As the map on this interval is developed, trends will appear of thin (drillable) areas and thick, off-structure areas.

There are additional indicators of paleostructures when the three conditions described above are limited.

1. *Oil shows*—If there is a known show or a surface oil seep, this is an indicator of migration and accumulation. The author is well acquainted with several areas that had reported shows that later turned out to be productive but were passed up at the time they were drilled. If a paleostructure is mapped and there is a report of an oil show it is important to evaluate the show very carefully.
2. *Fractures in limestones*—In one field in South Texas, production was from a fractured limestone that had no matrix porosity. A synchronous high could be mapped and used to pick additional locations. The fractured zone was on the crest of the paleostructure and did not extend to the flanks.
3. *Paleogeomorphology*—This can be applied in both sandstone and carbonate deposition. In sandstone areas the author has successfully reconstructed the paleo environment and traced the crossing of two channel sands which in turn created a paleostructure due to the non-compaction of the sands and the compaction of the surrounding shales (Gretener and Labute, 1969). In carbonate areas reconstruction of the classic undoform, clinoform, and fondoform (shelf, slope, basin) can help pinpoint the preferential habitat of carbonate reefs (Van Siclen, 1957, and Rich, 1951). The nonmarine Taching

field in the People's Republic of China is an example of a synchronous high. Once the compaction was taken out of the shales, correlations became much easier (Gray, 1980).
4. *Landsat tonal anomalies*—As experience is gained, this is a useful tool to make comparisons with the paleogeologic map. In older areas, there is an amazing correlation between the tonal anomalies and the paleostructures (Pyron, 1984).
5. *Paleontology and sample logs*—Where it is difficult to pick time markers, certain guide fossils can be used to identify time markers. It is also possible to utilize drillers' logs, particularly cable tool logs, to pick formation tops. This is only recommended when electric logs are not available.
6. *Surface features*—Topographic maps are very helpful to identify drainage patterns. An extension of this concept is now being utilized with the mapping of lineaments and circular anomalies using Landsat maps (Sabens, 1973).
7. *Diapirs and igneous plugs*—If well control is available, these features are readily identifiable if the movement was simultaneous with deposition. In areas where no control is available, gravity and magnetomic surveys can be useful reconnaisance tools for locating anomalies, but it should be remembered that such anomalies may not relate to the time of deposition. An excellent example of a field on a synchronous high over an igneous plug is Days Creek, Miller County, Arkansas.
8. *Recent developments*—Computers can be useful tools to speed up filing, sorting, and analyzing a greater amount of data than was heretofore possible with a small staff. On a spread sheet a large amount of well information can be stored and made available in a short period of time. It is also possible to add or delete data easily.

Summary

1. Stay close to control points.
2. Use all available control.
3. If a paleostructure is identified, do not give up on one well, as it is very probable that it is in the proximity of production.
4. Application of this technique requires a knowledge of subsurface geology, in particular geomorphology.
5. Confidence in this technique can be developed only through continued usage.

6. Drilling to provide well control is also the most direct approach to oil finding.
7. This method identifies areas to avoid, that is, areas to neither drill nor explore.
8. Use whatever data are available to upgrade the prospect, including oil shows, samples, paleontology, surface features, paleogeomorphology.
9. Finally, this technique is much more interpretative than structural mapping and the final results will be in proportion to the correctness of the interpretation.

Section II

Resource Occurrences

13
Survey of World Shallow Oil and Gas

Richard F. Meyer and Mary L. Fleming

Abstract

Shallow deposits are those in which hydrocarbons are first found at depths no greater than 5,000 ft (1,500 m). Ten percent of the world's shallow oil fields and 9 percent of the shallow nonassociated gas fields are found at depths of less than 500 ft (150 m). Nineteen percent of the shallow oil fields and 25 percent of the gas fields are within the depth range of 500-2,000 ft (150-600 m), and 71 percent of the shallow oil fields and 66 percent of the gas fields lie between 2,000 and 5,000 ft (600-1,500 m). The United States alone accounts for 79 percent of the world's shallow oil fields and 75 percent of such gas fields. Among the world's giant shallow oil fields having ultimate recoveries of at least 500 million barrels each, 78 are onshore and 16 are offshore; in all, the 94 fields contain 262 billion barrels of ultimately recoverable oil. Of 39 shallow giant gas fields in the world having ultimate recoveries of at least 3.5 trillion cu ft each, 2 are offshore; in all, these 39 fields account for 813 trillion cu ft of gas. In the United States alone, shallow oil fields of all sizes contain 49.7 billion barrels of ultimately recoverable oil or about one-third of the 140 billion barrels from all fields in the U.S. data base. The U.S. experience suggests that about 35 percent of the world's ulti-

---

We gratefully acknowledge the critical review of the manuscript by Emil Attanasi and Elizabeth Good (U.S. Geological Survey) and Gordon Zareski (Zinder Companies, Inc.) and the assistance of Kevin Laurent (U.S. Geological Survey) in programming retrievals from Petroconsultants data file.

mately recoverable oil resources and 13.5 percent of the ultimately recoverable nonassociated gas is in shallow fields. Most of such shallow oil and gas will be found by the exploration efforts of small operators.

Introduction

Enormous amounts of oil and gas occur at shallow depths, and the deposits are located on all continents and in many countries. This survey deals only with those shallow deposits that are actually known and that have been produced or may be producible. No attempt is made to assess the amount of presently unknown resources that may be discovered in the future.

Because actual production and reserve data for many parts of the world are scant, we have had to rely on numbers of fields to convey the idea of the importance of shallow hydrocarbon deposits. Enough information is available, however, to make clear that shallow deposits are not by any means necessarily small. Small deposits may occur at any depth, but most are economically producible only if they are shallow.

We are dealing with fields and not individual reservoirs. Thus, we attribute field totals on the basis of the shallowest producing reservoir. From available statistics on many multi-pay fields, we could not determine the depth of the principal production.

The aim of this article is to report on our survey of world shallow oil and nonassociated gas fields, to indicate the size of the known resource, and to emphasize that many shallow fields remain to be discovered in the world outside the United States. After defining terms and identifying some of our sources, we discuss and illustrate the geographic and depth distribution of shallow oil and gas fields in the world and in the United States, emphasizing the importance of giant shallow fields. We give statistics on exploration in the United States, explain why U.S. exploration has differed from that in the rest of the world, and extrapolate from the U.S. experience. In addition, we provide some information on bitumen and heavy-oil deposits.

Throughout this report, the following definitions apply.

Shallow occurrences are those in which the hydrocarbons are first found at depths no greater than 5,000 ft (1,500 m). This interval is divided into three categories:

1. Surface to 500 ft (150 m).
2. 500–2,000 ft (150–600 m).
3. 2,000–5,0000 ft (600–1,500 m).

These categories were selected because the ease of hydrocarbon recovery, by mining or conventional production methods, changes for each depth interval and because 5,000 ft is a depth within range of most small operating companies.

Heavy oil is that crude oil less than 20° gravity API (American Petroleum Institute) and less viscous than 10,000 cP. *Bitumen* refers to hydrocarbons that are more viscous than 10,000 cP and that are often referred to as *oil sand* or *tar sand*.

Natural-gas data are given for nonassociated gas only.

Our chief sources of information are the Petroleum Data System (PDS), an automated file of oil- and gas-field data for the United States, and Petroconsultants automated file on oil and gas fields in countries outside the United States and Canada. Information for the fields in Canada is available from Provincial data files and publications. Other information sources are cited in the "References."

Geographic and Depth Distribution of Shallow Oil and Gas Fields

The geographic distribution of oil fields, gas fields, and bitumen deposits that occur at depths of less than 500 ft is given in Figure 1. The distribution of oil and gas fields found at depths of 500–2,000 ft is shown in Figure 2, and that of fields found at 2,000–5,000 ft is shown in Figure 3.

Table 1 gives the depth distribution of known shallow fields by country, and Table 2 summarizes these data by region. The totals and the broad distribution are impressive. Shallow fields offshore constitute about 3 percent of the oil and 4 percent of the gas. Unfortunately, data on water depths for these fields are lacking.

In making an a priori judgment on depth distribution of all shallow fields, we might expect most of the fields to be deeper than 2,000 ft. If the United States is excepted, this expectation is not realized; most of the fields are at the shallower depths. The geographic distribution of fields is fairly even, although relatively few shallow oil fields are in the Middle East; the highest number outside North America is in Europe. The U.S.S.R. is included with Europe because we have no feasible means to subdivide it for statistical purposes; obviously, many of the Soviet fields are in Asia. When the United States is included in the world totals (Table 2), then 10 percent of the shallow oil fields are at less than 500 ft, 19 percent are between 500 and 2,000 ft, and 71 percent are between 2,000 and 5,000 ft. For shallow gas fields, 9 percent are at less than 500 ft, 25 percent are at 500–2,000 ft, and 66 percent are at 2,000–5,000 ft.

(Text continued on page 152.)

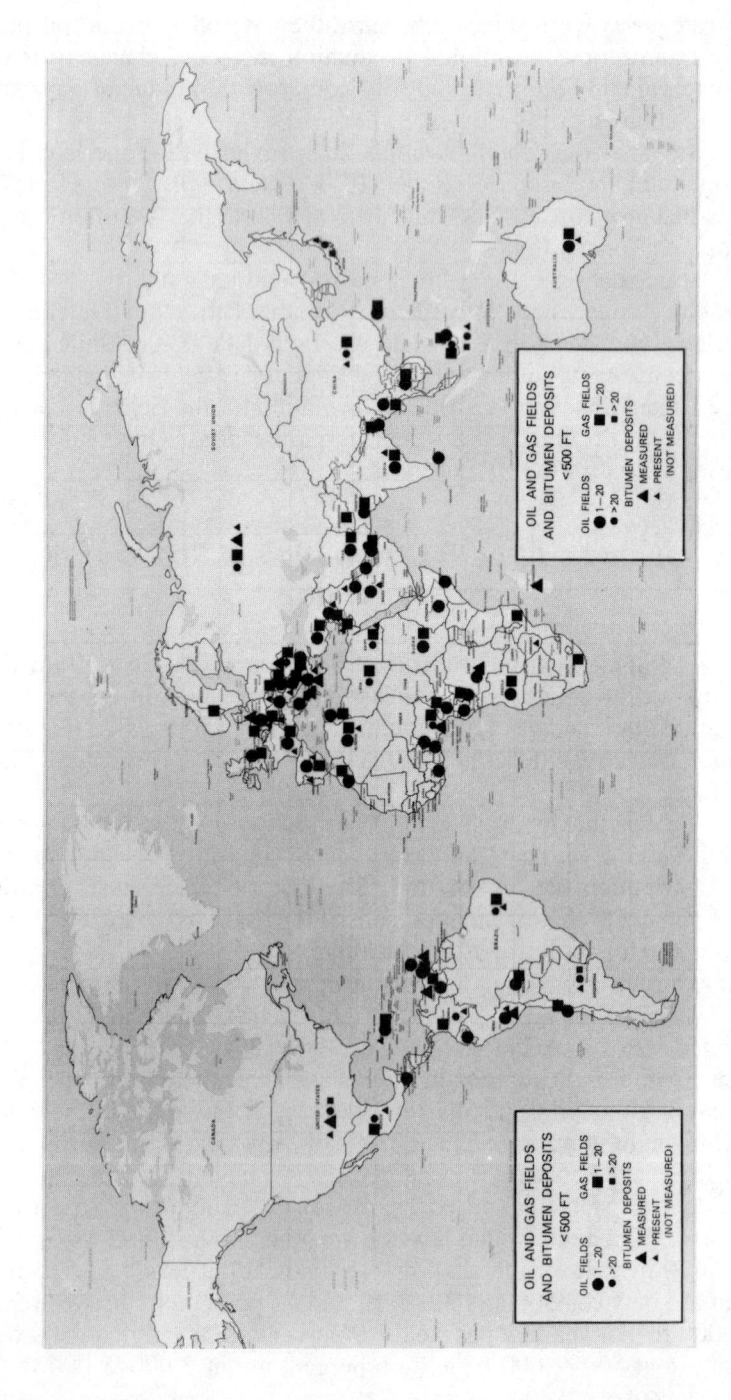

Figure 1. World oil fields, gas fields, and bitumen deposits less than 500 ft (150 m) below the surface.

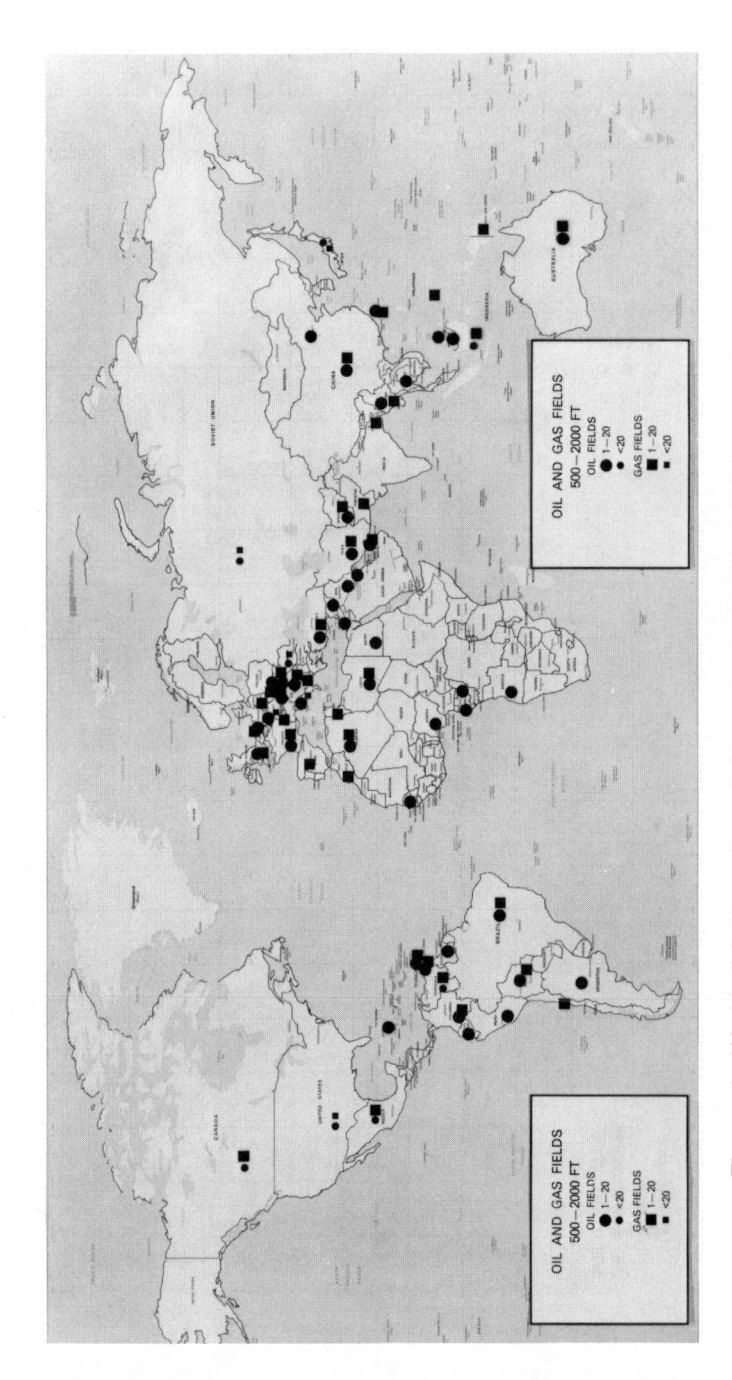

Figure 2. World oil fields and gas fields at depths of 500–2,000 ft (150–600 m).

138 Shallow Oil and Gas Resources

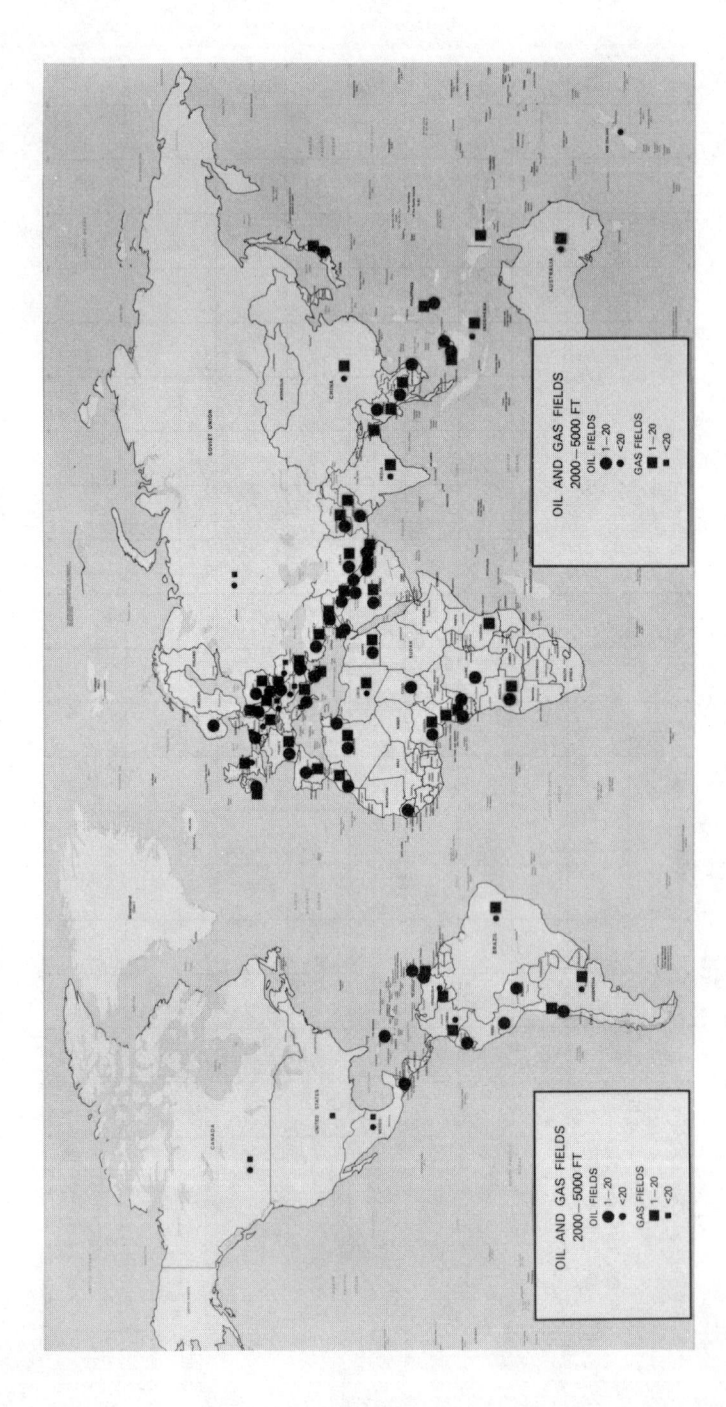

Figure 3. World oil fields and gas fields at depths of 2,000–5,000 ft (600–1,500 m).

Table 1
Depth Distribution of Shallow Oil and Gas Fields in the World by Country

Region and Country	Depth Category*	Onshore Oil	Onshore Gas	Offshore Oil	Offshore Gas
South and Central America and the Caribbean					
Cuba	a	3	1		
	b	4		1	
	c	2			
Guatemala	a	1			
	c	3			
Barbados	a	2			
	b	3	1		
	c	2	1		
Colombia	a	22	5		
	b	17	2		
	c	27	2		
Venezuela	a	18	1	2	1
	b	51	2	1	
	c	76	2	2	1
Trinidad	a	23		4	6
	b	15	1	1	1
	c	4	3	6	
Suriname	b	2			
Ecuador	a	16			
	b	1			
	c	3			
Peru	a	10		1	
	b	14		2	
	c	17		3	
Bolivia	a	4	8		
	b	6	1		
	c	11	2		
Brazil	a	77	7	7	3
	b	13	2	2	
	c	36	2	3	
Chile	a	8	7	2	
	b		2		
	c	1	2		1
Argentina	a	128	16	4	5
	b	10			
	c	62		3	

* [a, less than 500 ft (150 m); b, 500-2,000 ft (150-600 m); c, 2,000-5,000 ft (600-1,500 m).
Source: Petroleum Data System; Petroconsultants automated data file; Saskatchewan E and M, 1982; British Columbia MEM and PR, 1981; Alberta ERCB, 1982]

table continued

Table 1 continued

Region and Country	Depth Category*	Onshore Oil	Onshore Gas	Offshore Oil	Offshore Gas
Europe					
Norway	a			4	2
	c				1
Sweden	b	13			
Ireland	c			1	1
United Kingdom	a	2	6	10	4
	b	4	1	1	4
	c	21	2	4	7
The Netherlands	a	1	7		13
	b	1	6		
	c	13	29	5	2
West Germany	a	9	1		1
	b	30	7		
	c	86	28		
France	a	2	1		
	b	15	3		
	c	18	2		
Switzerland	b		1		
Austria	a	6	7		
	b	15	16		
	c	30	26		
Italy	a	15	53	4	36
	b	9	24		
	c	3	64	3	12
Yugoslavia	a	7	5		
	b	11	9		
	c	40	25	3	
Greece	c	1	1	1	
Spain	a	1	2	2	2
	b		1		
	c	3	2	3	5
U.S.S.R. (including parts in Asia)	a	30	3	2	
	b	73	39	1	1
	c	175	102	3	1
German Democratic Republic	b		2		
	c	2	4		
Poland	c	1	10		
Czechoslovakia	a		2		
	b	7	4		
	c	7	1		

* [a, less than 500 ft (150 m); b, 500–2,000 ft (150–600 m); c, 2,000–5,000 ft (600–1,500 m). Source: Petroleum Data System; Petroconsultants automated data file; Saskatchewan E and M, 1982; British Columbia MEM and PR, 1981; Alberta ERCB, 1982]

Table 1
continued

Region and Country	Depth Category*	Onshore Oil	Onshore Gas	Offshore Oil	Offshore Gas
Hungary	a	1	2		
	b	2	3		
	c	20	19		
Romania	a	38	6		
	b	28	39		
	c	43	42		
Bulgaria	a	1	1		
	c	1			
Albania	a	2			
	b	3	3		
	c	2	2		
Middle East					
Turkey	a	18	5		
	b	2	2		
	c	17			1
Syria	a	2	7		
	b	1			
	c	4	1		
Israel	a		1		
	b	1			
	c		5		
Iraq	b	11			
	c	6	1		
Iran	a	7	5	2	1
	b	5	1		
	c	15	3	2	1
Saudi Arabia	a	11		6	
	c	6	2	1	
Kuwait	b	2			
	c	4			
Neutral Zone	a	1			
	c	1			
Bahrain	a	1			
Qatar	c			2	1
Abu Dhabi	a	2		2	
Oman	a	32	2		
	b	12	1		
	c	17	2		

* *[a, less than 500 ft (150 m); b, 500–2,000 ft (150–600 m); c, 2,000–5,000 ft (600–1,500 m).*
Source: Petroleum Data System; Petroconsultants automated data file; Saskatchewan E and M, 1982; British Columbia MEM and PR, 1981; Alberta ERCB, 1982]

table continued

Table 1
continued

Region and Country	Depth Category*	Onshore Oil	Onshore Gas	Offshore Oil	Offshore Gas
Africa					
Morocco	a	3	1		
	c	11	11		
Algeria	a	14	7		
	b	10	9		
	c	20	18		
Tunisia	a	4	1	2	3
	b		3		
	c	2	1	4	
Libya	a	72	5		6
	b	9		1	1
	c	40	3		
Egypt	a	15	2	25	
	b	5			
	c	11	1	4	
Chad	c	1			
Sudan	a	5			
Ethiopia	a		1		
Somalia	a	1			
Senegal	b				
	c	3	1		
Ivory Coast	a	1		1	
Togo	a	1		1	
Benin	a	1		1	
Nigeria	a	143	14	53	6
	b	1			
	c	6	2	11	1
Cameroon	a	2	1	12	7
	c	2	1	20	5
Gabon	a	7		9	
	b	4		3	
	c	10	1	9	
Congo	a	1		7	3
	b			2	
	c	1		6	
Zaire	a	4		2	
	c	1			
Tanzania	c		1		
Angola	a	2	1	10	3
	b			2	
	c	8		3	

* [a, less than 500 ft (150 m); b, 500–2,000 ft (150–600 m); c, 2,000–5,000 ft (600–1,500 m). Source: Petroleum Data System; Petroconsultants automated data file; Saskatchewan E and M, 1982; British Columbia MEM and PR, 1981; Alberta ERCB, 1982]

Table 1 continued

Region and Country	Depth Category*	Onshore Oil	Onshore Gas	Offshore Oil	Offshore Gas
Mozambique	c		4		
South Africa	a				1
Far East					
China	a	20	4	1	1
	b	19	7		
	c	22	10	2	
Japan	a	50	37		
	b	40	33	2	
	c	15	16		
Afghanistan	a		1		
	b	1	2		
	c	5	3		
Pakistan	a	1	1		
	b		2		
	c	2	7		
India	a	18	5	2	1
	c	21	4	3	
Bangladesh	a	1	4		
	b		2		
	c		3		
Burma	a	4	2		
	b	4	2		
	c	4	2		
Taiwan	a	2	5		1
	b	3	3		
	c	6	4		
Viet Nam	c		1		
Thailand	a			2	1
	b	3			
	c	1			4
Philippines	b		1		1
	c		3	2	2
Sri Lanka	a	1			
Malaysia	a	1		27	23
	b			1	
	c			13	5
Brunei	a	2		9	3
	b	1			
	c	1		2	

* [a, less than 500 ft (150 m); b, 500-2,000 ft (150-600 m); c, 2,000-5,000 ft (600-1,500 m).
Source: Petroleum Data System; Petroconsultants automated data file; Saskatchewan E and M, 1982; British Columbia MEM and PR, 1981; Alberta ERCB, 1982]

table continued

Table 1 continued

Region and Country	Depth Category[*]	Onshore Oil	Onshore Gas	Offshore Oil	Offshore Gas
Indonesia	a	138	45	48	18
	b	63	2	3	
	c	104	11	40	5
Papua New Guinea	b		1		
	c		2		
Australia	a	1	2		1
	b	1	11	1	
	c	23	47	11	6
New Zealand	c	1			
North America					
Mexico	a	25	1	6	
	b	22	7		
	c	95	40	1	
Canada					
British Columbia[1,2]	b	29			
	c	23			
Saskatchewan[1]	b	36	11		
	c	153	27		
Alberta[1]	b	15	108		
	c	182	145		
United States					
PAD[3] district 1	a	15	2		
	b	142	24		
	c	27	65		
2	a	303	100		
	b	1,592	652		
	c	5405	1,138		
3	a	72	28	1	
	b	935	548	2	
	c	5,014	2,186	25	49
4	a	10	1		
	b	10	22		
	c	217	96		
5	a	7			
	b	75	21		
	c	115	97	5	
United States Totals		13,939	4,980	33	49

[a, less than 500 ft (150 m); b, 500–2,000 ft (150–600 m); c, 2,000–5,000 ft (600–1,500 m). Source: Petroleum Data System; Petroconsultants automated data file; Saskatchewan E and M, 1982; British Columbia MEM and PR, 1981; Alberta ERCB, 1982]

[1] Miscellaneous unnamed other areas in Alberta, British Columbia, and Saskatchewan are considered to be one field in each depth category.
[2] Natural gas data not compiled.
[3] PAD, Petroleum Administration for Defense.

Table 2
Depth Distribution of Shallow Oil and Gas Fields in the World by Region—A Summary of Data in Table 1*

		Numbers of Fields			
		Onshore		Offshore	
Continent	Depth Category*	Oil	Gas	Oil	Gas
South and Central America and the Caribbean	a	312	45	20	15
	b	136	11	7	1
	c	244	14	17	2
	Total	692	70	44	18
Europe	a	115	96	22	58
	b	211	158	2	5
	c	466	359	23	29
	Total	792	613	47	92
Middle East	a	74	20	10	1
	b	34	4	0	0
	c	70	14	5	3
	Total	178	38	15	4
Africa	a	276	33	123	29
	b	29	12	8	1
	c	116	44	57	6
	Total	421	89	188	36
Far East	a	239	106	89	49
	b	135	66	7	1
	c	205	113	73	22
	Total	579	285	169	72
North America[1]	a	432	132	7	0
	b	2,856	1,393	2	0
	c	11,231	3,794	31	49
	Total	14,519	5,319	40	49
World	a	1,448	432	271	152
	b	3,401	1,644	26	8
	c	12,332	4,338	206	111
	Total	17,181	6,414	503	271

* [a, less than 500 ft (150 m); b, 500–2,000 ft (150–600 m); c, 2,000–5,000 ft (600–1,500 m)]
[1] Gas totals for British Columbia excluded.

Thus, as usual in dealing with oil and gas data, the statistics are dominated by the United States, which accounts for 79 percent of the world shallow oil fields and 75 percent of the shallow gas fields (Tables 1 and 2). U.S. PAD (Petroleum Administration for Defense) districts 2 and 3 alone contain 75 percent of the shallow oil and 70 percent of the shallow gas fields of the world.

The shallow giant oil fields of the world (Figure 4) are listed in Table 3. Among those onshore, 26 are less than 2,000 ft deep, and another 52 are 2,000–5,000 ft deep. Offshore, 7 are less than 2,000 ft deep, and 9 are between 2,000 ft and 5,000 ft. The 94 shallow giant oil fields repre-

146   Shallow Oil and Gas Resources

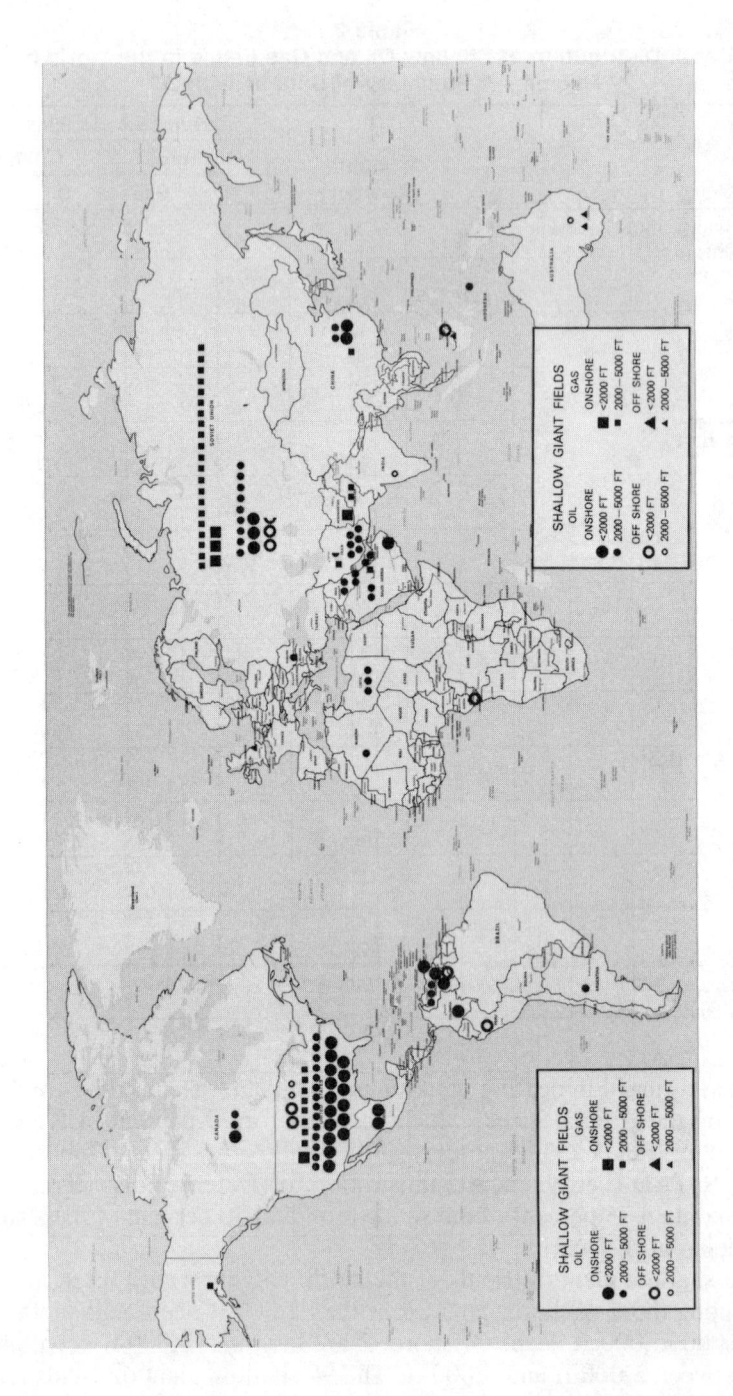

Figure 4. World giant oil fields and gas fields less than 5,000 ft (1,500 m) below the surface. Giant oil fields have ultimate recoveries of at least 500 million barrels; giant gas fields have ultimate recoveries of at least 3.5 trillion cu ft.

Table 3
Locations, Depths, and Ultimate Recoveries of Shallow Giant Oil Fields in the World*

Field	Country	State	Depth (ft)	Ultimate Recovery (millions of bbl)
		Onshore		
Burgan complex	Kuwait		4,800	66,000
Kirkuk	Iraq		2,800– 4,200	15,000
Romashkino	U.S.S.R.		4,800– 5,000	14,310
Gach Saran	Iran		4,200– 7,000	8,000
East Texas	United States	Texas	3,450– 3,850	6,000
Dammam	Saudi Arabia		4,490– 4,900	5,500
Amal-Nafoora-Augila	Libya		3,500– 11,000	5,200
Bibi Hakimeh	Iran		3,350– 6,800	4,500
Wafra	Neutral Zone		3,600	4,500
Arlan complex	U.S.S.R.		4,080– 4,305	4,100
Minas	Indonesia		2,400	4,000
Kotur-Tepe (Leninskoye)	U.S.S.R.		4,100– 12,730	4,000
Uzen	U.S.S.R.		3,300– 8,500	3,650
Pazanan	Iran		3,500– 9,000	3,500
Novoyelkhov-Aktash	U.S.S.R.		3,940– 5,580	3,240
Malgobek-Voznesenka-Aliyurt	U.S.S.R.		1,640– 10,830	3,000
Balakhany-Sabunchi-Ramany	U.S.S.R.		3,280– 6,560	2,400
Tuymazy	U.S.S.R.		3,340– 5,900	2,236
Haft Kel	Iran		3,500	2,000
Gialo	Libya		2,200– 6,300	2,000
Comodoro Rivadavia	Argentina		2,500– 9,000	2,000
Masjid-i-Suleiman	Iran		3,500	1,900
Kelly-Snyder-Diamond M	United States	Texas	2,570– 6,500	1,683
Panhandle	United States	Texas	2,050– 3,500	1,650
Naranjos-Cerro-Azul	Mexico		1,800– 2,300	1,400
Zarzaitine	Algeria		3,200– 4,600	1,300
Elk Hills	United States	California	2,400– 9,500	1,300
Yates	United States	Texas	1,150– 1,300	1,300
Midway-Sunset	United States	California	590– 5,300	1,200
Bai Hassan	Iraq		4,800– 5,400	1,200
Bahrain	Bahrain		1,850– 9,000	1,050
La Paz	Venezuela		4,250– 8,000	1,000
Quiriquire	Venezuela		1,800– 4,100	1,000
Fahud	Oman		1,250– 2,800	1,000
Shaybah	Saudi Arabia		5,000	1,000
Bushgan	Iran		4,500	1,000

* [a, less than 500 ft (150 m); b, 500–2,000 ft (150–600 m); c, 2,000–5,000 ft (600–1,500 m). Source: Petroleum Data System; Petroconsultants automated data file; Saskatchewan E and M, 1982; British Columbia MEM and PR, 1981; Alberta ERCB, 1982]

Table 3
continued

Field	Country	State	Depth (ft)	Ultimate Recovery (millions of bbl)
Ebano-Panuco	Mexico		1,450	971
Sho-Vel-Tum	United States	Oklahoma	1,000– 9,780	901
Seminole	United States	Oklahoma	3,550– 4,600	822
Ventura Avenue	United States	California	300– 7,180	812
Moreni-Gura-Ocnitei	Romania		2,950– 5,250	800
Oklahoma City	United States	Oklahoma	2,600– 5,200	771
Seeligson complex	United States	Texas	3,200– 6,700	752
Karamai	China		328– 7,545	730
Rainbow	Canada	Alberta	2,000– 3,200	720
Redwater	Canada	Alberta	2,600– 4,000	700
Dahra-Hofra	Libya		3,100– 3,700	700
Goldsmith-Andector	United States	Texas	3,700– 9,500	693
Illinois old fields	United States	Illinois	1,000– 1,700	675
Coalinga	United States	California	1,890– 2,750	664
Infantas-La Cira	Colombia		1,000– 3,250	660
Bradford	United States	Penn.	1,700– 2,300	658
Starogrozny	U.S.S.R.		660– 13,500	650
Hastings	United States	Texas	2,925– 6,130	630
Ta-Ching (Daqing)	China		2,230– 4,265	624
Buena Vista Hills	United States	California	1,450– 5,300	615
Santa Fe Springs	United States	California	1,145– 9,100	615
Kern River	United States	California	100– 1,500	615
Usa	U.S.S.R.		4,400– 7,120	614
Oficina	Venezuela		4,100– 6,550	610
Lung-nussu	China		3,610– 4,920	606
Conroe	United States	Texas	2,000– 5,050	605
Wasson	United States	Texas	4,890– 8,500	603
Karachukhur-Zykh	U.S.S.R.		2,690– 8,860	600
Rangely	United States	Colorado	1,700– 6,150	600
Mene Grande	Venezuela		500– 5,000	600
Cowden complex	United States	Texas	3,800– 9,300	580
Leng-hu complex	China		1,400– 4,265	528
Hawkins	United States	Texas	3,700– 4,900	526
Smackover	United States	Arkansas	2,000– 4,900	525
Oktyabr'skoye	U.S.S.R.		1,390– 3,700	520
Slaughter	United States	Texas	4,950– 9,250	518
Lima-Indiana	United States	Ohio-Ind.	1,250	514
Fyzabad group	Trinidad		50– 8,000	510
Salt Creek	United States	Wyoming	900– 3,800	510
Burbank	United States	Oklahoma	2,500– 3,200	500
Leduc-Woodbend	Canada	Alberta	5,100– 6,400	500
Tom O'Connor	United States	Texas	2,000– 5,900	500
Bolivar Coastal	Venezuela		560– 8,500	30,000

**Table 3
continued**

Field	Country	State	Depth (ft)	Ultimate Recovery (millions of bbl)
Bay Marchand-Timbalier-Caillou Island	United States	Louisiana	1,000– 20,000	3,400
Wilmington	United States	California	2,200– 6,600	2,600
Neftyanyye-Kamni	U.S.S.R.		650– 5,900	2,500
Idd-el-Shargi	Qatar		4,800– 8,250	2,100
Bibi Eybat	U.S.S.R.		330– 7,550	2,000
Seria	Brunei		834– 10,450	1,730
Bombay High	India		4,500	1,500
La Brea-Parinas-Talara	Peru		2,600– 8,700	1,000
Rostam	Iran		4,000– 6,700	1,000
Huntington Beach	United States	California	1,900– 6,700	970
Long Beach	United States	California	2,500– 9,500	892
South Pass Block 24	United States	Louisiana	2,000– 12,000	750
Cheleken	U.S.S.R.		3,610– 8,200	640
Emeraude Marin	Congo		1,820– 1,970	500
Barracouta	Australia		3,854– 3,900	500

* *[Ultimate recovery at least 500 million barrels; shallowest depth less than 5,000 ft (1,500 m).*
Source: Halbouty and others, 1970]

sent a recoverable resource of about 210 billion barrels of oil onshore and another 52 billion barrels offshore. This is an enormous amount of oil at relatively shallow depth. These data also do not take into account the greatest oil deposit in the world, the Faja Petrolifera del Orinoco or Orinoco Oil Belt (Fiorillo, 1984). Still in the evaluation stage, this deposit is estimated to contain 1.2 trillion barrels of oil in place, of which 217 billion barrels in priority areas and 50 billion barrels in nonpriority areas are deemed to be recoverable. The oil density ranges from 4° to 17° gravity API; depth to the top of the reservoir ranges from 500 to 4,300 ft and averages 2,000 ft.

A similar story is told by the 39 shallow giant gas fields (Table 4). They represent a recoverable resource of about 813 trillion cu ft (TCF), of which 7.6 TCF is offshore. Four of the fields, representing 58.4 TCF, are at depths of less than 2,000 ft. Nine of the fields, having resources of 112 TCF, are in the United States.

With respect to the United States (Table 5), shallow fields are a very significant factor in terms of recoverable resources of oil and gas. The data are derived from the Petroleum Data System (PDS) and compiled for convenience into PAD districts (Figure 5). About 627 million barrels is

Table 4
Locations, Depths, and Ultimate Recoveries of Shallow Giant Gas Fields in the World*

Field	Country	State	Depth (ft)	Ultimate Recovery (trillions of cu ft)
Onshore				
Urengoy	U.S.S.R.		3,450– 10,500	210.0
Yubileynyy	U.S.S.R.		4,000	70.0
Zapolyarnoye	U.S.S.R.		3,400– 4,400	54.3
Pazanan	Iran		3,500– 6,000	50.0
Tax	U.S.S.R.		3,600– 3,950	40.4
Hugoton	United States	Kans., Okla. and Texas	2,500– 3,200	39.5
Medvezh'ye	U.S.S.R.		4,000– 4,350	35.3
Panhandle	United States	Texas	1,450– 3,500	30.5
Yamburg	U.S.S.R.		3,600	30.0
Krasny Kholm (Orenburg)	U.S.S.R.		4,590– 5,900	26.5
Bahrain	Bahrain		4,200– 9,000	20.0
Gazii	U.S.S.R.		1,250– 3,840	17.0
Shebelinka	U.S.S.R.		4,135– 9,840	16.3
Sredne-Vilyuy	U.S.S.R.		4,700– 9,515	15.9
Messoyakha	U.S.S.R.		2,395– 5,085	14.0
Gubkin	U.S.S.R.		2,130– 2,400	12.3
Blanco-Basin	United States	New Mexico	4,600– 7,530	11.0
Vyngapur	U.S.S.R.		2,900	10.6
Russkoye	U.S.S.R.		3,300– 3,400	10.6
Jalmat	United States	New Mexico	2,900– 3,600	8.1
Shih-you-kou-Tung-hai	China		4,920– 6,560	7.8
Severo-Stavropol-Pelagioda	U.S.S.R.		1,500– 4,000	7.3
Monroe	United States	Louisiana	2,125	7.0
Sui	Pakistan		4,500– 5,000	6.0–9.0
Novyy-Port	U.S.S.R.		3,650– 6,750	5.1
Achak	U.S.S.R.		4,265– 6,856	5.0
Kenai	United States	Alaska	4,200	5.0
Mari	Pakistan		2,260	5.0
Mocaine-Laverne	United States	Oklahoma	4,250– 7,050	3.8
Buyerdeshik	U.S.S.R.		4,920	3.6
Shibargan	Afghanistan		1,340– 7,250	3.6
Rio Vista	United States	California	3,800– 5,750	3.5
Komsonol'	U.S.S.R.		2,760– 2,970	3.5
Nyda	U.S.S.R.		3,900– 4,000	3.5
Solenaya	U.S.S.R.		2,939– 9,680	3.5
Bayou Sale	United States	Louisiana	4,500– 14,500	3.5
Gugurtli	U.S.S.R.		3,600– 6,900	3.5
Offshore				
Hewett	United Kingdom		4,500– 6,500	4.0
Marlin	Australia		4,532– 10,000	3.6

*[Ultimate recovery at least 3.5 trillion cu ft; shallowest depth less than 5,000 ft (1,500 m). Source: Halbouty and others, 1970]

Table 5
Shallow Oil and Gas in the United States*

PAD[1] district	Depth Category	Ultimate Recovery Oil (millions of bbl)	Gas (billions of cu ft)
\multicolumn{4}{c}{Onshore}			
1	a	4.2	18.9
	b	839.3	0.1
	c	31.5	1,093.4
2	a	7.4	29.5
	b	327.8	316.9
	c	1,054.4	1,369.0
3	a	609.5	155.7
	b	2,999.0	4,246.2
	c	28,165.0	56,702.3
4	a	0.3	
	b	25.6	158.2
	c	240.0	1.2
5	a	2.3	2.0
	b	5,537.2	697.3
	c	9,116.0	10,279.1
\multicolumn{4}{c}{Offshore}			
3	a	2.7	
	b		1.0
	c	229.2	316.7
5	a		
	b		
	c	570.0	226.8

* [a, less than 500 ft (150 m); b, 500–2,000 ft (150–600 m); c, 2,000–5,000 ft (600–1,500 m). Source: Petroleum Data System]

[1] PAD, Petroleum Administration for Defense.

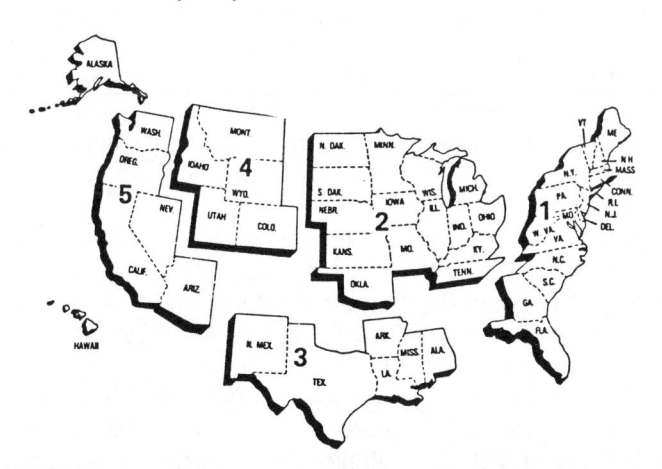

Figure 5. Index map of Petroleum Administration for Defense (PAD) districts in the United States.

less than 500 ft deep, 9.7 billion barrels is between 500 and 2,000 ft, and another 39.4 billion barrels is at depths from 2,000 to 5,000 ft. Approximately half the U.S. shallow oil is contained in the giant fields (Tables 3 and 5), leaving a significant amount in the smaller ones.

For shallow gas in the United States (Table 5), the resource also is large. The total of 75.6 TCF from Table 5 is less than the amount given in Table 4 for shallow giant fields alone; this difference is the result of several factors. The data for Table 5 were computed from latest annual production, and the decline factor applied was probably too small. In addition, cumulative gas totals are often understated in the PDS. Nevertheless, Table 5 gives a very good idea of the depth distribution of shallow gas; nearly 93 percent is from 2,000 to 5,000 ft in depth, essentially all the rest is between 500 and 2,000 ft, and almost none is from depths shallower than 500 ft.

Oil and Gas in the United States by Depth and Year of Discovery

A perspective on the distribution of U.S. ultimately recoverable oil and gas by depth and year of discovery is provided by Tables 6 and 7. Table 6 indicates a total of 140.0 billion barrels to be recovered in known fields. Of this amount, barely 7 percent is shallower than 2,000 ft, about 26 percent occurs between 2,000 and 5,000 ft, and two-thirds is below 5,000 ft. Almost from the industry's beginning, oil being found below 5,000 ft equaled or exceeded that being found at the shallower depths. At no time has a significant proportion of the oil been derived from depths of less than 2,000 ft. Since 1930, the share found below 5,000 ft has been entirely dominant.

Similarly for U.S. natural gas (Table 7), the discoveries have been overwhelmingly from the depths below 5,000 ft. There has never been a significant amount of gas found at depths shallower than 2,000 ft. The interval from 2,001 to 5,000 ft accounts for about 12.5 percent of natural gas.

The intense level of drilling in the United States has resulted in major amounts of oil and gas being found at shallow depths. Most of these hydrocarbons are found at depths between 2,000 and 5,000 ft; the greatest amounts by far are in PAD district 3, and lesser but important resources are in PAD district 5 (Table 5).

Most of the world's shallow oil and gas fields are found in the United States because of the way business is conducted in that country and not because the United States is geologically or geochemically unique in the

Table 6
Amount and Percentage of Ultimately Recoverable Crude Oil in the United States, by Decade in Which Discovered

Depth class (ft)	Through 1900	1901–1910	1911–1920	1921–1930	1931–1940	1941–1950	1951–1960	1961–1970	1971–1980	1981–1983	Total
Year-of-Discovery Class (billions of barrels)												
0–500	0.007	0.000	0.000	0.530	0.005	0.007	0.021	0.003	0.001	0.000	0.580
501–2,000	2.800	2.300	.880	2.100	.480	.340	.280	.073	.020	.003	9.200
2,001–5,000	.380	1.500	3.200	13.000	8.600	4.000	3.100	1.600	.320	.022	36.000
5,001+	2.500	3.900	5.700	8.000	17.000	18.000	17.000	12.000	3.700	7.700	95.000
Totals	5.600	7.700	9.900	24.000	26.000	22.000	20.000	14.000	4.000	7.700	140.000
Year-of-Discovery Class (% of total)												
0–500	0.00	0.00	0.00	0.38	0.00	0.01	0.01	0.00	0.00	0.00	0.41
501–2,000	1.98	1.60	.63	1.48	.34	0.24	.20	.05	.01	.00	6.54
2,001–5,000	.27	1.05	2.29	9.41	6.13	2.83	2.22	1.11	.22	.02	25.55
5,001+	1.74	2.78	4.07	5.69	11.94	12.62	11.91	8.73	2.60	5.43	67.49
Totals	3.99	5.43	6.99	16.97	18.42	15.69	14.34	9.89	2.83	5.44	100.00

[*Source: Petroleum Data System*]

Table 7
Amount and Percentage of Ultimately Recoverable Natural Gas in the United States, by Decade in Which Discovered

Depth class (ft)	Through 1900	1901–1910	1911–1920	1921–1930	1931–1940	1941–1950	1951–1960	1961–1970	1971–1980	1981–1983	Total
\multicolumn{12}{c}{Year-of-Discovery Class (trillion of cu ft)}
0–500	0.002	0.000	0.000	0.064	0.003	0.001	0.100	0.002	0.053	0.001	0.230
501–2,000	.180	.340	.220	.560	1.300	.360	1.300	.300	.370	.019	5.000
2,001–5,000	.450	1.400	.880	10.000	28.000	5.600	16.000	4.600	2.900	.220	70.000
5,001+	1.200	2.600	11.000	42.000	51.000	77.000	120.000	85.000	47.000	50.000	490.000
Totals	1.800	4.400	12.000	53.000	80.000	83.000	140.000	90.000	51.000	50.000	560.000
\multicolumn{12}{c}{Year-of-Discovery Class (% of total)}
0–500	0.00	0.00	0.00	0.01	0.00	0.00	0.02	0.00	0.01	0.00	0.04
501–2,000	.03	.06	.04	.10	.23	.06	.23	.05	.07	.00	.88
2,001–5,000	.08	.26	.16	1.81	4.88	.99	2.91	.82	.51	.04	12.46
5,001+	.21	.46	1.98	7.38	9.10	13.72	21.52	15.04	8.40	8.81	86.62
Totals	.32	.77	2.17	9.31	14.21	14.77	24.68	15.92	8.99	8.85	100.00

[Source: Petroleum Data System]

world. Outside the United States and Canada, oil and gas exploration historically has been conducted by major companies or state-owned companies whose obligations cannot be met by shallow fields, which are perceived, often incorrectly, to be small.

Bitumen and Heavy-Oil Deposits

Although the number of bitumen deposits is not large compared with the number of shallow oil fields, the resource contained in the world's bitumen deposits is enormous (Table 8). The deposits in Alberta, Canada, alone contain at least 1.2 trillion barrels of bitumen in place. The total for the deposits in the U.S.S.R. may be about as large, although information is not nearly so precise. In the United States, none of the deposits approach the size of the giant deposits of the world, but many are very large and potentially exploitable. Those in Utah have long been described and are the subject of current efforts at development. The Kentucky deposits were vastly underestimated for many years, but following recent research by the Kentucky Geological Survey, the true resource level is known to be at least 3 billion barrels of bitumen in place. Table 9 lists some of the numerous occurrences of bitumen around the world that either are very small or else have never been evaluated in detail.

The United States has 630 shallow heavy-oil fields (Table 10). More than half of these are at depths of 2,000–5,000 ft, and 43 are at less than 500 ft.

Table 8
Measured Bitumen Deposits in the World

Country	Deposit Name	Bitumen in Place (millions of bbls)	Density (°API)
Albania	Selenizza	371	4.6–13.2
Canada	Athabasca/Wabasca, Alberta	918,981	8–10
	Athabasca deep, Alberta	42,651	8–10
	Buffalo Head Hills, Alberta	5,796	10
	Peace River, Alberta	75,072	8–9
	Cold Lake, Alberta	205,317	8–13
	Carbonate Trend, Alberta	315	7
		1,248,132	

Table 8
Continued

Country	Deposit Name	Bitumen in Place (millions of bbls)	Density (°API)
Italy	Ragusa area	14,000	
Madagascar	Bemolanga	21,000	3.4-12.6
Nigeria	Dahomey embayment	41,000	5.3-14.6
Peru	Jibaro	7	7.5-10.0
Romania	Derna	25	
Trinidad/Tobago	Asphalt Lake	60	1-2
U.S.S.R.	Melekess	127,000	
	Siligir	13,000	
	Olenek	600,000	
	Cheildag (near Baku)	24	
	North Caspian	280,000	
	Tunguska	17,500	
	Lena-Anubarsky	70,000	
	Timan-Pechora	45,500	
		1,153,024	
United States	North area, Alabama	4,300	
	Edna, California	175	
	McKittrick tar sand, California	9	
	Oxnard Vaca, California	400	5-8
	Oxnard Lower, California	165	
	Paris Valley, California	100	
	Point Arena, California	1	
	Santa Cruz, California	10	
	Sisquoc, California	106	4-8
	Santa Maria (Foxen), Calif.	000	9-14
	Richfield, California	40	12
	Asphalt area, Kentucky	3,000	
	Santa Rosa, New Mexico	91	
	Ohio	1	
	South-Central Oklahoma	800	
	Tri-State area, Utah	287	
	Asphalt Ridge, Utah	873	10.4
	Asphalt Ridge, NW, Utah	100	14
	P.R. Spring, Utah	3,700	9.5
	Hill Creek, Utah	830	9.1
	Sunnyside, Utah	2,000	8.6
	Tar Sand Triangle, Utah	6,100	4.3
	Circle Cliffs, Utah	1,137	−3
	Uvalde area, Texas	3,000	−2
	Wyoming	1,000	
		30,225	
Venezuela	Guanoco Pitch Lake	12	
West Germany	Eschershausen	Not known	
Zaire	Zaire-Cabinda border area	1,500	

[Source: Meyer, Fulton, and Dietzman, 1984]

Table 9
Poorly Known Bitumen Deposits in the World

Country	Area	Remarks
Deposits that Are Less than 10% Mineral Matter		
Mexico	Tamaulipas	Asphalt springs, Tamesi R.; Chijol
	Vera Cruz	Asphalt springs, Tuxpan; Chapapote
Cuba	Matanzas	20 barrels/day semiliquid asphalt
	Santa Clara	Hard asphalt
	Camaguey	Pure soft asphalt
	Santiago de Cuba	Soft asphalt, small
Venezuela	Maracaibo	Springs; 100,000 tons produced 1901–1905
	Tachira	Asphalt utilized for paving
France	Auvergne	Asphalt reefs; 100 tons/yr prior to 1914
Greece	Zante	Springs and seeps; 10° API
U.S.S.R.	Sakhalin	Great Okha Asphalt Lake and area; 3.5 million barrels of nearly pure asphalt
Philippines	Leyte	Several deposits of hard and soft asphalt
Deposits that Are More than 10% Mineral Matter		
United States	Missouri	Several large deposits
	Kansas	Linn County; 20 million tons; quarried at times
	Indiana	Princeton; liquid asphalt below coal bed
	Arkansas	Southwest; quarried at times
	Louisiana	Lafayette; 50 acres on surface
Cuba	Matanzas	Bottom of Cardenas Harbor; mined in past
	Havana	Mined in past
Brazil	Parana, Sao Paulo and Bahia	Small occurrences
Argentina	Laguna de la Brea	Asphalt Lake
Colombia	Bolivar	Seepages used to calk ships
	Antioquia	Nare area
	Santander	Many seepages
	Boyaca	Formerly used for paving
Ecuador	Guayas	Seepages reported in oil-prospecting pits
Peru	Junin and Puno	Asphalt used for paving
France	Landes	Asphalt mines at Guajacj and Bastennes
	Gard	Ales basin; asphalt formerly distilled for motor fuel
	Haute-Savoie and Ain	Asphalt mines
	Auvergne	Asphalt mines
Switzerland	Lac de Neuchatel	Asphalt mines in past
West Germany	Hanover	Near Limmer; 3 million tons; other deposits in area
Yugoslavia	Montenegro, Herzegovina	Asphalt limestones
Czechoslovakia	Trencsen, Moravia	Asphalt limestones
Greece		Various asphalt deposits
Spain		Asphalt deposits and pits
Portugal	Estremadura	Soft asphalt

table continued

Table 9
continued

Country	Area	Remarks
Deposits that Are More than 10% Mineral Matter		
U.S.S.R.	Simbirsk	Asphaltic sand (garj)
(Europe)	Caucasia	Soft asphalt (kir), fairly pure
	Transcaucasia	Numerous deposits
Syria		Many small deposits
Israel		Many small deposits
Iraq		Many small deposits
U.S.S.R. (Asia)		Many small deposits
Saudi Arabia		Many small deposits
Egypt		Many small deposits
India		Many small deposits
China		Many small deposits
Philippines	Leyte	650,000 tons rock asphalt
Japan		Small deposits formerly mined
Australia		Small deposits
Tasmania		Small deposits
New Zealand		Small deposits
Indonesia	Buton Island	Mine with 1.3 million tons of asphaltic marl
Algeria	Oran	Seepages
Zambia		Rock asphalt

[Source: Meyer, Fulton, and Dietzman, 1984]

In Canada, British Columbia has very little heavy oil. In Saskatchewan, virtually all the heavy oil is in the western part of the province, where 35 fields contain nearly 299 million barrels of heavy oil at depths of less than 2,000 ft (none are as shallow as 500 ft); another 13 fields having an ultimate recovery of 103 million barrels are between 2,000 and 5,000 ft. In Alberta, about 56 fields, many multi-pay, produce heavy oil. These are mostly at depths of 2,000–5,000 ft. The initial established heavy-crude reserves of Alberta are about 600 million barrels, most but not all of which is in the shallow fields.

Effect of Drilling Technology in the United States

Smith International Inc. has introduced a new statistical series embodied in its *Smith International Inc. Drilling Activity Correlator (Sii-DAC)*. This measure delineates activity on the basis of rig capability (Figure 6).

Table 10
Numbers of Shallow Heavy-Oil Fields in the United States by Petroleum Administration for Defense District and Depth Category*

PAD[1] district	Depth Category	Number of Fields
2	a	3
	b	20
	c	42
	Total	65
3	a	27
	b	152
	c	167
	Total	346
4	a	2
	b	14
	c	51
	Total	67
5	a	11
	b	65
	c	76
	Total	152
Entire United States	a	43
	b	251
	c	336
	Total	630

* *[oil less than 20° API; a, less than 500 ft (150 m); b, 500-2,000 ft (150-600 m); c, 2,000-5,000 ft (600-1,500 m). Source: Lewin and Associates, 1984]*
[1]*PAD, Petroleum Administration for Defense.*

Rigs of the three types may have different depth capabilities from basin to basin, so that in practice Smith International compiles its data on the basis of geologic provinces (Meyer, 1970). In general, Type I rigs may drill as deeply as 4,000 ft, Type II, 10,000 ft, and Type III, below 10,000 ft. The actual number of U.S. wells drilled by each type is given in Figure 7. The number of Type III wells has increased slowly relative to the number drilled by shallower-capability rigs, but this increase is misleading. Figure 8 demonstrates that the Type II well is the heart of the business of U.S. oil and gas drilling. However, although only 10 percent of the wells are Type III, these account for about 25 percent of the footage drilled. The simple Type I rig now accounts for more than 35 percent of the wells drilled but only 18 percent of the footage. Economic forces are the only obvious explanation for the solid improvements in both rig efficiency (the number of wells drilled per active rig) and drilling efficiency (the

160   Shallow Oil and Gas Resources

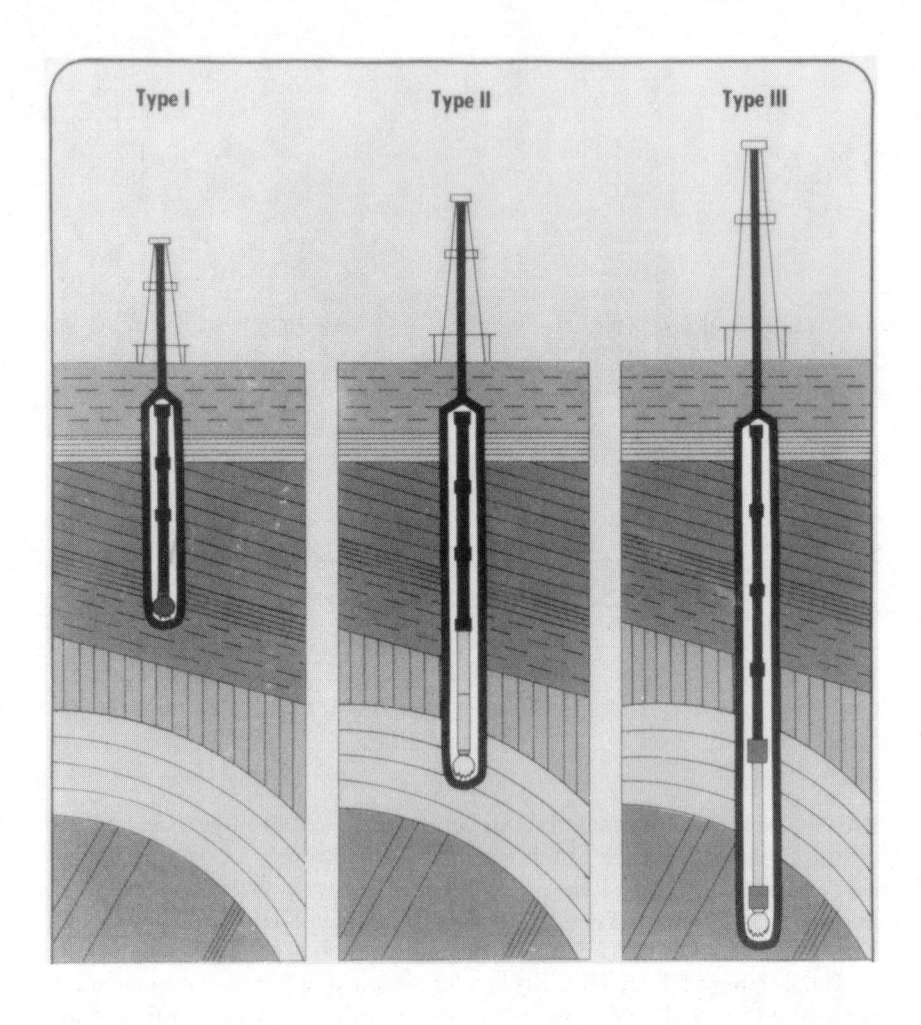

Figure 6. Sii-DAC (Smith International Inc. Drilling Activity Correlator) rig categories. Type I, small rig with simple drill string and open-bearing, milled-tooth bit used to drill to total depth; Type II, moderate-size rig with either a sophisticated drill bit or drill-string components, but not both, used to drill to total depth; and Type III, sophisticated rig using advanced drill-string components with premium-quality bit to drill to total depth. (Data from Smith International, Inc., 1974–83.)

Survey of World Shallow Oil and Gas 161

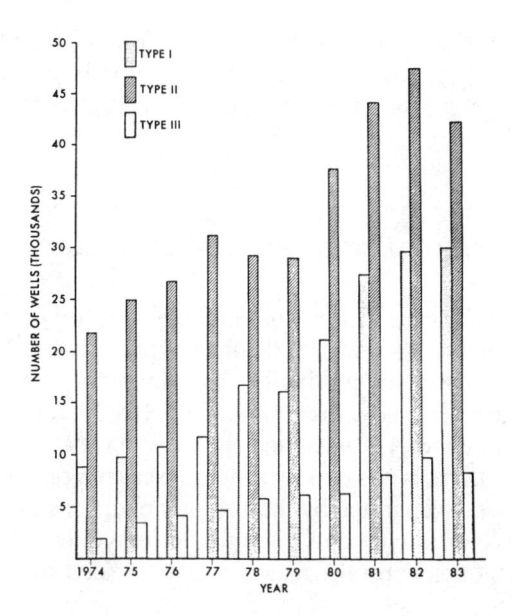

Figure 7. Numbers of U.S. wells drilled by rig type, 1974–83. (Data from Smith International, Inc., 1974–83.)

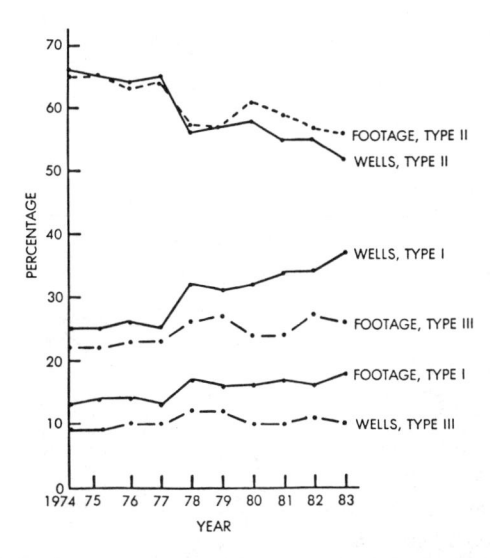

Figure 8. Percentage of U.S. wells and footage drilled by rig type, 1974–83. (Data from Smith International, Inc., 1974–83.)

footage per rig) during 1981–83 (Figure 9). With effort now being concentrated on drilling in and around known U.S. fields and in finding and drilling shallow fields, rig and drilling efficiencies are of paramount importance.

Exploration in the United States

In Figure 10 all data are for the United States and are indexed to 1967, taken as 100 for the wells and $1.00 for wellhead prices. From 1935 to about 1972, prices rose only gradually, from about $0.25 to about $1.15. In the ensuing decade, they rose rapidly to $15.20 for gas and about $10.00 for oil, or about 13 times and 9 times, respectively. During the first price phase, there were two drilling phases, the peaks of which, in 1937 and 1955, were separated by the special conditions of World War II. Whether there would have been a single perhaps more subdued phase without the war is conjectural because the war ended the economic de-

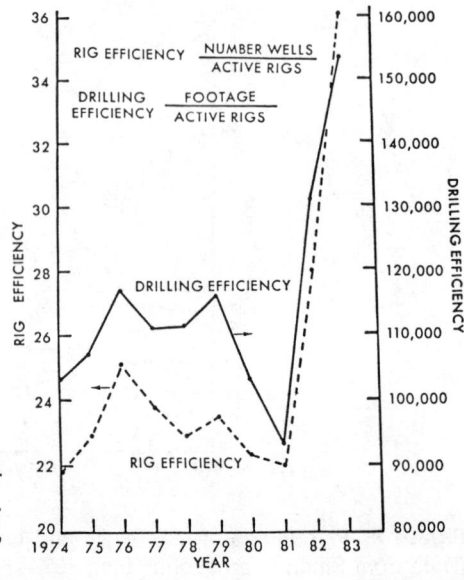

Figure 9. Rig and drilling efficiencies in the United States, 1974–83. (Data from Smith International, Inc., 1974–83.)

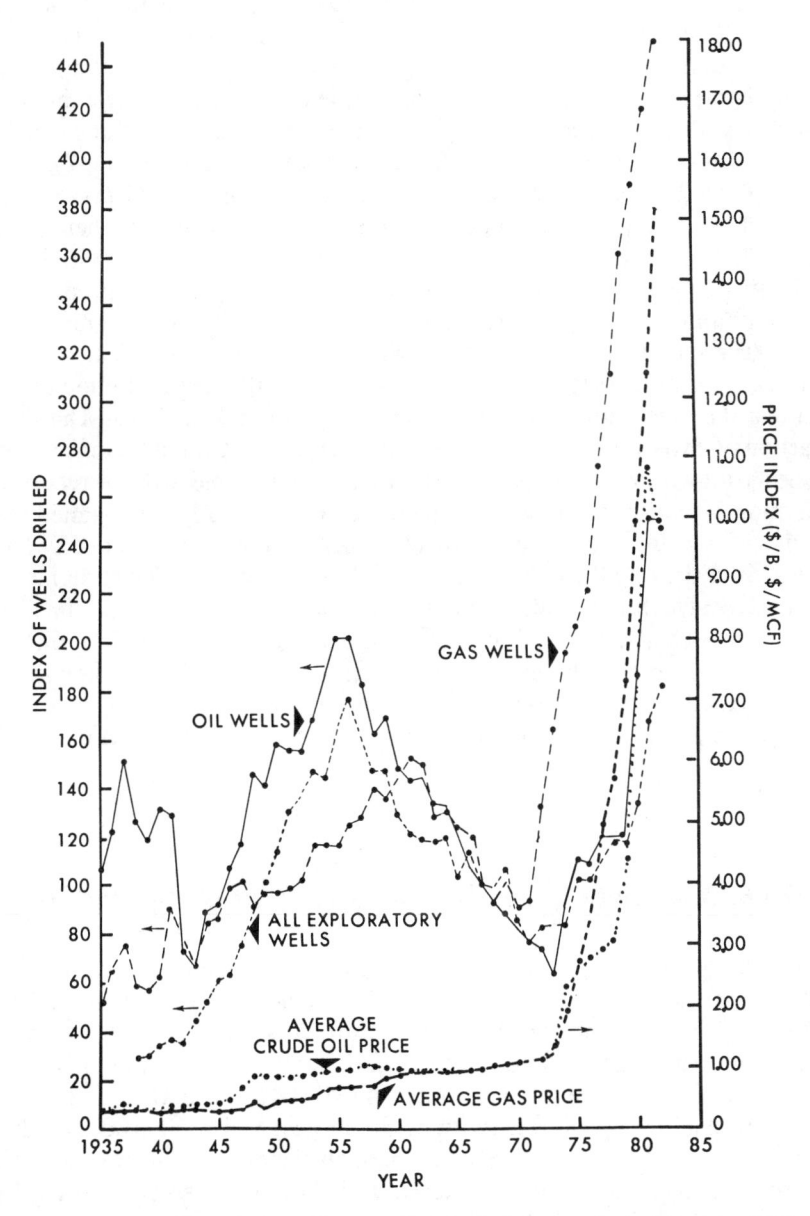

Figure 10. Numbers of U.S. oil, gas, and exploration wells indexed to 1967 = 100 and wellhead oil and gas prices indexed to 1967 = $1.00, 1935–82. (Data from DeGolyer and MacNaughton, 1983.)

pression. In any case, these peaks clearly were not price driven but resulted from resource availability, with volume production offsetting low price. The second phase, beginning in 1972, was instigated by strong price increase, due to non-national factors coupled with strong national demand and resource scarcity. The resultant increase in drilling definitely is price driven. Figure 10 shows little evidence to suggest that exploratory drilling leads development drilling. Indeed, nothing really seems to lead anything consistently. This lack of pattern can most easily be explained in terms of economic factors, regulation, and tax policy. The net effect is that when times are good, everybody drills; when times are bad, everybody retreats.

Even though additional wells drilled in fields can add to production capacity and reserves, new discoveries are the only means to add to future reserves. Of the five classes of exploration wells, one, the new-field wildcat, normally offers the opportunity for a really large addition to reserves; the others are drilled in association with known fields. The proportion of exploration wells drilled each year in the United States in each class is shown in Figure 11. Historically, the new-field wildcat received the most attention because the potential rewards greatly outweighed the higher risks. In recent years, however, perceptions have changed. The rewards of drilling the more certain well have increased. The recognition that many structures producing at shallow depths persist at depth has led to deep-pool tests, fostered by improved drilling technology. In terms of success (Figure 12), though not necessarily of rewards, the shallow-pool test is still the most certain; declining success of the shallow tests in recent years suggests that the most obvious prospects have been drilled.

Drilling Activity in the United States by Company Size

In attempts to compare activity of large and small companies, the U.S. Bureau of the Census (1984) does not name individual companies. One grouping of companies by the Census Bureau is based upon lease revenues on a net-company-interest basis, a desirable approach in analyzing exploration and development in the United States. The U.S. Bureau of the Census (1984) groups activity by the largest 8, 16, 50, 100, and 200 companies, and by all others combined, in terms of lease revenues. If these groupings are roughly related to total company assets, as listed by the *Oil & Gas Journal* (1983) for the largest 400 companies, then the largest 8 would equate to the 8 internationals, the largest 16 would include most of the large, integrated companies having assets of about $9

Survey of World Shallow Oil and Gas   165

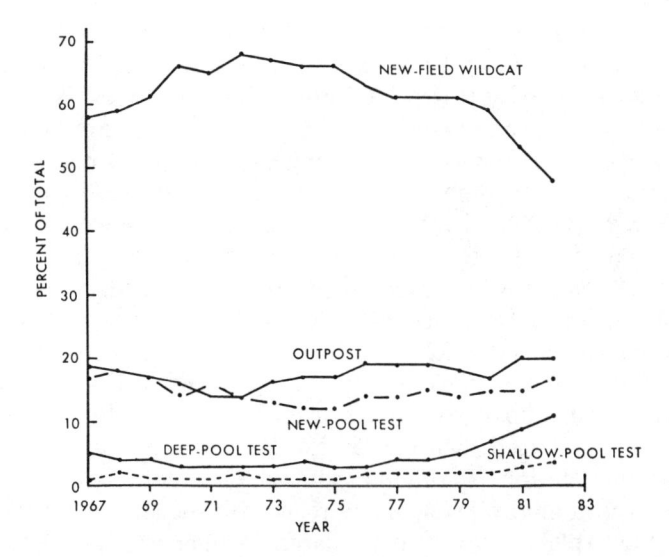

Figure 11. Percentage of U.S. exploration wells by class, 1967–82. (Data from Johnston, 1983.)

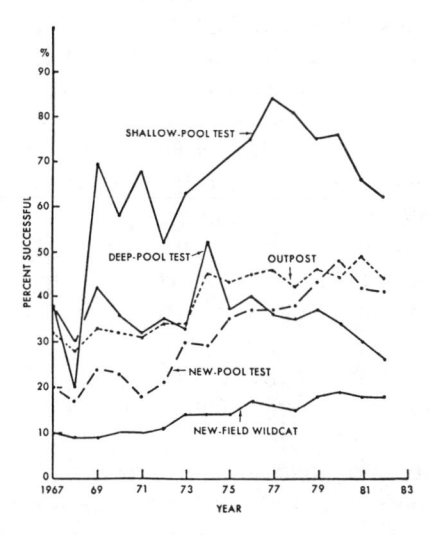

Figure 12. Percentage of successful U.S. exploration wells, by class, 1967–82. (Data are from Johnston, 1983.)

billion, and the largest 50 would be as large as Mitchell Energy and Louisiana Land, which have assets of about $1.7 billion. The largest 100 would have assets of at least $550 million, and the largest 200 all would have assets greater than $40 million. Figure 13 shows a progressive increase in well and footage drilling and expenditures for this purpose, through the first 200 companies. All the other companies together, totaling probably 16,000 to 18,000 enterprises, drill more wells, drill more footage, and spend far less money for the purpose.

Figure 14 compares the ratio of expenditures for production drilling to expenditures for exploration drilling for the various groups of companies for the years 1978 and 1982. For 1978, the largest 200 companies spent as much as $1.70 for development drilling for each $1.00 for exploration, whereas the small companies spent only $1.05. But in 1982, this situation was reversed, and the largest companies spent little more for development than for exploration, while the smallest companies greatly increased spending for development relative to spending for exploration. A probable explanation is that the larger companies mostly caught up their exploitation of past discoveries and moved back into exploration while the smaller companies, with less venture capital available in the recession economy, moved into a more conservative position. The popularity of development-drilling programs as tax shelters also influenced change, as each partnership is treated as a separate firm.

Table 11 gives drilling statistics for 1982 for the petroleum industry in the United States. The data make clear that company size has an enormous influence on company activity. The effort of the small enterprises is clearly focused on shallow drilling. In only one category—the drilling of development gas wells—does the average well depth significantly exceed 5,000 ft for the small companies. This focus on shallow drilling will change in the future because the average depth of exploration gas wells drilled by small companies is about 4,500 ft.

Among exploration wells, whether dry, oil, or gas, per-well costs for the 200 largest companies are on average not less than $1.4 million. For the small companies, the average well costs hardly exceed $200,000. This cost difference is not merely a function of depth, although depth is an important factor. The exploratory wells drilled by large companies tend to be in remote and environmentally difficult locations and rely heavily on the more sophisticated and costly Type II and Type III rigs. Similar comments apply to development wells, but the average costs and average depths are much less than those for exploration wells. The sharp division between the large and small companies remains. Although the differences in average well depths are much narrower for development wells than they are for exploratory wells, the cost differentials remain high. Again, this difference is not due simply to depth.

Survey of World Shallow Oil and Gas

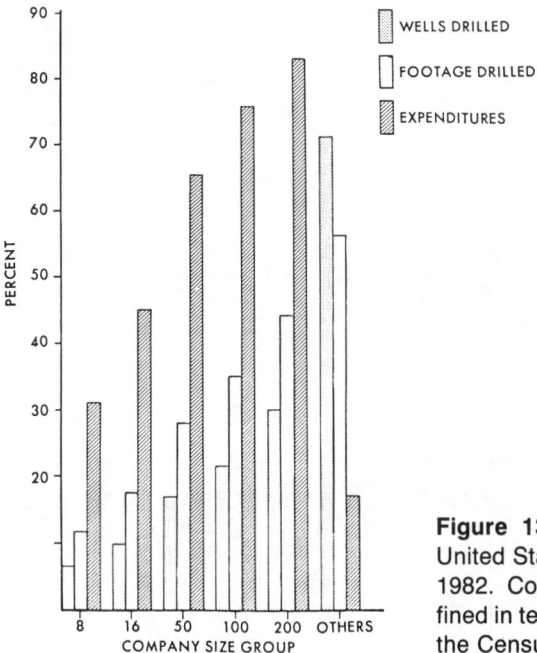

Figure 13. Exploration activity in the United States by company size group, 1982. Company size groups are defined in text. (Data from U.S. Bureau of the Census, 1984.)

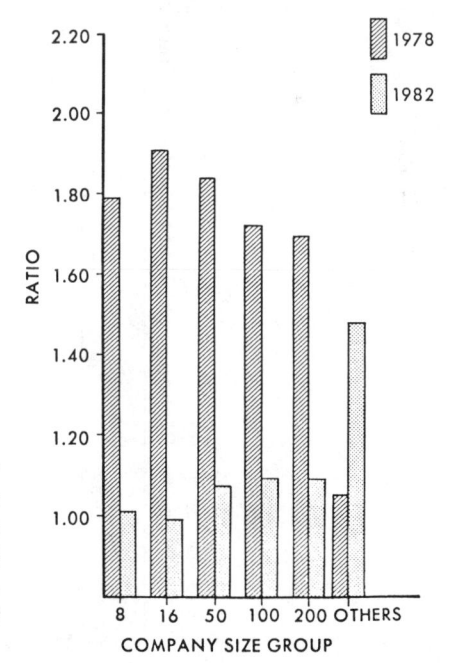

Figure 14. Ratio of expenditures for production drilling to expenditures for exploration drilling in the United States by company size group, 1978 and 1982. Company size groups are defined in text. (Data from U.S. Bureau of the Census, 1980 and 1984.)

Table 11
Statistics for Drilling in the United States by Companies Ranked by Size (Net Company Lease Interest), 1982

Item	Unit	1st 50	1st 100	1st 200	All Others
Exploration dry holes					
Number of wells	number	1,583	2,098	2,904	6,084
Total expenditures	mil. dol.	4,492.3	5,192.4	5,692.4	883.8
Average expenditure per well	mil. dol.	2.8	2.5	2.0	0.145
Total footage	thous. ft	15,324	19,425	25,492	27,902
Average footage per well	ft	9,680	9,259	8,778	4,586
Exploration oil wells					
Number of wells	number	681	827	1,073	4,230
Total expenditures	mil. dol.	1,153.7	1,362.3	1,527.7	760.1
Average expenditure per well	mil. dol.	1.7	1.6	1.4	0.180
Total footage	thous. ft	5,763	6,829	8,505	1,910.3
Average footage per well	ft	8,463	8,258	7,926	4,516
Exploration gas wells					
Number of wells	number	761	978	1,279	2,472
Total expenditures	mil. dol.	2,654.8	3,041.1	3,307.9	510.7
Average expenditure per well	mil. dol.	3.5	3.1	2.6	0.207
Total footage	thous. ft	8,032	9,891	11,983	10,994
Average footage per well	ft	10,554	10,113	9,369	4,447

Development dry holes						
Number of wells	number	879	1,177	1,559	2,750	
Total expenditures	mil. dol.	929.6	1,183.6	1,360.5	500.7	
Average expenditure per well	mil. dol.	1.1	1.0	0.873	0.182	
Total footage	thous. ft	6,545	8,679	11,298	14,035	
Average footage per well	ft	7,446	7,374	7,247	5,104	
Development oil wells						
Number of wells	number	6,162	6,798	8,020	6,655	
Total expenditures	mil. dol.	3,915.6	4,389.4	4,752.7	1,322.7	
Average expenditure per well	mil. dol.	0.635	0.646	0.593	0.199	
Total footage	thous. ft	31,957	35,693	41,620	30,648	
Average footage per well	ft	5,186	5,251	5,190	4,605	
Development gas wells						
Number of wells	number	2,269	3,115	4,122	5,181	
Total expenditures	mil. dol.	4,037.0	4,850.8	5,377.1	1,368.3	
Average expenditure per well	mil. dol.	1.8	1.6	1.3	0.264	
Total footage	thous. ft	19,559	26,009	31,436	29,882	
Average footage per well	ft	8,620	8,350	7,626	5,768	

[Source: U.S. Bureau of the Census, 1984]

Conclusions

Masses of statistical data can be intimidating and can be variously interpreted. Used judiciously, the data make possible an understanding of past activity and future trends.

The shallow fields discovered to date in the world were found by companies of all sizes, but most were found by large companies. This is due in great part to the fact that in many countries, only large companies can and do operate, mostly for economic reasons. The shallow fields, and especially the large ones, are normally found early in the exploration process, and many small shallow ones are early indicators of larger and deeper reservoirs.

In the United States, certain regions have largely been the domain of small companies, and as a consequence, such companies found most of the shallow oil. As finding shallow, large fields having large reserves became more and more difficult, the large companies came to concentrate on deep, costly prospects. Now, new fields being discovered are mostly small or very small, and have less than 10 million barrels of ultimately recoverable oil. These small fields are not necessarily shallow, because increased oil and gas prices have altered petroleum production economics significantly. The data in Figures 13 and 14 and Table 11 all point to the fact that the small companies today are emphasizing production drilling at shallow depths. This emphasis means that fewer shallow fields are available for discovery by current new-field wildcat exploration; therefore, future production will be from greater depths and will be increasingly concentrated in fewer and larger companies. Such changes will not come quickly, but they appear to be inevitable because the U.S. petroleum industry has been too successful. The number of prospects at shallow depths must diminish, except for the very small fields.

If we extrapolate from the U.S. experience and assume that the United States is in no way unique except in exploitation effort, then we estimate that about 35 percent of the world's ultimately recoverable oil resources and 13.5 percent of the ultimately recoverable nonassociated gas resource should be in shallow fields. Should the Orinoco Oil Belt be as productive as appears to be likely, then the shallow oil proportion will be vastly greater. Probably some basins are characterized by many small fields, as in U.S. PAD district 2, and others by fewer and larger fields, as in U.S. PAD districts 3 and 5. It is unlikely that any basin will contain only or even mostly deep fields or large fields.

Most of the shallow oil and gas remaining in the world will be found and exploited by small operators. Although shallow fields do not equate with small fields, most shallow fields to be found in the future will be small. In an evolving exploration process in relatively unexplored basins, large shallow fields will be found before small shallow fields, if large ones are present.

Discovery of large, shallow, readily exploitable fields leads to the development of a production and transportation infrastructure that facilitates development of small fields either found later or else shut in to await a transportation network.

In general, we can safely assume that most of the large, shallow fields in the world have already been found. Exploration, at least in the free world, has been intensive for most of this century.

Most of the world's bitumen deposits already are known. They are numerous and have commonly signalled the presence at depth of oil fields; some oil fields associated with bitumen deposits are very large, for example, Burgan and Kirkuk (Table 3). The exploitation of bitumen deposits as a source of oil is a matter of cost relative to costs of alternative sources of conventional crude oil.

References

Alberta Energy Resources Conservation Board, 1982, Alberta's reserves of crude oil, gas, natural gas liquids, and sulphur at 31 December 1981: Alberta Energy Resources Conservation Board ERCB 82-18, 9 chapters.

British Columbia Ministry of Energy, Mines and Petroleum Resources, 1981, Engineering and geological reference book, oil and gas pools in British Columbia: Victoria, B.C., unpaged.

DeGolyer and MacNaughton, 1983, Twentieth century petroleum statistics 1983: Dallas, DeGolyer and MacNaughton, 126 p.

Fiorillo, G. J., 1984, Exploration and evaluation of the Orinoco Oil Belt: American Association of Petroleum Geologists Research Conference, Exploration for Heavy Crude Oil and Bitumen, Santa Maria, CA, Oct. 29-Nov. 2, 1984, 26 p., preprint.

Halbouty, M. T., et al., 1970, World's giant oil and gas fields (Tables 1 and 2), in (Halbouty, M. T., Ed.), Geology of giant petroleum fields: American Association of Petroleum Geologists Memoir 14, pp. 502-528.

Johnston, R. R., 1983, North American drilling activity in 1982: American Association of Petroleum Geologists Bulletin, v. 67, no. 10, pp. 1495-1528.

Lewin and Associates, 1984, Major tar sand and heavy oil deposits of the United States: Oklahoma City, Interstate Oil Compact Commission, 272 p.

Meyer, R. F., 1970, Geologic provinces code map for computer use: American Association of Petroleum Geologists Bulletin, v. 54, no. 7, pp. 1301–1305.

Meyer, R. F., Fulton, P. A., and Dietzman, W. D., 1984, A preliminary estimate of world heavy crude oil and bitumen resources, *in* (Meyer, R. F., Wynn, J. C., and Olson, J. C., Eds.), Heavy crude and tar sands: New York, McGraw-Hill, pp. 97–158.

Oil & Gas Journal, 1983, OGJ Report, the OGJ 400: Oil & Gas Journal, v. 81, no. 42, pp. 75–104.

Saskatchewan Energy and Mines, 1982, Reservoir annual 1981: Saskatchewan Energy and Mines Miscellaneous Report 82-1, 11 sections.

Smith International Inc., 1974–83, Smith International Inc. Drilling Activity Correlator (Sii-DAC): Newport Beach, California, issued quarterly.

U.S. Bureau of the Census, 1980, Annual survey of oil and gas, 1978: U.S. Bureau of the Census MA-13K (78)-1, 47 p.

U.S. Bureau of the Census, 1984, Annual survey of oil and gas, 1982: U.S. Bureau of the Census MA-13K (82)-1, 31 p.

14
Summary of Exploration History and Hydrocarbon Potential of Morocco

A. El Morabet

Introduction

Oil exploration in Morocco started early in this century on the basis of the many oil seepages along the front of the Pre-Rif nappe. Several foreign oil exploration companies became active during that period in the Pre-Rif area, but the results were negative, except for modest oil discoveries in 1919 in Tselfat, Sidi Moussa, Ben Zered (1921–23), and Ain Hamra (1923) from Miocene reservoirs. The complexity of the geology of the Rif and Pre-Rif area, where most of the oil shows are present, was an obstacle towards any serious exploration efforts in that area for many years. For this reason, it was decided to unify the action both financially and technically. The creation of "Bureau des Recherches et des Participations Minieres—BRPM" in 1928 and "La Societé Cherifienne de Petroles SCP" in 1929 was the result of the interaction of all experiences and thoughts acquired in the early phase of exploration in Morocco.

SCP Activity 1929–1945

Because of the diversity of opinion on the origin of oil in the Gharb basin (Sidi Moussa, Ben Zered, Ain Hamra and Tselfat) the SCP put a group of geologists, among them the eminent diapir specialist of that time, Mrazec, to study the region and give a general synthesis of the geology and tectonics, despite differences of opinion regarding the complex geology and structure.The group recommended that the exploration effort be concentrated in the northern part of the basin (Ain Hamra region) and on the Tselfat anticline, close to production.

Research in the northern part of the basin stopped as the geological complexity of the region, due to Alpine thrusting, became more and more evident. The discovery in 1934 of oil in the Middle Lias of the Tselfat anticline, with a flow rate of 250 tons/day (1750 b/d), gave new impetus to research in Morocco and particularly on the Lias reservoir. Exploration was extended to the Jurassic structures resembling Tselfat. Several wells were drilled on Tratt, to the North of Fes, Ari, near Mekenes, and Outita and Bab Tissra, near Sidi Kacem. The results were discouraging, because the reservoirs were flushed. An exception was the Bou Draa structure, which produced no more than 20,000 tons (140,000 barrels) from Middle Lias.

The SCP concentrated its activity in the plain of the Gharb basin and its border. The wells drilled in this region confirmed the presence of a thick, heterogeneous complex named "La nappe Prerifaine" overlying the structurally complex Mesozoic substratum. The Pre-Rif nappe in the Gharb Basin is covered by the Mio-Pliocene marls; methane gas was discovered in sand lenses in the marls. Geophysics, particularly seismic reflection, was at that time unable to penetrate the Pre-Rif nappe, which constituted an obstacle towards the exploration of the infra-nappe substratum.

SCP Activity 1946-1958

After the second world war SCP renewed its drilling. The activity now was oriented toward the external margin of the Mesozoic Basin of the Pre-Rif area. The first well (Oued Beht No. 1), drilled in 1946, confirmed the presence here of oil in Miocene sand lenses and Triassic breccia. Intensive exploration of this zone was undertaken. Detailed seismic coverage was run and several hundred exploration and development wells were drilled in the following decade. This resulted in the discovery of the Oued Beht oil field. Oil was produced from several reservoirs in either weathered granites and metamorphosed Paleozoic rocks, or Jurassic sands and carbonates.

Other oil fields of the Oued Beht trend were discovered in the following years: Baton (1948), Paleozoic; Tisserand (1950), Paleozoic; Oued Mellah (1948), Paleozoic; Sidi Filli (1950), Jurassic and Paleozoic; Mers El Kharez (1951), Paleozoic; Bled Eddoum (1952), Jurassic and Paleozoic; Bled Khatara (1953), Jurassic and Paleozoic; and Zrar (1955),

Jurassic and Paleozoic. At the end of 1957 the total production of the Oued Beht oil fields and Haricha did not exceed 700,000 tons (4,900,000 bbls).

In 1957, a well was drilled on seismic data on the Haricha structure, discovering commercial oil in a narrow Jurassic block injected in the pre-Rifaine nappe. In order to find new targets, the SCP conducted seismic surveys and exploratory drilling in the Sais Basin, but the negative results led to the abandonment of any further activity in the Northwest of Morocco.

SCP Activity in the Essaouira Basin

The Mesozoic Essaouira Basin, in west-central Morocco, is composed of more than 4,000 m of red bed continental sands and clays of Permo-Triassic age, Jurassic carbonates, and Cretaceous marls and carbonates, folded and uplifted in the early Tertiary. The first well, drilled in 1957, was on Kechoula anticline and discovered gas in Upper Jurassic sandy dolomites. In 1958, another well was drilled on the Jebel Jeer anticline, to the Southeast of Kechoula, finding commercial reserves in the same reservoir as in Kechoula.

In 1961, the first well drilled on top of the Sidi Rhalem diapir structure led to the discovery of an oil field in the upper Jurassic and Dogger limestones. The success of SCP in the Essaouira basin gave the impulse to drill most of the surface structures in the basin. Twelve more wells were drilled, but none was commercial. Seismic reflection had poor penetration through the Triassic basalt, which hindered any further exploration activity in the region.

Bureau de Recherches et de Participations Minieres 1958–1969

After the independence of Morocco in 1956 and the promulgation of the hydrocarbon law in 1958, BRPM became active in the field of hydrocarbon exploration, as the government representative with foreign companies. Activity at first was characterized mainly by the participation of BRPM with the international oil companies, according to the newly published hydrocarbons law for oil exploration either offshore or onshore Morocco.

BRPM signed an agreement with AGIP in 1958, to explore the Tarfaya onshore basin and in 1963 the High Plateaux. Several other agreements were signed during this period which resulted in the discovery of methane gas in sand lenses of the Miocene of the Gharb Basin by APEX. On the other hand ESSO drilled several wells offshore the Tarfaya Basin, still further south. Their second well, MO2, discovered heavy oil (13° API) in the upper Jurassic (14 million barrels). The rest of the wells, drilled both offshore or onshore by such companies as Texas Eastern, Petrofina, Sun Oil Company, Burma and Phillips, were discouraging, but nonetheless the results were important for their valuable new information. In 1969, BRPM started its own exploration with a seismic survey in the Guercif Basin where a Jurassic section exceeding 6,000 m is present. This was followed by work in the plain of Beni Mellal and Boujaad. Two wells, KAT-1 and KAT-2, were drilled in Beni Mellal Basin to evaluate the Carboniferous carbonate lenses. Oil shows were encountered in the Carboniferous, but the difficulty of mappping the deeper horizons because of intense faulting caused exploration to be abandoned in this area.

In 1973, BRPM conducted a seismic survey in the Essaouira Basin, and due to new developments in data acquisition and processing, it was possible for the first time to map deeper horizons than the Lias, initiating a new phase of exploration. As a result, the Toukimt Field was discovered in 1975. It has an estimated 250 million cubic meters of gas and 110,000 cubic meters of gas condensate in place. The reservoir is in fractured dolomites of Upper Jurassic age, overlain by more than 1,500 m of anhydrite. In 1976, a small gas field was discovered in N'Dark anticline, in the same reservoir as at Toukimt. Meskala No. 1, drilled to east of Toukimt, encountered important gas shows under high pressure in the Triassic and Paleozoic, confirming a commercial reservoir. An intensive first phase of exploration of the Essaouira Basin and the appraisal of the Meskala field was undertaken by BRPM with a loan from the World Bank. Meanwhile, exploration by BRPM continued in other basins in Morocco, where a vast seismic program was run, but drilling took place in the Gharb Basin and the Pre-Rif hills.

Creation of ONAREP

The Petroleum Exploration Division and the Drilling Division were separated from BRPM and the Office Nationale de Recherches et de l'Exploitations Pétrolières (ONAREP) was created in 1981. Thus, all rights of hydrocarbons exploration and exploitation were transferred to ONAREP. The new Office continued the negotiation of a loan to finance the second phase of the appraisal of Meskala Field, while trying to attract foreign investment in oil exploration. Several agreements were signed with exploration companies, including ARCO, AMOCO, Elf Aquitaine, and Mobil, to explore for oil and gas both offshore and onshore.

In 1983, presentations were made in Morocco, Europe, Canada, and the United States to encourage international oil companies to invest in oil exploration in Morocco. ONAREP is actively pursuing liberalization of the Moroccan hydrocarbon law and promoting a new law which will be more attractive to foreign investment.

Hydrocarbon Potential of Morocco

The surface of the offshore basins, on the Atlantic and the Mediterranean platforms, exceeds 100,000 km^2. The total surface area is over 150,000 km^2. The thickness of the sedimentary section can be more than 10,000 m, ranging from Precambrian to Recent.

Cambrian shales and carbonates are well exposed in the Anti-Atlas chain, with thick Archaeocyathid reefs which could serve as excellent reservoirs. Ordovician quartzites and sands of the southern High Atlas show good fracture porosity, forming a good reservoir for oil from the Silurian organic-rich black shales, as known in the Anti-Atlas, Skoura, and Doukkala regions.

The Devonian has both a good potential source rock in the shales of the Lower and Upper Devonian, and a widespread reservoir in the middle Devonian reefs.

Thick Carboniferous sediments are well represented in Morocco in the Beni Mellal basin, the high Plateaux, at Boudnib and in the Draa basin. There have been interesting shows in the Boujaad Beni Mellal basin. The geochemical analysis done on the Upper Carboniferous shales of high Plateaux well (TD 2) revealed a high organic-carbon content.

The basal sands of the Triassic have shown their potential as reservoir rocks in the Essaouira basin.

The Mesozoic section, particularly the Jurassic, has been the target of exploration pioneers since the beginning of the century. All the geochemical analyses, scanty though they are, and done mostly on reservoir cores rather than the shaly part of the section, demonstrate the good qualities of the Lias as source rock, especially in the Pre-Rif area. The Middle Lias reefs and dolomitized sandy limestones are well known as good reservoir rock in the Pre-Rif hills and the Essaouira Basin.

The thick Cretaceous sequence, particularly in the offshore Atlantic, has demonstrated excellent source-rock potential where its upper part could be found sufficiently buried.

Geochemical analysis of the Miocene shales indicates rich marine organic matter to be present in the offshore Atlantic and the Pre-Rif, but due to the lack of adequate burial, the organic matter is still immature, except in the eastern Pre-Rif area (well BFS1), where the infranappe Miocene is in the oil zone. Extremely dry gas (methane 99%) is being produced from lenses embedded in the Miocene marls of the Gharb Basin at a depth between 1,000 m and 1,700 m. If the geothermal gradient were high enough, these sediments could generate oil at depths of about 2,000 m.

Conclusion

Exploration started early in the century in Morocco, in a region which is well known to geologists and explorationists for its geological and structural complexity. Oil seepages along the front of the Pre-Rif nappe attracted the first explorers. The result was the discovery of several oil fields in the Pre-Rif hills. This sustained a heavy exploration program after the second world war.

The Essaouira Basin has demonstrated good hydrocarbon potential. The discovery of oil at Sidi Ghalem and gas at Kechoula and Jeer in the 1960s was followed more recently by the gas and condensate discoveries in the Triassic of Meskala. Most of the sedimentary basins could still be considered as not well explored, due to insufficient drilling. The exploration activity has been concentrated mainly in the Pre-Rif hills and the Essaouira Basin. The targets drilled have been at shallow depths, while the deeper horizons are still waiting to be explored.

The Atlantic and Mediterranean platforms, at water depths not exceeding 300 m, are over 60,000 km^2. In this vast area only 20 wells have been drilled, most of them concentrated in the offshore Tarfaya Basin. It is evident that more drilling is necessary in order to fully evaluate the potential of this large and prospective area.

References

Alem, A., Exploration petroliere du Maroc: de 1970–1983.
Beicip, Analyse geochimique de 10 puits dans les régions des Doukkala, Essaouira et Ksar Essouk, 1980.
El Morabet, A., General outlook on the geology of Morocco and exploration activity, 1980.
Gaffney Cline and Associates, Petroleum exploration prospects in Morocco, a summary report for management.

Lardenois, J., Le gisement de petrole de la region de l'Oued Beht de Mers: El Kharez et Sidi Filli, 3rd International Petroleum congress, La Haye, 1951.

Lardenois, et al., Gisement de petrole du Maroc, 1956.

Le Tran, K., et al., Etude geochimique de divers fluides Marocains, 1980.

Levy, R. G., Les travaux de la Societe Cherifienne des Petroles dans l'exploration petroliere au Maroc: *Journée Geologique du Maroc,* 1981.

Michard, A., Elements de la geologie Marocain. Notes et Memoires du Service geologique, MEM 1976.

Saadi, M., Schema structural du Maroc.

15
Shallow Oil Fields of the Denver Basin, Colorado and Nebraska, U.S.A.

J. F. deChadenedes

Introduction

Shallow oil has been produced in eastern Colorado since the year 1862. The earliest discoveries were found by drilling near surface seepages of oil and tar. Subsequently, the more obvious closed structures mapped on the surface along the eastern front of the Rocky Mountains were proved by drilling to be productive.

The oil occurs in sandstones within and beneath the Cretaceous Pierre shale (Figure 1). Underlying the Pierre is the Niobrara formation, a chalky shale and impure limestone, rich in organic material, which is the principal oil and gas source rock in the basin. The Niobrara shale, where fractured, also is one of the reservoir rocks. The D and J sandstones, in the Dakota group at the base of the Cretaceous, contain the principal reserves. Second in reserves are the Terry and Hygiene sandstones in the middle of the Pierre shale at depths 4,500 ft (1,390 m) shallower than the D and J sandstones. Through 1978, 885 fields had been discovered in the Denver Basin, with a cumulative production of 454 million barrels of oil and almost a trillion cubic feet of natural gas. The producing trend is now 160 miles in length, from north to south, and 40 miles wide (259 by 65 km).

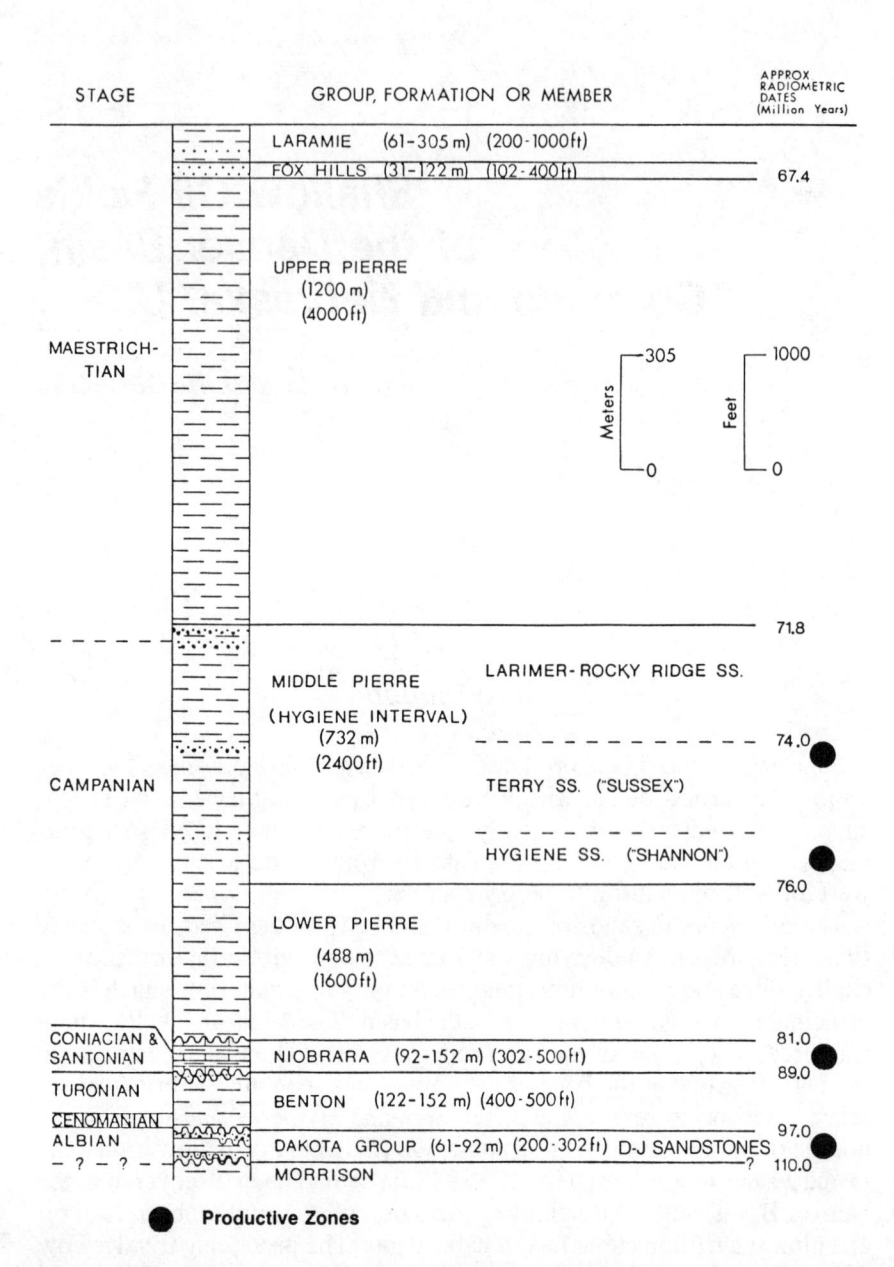

Figure 1. Cretaceous formations of the Denver Basin (after Porter and Weimer, 1982).

History of Development

Discovery has historically moved from the shallowest to the deepest portions of the Basin. The 1862 discovery of the Florence Oil Field at the extreme south end resulted from shallow pits dug into gravel banks, and water wells. Ultimately cable tool wells were drilled, and the field was defined between 1,300 and 3,100 ft in depth (396-945 m). Though the fractured shale is a reservoir of poor quality, cumulative production to 1975 exceeded 14 million barrels from a total of 600 wells. In 1974 the 26 wells still active produced 19,000 barrels. The J sandstone contains the largest reserves of the basin. The earliest discoveries, beginning in 1923, were structures on the steep west flank of the basin at depths from 2,000 to 3,000 ft (620 to 920 m). In 1949, development of the east flank of the basin was initiated by the Ohio Oil Company's discovery of the Gurley Field in western Nebraska, at a depth of about 4,000 ft (1,235 m). The field was on a seismic structure with closure, one of the few on the east flank. Subsequent development drilling down the anticlinal flank beyond closure showed the stratigraphic element of the trap. Within the supposed "trend," as it came to be called, major companies and independents initiated large-scale lease acquisition on the privately-owned lands between western Nebraska and Denver. The larger companies carried out regional seismic mapping, which ultimately proved to be mostly wasted due to the low relief of the rare closed structures. However, enough discoveries were made, and encouraging shows found, to justify continued enthusiasm. The early wells gave rough definition to a pattern of lenticular sandstone bodies in the J interval. These were interpreted to have originated within and on the seaward front of a major Cretaceous delta having its source to the northeast. However, marine transgression over the delta had redistributed the channel sands to an almost random pattern. Only near the original seaward slope of the delta were offshore bars preserved with clearly defined shapes and predictable trends.

The discovery of several large oil fields, such as Adena, with cumulative production to 1983 of 61 million barrels of oil and 88 billion cubic feet of gas, Little Beaver, 17 million barrels, and Plum Bush, 19 million barrels, greatly accelerated exploration. These are in the delta-front bar-sand areas, with up to 70 ft (22 m) of sandstone pay compared with the more common 5 to 25 ft (2 to 8 m) for such areas. Most of the fields had 1-10 wells with an average of 5, and these targets were too small to be attractive to the major companies. The large companies, especially Amoco Production Co. (the largest leaseholder), established a policy of farming out interest. The usual deal, with variations, was to give up to three 80-acre tracts in a checkerboard pattern, and contribute dry-hole money ranging from $1.00 to $2.00 per foot at J-sandstone depth. Independents cooperated with small leaseholders to put together multiple drilling deals, and the owners of drilling rigs sometimes bought interest in the wells by contributing drilling time. Drilling costs were low in the 1950s and 1960s, from $10,000 to $15,000 for wells of 4,000 to 6,000 ft (1,230 to 1,850 m), and drilling time was only five or six days. This low cost, and the continued discoveries, encouraged many thousands of wells to be drilled in the two decades. Though the J pay sand thickness was only 5-20 ft (6 m), averaging about 9 ft (3 m), the quality of the oil (28° to 45° gravity API) was high, and with initial production rates of 100 to 200 barrels per day, payout times were as low as one to two months. Costs were further minimized by eliminating diamond cores, and restricting electrically logged intervals to the D and J zones. Production rates rapidly declined after the first few months, due to lowering of gas pressure in the closed reservoirs. But with oil prices as low as $2.00 to $3.00 per barrel, production lasted for many years even at rates of 10 barrels per day.

Rapid development tapered off in the 1970s, as drilling density in the most productive areas grew to 1-4 wells per square mile. Secondary recovery by water injection was generally successful where justified, in some cases almost doubling the reserves produced by primary methods (from 30% to 40% of the oil in place). Exploration moved to deeper portions of the basin, to 8,000 ft (2,500 m) and deeper, near the regional syncline in the vicinity of Denver. In this area the J sandstone is continuous, but clay-filled and tightly compacted, with porosities from 5% to 10%, and permeabilities from a fraction of a millidarcy to 5 millidarcies. It is, however, saturated with gas, and with the help of the recently developed massive sand fracturing, Amoco found it economically feasible to produce what became the Wattenberg Gas Field. This field now covers more than 600 square miles (1,670 km^2) and has ultimate reserves of over one trillion cubic feet of gas and 70 million barrels of oil.

In the western portion of Wattenberg it was discovered that the Terry and Hygiene sandstones (Figure 2), which at 4,000 ft (130 m) had been bypassed in drilling to the J, were potentially oil productive. In this manner the Spindle Oil Field (Figure 3) was discovered in 1972. Porosities and permeabilities were better than those in the J, and initial production after fracturing was from 50 to 200 barrels per day. Spindle has attained a daily production of over 25,000 barrels per day, and has contributed substantially to recent production in the basin.

Exploration continues in the Denver Basin in mid-1984. The ultimate limits of the producing trend have not determined in any direction. Four large fields have been discovered since 1970, in a southwest extension of the J trend. These are Peoria (15 to 20 million barrels), Kachina, Shelda, and Third Creek (each over two million barrels). The productive area

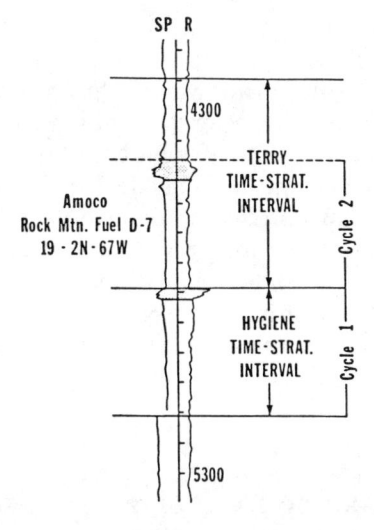

Figure 2. Electric log of Terry and Hygiene sandstones (after Porter and Weimer, 1982).

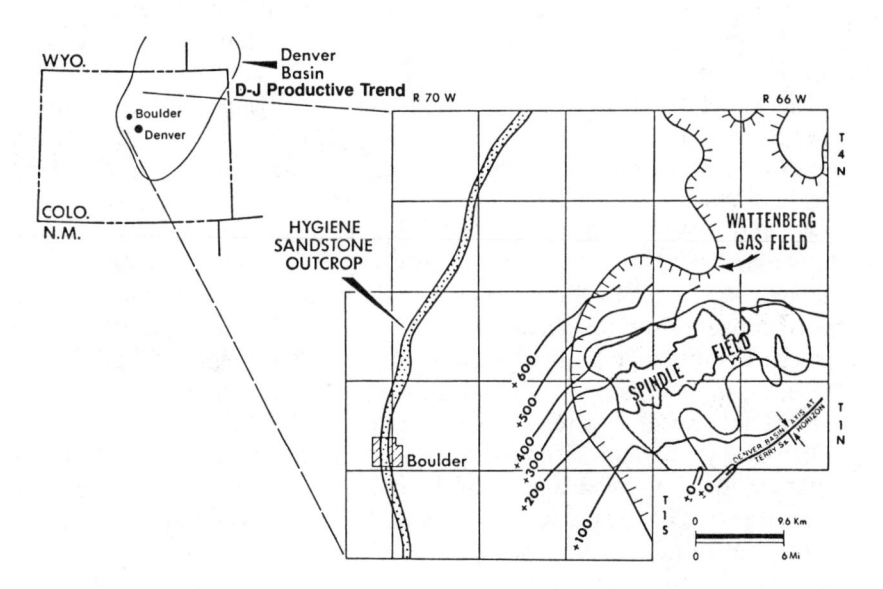

Figure 3. Location map of Spindle Field (after Porter and Weimer, 1982).

may extend even further southwest, and also into the more sparsely drilled areas within the general trend. It will also extend eastward onto the shallower flanks of the basin. Spindle has given impetus to drilling for shallow objectives in the western Basin. On the extreme east, along the Kansas-Colorado border, shallow gas has been discovered in the Niobrara. The Niobrara and the associated Codell sandstone in the area northwest of Denver have proved productive in fractured shale traps.

Though drilling costs have multiplied, now ranging from $30,000 to $50,000 for wells to the J sandstone in the trend, the price of oil has also increased. Although the drilling rate has slowed with the recent decrease in the world price of oil, any increase in oil price should stimulate exploration in the Denver Basin.

Prospecting for D and J Sandstones

Locating wells in this play has proved to be very risky because of the generally unpredictable and patchy occurrence of the sandstones. Many an independent has gone broke after drilling 10 to 20 consecutive dry holes. In some areas the J may contain multiple sandstones separated by thin shales. Most of the sands are likely to be water bearing, and the driller who drills near or beyond the oil-water contract may find that a water-free completion is impossible. Fracturing an individual sandstone in the sequence may cause fluid communication with the water-wet sandstones.

The structure of the basin is simple. The west flank dips eastward at rates from 5° to 30°. The east flank is a gentle ramp tilted to the west and dipping regionally at an average of 40 ft per mile. Commonly, there are crenulations in the contours, and rarely even reversals of regional dip or low-relief closures. On the top of the J sandstone, structural closures on the order of 20-50 ft (3-15 m) may occur, too small to be reliably mapped by seismic methods. These are the result of draping and differential compaction of the enclosing shales over J sands. Because of this structural unpredictability as well as the rapidly varying sand thickness, there are numerous one-well fields offset in four directions, in the official 40-acre spacing pattern, by dry holes.

For every operator discouraged by dry holes, there have been many others ready to take advantage of the information from these wells. The free exchange of data from all wells has greatly expedited the rapid development of the Denver Basin.

Despite the general unpredictability of the sand bodies, geologic study has contributed much to the discovery of most of the reserves. A typical simple "play" might be described as follows: an area of a square mile or less lying between an updip well having an impermeable J sandstone that is oil stained, and a downdip well that found a porous and water-wet J sandstone 10–20 ft (3–6 m) thick. One or more of these elements may be lacking or the updip well may be porous with no shows of oil, and the downdip sandstone may be porous and oil stained. In about one out of 12 instances a well drilled on this play will be a discovery, and with luck this will develop into a field of 5 or more producing wells. Amoco, with intensive use of geologic interpretation, achieved a success rate of one discovery in eight wildcats drilled.

In the area along the original delta front, the J bar sands are easily identifiable, with lagoonal shales forming an updip seal to the traps, as at Adena. Analysis of the environment of deposition from logs and drill cuttings contributed substantially to successful exploration. In some J bar- and channel-sand deposits, interpretation of depositional trend from crossbedding, as measured in oriented cores or with dipmeter, aided predictions of trend and shape of the sand body. These tools were too expensive for the average independent operator, who used only the most basic information and inexpensive drilling tools.

Spindle Field

This is the largest field at depths of less than 5,000 ft (1,540 m) in the Denver Basin. It is located a few miles north of Denver, adjacent to the Basin synclinal axis. It was not discovered until 1972 because the Terry and Hygiene sandstones were not considered to be proved exploration objectives. Spindle overlies the western portion of the Wattenberg Gas Field (Figure 4), and was discovered by plugging back a failed J sand well, with recompletion in the Terry interval.

As Porter and Weimer (1982) and Kiteley (1977) demonstrated, the Terry and Hygiene were deposited as offshore bar complexes (Figures 5–6) along a delta front whose source was to the northwest, in Wyoming. Each reservoir sand lies at the top of a thick, silty sand sequence within the Pierre shale. In each zone the texture coarsens upward and is capped by a fine grained, fairly well sorted, silty sandstone. These sequences are interpreted as due to a shallowing marine environment of deposition, which resulted in greater exposure of the bottom to wave action. Both

188 Shallow Oil and Gas Resources

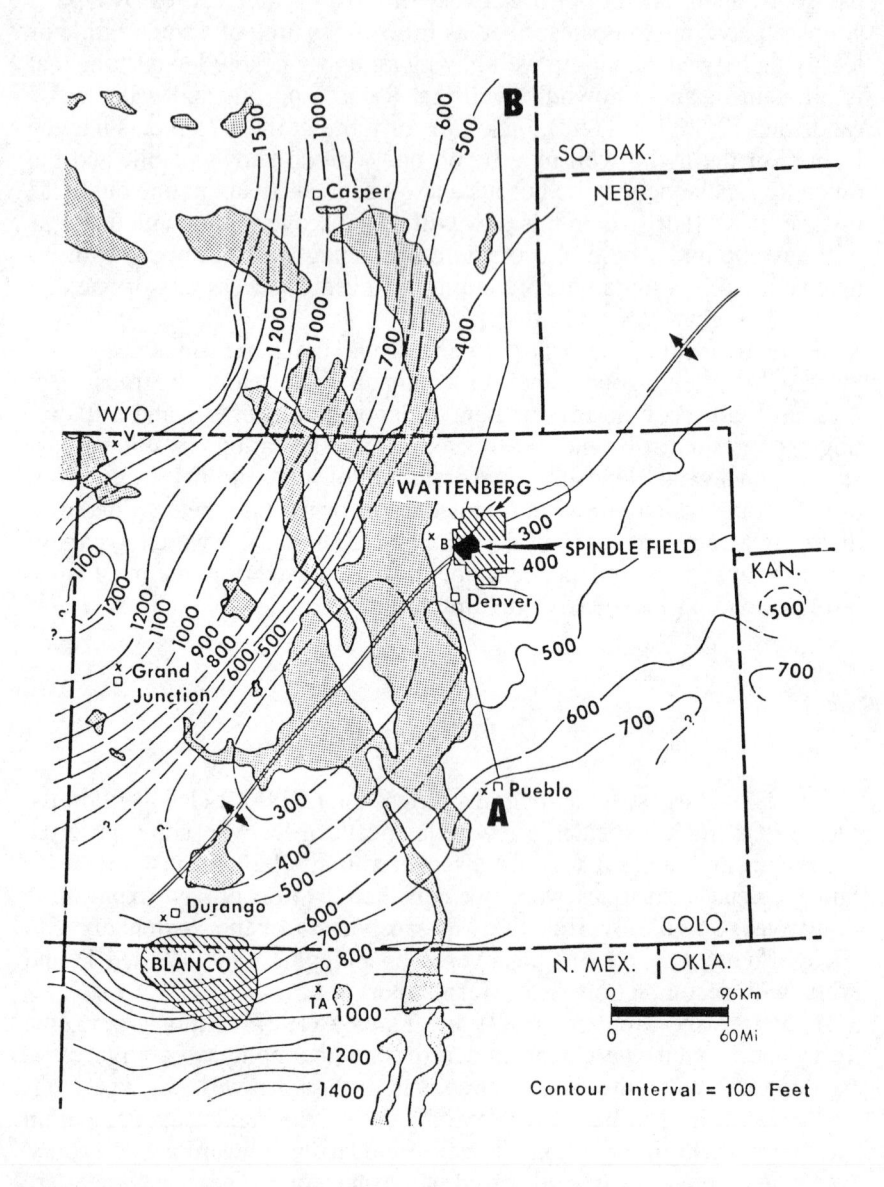

Figure 4. Isopach map of Niobrara showing Cretaceous paleoarch (after Porter and Weimer, 1982)

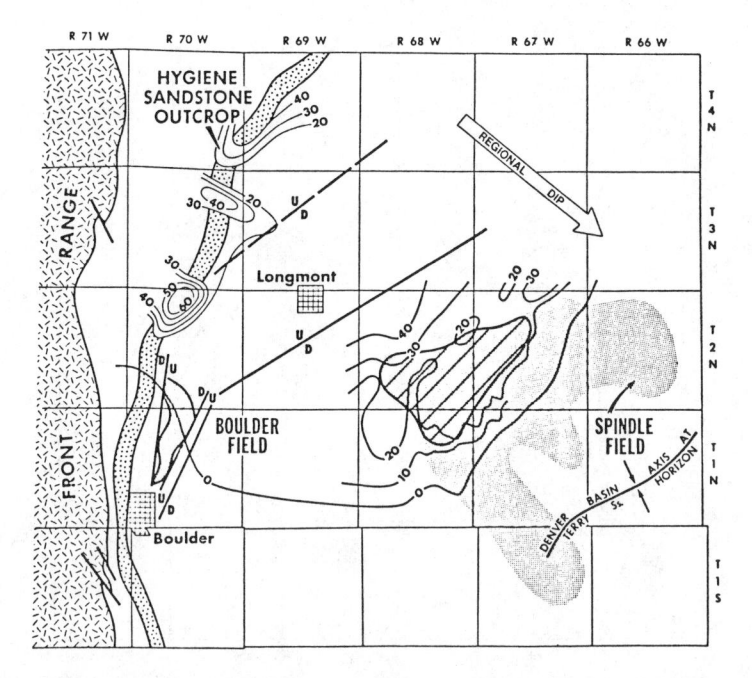

Figure 5. Isopach map of Terry bar sandstone, Spindle Field (after Porter and Weimer, 1982).

Wattenberg and Spindle are on the crest of a regional paleo-arch trending northeastward across northeast Colorado. This arch is older than the Laramide folding and faulting that produced the present Rocky Mountain structure and the Denver Basin. Porter and Weimer (1982) deduce that it was present at least as a topographic high in the shallow Cretaceous sea, and thus influenced sand deposition. In my own study of the Terry and Hygiene, I found pre-Laramide growth faults parallel to the axis of the arch across the Spindle Field. These strongly influenced sand deposition, and may form trap elements. These effects are shown by abrupt thickness and quality changes in the sandstones across the faults, and strong salinity changes in the formation waters on either side of the major fault (Figures 5-6).

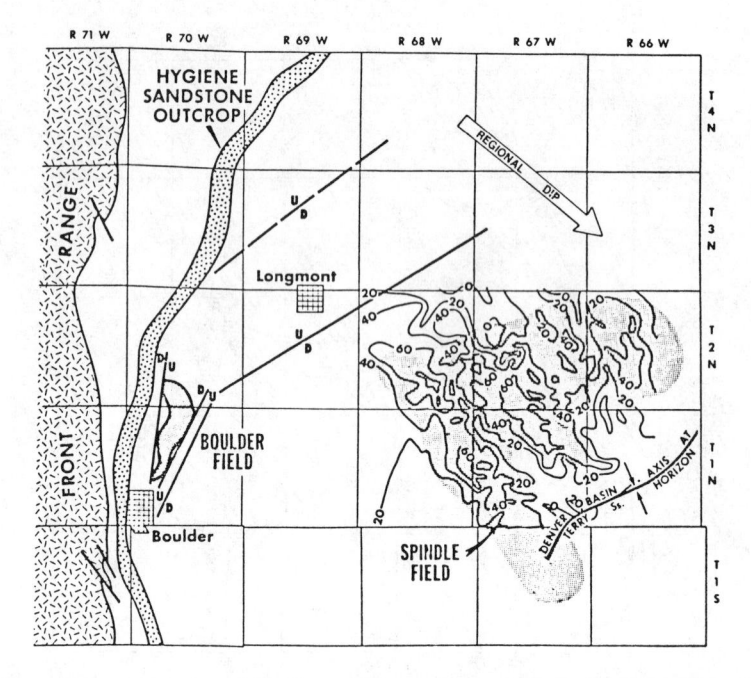

Figure 6. Isopach map of Hygiene bar sandstones, Spindle Field (after Porter and Weimer, 1982).

Porosity in the producing sandstones varies from 8% to 16%, averaging near 12%, and permeabilities range from 5 to 50 millidarcies. The original sandstone had greater porosity, but this was reduced by such diagenetic processes as compaction, chloritization and calcite deposition. Much of the present pore space is secondary, caused by solution subsequent to diagenesis. Low permeabilities are improved by large-scale sand-oil fracturing. Initial production rates range from 50 to 200 barrels of oil per day.

Cumulative production to 1983 was 37 million barrels of oil and 164 billion cu ft of gas, from over 700 wells. Secondary recovery methods are being considered for the more uniformly porous areas of the field, where pay sections range from 20 to 60 ft (6 to 18 m). Ultimate reserves by primary methods are estimated by Pruitt (1978) to be over 50 million barrels of oil.

Niobrara Gas Fields

Increased gas prices encouraged the development in the 1970s of shallow gas fields in the Niobrara chalk in the extreme eastern Denver Basin (Figure 7). Over 30 fields have been discovered at depths from 900 to 2,800 ft (275 to 860 m) on both sides of the border between Colorado and Kansas.

Regional dip is to the northwest at the extremely gentle rate of 15 to 20 ft (5 to 6 m) per mile. There are many low-relief structures with closure of 50 to 200 ft (15 to 60 m) on the top of the Niobrara. The gas is entirely within structural closure.

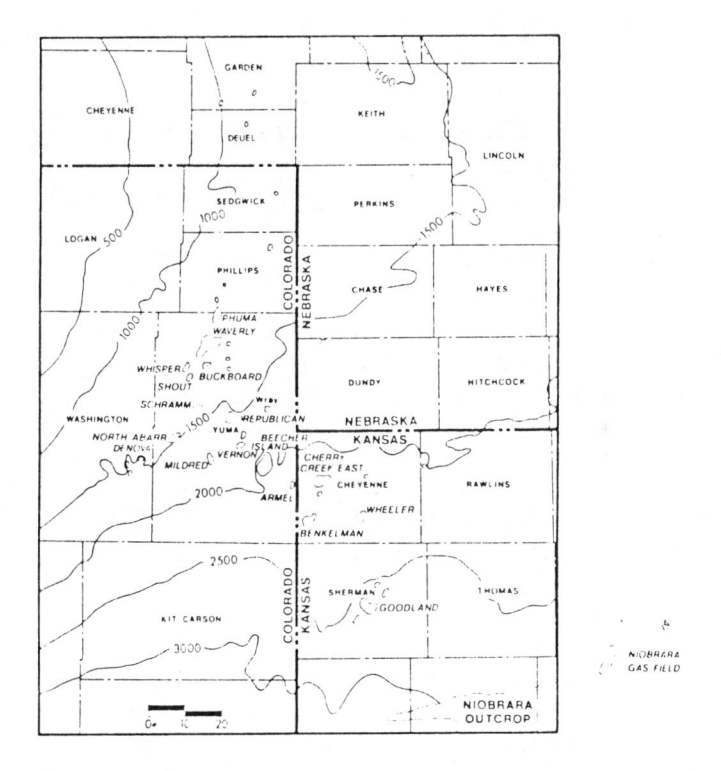

Figure 7. Location map of Niobrara gas fields; structure contours on top Niobrara (after Lockridge, 1978).

The Niobrara, which in the western Basin is an impure chalky limestone, is here a relatively pure chalk. Reservoir porosity ranges from 30% to 45%, decreasing with depth. Permeabilities vary from 0.1 to 16 millidarcies. Reservoir pressures are below normal, from 60 psi at 900 ft (275 m) to 780 psi at 2,800 ft (860 m).

In spite of low deliverability, with initial production rates from 20,000 to 300,000 cubic feet per day, the low cost of completed wells gives favorable economics. Well cost, including fracture stimulation by sand and nitrogen foam, or sand and water, is about $50,000. The gas averages about 1,000 Btu per standard cubic foot, and sells for $1.48 per thousand cubic feet. Payout is estimated at 15 months. Ultimate reserves cannot yet be calculated because of short production history, but are estimated to be 11 billion cubic feet in the present fields. Exploration is continuing in extended areas, and to deeper chalk reservoirs known to exist.

Comparison with Other Regions

The Denver Basin is a structural subdivision of a Cretaceous depositional geosyncline that extended from the Arctic Ocean to the Gulf of Mexico. A similar Cretaceous geosyncline exists in South America, to the east and along the full length of the Andes Mountain chain, where the stratigraphic section is similar to that of the Denver Basin, including source rocks underlying a thick marine shale section. Thick sandstone reservoirs are widespread in the lower shale section, as they are in the Denver Basin. Along the east flank of the Andes, these outcropping sandstones are commonly saturated with oil. Several oil fields have been discovered on large, closed structures in Colombia, Ecuador, and Peru. It seems probable that major, or even giant, oil fields remain to be discovered in stratigraphic traps related to Cretaceous delta systems.

Other basins of marine deposition are well known in the interior of, or fringing other continents. The abundant stratigraphic traps of the Denver Basin show that it is necessary to analyze patterns of sand deposition in relation to ancient deltas, and that many small to major oil fields can be found as a result. Under these conditions, multiple exploration approaches, such as are achieved with many competing operators, are necessary to achieve optimum results in exploration.

References

Clayton, J. L., and Swetland, P. J., 1980, Petroleum generation and migration in Denver Basin: AAPG Bull., v. 64, pp. 1613–1633.

Kiteley, L. W., 1977, Shallow marine deposits in the Upper Cretaceous Pierre shale of the northern Denver Basin and their relation to hydrocarbon accumulation: Rocky Mt. Assoc. Geologists 1977 Symposium, Exploration Frontiers of the Central and Southern Rockies, pp. 197–211.

Lockridge, J. P., and Scholle, P. A., 1978, Niobrara Gas in eastern Colorado and northwestern Kansas: Rocky Mt. Assoc. Geologists 1978 Symposium, Energy Resources of the Denver Basin, pp. 35–49.

Mallory, W. W., 1977, Oil and gas from fractured shale reservoirs in Colorado and northwest New Mexico: Rocky Mt. Assoc. Geologists Special Publication No. 1, pp. 1–38.

Porter, K. W., and Weimer, R. J., 1982, Diagenetic sequence related to structural history and petroleum accumulation: Spindle field, Colorado: AAPG Bull., v. 66, pp. 2543–2560.

Pruitt, J. D., 1978, Statistical and geological evaluation of oil and gas production from the J sandstone, Denver Basin, Colorado, Nebraska and Wyoming: Rocky Mt. Assoc. Geologists, 1978 Symposium, Energy Resources of the Denver Basin, pp. 9–34.

16
Petroleum Exploration Strategy for Zambia

Chisengu Mdala

Abstract

A high-sensitivity aeromagnetic survey totaling over 100,000 line kilometers was flown over the major sedimentary basins of Zambia to assess their oil and gas potential.

Although the sediments are thought to be largely of continental origin, from the tectono-stratigraphic considerations, the Karroo sediments within some of the valleys would appear to possess potential petroleum source rocks, reservoir rocks, and possible cap rocks and trapping mechanisms.

From aeromagnetic considerations, it is evident that sufficient thickness of sediments exists for the maturation and generation of potential hydrocarbons within some of the valleys.

Although the road to the first exploration drill hole is still a long way, aeromagnetic data obtained so far have raised enough interest to proceed with the gravity survey in 1984 as originally planned. Because some of the sedimentary basins of Zambia have basaltic flows overlying the Karroo sediments, it is proposed to supplement the gravity and aeromagnetic information with that of magnetotelluric soundings. It is also envisaged to use Landsat data for geological basin study in Zambia.

In a country like Zambia where most geological work has been biased more towards exploration for base metals and precious and semi-precious minerals, a need arises to reexamine the geological data and information with oil and gas resources in mind. The services of a petroleum geologist will be sought for this purpose.

Zambia does not have any petroleum legislation at the moment. Petroleum legislation will, therefore, be prepared by legal consultants. A number of incentives are expected to be included in the legislation.

An integrated report will be prepared covering the findings of the geophysical, Landsat imagery, and geological programs. The reports will be sold to interested companies to promote hydrocarbon exploration by attracting foreign companies to undertake any required follow-up investigations for petroleum.

Introduction

Zambia has no known petroleum deposits. All products must, therefore, be imported. In 1981 the petroleum import bill was equivalent to above 18% of the total import bill. While the total volume of petroleum imports declined by 1.7% per year (from 732,000 metric tons in 1976 to 683,000 metric tons in 1980), the cost of petroleum imports increased by more than 15% (from 72 million Kwacha to 155 million Kwacha).

The country has become increasingly concerned about the high cost of imported petroleum products. Measures taken to deal with this petroleum import bill now include efforts to explore for oil and gas in the country's sedimentary basins. In May 1982 Zambia obtained a World Bank Loan of US $6.6 million for a Petroleum Exploration Promotion Project (PEPP). Some of the categories of items which were to be financed out of the proceeds of the loan included aeromagnetic surveys totaling about 90,000 line-kilometers of the prospective sedimentary areas, gravity survey to cover selected traverses in areas identified in the airmag survey, and reprocessing of mineral aeromagnetic data totalling about 50,000 line-kilometers. These items constitute the first part of the project.

The second part of the project consists of services of exploration consultants, legal consultants, and consultants' services for training of Zambian staff in petroleum geology and interpretation of geophysical survey results and other items.

The Hydrocarbon Unit

In order to execute the project effectively the Government established and maintains in the Ministry of Mines, a petroleum unit known as the "Hydrocarbon Unit," whose function is to coordinate the execution of the project. Although the original terms required the unit to be headed by a project coordinator who was supposed to be a senior professional officer, practical experience has shown that a project administrator is the more suitable officer to head and administer the unit. This allows the project coordinator more time to attend to technical matters which include field visits. Apart from the project administrator and the project coordinator other officers of the unit include, at the moment, one geologist, one geophysicist, and an administrator.

In 1982, two separate contracts were signed with two international companies, namely, Exploration Data Consultants (EDCON) of Denver, Colorado, U.S.A. and Geo-Survey International Ltd. based in Nairobi but with headquarters in London. Geosurvey International Ltd. undertook the aeromagnetic data acquisition contract whereas EDCON was the government's consultant for quality control during data acquisition and processing. EDCON also carried out the final interpretation of the processed data. In addition EDCON was contracted to reprocess and interpret existing mineral airmag data, obtained in the basins, in the perspective of petroleum resources.

Brief Geologic Structure of Zambia

Geologically, Zambia lies between the Congo and Zimbabwe cratons and therefore contains elements of mobile belts as well as aspects of ancient cratons (Money, 1981). Apart from limited areas underlain by Karroo (Permian to Jurassic) sedimentary and volcanic rocks, and the poorly exposed Cretaceous beds and Pleistocene Kalahari sands which obscure much of the solid geology in the West, Zambia is underlain almost entirely by pre-Silurian, mainly Precambrian rocks.

Zambia can be subdivided into structural-stratigraphic provinces as follows:

1. Bangweulu Block (over 1,800 million years)
2. Kibaran Belt (approximately 1,300 million years)
3. Lufilian Arc (840–465 million years)
4. Mozambique Belt (approximately 460 million years)
5. Paleozoic sediments—the unmetamorphosed sedimentary succession beneath the basal Karroo rocks belonging to Ordovician through Devonian systems.
6. Upper Paleozoic, Mesozoic and Recent rocks. These are nearly horizontal, unmetamorphosed continental sediments, locally overlying basalts of the Karroo system, the nonmarine Cretaceous beds of Western Zambia, and Pleistocene Kalahari sands.

Sedimentary Basins of Zambia

Zambia has at least seven sedimentary basins (Figure 1). These are:

1. Western Zambia Basin
2. Luangwa Basin
3. Mid-Zambezi Basin (was not flown during the airmag survey)
4. Kafue trough Basin
5. Luano-Lukusashi Basin
6. Bangweulu Swamp Basin
7. Lukanga Swamp Basin

With the exception of the Bangweulu Swamp and Lukanga Swamp basins, all these basins have sediments dating from Permian time (over 200 million years).

The three most important basins as far as petroleum is concerned are the Luangwa Basin, the Western Basin, and the Luano-Lukusashi Basin.

The Luangwa Basin is a linear depression up to 95 km wide, extending for over 600 km, that is, an area about 50,000 sq km. Aeromagnetic interpretative results have revealed interesting tectonic aspects of this basin to place it as the priority area.

The Western Zambia Basin constitutes the largest sedimentary basin in Zambia, approximately 150,000 sq km. Although placed second in priority, this basin, because of its large size, might be of top priority if some of the outstanding geological questions were answered.

Aeromagnetic Survey

Aeromagnetic maps indicate the distribution of magnetic material within a survey area. Igneous rocks, especially of the mafic-ultramafic variety, produce strong magnetic responses, whereas recent sedimentary rocks, which are virtually devoid of magnetite, the most dominant magnetic material, produce few significant magnetic anomalies.

From June to November 1982, a high-sensitivity aeromagnetic survey was flown under contract to Geosurvey International, over the major sedimentary basins of Zambia to assess their oil and gas potential. Original areas to be flown were A, B, C, D, E, and F, but later included G, H, and J. The survey was flown using a Cessna Titan 404 aircraft equipped with Doppler navigation and a Texas Instruments AN ASQ81 Helium magnetometer. Data were recorded in both analogue and digital form, and the survey precision was about 0.1 gamma.

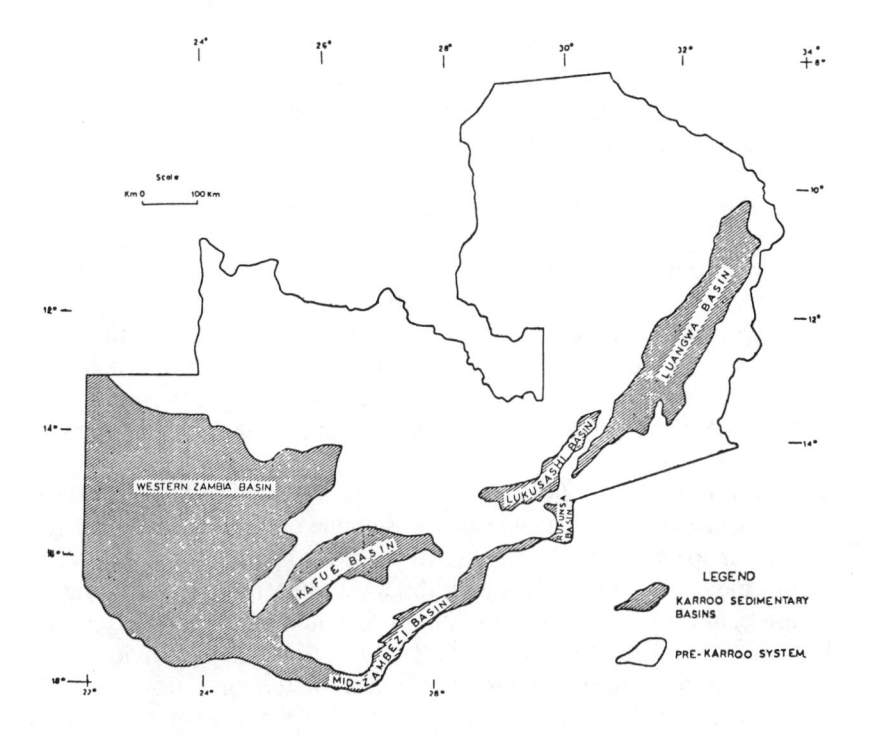

Figure 1. Karroo sedimentary basins in Zambia.

The flight line spacing was 5 km, with tie line spacing of 20 km. The survey altitude was 3.35 km in some cases and 2.44 km in others. A number of factors influenced the decision on the survey altitude. In the Western Basin, although lying flat, it was found necessary to fly at a higher altitude of 3.35 km (compared to 2.44 km for most areas) in order to filter off the "screening effects" of the basalts. This value of 3.35 km was determined mathematically, based on the physics of the problem. In other parts of the survey area effects of weather determined the flying height.

Quality control of both acquisition and data processing was performed by EDCON. Prior to 1982 Zambia had been covered by an aeromagnetic survey, leaving only a few areas unexplored. This was mineral-based coverage and therefore had been flown at a much lower altitude. These magnetic data, acquired about 1975, had been recorded both digitally and in analogue form. It was therefore originally decided to reprocess it for areas G, H, and J and interpret it in terms of petroleum resources. This was part of the contract entered with EDCON. Unfortunately degradation of digital data, because of storage in an unsuitable environment, required that the areas be flown again. This was an extension to the contract with Geosurvey International.

Deliverable items to the Government of Zambia include:

1. *Total intensity maps*: where the magnetic field is proportional to the Earth's inducing field at that location.
2. *Downward continuation maps*: by downward-continuing the data, greater resolution and anomaly separation is achieved. (It is like looking at a forest of trees which looks a general green from the air but where you can distinguish the individual trees when the plane is landing.)
3. *Reduction-to-pole (RTP) maps*: In the reduction-to-pole (RTP) maps, the field is created as if it was at one of the Earth's poles. The RTP maps give the broad overall picture of magnetic source without the distortion displayed by the total intensity maps. Anomalies are better resolved in the RTP maps than in the total intensity maps.
4. *Second-vertical derivative (SVD) maps*: SVD maps provide valuable information on geological contacts including faults.
5. *Interpretative overlays*: which summarize the quantitative details derived from the data, including contours of the estimated depth to the magnetic basement, regions of quiet magnetic signature, which are equivalent to areas underlain by notable thicknesses of Karroo sediments, possible faults and contacts, and regions of active magnetic signature, mostly indicative of igneous activity.

The aeromagnetic phase of the Petroleum Exploration Promotion Project is nearly completed, resulting in data of sufficient interest to warrant follow-up work. Areas F, E, and A (northwestern corner) have been classified as prospective. Some of the areas have been classified as possibly prospective, depending on whether the outstanding geological unknown quantities are solved.

From tectono-stratigraphic consideration, Karroo sediments of certain basins, such as within the Luangwa and Luano Valleys, would appear to possess potential petroleum source rocks, reservoir rocks, and possible cap rock and trapping mechanisms.

In the Luangwa Valley, generally, the depth to magnetic basement varies considerably from less than 1,000 to 6,500 m in its deepest sections. Assuming a continental geothermal gradient of $3.5°C$ per 100 m depth, and assuming all things being equal, the depths obtained in the Luangwa Valley fall within the desired range. Ridges or linear uplifts which occur in certain parts of Luangwa Valley are of interest. Takutu graben, in the southwestern Guyana, which consists of nonmarine sediments and has very similar tectonic and statigraphic set-up as the Luangwa Valley, recently was the site of an oil discovery (Crawford et al., 1984).

Follow-Up

The aeromagnetic survey has revealed tectonic information in some of the sedimentary basins of Zambia of sufficient interest to warrant follow-up work.

Gravity traverses will be made in selected areas to confirm or disprove the results of the aeromagnetic survey. Preparations of terms of reference and contract specifications are already underway. It is envisaged that the gravity survey will be helicopter supported.

Magnetotelluric (MT) sounding investigations were not in the original contract between the World Bank and the Ministry of Mines. It is strongly believed that a well-executed MT survey will provide additional information on the sedimentary basins, especially regarding the thickness of the different rock formations. This is especially important in the case of the Western Basin, where basaltic flows exist. Negotiations with the World Bank for the inclusion of the MT soundings are expected to be made.

Evaluation of existing geological data and acquisition of new geological data as it relates to petroleum resources forms one of the components of PEPP. Services of petroleum geologists will be sought on a consultative basis. Studies will include paleontology and palynology.

PEPP has provisions to use Landsat data for geological studies. This effectively will be the fourth control on the tectonic data after aeromagnetics, gravity, and MT.

Promotion Phase

This is the last but one of the most crucial phases of PEPP. Information from aeromagnetics, gravity, MT, and Landsat imagery will all be integrated into one report. Available at the same time will be the Petroleum Legislation in Zambia. All this information will be available for sale to interested companies.

Conclusion

It is hoped that the release of the integrated promotion report will herald the start of petroleum exploration, drilling, and production in Zambia.

References

Crawford, F. D., 1984, Geology and exploration in the Takutu graben of Guyana: Oil and Gas Journal, March 5, 1984.

EDCON, 1984, Interpretation of areas E and F Zambia: for Ministry of Mines, Republic of Zambia (unpublished).

Money, N. J., 1981, Hydrocarbon potential of Zambia: Occasional paper, No. 06-1981.

17
Note Concerning the Research and Development of Shallow Petroleum Fields in Senegal

Babacar Faye

Senegal, a country situated in West Africa, south of the Sahara, has two shallow oil fields which are of great interest and about which studies in progress could result in these fields being brought into production.

Dome Flore Field was discovered in 1968 by Total Petroleum Co., 60 kilometers offshore Casamance in the southern region of Senegal, at a water depth of about 70 meters. This field comprises two reservoirs at different levels. The first, between 250–300 m deep, consists of an accumulation of heavy oil in Oligocene sandy layers with reserves estimated to be 200 million tons. This heavy oil is very viscous, has a density of 10° API, and is composed essentially of paraffin and naphthene hydrocarbons. Putting this deposit into production appears to be difficult because of the nature of the nonconsolidated formations containing the oil and the method needed to recover it.

The second reservoir in this field contains an accumulation of light oil (23° API) with reserves estimated at 5 million tons. This accumulation, found between 700–800 meters in the Paleocene and at the base of Maastrichtian, is now the subject of a legal and economic study, which would permit its economic exploitation by the National Petroleum Company of Senegal (PETROSEN) in association with foreign partners. The details

---

*Note: This is an unofficial translation from the French original—Ed.

of such an association are being studied in order to carry out borings before bringing the reservoir into production. The studies concerning the Dome Flore Field are being financed by the World Bank at the request of the Government of Senegal. These studies are being followed by one of the sedimentary basin, financed by the World Bank, and a complementary study financed by Petro-Canada.

The second field, called Niam Niadio, is situated about 40 kilometers north of Dakar, the capital of Senegal. It consists of three small accumulations of gas discovered in 1965 by the African Petroleum Company (SAP) and two small oil accumulations located in the same sector over a surface area of 1,200 square kilometers and at a depth averaging 1,000 meters. Putting them into production poses some problems due to the very limited quantity of hydrocarbons found. However, an economic study with a view to decreasing the oil import bill of Senegal has permitted the initiation of production ot two small deposits of gas equivalent to 30 million calories to supply a central electrical power plant located nearby.

The increased production offtake allowed water to enter the reservoirs and thus their exploitation had to be stopped. Some seismic studies and drilling (four wells) are foreseen in order to, on the one hand, restart production of the flooded gas deposits and to find other deposits of gas, and on the other hand to gather evidence of oil deposits in this region. Wells already drilled have lead to the discovery of both oil and gas. An economic study and a seismic study would permit PETROSEN to undertake the drilling of wells as planned. For this, financing has been obtained from the World Bank, and Petro-Canada has been asked to take part.

Section III

Shallow Oil Production and Technology

18
Aspects and Specifics of Shallow Oil Production: Field Development Techniques, Equipment, and Economics of Production

C. G. Rosaire, Jr., and Kenneth J. Schmitt

Introduction

We divide shallow oil production into two categories, wells which produce more than 50 barrels per day and those which produce less than 50 barrels daily. The two categories differ in some respects with methods of completion and production sometimes differing. Both eventually become what is referred to as "stripper oil," producing less than 10 barrels per day. We concentrate primarily on producing wells of less than 50 barrels per day, but the methods can be applied to greater production as well.

Economics divides the two categories; a 50-barrel-per-day well or greater will not carry the same set of economically critical considerations as one producing less than the 50 barrels per day.

Development Planning

We assume that a field has been discovered and several test wells have been drilled, logged, and cored.

It is of prime importance to determine the type of formation and its characteristics. From the logs and cores we learn the type of drilling fluid best suited to cause the least formation damage, the permeability, porosity, and the likely components which will affect the end method of completion and stimulation. We will also learn how best to begin field development. Once this information is assimilated and a drilling program determined, we look to the best and most economical methods of drilling, rotary air or fluid.

Selection of Proper Drilling Equipment

A number of small shallow rigs are available capable of drilling to 2,500 ft or deeper with either air or fluid. These units are generally truck mounted, highly portable, and require a three-man crew. The units are basically identical to larger land rigs but sized down in capacity. In remote areas, helicopter units are available. Drilling components can be air lifted by helicopter in two or three loads, together with their own portable mud pits. Both types of units are capable of reaching target depth in two to five days, depending on the formation, and usually work daylight hours only. They also are capable of setting surface casing, running the long drill string, and cementing both.

Selection of Completion Units

A number of types of completion units are available, ranging from sophisticated straight hydraulic units to single-pole, shallow-well workover/completion units. Simple drilling and completion units are easiest to maintain; maximum maintenance means minimum breakdown time. For downhole correlation purposes, a number of small, inexpensive gamma ray-neutron logging tools are available which can be run by rig personnel. It is not necessary to run expensive logs on each well.

Selection of Production Equiment

Establishing the type of formation and gravity of the oil to be produced to a great extent will determine the method and type of production equipment needed, air lift, gas lift, beam pump, positive displacement, bottom hole pump, or other. In order for the project to be most economical, proper research and planning must go into the initial stages of equipment specification and procurement. Secondary stages of planning are concurrent with actual field development. Time bar charting may be utilized, consisting of:

1. Well location
2. Site preparation
3. Storage location
4. Flowline schematic drawings
5. Energy type (pumping)
6. Equipment procurement
7. Roustabout work
8. Drilling operations
9. Logging and coring
10. Completion and stimulation

Production Maintenance

This is of prime concern at low barrel-per-day production, which is characteristic of much shallow oil. Constraining factors, such as paraffin build-up, sanding, emulsion problems, and gas locking, must be dealt with effectively on a daily basis. Flush production may decline rapidly to 50% of its potential without proper maintenance. In many cases consistent, effective production maintenance determines whether a well is economically successful.

Proper planning and operation management, or their lack, impacts heavily on the economical operation of a project.

Experience and Methods for the Navarro Formation

To this point, we have been examining shallow oil production in general. We will now focus attention on the Navarro Formation of central Texas.

The Navarro Sands belong to the Upper Cretaceous Gulf Series and consist of a laminated sequence of sand and silty shale, deposited in a tidal marsh environment. These sands and shales are contaminated with anhydrous clays which act as a cementing agent and are highly susceptible to swelling when in contact with fresh water. This susceptibility can lead to extensive formation damage if improper procedures are followed during well development.

Over the past several years, we have drilled about 125 wells, with a success ratio of over 90%, and have learned that well and lease planning and economics go hand in hand. During this period we have determined what we feel are the best approaches and methods of accomplishment.

Through "hands on" operations, one gains much of the experience in field development necessary to achieve optimal production. Now let us examine some of the methods we use, many gained through experience.

Drilling Methods and Determinations

Drilling unit: Portable truck-mounted rig, normally designated as 2,000/2,500 with Gardner-Denver duplex $5^{1}/_{2} \times {}^{8}/_{10}$ fluid pump.

Hole size: $6^{3}/_{4}$ in.

Pit system: 250–300 bbl, 2-pit earthen system. There are two reasons to use the 2-pit configuration.

1. It is economical to use native mud until marked depth is reached, usually about 75 to 100 ft above the projected producing zone, at which time we jet the suction side and mix a polymer/salt system for drilling the formation.
2. It makes it possible to maintain a sufficient amount of heavy mud on location in the event that higher bottom-hole pressures than expected are encountered.

As a matter of interest, we have had three wells blow out in the past 2 years; it is not uncommon to experience bottom-hole pressures of 700 psi or greater.

Proper planning leads to:

1. Better wells by preventing undue formation damage during drilling and completion;
2. Reduction by 75% of drilling mud bills; and
3. Preparation for unexpected events without additional cost.

Fluid system: Native mud/polymer-salt system. The Navarro sands are highly susceptible to formation damage when they come in contact with fresh water. A polymer-salt system prevents the formation clays from swelling.

Logging principles: A good log with sidewall cores is mandatory when drilling field-limitation development wells. During drilling of infield wells, it's a matter or operator choice; however, we feel that a simple correlation log is all that is required.

Casing: For casing, we utilize 4½ in., 9 lb X 42 line pipe, tested to 3,000 lbs, threaded and collared with J-55 short collars. Savings have been approximately 25% over 9 lb A.P.I. J-55 pipe.

Cementing procedure: Under normal conditions we have found that it is less expensive to cement from top to bottom. For economic reasons we use a 50/50 Poz and Class H cement, with the "H" covering the formation. This system generally runs about 25% to 30% less expensive than straight "H." Our pad is 2% to 4% KCl solution, as is the displacement fluid, the purpose being to protect the formation from fresh water.

Perforating: Four shots per foot with casing gun preferred, due to the laminated nature of the formation.

Completion: More dry holes are created through improper completion methods and mechanical failure than are plugged prior to setting pipe. One must understand as much as possible about the formation and how much abuse it can stand during well completion. More than 79% of the Navarro producing zone runs 140 mesh or greater; during the swabbing process, should you put too "hard a pull" (vacuum) on the formation, you could start it to "coming in" (sanding). This creates all sorts of problems, up to the point of losing the well. The least problem would be a "sand pumping process." We have developed a special procedure for this period using an incremental approach to swabbing and the input of much patience.

Stimulation procedure: Some operators will fracture the well while perforating it; others wait thirty days. Each well is different. Generally, we use a "bucket" or "Freeman frac," named after the man who first designed it. This commonly consists of 1,800 lbs of 20/40 sand and 500 lbs of 12/20 sand, using 120 bbls of either lease water, lease diesel oil, or KCl solution as the carrying agent. This gives good wellbore cleanup with at least 2 ft of penetration. The well is shut in for 36 to 48 hours, allowing the formation to heal itself. If the well is brought back too soon or fast, sanding may result. This fracing method is appropriate and economical for the Navarro.

Production equipment for non flowing well: Through proper evaluation and experience, we have found the beam system with a downhole insert pump to be the most effective. Commonly we use a bottom hold-down, 7-cup pump with sand and gas checks, but the downhole configuration varies from well to well. We have tried such pumping systems as air lift, submersible, auger, and a variety of others, with varying degrees of success, depending on well conditions, yet find the beam system preferable. Overall service reliability and economics are the most important considerations.

By utilizing the techniques and equipment described here, we currently hold our cost to drill and complete the 1,500-ft initial well of a lease at $28,000 to $31,000, using new equipment.

Production maintenance: The major maintenance problems in our area are sanding, gas locking, and paraffin deposition. Sanding can occur for a number of reasons, the most common being improper well completion or servicing. Sand checks are used on the pump to assist in control. Gas locking is prevented with gas checks on the pump and by holding a predetermined back-pressure on the casing. Varying the bottom hole configuration helps to alleviate both sanding and gas locking problems.

Paraffin is a naturally occurring constituent of the oil produced from the Navarro sands. The paraffin moves with the oil flow to the well and solidifies on the downhole equipment, in particular the inside of the pump tubing string, restricting flow. Without preventive maintenance, our wells would shut themselves down in about 21 days due to paraffin. Paraffin is a significant production and economic constraint. We have developed, through computer analysis, a service schedule for each well using chemicals, mainly solvents, and hot oil treatment for paraffin control. Production maintenance problems require service by a competent pumper each day, for optimal production.

The Future of Shallow Oil Production

The economic outlook for shallow oil production is bright. Much of it worldwide is yet to be developed.

Shallow production allows one to "spread the risk." Wells require less expensive drilling and production equipment and are generally completed in a few days at a cost allowing one to drill ten or more wells for the price of a single deep well. The single deep well may produce more than the ten shallow wells combined, yet it may also be a dry hole. It is possible to explore and develop a potential area at a low cost per well, allowing an incremental program.

Perhaps the most important factor in shallow oil production is simple economics. Careful planning, retention of knowledgable project personnel, and application of modern technology and equipment are the keys to successful projects.

19
Producing Shallow Heavy Crudes Having High Water Content

Charles D. Haynes

Introduction and Scope

The common concept of conventional heavy crude oil production is that of a slow-moving pumpjack lifting a viscous, tarry mass with much difficulty and with poor efficiency. This paper concerns the opposite extreme from this concept, whereby the heavy crude oil content may be only one percent of the total fluid stream, with the remainder a concentrated brine having nearly 100,000 parts per million (ppm) total dissolved solids.

Belden & Blake Corporation has accumulated several years experience in producing heavy crude oil accompanied by significant produced water. Our philosophy of production, oil dehydration, and produced water disposal in two fields will be reviewed in this paper, with cost data given where appropriate.

Background

Belden & Blake Corporation operates two oil fields in south Alabama (USA). These are the Gilbertown and South Carlton fields, located as shown in Figure 1. The two fields are similar in that they both produce heavy crude oil between 15° and 18° API from relatively shallow depths, but have several prominent differences that warrant separate discussion.

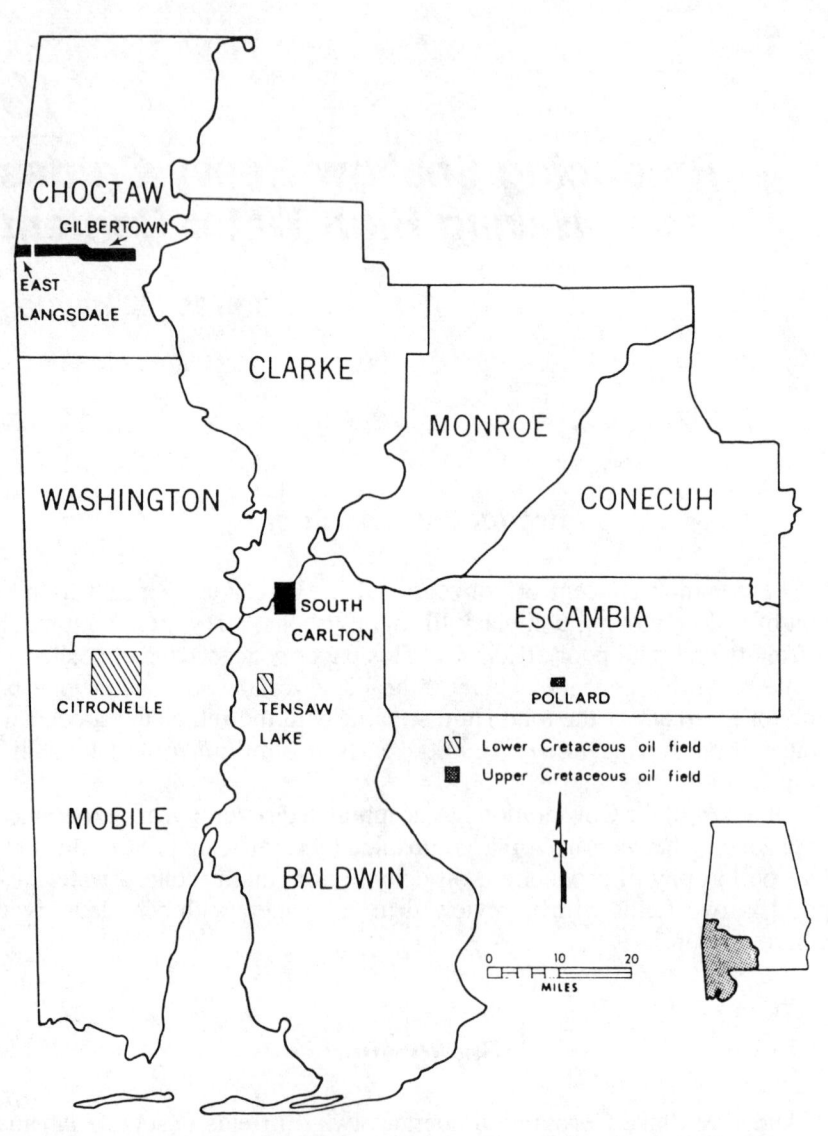

Figure 1. Location of Cretaceous oil fields in southwest Alabama, U.S.A.

Gilbertown Field

The Gilbertown field, discovered by the Hunt Oil Company in 1944, was the first oil field in Alabama. Initial production was found in fracture zones of the Selma Group chalks, but was subsequently found in the Eutaw Formation sands immediately underlying the Selma. Productive depths vary between 2,800–3,600 ft (854–1,097 m). The basic entrapment feature at Gilbertown is a series of faults oriented approximately east-west, with a down-to-the-north sealing fault. The Gilbertown field, through rapid development, eventually reached an annual production of 898,075 barrels (bbl) in 1951. An exceptionally strong water drive has resulted in significant water production at Gilbertown. Oil production remained profitable, by combining low operating costs with surface disposal of the produced water in the numerous streams flowing through the area.

In 1969, a statewide no-pit order virtually shut down the entire field. Production resumed, but only in the best wells, which could support subsurface water disposal. By 1971, production had declined to 86,946 bbl annually, and the field began to pass through a succession of operators. Belden & Blake acquired 33 of 35 available producing wells in 1976. At that time, the field had produced 10.9 million bbl of oil and was considered by many to have reached its economic limit.

Operating Philosophy

During the engineering evaluation prior to purchase of the Gilbertown wells, the author determined that two factors would be necessary for successful operation of the field:

1. High-volume fluid withdrawal from producing wells.
2. Low-cost subsurface disposal of produced water.

Most of the wells produced 5 percent or less oil of the total fluid, with many around 1 percent. Assuming that the oil-water content remains in the same proportion with increased total fluid withdrawal, a "spread" would develop between the value of the increased oil produced *if* the unit water disposal cost remained minimal. For example, a well having an oil percentage of total fluid of 2 percent would produce 10 bbl oil/day, and 490 bbl water/day at 500 bbl/day output of total fluid. If the net value of the oil to the producer is $12/bbl after operating expenses, and the water disposal cost is $0.05/bbl, there is a "spread" of $95.50/day between the value of the oil and the cost of water disposal. If the total fluid is increased to 1,000 bbl/day and the proportion of oil to total fluid remains constant, then the new productivity levels would be 20 bbl oil/day, and 980 bbl water/day. If the cost of producing the additional total fluid is $0.03/bbl and the disposal cost remains the same, then the "spread" between the value of the oil and the incremental cost of producing the additional fluid and disposing of it increases to:

$$(20 \times \$12) - [(980 \times \$0.05) + (\$0.03 \times 500)] = \$176.00/\text{day}.$$

In practice, the "spread" has been somewhat greater than this because the unit cost of water disposal has been reduced to $0.02/bbl in low-pressure systems having simple water-treating facilities.

Productivity Analysis

When Belden & Blake made its initial purchase in the Gilbertown field, each well was analyzed to determine its candidacy for high-volume lift. Fluid level and pumping unit dynamometer data were taken on each well. Typically, a well was then being produced by a beam pump-sucker rod (BPSR) system which was overstressed in attempting to produce a greater volume of fluid than originally intended. Also, it was found that many of these pumping units were being overstressed simply by having the downhole pump set deeper than necessary. For example, some pumps were set at 3,000 ft (914 m) in a 3,600 ft (1,097 m) well when the pumping fluid level was only 600 ft (183 m) from the surface. In many cases, output was increased by resetting the pump to a fluid submergence of 300 ft (91 m) or less, placing the surplus rods and tubing in inventory, increasing the downhole pump diameter for additional fluid production, and reducing the stress on the pumping unit. The additional fluid production brought with it the "spread" just mentioned.

The fine-tuning of existing pumping capacity had its practical limits, beyond which other measures were investigated. Typically, the existing pumping units were API 80 or 114 series, having practical limits of 1,000 bbl/day of total fluid at their maximum stroke, pumping speed, and gearbox rating. By analyzing dynamometer and fluid-level data, those wells having 1,000 bbl/day total fluid production at pumping fluid levels less than 1,000 ft (305 m) from the surface were candidates for high-volume artificial lift. Available systems were studied, with the following results:

1. Larger BPSR systems were acceptable if the pumping speed was kept within the manufacturer's recommendation for rod and tubing wear. Since the oil at Gilbertown is nonlubricating, wear considerations were extremely important.
2. Hydraulic downhole pumping (HDP) systems were discarded because of the decentralized nature of the field and the relatively shallow producing formations.
3. Electric submergible pumping (ESP) systems were acceptable as an alternative to the BPSR system if increased pump life would offset the additional cost of pump and motor overhaul.

After initial experimentation, several electric submergible pumps were installed where total fluid production would be increased to 2,000–2,500 bbl/day without pump-off.

Operating costs before and after ESP installations were analyzed to determine if these units were comparable to the previously used BPSR systems. Table 1 shows the comparison in operating costs before and after conversion to ESP. In addition to the lower operating costs, note also that the on-line time of each well improved sharply. Typical ESP life now averages between 1–2 years. Clearly, ESP units have proven to be cost effective in the higher fluid output wells at Gilbertown.

Oil Dehydration and Storage Facilities

The decentralized oil dehydration and storage facilities are designed for high-water production. For example, each tank battery utilizes a free-water knockout to remove as much as 95 percent of the total fluid stream. An ambient-temperature, float-controlled, horizontal vessel is used which "fails-safe" to either the stock tanks or skim-oil tank at the dis-

Table 1
Gilbertown Field Operating-Cost Comparison Beam Pump-Sucker Rod System (BPSR) and Electric Submergible Pumps (ESP)

Well	BPSR Previous 6-Month Average On-line (%)	Op. Cost ($/bbl)	ESP Post-Conversion 6-Month Average On-line (%)	Op. Cost ($/bbl)
Rex Alman 3	89	4.15	94	3.16
Rex Alman 4	81	4.96	97	2.55
F. Stewart 1	89	2.56	96	1.56
W.C. Abston 1	69	4.84	98	2.26

posal facility. Other than occasional recycling or oil skimming operations, no problems have resulted from this design philosophy. The use of the free-water knockout enables the other tank battery equipment to be reduced in size to that of the expected oil-water emulsion volume. An automatic shut-down system prevents tank overflow.

Ordinarily, the total fluid stream has enough inherent heat (about 27°C) to provide adequate dehydration after the free water is stripped. The moderate South Alabama climate aids in this dehydration during most of the year. Vessels are painted a flat-black color to absorb as much heat as possible. De-emulsion chemical or liquid petroleum gas (LPG) heat is used in extreme situations, such as in treating a batch of nonspecification oil or during cold weather. A home-designed, hot-oil unit is used in these situations. Sufficient gas to operate emulsion treaters is occasionally found in Selma completions. However, the usual gas-oil ratios are too low to provide an adequate long-term gas supply for surface equipment utilization.

Produced Water Disposal

It was mentioned earlier that the philosophy of produced water disposal (PWD) was to simplify the systems as much as possible to ensure a minimal disposal cost. One of the first requirements was to assure compatibility of the disposed water with the disposal formation water. Produced water cannot be injected into a producing zone in Alabama unless the zone is in a secondary recovery or pressure-maintenance unit. Certain sandstones immediately below the producing Eutaw sands were found to be excellent disposal zones. Candidate wells for disposal were selected from:

1. Abandoned oil wells cased through the potential disposal zones.
2. Abandoned wells drilled but not cased through the potential disposal zones.
3. Sites convenient to existing disposal facilities that could be drilled for PWD.

The selection procedure conformed to the preceding order. A typical 5,000-ft (1,524-m) PWD well at Gilbertown is estimated to cost as follows:

1. $10,000–20,000.
2. $25,000–50,000.
3. $100,000–120,000.

When converting or drilling a PWD well, it is extremely important to prevent contamination of the disposal zones. An independent operator typically cannot afford "operating room" conditions at the disposal well. At least, fresh water is kept off the formation face as much as possible and native formation fluids or inert gases are used whenever possible for downhole fluid circulation. When perforating the disposal zone, the fluid level in the well is adjusted to ensure an "underpressure" situation, where the formation pressure exceeds the pressure exerted by the fluid column in the wellbore. This causes the formation, when perforated, to expel contaminants in the perforation tunnels back into the wellbore, where they are bailed or circulated out of the well.

The nature of the disposal pumping system will greatly affect disposal cost. A system can range from gravity disposal (well on vacuum) where no pump is needed, to high pressure disposal requiring a positive displacement pump. Typical Belden & Blake disposal pressures range from 250–1,000 psi, with one gravity system.

At the time of purchase, the Gilbertown field had three PWD systems operating, all on triplex plunger pumps. Injection pressures corresponding to these systems suggested that centrifugal pumps could be substituted for the plunger pumps at considerable savings in purchase price and maintenance costs. This decision has been borne out through the continued use of centrifugal pumps in the two converted systems, the Chestnut-Rudder and Alman-Johnson, which dispose about 15,000 bbl/day. Typically, two-state centrifugal pumps can provide up to 300 pounds per square inch gauge (psig) injection pressure at 15,000 bbl/day disposal capacity without resorting to exotic designs. These pumps, such as the Goulds 33-16, are priced around $7,000 each for cast iron bodies and stainless steel trim, with a surcharge of about 100 percent for all-stainless steel versions. A comparable triplex pump will cost around $20,000. In general, if a centrifugal pump is capable of providing needed injection pressure, we have found it less expensive to purchase, operate, and maintain. Only in the pressure ranges above 300 psig is it less effective than the plunger pump. Multistage centrifugal pumps are capable of higher pressures, but generally incur proportionately higher purchase and operating costs.

Produced-Water Disposal Surface Facilities

Surface facility design is simple and straightforward. No filters are used, and storage tank size is selected to provide a minimum solids-settling time of eight hours. In addition, a minimum two-tank facility, connected by a water leg, prevents oil migration from the skim tank to the pump tank. Since the heavy crude oil, when carrying fine silt, can approximate salt water density in cold weather, the tanks are sampled periodically for floating oil below the surface of the fluid. Oil skimmers are included with each disposal system to remove crude oil carryover. Periodic checks of oil content in the produced water are made to monitor any increases in carryover. Typically, oil content is kept below 100 ppm in the water system. This is an acceptably low content consistent with an economic facility design.

Table 2 shows the current Belden & Blake PWD systems in Gilbertown, their operating characteristics, and typical disposal costs in 1983.

Planning additional field development has, as its first priority, the expansion of PWD facilities. Recently, the central Gilbertown area has begun development around an abandoned wildcat well converted to PWD. This development will include reentries of abandoned wells with high water contents, as well as new wells drilled from recent geological interpretations.

Operating Costs

Operating or "lifting" costs vary throughout the Gilbertown field, depending on the type of pumping system, whether electric power is purchased wholesale or retail, and the volume of produced fluid. Selma wells having low-maintenance API 114 BPSR systems may be operated for as little as $1.00/bbl oil when producing after initial completion. As the water content increases, increased total fluid production will increase

Table 2
Gilbertown Field Produced-Water Disposal Facilities and Costs (1983)

Area	System	Injection (bbl)	Operating cost ($)	Injection Pressure (psig)	Disposal Cost ($/bbl)
East	Chestnut-Rudder	591,248	$ 32,528	250	0.06
E. Central	Alman-Johnson	3,883,492	226,333	300	0.06
E. Central	Stewart 2	2,572,195	90,079	1,000	0.04
E. Central	Stewart 3-15	699,680	35,616	1,000	0.05
Central	Singley	245,884	4,812	1,000	0.02
Central	Trice 34-6	(Under construction)			
W. Central	Scruggs	842,931	31,244	300	0.04
West	Bonner	51,068	997	Vacuum	0.02
		8,886,498	$421,608		$0.05

Table 3
Gilbertown Field Operating Cost and Oil Price Data

Year	Operating Cost ($/bbl)	1977 Adjusted Price ($/bbl)	Average Oil Price ($/bbl)	Ratio Price/Cost
1977	4.71	–	10.79	2.29
1978	4.56	4.18	11.34	2.49
1979	5.00	4.05	14.90	2.98
1980	6.34	4.57	25.79	4.06
1981	6.30	4.16	28.06	4.45
1982	7.42	4.72	23.48	3.16
1983	7.68	4.71	21.36	2.78

this unit cost to $4–$8/bbl oil. Continual increases in efficiency have held operating cost increases at or below the normal inflationary growth since 1977. For example, in 1977, operating costs averaged $4.71/bbl oil. In 1983, these costs had increased to $7.68/bbl oil. However, when adjusting the 1983 costs for inflation, the cost for 1983 is reduced to $4.71/bbl oil, identical to the 1977 cost (Table 3).

South Carlton Field

Belden & Blake Corporation has drilled 34 wells in the South Carlton field since 1973, 33 of which produce asphaltic heavy crude oil averaging 15° API. These wells were drilled in fringe acreage around the original 22 wells in the field.

South Carlton was discovered by Humble Oil and Refining Company in 1950, and has produced steadily since that time. The primary productive zones are in the lower Tuscaloosa Group of Cretaceous age at a depth of 5,300 feet (1,616 m). Two intervals are prominent: the "massive" (lower) and the Pilot (upper). The Pilot and "massive" are productive on the flanks.

This field presents unusual drilling and operating problems since it is located on an isolated flood plain near the confluence of the Tombigbee and Alabama Rivers. Seasonal flooding isolates the field, and dictates that all well locations be raised above the expected high water level. Also, several wells are directionally drilled at angles up to 34° from the vertical to drain drilling units under permanent bodies of water or to minimize surface preparation.

Operating Philosophy

Belden & Blake Corporation originally developed South Carlton as a quasi-offshore field, with marine-oriented logistics. Since site preparation and roadwork were restricted by the surface owner, it was necessary to concentrate the early wells on multiple-well "drilling pads," earthen mounds built above the seasonal high-water level. The isolated nature of the wells encouraged a production system that would not require a mobile workover rig. The initial 23 wells were produced by casing-free hydraulic pumps using produced brine as the power fluid. Two central tank batteries were constructed to dehydrate and store the crude oil, clean the produced water, and transfer it to triplex pumps for return to the producing wells. Surplus produced water was disposed in two wells located near the tank batteries.

Unfortunately, the hydraulic pumps did not perform as anticipated. The relatively high water content of the produced fluid, around 94 percent, together with the nonlubricating asphaltic crude oil and entrained fine sand, greatly decreased the operating life of pumps. Distribution of the power fluid to remote wells was hampered by inadequate working pressures and design of the power fluid lines. An alternative producing system was clearly needed.

The author, upon assuming operating responsibilities for this property, began testing alternative pumping systems in the directional wells. Wear of the sucker rods, tubing, and casing in these wells was a primary concern in the possible conversion to BPSR units. Hands-on experience in the Gulf Coast area was not readily available, so several California-based heavy crude operators were interviewed for their operating practices. As a result of these interviews, a plan was made for operating BPSR units in the directional wells. This plan essentially called for the following procedures:

1. Run J-55 tubing on a tubing anchor.
2. Anchor tubing, pulling 10,000# over string weight.
3. Run rod guides as determined by empirical design.
4. Install a rod rotator on the carrier bar of the surface pumping unit.

In addition, a pumping unit one size larger in API designation than that necessary for minimum anticipated straight-hole requirements was specified. This wasn't strictly necessary, but was a matter of conservative design in anticipating the need for increased total fluid withdrawal over the life of the field.

The BPSR units installed in the directional wells have proven to be a good investment over the past 6 years of use. Rod break and tubing leak frequency on these wells is no worse than "straight-hole" wells operating in the same environment. Although most new drilling in this field is being done using straight holes, our experience with BPSR units in directional wells would not prevent this type of drilling where dictated by physical location.

ESP units were installed in three wells having high total fluid outputs. These units went through design evolution before reaching an acceptable longevity. Early units suffered frequent shaft breakage from vibration, later traced to radial impeller wear. Shaft bearing redistribution increased pump runs from several weeks to over one year. The one-year runs are considerd optimal under existing operating conditions.

Oil Dehydration and Storage Facilities

Two centrally located tank batteries serve the 33 producing wells. Since the produced fluid contains about 94 percent water, stripping the free water permits smaller downstream equipment and reduces energy consumption for unnecessary heating of the free water. Normally, over 90 percent of the produced fluid is stripped as free water in the knockout vessel, passing the remaining emulsion to the gunbarrel or settling tank. Heating coils are installed in the gunbarrel and stock tanks to aid in oil dehydration. A steam generator burns natural gas condensate purchased in bulk quantities and stored at the tank battery location. Heat loss calculations were performed on each tank battery for wide ranges of temperatures and wind velocity to optimize fuel consumption.

Dehydration is usually effective to about 2 percent basic sediment and water (BS&W) in the gunbarrel. The large storage capacity, necessitated by monthly barge shipments of the crude oil, allows additional gravity dehydration between oil shipments. The stock tanks are heated to 110°–120°F beginning about three days ahead of shipment, further dehydrating the oil to levels between 0.5–1.0 percent BS&W. This also lowers oil viscosity to suitable levels for pumping. During winter, a small amount of dehydration chemical is used at flowline locations to improve water-oil separation.

Produced Water Disposal

The volume of produced water at South Carlton averages about 2.2 million bbl/yr, low compared with Gilbertown. The injection pressure is around 500 psig, slightly above the capacity of centrifugal pumps. Even though triplex pumps are used, the disposal cost still remains under $0.10/bbl. Three disposal wells are used, one of which is a standby. This precaution is taken because of the isolated location of the field, preventing convenient access by service equipment during the high-water periods.

Operating Costs

The initial operating cost of the hydraulic pumping units was quite high, about $14/bbl oil. This cost, coupled with an oil sales price of less than $10/bbl in 1977, had the operation firmly "in the red." The conversion to BPSR and ESP systems, begun in 1978, lowered the operating costs almost immediately. Table 4 shows the oil price-operating cost history since the conversion to the new production systems. Rapid increases in the crude oil price occurred from 1978-1981, peaking at $26.79/bbl. At the same time, the new systems, operating essentially trouble-free, drastically lowered the operating cost in 1979 to $8.04/bbl oil. The inflationary spiral, coupled with the expected increase in rod parts and tubing leaks as the systems aged, produced an increase in the operating cost, finally reaching $12.78/bbl oil in 1981. Some of this increase was the result of the tendency of oil prices to increase, which, in turn, set an environment for spending. The downturn in oil prices in 1982 brought with it a reduction in expenditures, reaching $9.88/bbl oil in 1983.

An indication of operating efficiency with the new production systems is shown by adjusting the operating costs from 1977 onward back to a 1977 basis. The "adjusted" operating cost for 1983 was $6.06/bbl using a 1977 basis. This is a reduction of 57 percent from the 1977 operating cost.

Table 4 also compares the ratio of oil sales price to operating cost per barrel of oil. This ratio reached 1.0, or breakeven, in early 1978. Through 1981, the rapid increase in the oil sales price resulted in a maximum ratio of 2.27 reached in 1980. Since that time, the ratio has remained near 2.0.

Summary

The Belden & Blake operations at Gilbertown and South Carlton include several facets unique to the production of heavy crude oil from shallow depths. High oil viscosity, corrosive salt water, and directionally-drilled wells in isolated locations are common. These condi-

Table 4
South Carlton Field Operating Cost and Oil Price Data

Year	Operating Cost ($/bbl)	1977 Adjusted Cost ($/bbl)	Average Oil Price ($/bbl)	Ratio Price/Cost
1977	14.21	14.21	9.97	0.70
1978	12.66	11.61	10.89	0.86
1979	8.04	6.50	13.60	1.69
1980	10.74	7.73	24.42	2.27
1981	12.78	8.45	26.79	2.10
1982	12.47	7.94	22.51	1.80
1983	9.88	6.06	21.43	2.17

tions, coupled with an extremely high produced-water content, present an ongoing operating challenge. The continuing profitability of these operations attests to the sound technical planning, and most importantly, the careful and conscientious field management by Belden & Blake personnel.

Acknowledgments

Thanks are due the following persons for their assistance in the preparation of this paper: Duane Chase, Lorraine Hayes, Judith Haynes, Betty Irby, Dian Turner, and Juanell Watkins.

20

How to Predict Stability of Naturally Flowing Oil Wells

Edward F. Blick

Introduction

Quite often a naturally flowing oil well (a well without artificial lift) will exhibit a strange behavior called heading, producing oil in an intermittent pulsing fashion (Nind, 1981). Once this phenomenon starts, the well usually dies; that is, it stops flowing naturally. Sometimes the well owner can return the well to natural flow by pulling the tubing and replacing it with smaller-diameter tubing. If that doesn't work the well owner may elect to install some means of artificial lift, such as a sucker rod pump, a gas lift system, or an electrical submersible pump. Such artificial lift systems are expensive and greatly increase the lifting cost of each barrel of oil.

The solution that follows allows an engineer to predict when a well will become unstable, and it allows him to determine what well parameters (tubing size, choke size, etc.) can be changed to make the well stable again. This solution is limited to an incompressible reservoir fluid.

The Mathematical Model

Assume the oil well is flowing naturally and contains no tubing packer (Figure 1). The steady-state equations which relate pressure drop to flow rate are

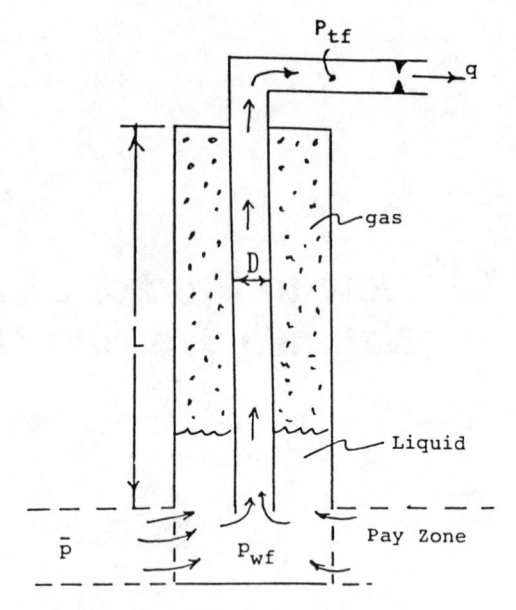

Figure 1. A naturally flowing well without a casing packer.

Steady-State Equations

Pressure drop reservoir to well bore

$$\bar{p} - p_{wfo} = q_o/J \tag{1}$$

Pressure drop in tubing

$$\Delta p_o = p_{wfo} - p_{tfo} = f(q_o, D, GLR_o, L, p_{tfo}) \tag{2}$$

Pressure drop through choke (sonic flow)

$$p_{tfo} = K \frac{(GLR_o)^{1/2}}{(d_o)^2} q_o \tag{3}$$

Subscript zero indicates a steady-state variable.

Unsteady-State Equations

Define instantaneous variables as:

$p_{wf} = p_{wfo} + p_{wfl}$

(instantaneous) = (steady state) + (fluctuating)

$\Delta p = \Delta p_o + \Delta p_1$

(etc.)

For small deviations from steady state it can be shown

$$\Delta p_1 = \left(\frac{\partial \Delta p}{\partial GLR_o}\right)(GLR)_1 + \left(\frac{\partial \Delta p}{\partial q_o}\right) q_1 + M \frac{dq_1}{dt} * \tag{4}$$

$$p_{tf1} = \left(\frac{\partial p_{tf}}{\partial GLR_o}\right)(GLR)_1 + \left(\frac{\partial p_{tf}}{\partial q_o}\right) q_1 + \left(\frac{\partial p_{tf}}{\partial d}\right)_o d_1 \tag{5}$$

$$\Delta p_1 = p_{wf1} - p_{tf1} \tag{6}$$

Reservoir flow perturbation (incompressible reservoir fluid)

$$q_r = -J p_{wfi} \tag{7}$$

* Annulus flow perturbation (capacitance effect)

$$q_a = -C_a \frac{dp_{wf1}}{dt} \tag{8}$$

* Tubing flow perturbation (capacitance effect)

$$q_t = C_t \frac{d\Delta p}{dt} \tag{9}$$

Total flow perturbation

$$q_1 = q_r + q_a + q_r \tag{10}$$

---

* See the notation for the derivation of M, C_a, and C_t.

By taking the Laplace transform of Equations 4–10 one obtains

$$\Delta p_1(s) = a_1(GLR)_1(s) + q_1(s) a_2 + Ms \quad (11)$$

$$p_{tf1}(s) = b_1(GLR_1)(s) + b_2 q_1(s) + b_3 d_1(s) \quad (12)$$

$$\Delta p_1(s) = p_{wf1}(s) - p_{tf1}(s) \quad (13)$$

$$q_r(s) = -J\, p_{wf1}(s) \quad (14)$$

$$q_a(s) = -C_a s\, P_{wf1}(s) \quad (15)$$

$$q_t(s) = C_t S\, \Delta P(s) \quad (16)$$

$$q_1(s) = q_r(s) + q_a(s) + q_t(s) \quad (17)$$

where $a_1 = \left(\dfrac{\partial \Delta P}{\partial GLR}\right)_o$

$a_2 = \left(\dfrac{\partial \Delta P}{\partial q}\right)_o$

$b_1 = \left(\dfrac{\partial P_{tf}}{\partial GLR}\right)_o^*$

$b_2 = \left(\dfrac{\partial P_{tf}}{\partial q}\right)_o^*$

$b_3 = \left(\dfrac{\partial P_{tf}}{\partial d}\right)_o^*$

Equations 11–17 represent seven equations with nine variables, $\Delta P_1(s)$, $p_{tf1}(s)$, $p_{wf1}(s)$, $q_1(s)$, $q_r(s)$, $q_a(s)$, $q_t(s)$, $GLR_1(s)$ and $d_1(s)$. Assume that two of these variables $d_1(s)$ and $GLR_1(s)$ are known. Hence

---

* Using Equation 3 it is possible to show that

$$b_1 = \dfrac{1}{2}\left(\dfrac{P_{tfo}}{GLR_o}\right),\ b_3 = \dfrac{P_{tfo}}{q_o},\ b_3 = \dfrac{-2P_{tfo}}{d_o}$$

Equations 11–17 can be solved by any one of several methods in terms of $d_1(s)$ and $GLR_1(s)$. The solutions take the form of

$$p_{tf_1}(S) = \left[\frac{ES^2 + FS + G}{AS^2 + BS + C}\right] d_1(S) + \left[\frac{HS^2 + IS + L}{AS^2 + BS + C}\right] GLR_1(S)$$

$$p_{wf_1}(S) = \left[\frac{f_1(S)}{AS^2 + BS + C}\right] d_1(S) + \left[\frac{f_2(S)}{AS^2 + BS + C}\right] GLR_1(S)$$

$$q_1(S) = \left[\frac{f_3(S)}{AS^2 + BS + C}\right] d_1(S) + \left[\frac{f_4(S)}{AS^2 + BS + C}\right] GLR_1(S)$$

where $A = M(Ca - Ct)$

$$B = \left(\frac{P_{tfo}}{q_0} + \left(\frac{\partial \Delta P}{\partial q_0}\right)_0\right) Ca - Ct \left(\frac{\partial \Delta P}{\partial q_0}\right)_0 + MJ$$

$$C = 1 + J\left(\frac{P_{tfo}}{q_0} + \left(\frac{\partial \Delta P}{\partial q}\right)_0\right)$$

The roots of the *characteristic equation*

$$AS^2 + BS + C = 0$$

Determine the time domain solution. Assuming the characteristic equation has roots

$S_1 = m$

$S_2 = n$

then it is possible to show for unit impulse inputs,

$p_{tf_1}(t) = c_1 e^{mt} + c_2 e^{nt}$

$p_{wf_1}(t) = c_3 e^{mt} + c_4 e^{nt}$

$q_1(T) = c_5 e^{mt} + c_6 e^{nt}$

m and n can be real or complex conjugate.

If m and n are *complex conjugate,* then the solutions take the form

$$p_{tf_1}(t) = A_1 e^{at} \sin wt$$

$$p_{wf_1}(t) = A_2 e^{at} \sin(wt - \gamma_1)$$

$$q_1(t) = A_3 e^{at} \sin(wt - \gamma_2)$$

In either case the solutions converge to zero (stable well) for negative values of the roots of the characteristic equation (or negative real part of complex conjugate) and the solutions diverge (approach ∞) for positive (real parts of roots) roots of characteristic equation.

A well will head up and die if a root or both roots are positive. For second-order characteristic equations it is easy to show that the system is stable if any of the following occur

$$A > 0$$

or

$$B > 0$$

or

$$C > 0$$

In other words the well is stable if

$$A = M(Ca - Ct) > 0$$

$$B = \left[\frac{P_{tf0}}{q_0} + \left(\frac{\partial \Delta P}{\partial q}\right)_0\right] Ca - Ct \left(\frac{\partial \Delta P}{\partial q}\right)_0 + MJ > 0$$

$$C = 1 + J\left[\frac{P_{tf0}}{q_0} + \left(\frac{\partial \Delta P}{\partial q}\right)_0\right] > 0$$

The most variable term in the "A," "B," and "C" equation is

$$\left(\frac{\partial \Delta P}{\partial q}\right)_0$$

This term is most important in determining the stability of the well. (See Figure 2.)

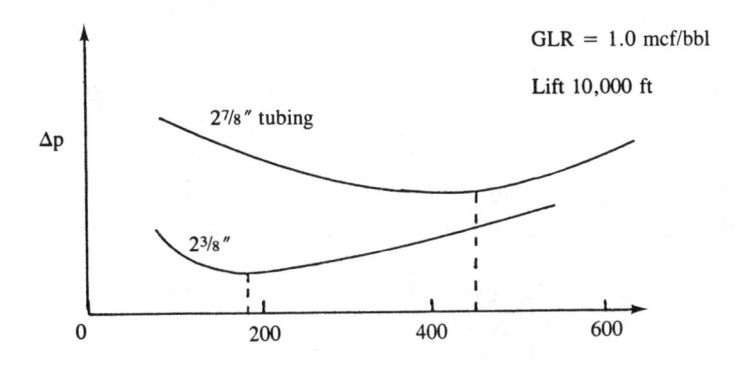

How to Predict Stability of Naturally Flowing Oil Wells 233

GLR = 1.0 mcf/bbl
Lift 10,000 ft

From the figure:

Tubing Size	$\left(\dfrac{\partial \Delta P}{\partial q}\right)_0$ Negative for q_0
2⅞"	< 450 bbl/day
2⅜"	< 190 bbl/day

Figure 2. Typical variation of $(\partial \Delta P/\partial q)_0$.

Hence, a well with 2⅞-in. tubing tends to go unstable for $q_0 < 450$ bbl/day. If 2⅜-in. tubing is used, the well tends to go unstable for $q_0 < 190$ bbl/day. These theoretical observations on stability concerning the effect of tubing size and flow have been confirmed many times in field operations, where the substitution of smaller tubing stabilized a heading well.

Notation

- A   annulus area
- A_t area of tubing
- C_a $\left[\dfrac{\rho g}{A} + \dfrac{Pa}{Va}\right]^{-1}$ = annulus capacitance
- C_t $V_t\left[\dfrac{1}{\beta c} + \dfrac{1}{\beta \ell} + \dfrac{V_g}{V_t}\left(\dfrac{1}{\beta g}\right)\right]$ = tubing capacitance
- D   tubing diameter
- d   choke diameter
- GLR   gas/liquid ratio

234 Shallow Oil and Gas Resources

g   acceleration of gravity
J   productivity index $q/(\bar{p} - p_{wf})$
L   well depth
M $\rho L/A_t$ = tubing inertance
P_a annulus gas pressure
P_{tf} tubing head pressure
p_{wf} flowing bottom hole pressure
$\bar{p}$ reservoir pressure
q   flow rate, bbl/day
V_a annulus gas volume
V_g volume of gas in tubing
V_t tubing volume
$\beta_c, \beta_\ell, \beta_g$ casing, liquid and gas modulus of elasticity
Δp pressure drop in tubing
ρ density of flowing fluid (oil, water, gas mixture)

References

Nind, T. E. W., 1981, Principles of oil well production: McGraw-Hill Book Co., New York, pp. 159-165.

21
Enhanced Oil Recovery and Stimulation by High-Energy Gas Fracturing

Dale A. Eastman

Abstract

This paper deals with the development, application, and results of using Radialfrac (Reg. U.S. Pat. & Tm. Off.), a downhole, self-contained, high-energy gas generating device. It has been conclusively demonstrated that high-energy gas fracturing initiates and propagates fractures in all directions from the wellbore, dramatically increasing permeability (Ford and Mohaupt, 1981). Accurate fracture lengths have not yet been scientifically documented. Communication between wells at distances of up to 450 feet has been observed.

Introduction

Oil and gas wells have been stimulated with high-energy explosives for a number of years. Problems of wellbore damage, safety, and unpredictable results have greatly limited the number of wells successfully stimulated by high-grade explosives. However, new, tailored, pulse-loading techniques which involve the use of controlled propellant burns are rapidly receiving acceptance in the industry (Schmidt et al., 1980, Moore et al., 1977, and Fitzgerald and Anderson, 1978).

Both explosives and propellants may decompose chemically ("burn") and, in doing so, provide energy in the form of pressurized gases and heat.

There are three important differences between explosives and propellants:

1. Generally, the burn rate of explosives is 1,000 or more times greater than that of propellants.
2. The burn rate of propellants can be controlled, whereas explosives "detonate" at a rate depending only on the nature of the explosive itself. Both eventually deliver about the same energy per pound of material.
3. The energy from propellants (mainly hot gases) can readily be transformed into mechanical work, e.g., to drive a rocket or to push a projectile through a barrel, while the energy from an explosive (mainly a short, sharp pressure "shock") will be destructive to all known materials, as seen in its applications in warheads or for demolition.

Propellants are used to launch space vehicles gently enough for man to be on board. The same amount of energy release, in the form of explosives, would break windows as far as 1,000 ft away and dig a crater at the launch site about 100 to 500 ft across. The space vehicle would be destroyed.

The bazooka was an early application of propellants, for launching an explosive warhead at a tank. During launch by the propellants, the bazooka was safely hand-held by a soldier; but on impact the warhead could destroy a tank. The energy in the propellant needed to launch the warhead was roughly equal to the explosive energy of the warhead.

Radialfrac is such a controlled propellant combustion process. Both pressure build-up rate and peak pressure are reached at the selected zone in a small fraction of a second. The energy transfer rate is considerably less than that observed in most explosive stimulation devices, yet it is enough to permit operation of the Radialfrac tool without packers or plugs for most well-stimulation activities. The slower pressure pulse produced by the Radialfrac tool eliminates the possibility of casing damage (Warpenski et al., 1979, Schmidt, 1979, Schmidt et al., 1979). In contrast to hydraulic fracturing (Howard and Fast, 1978), which is conducted at pressures that are only slightly higher than the minimum in-situ fracture stress, on time scales involving hundreds of seconds and resulting in a single fracture whose orientation is perpendicular to the minimum principal stress, the Radialfrac tool, by producing higher wellbore pressures, can initiate and drive fractures in less obvious directions with respect to the in-situ stresses in order to produce multiple fractures (Warpenski et al., 1979, and Northrup et al., 1978). While these multiple fractures may not extend as far as a hydraulic fracture, they may link the well to natural fractures which run parallel to the preferred in-situ stress pattern of the natural fractures. Other advantages of the Radialfrac technique are the low cost associated with well treatment, the unique, insensitive, and safe propellant ingredients, and the ability to select a tool suitable to a particular class of formation from a family of propellant-actuated tools which can easily and safely produce the pressure-rise rates and the peak pressures required to produce the multiple fractures. Figure 1 contrasts fracturing effects of three well-stimulation methods. The hydraulic region is affected by artificial fracturing, the zone of multiple fractures by the Radialfrac tool, and the crushed zone by explosives. Figure 2 depicts the region around the wellbore after the three types of well stimulation.

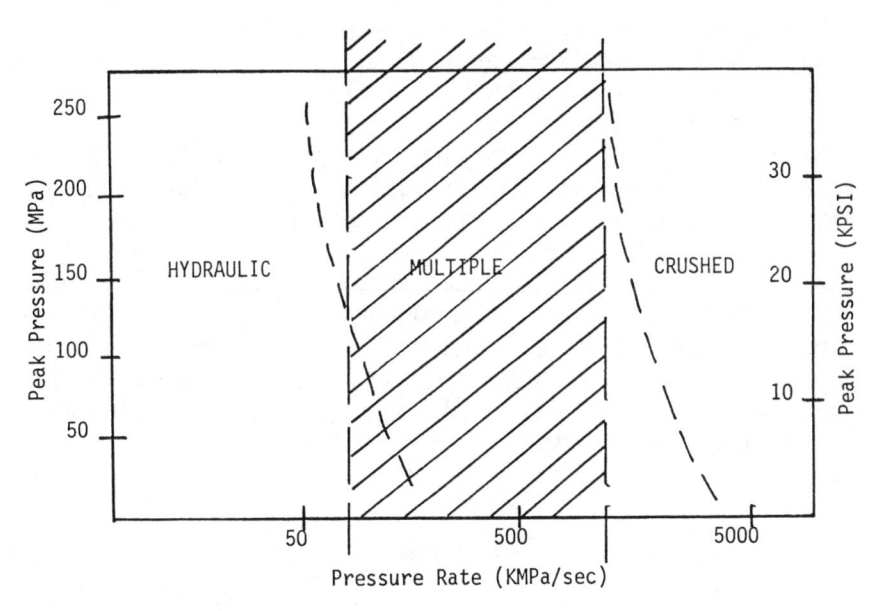

Figure 1. Regions affected by artificial hydraulic fracturing, Radialfrac tool (cross-hatched), and explosives (crushed).

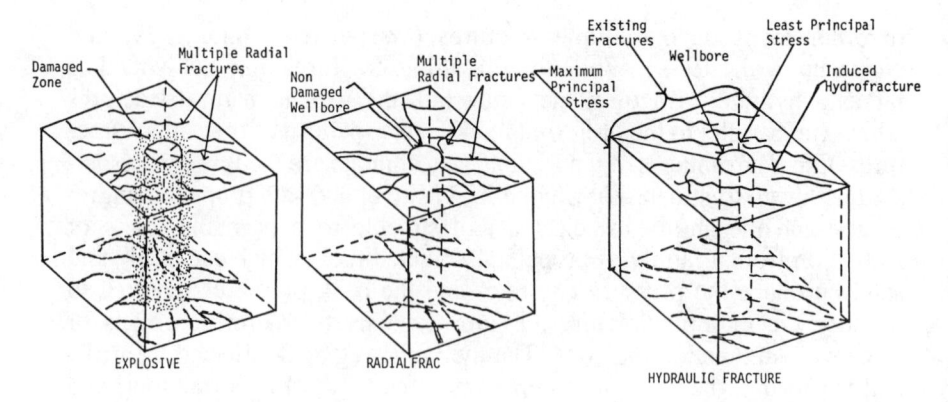

Figure 2. Effects of three methods of well stimulation: explosive, Radialfrac, and hydraulic fracturing.

Development History

As characteristics and possibilities of propellants were recognized, the application of this technology to well stimulation became obvious.

It then became a question of devising a safe, effective, economical procedure to spot the propellants adjacent to the desired formation and ignite them so that a controlled deflagration would achieve the desired predetermined dynamic pressure loading.

Numerous proven oil field methods were available to isolate specific formations at desired depths for standard static-loading procedures. All required the man-hours and equipment necessary to set packers or retrievable plugs. Radialfrac produces effective fracturing while eliminating the manpower, equipment, and time requirements.

The current state-of-the-art tool consists of a propellant design prepared in a controlled laboratory setting, loaded into a pre-fabricated casing with a custom-designed ignition system (Figure 3). The tool is then shipped to location fully operational, requiring only connection to a wireline and final arming of the ignition system.

Operations

Cased-hole application requires four perforations per foot. Pressure containment is accomplished by the use of a hydrostatic head using fluids

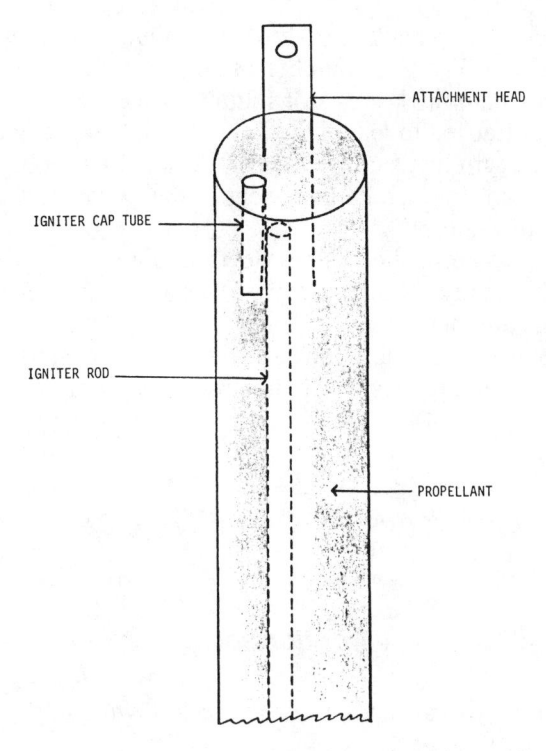

Figure 3. Propellant-type well-stimulation tool.

known to be compatible with the particular formation to be treated. This "fluid tamp" provides the following functions, when used in conjunction with the process:

1. An inexpensive pressure-control agent, which through mass and friction, forces the gases at sonic velocities into the formation at desired intervals.
2. A compatible medium for "spotting" various solutions for formation enhancement, such as surfactants, demulsifing agents, acid, etc.
3. A temperature-control agent. Open-hole completion eliminates the need for shaped-charge perforating or notching procedures.

Radialfrac tools are manufactured to a maximum length of 12 ft and a minimum length of 3 ft. Any number of runs can be made in the same well to accommodate thick zones or multiple-zone wells.

The tools are attached to lead line of any desired length, which in turn is attached to the wireline's collar locater. A radial-firing blasting cap is inserted into a 6-in. molded cap adjacent to the igniter rod, which runs the full length of the tool.

The tool is lowered to the desired depth at approximately 100 ft per min. Casing collars are located on the way down and correlated with existing logs to ensure accurate placement.

Once the tool is in place, it is activated by an electric signal to the blasting cap which in turn ignites the ignitor rod resulting in the full propellant deflagration.

Application

The Radialfrac process lends itself to the following applications:

1. A secondary recovery method for previously produced wells or wells with bore damage, such as emulsification blocks, plugged perforations, or healed fractures.
2. A primary stimulation technique requiring no additional treatments. This application is limited to wells to depths of approximately 6,000 ft. The limitation is due to overburden pressures substantial enough to collapse the scoured fractures created by Radialfrac; it can be overcome by the use of additional proppants.
3. A pre-frac treatment to initiate formation breakdown and a radial fracture pattern which will greatly enhance any static-loading process utilizing acids and/or proppants.

Radialfrac tools can be designed to function with virtually no pressure, temperature, or depth limitations. To date, wells have been successfully fractured at depths ranging from 100 ft to 12,000 ft. Because the tool generates inert gases in a formation-compatible fluid, it has universal application in most formations. Only formations which are known to be extremely soft, such as ash, unconsolidated sands or highly-frangible shales with poor cementing between grains, are not recommended for Radialfrac.

Performance

Radialfrac tools have successfully stimulated oil and gas wells under diverse conditions. Formation conditions have included open and cased holes in sandstones, limestones, dolomites, and shales, with wide ranges in porosity and permeability.

Radialfrac creates open fractures substantially wider than those created by static-loading processes. The wider fractures are a result of the scouring action which takes place when the gases are injected into the formation at sonic speeds. It is the turbulent action of these gases eroding the walls of each fracture and irregularly depositing the eroded grains that provides a natural proppant. The resulting dramatic increase in permeability is particularly effective in shallow, low-pressure formations, which rely heavily on gravity flow for production.

Application of the Radialfrac process to production problems resulting from paraffin or low-viscosity crude are addressed through the chemical "spotting" capability in the fluid tamp. The combination of chemical injection and increased wellbore permeability is an effective solution in single-well application.

While no multiple-well tertiary injection program has been attempted to stimulate heavy crude production, the fracture communication capabilities of Radialfrac should greatly enhance a chemical injection system.

Conclusions

High Energy Gas Fracturing (HEGF) in general and Radialfrac in particular are receiving attention not only in field application but in continued laboratory research. Currently the Gas Research Institute has undertaken the task of designing a downhole monitor and associated surface equipment to measure and record the pressure transient generated during the HEGF stimulation event.

The information gathered from these readings under field conditions and used in conjunction with known mathematical well models and reservoir engineering techniques will further the understanding of the dynamic-loading process in fracture mechanics. These studies will also investigate the effect of the viscosity and mass of the fluid tamp on various formation-treatment agents.

The process is not considered a replacement for hydraulic fracturing, acidizing, or other stimulation techniques; rather it is a new technology to be used with existing technology to improve production.

References

Fitzgerald, R., and Anderson, R., 1978, Kinefrac, a new approach to well stimulation: ASME Paper 78-Pet-25, ASME Energy Technology Conference and Exhibition, Houston, Texas.

Ford, F. C., and Mohaupt, H., 1981, Stressfrac stimulation using tailored gas pulse techniques: presented at Explosives Conference of the International Association of Drilling Contractors, June.

Howard, G. C., and Fast, C. R., 1970, Hydraulic fracturing: Society of Petroleum Engineers of AIME Monograph Series, Dallas, Texas.

Moore, E. T., Mumma, D. M., and Seifer, K. D., 1977, Dynafrac application of a novel rock fracturing method to oil and gas recovery: Physics International Final Report 827, April.

Northrup, D. A., et al., 1978, Stimulation and mine back experiment project—a direct observation of hydraulic and explosive fracturing tests: Proceedings of the Fourth Annual DOE Symposium on Enhanced Oil and Gas Recovery, Tulsa, Okla., August.

Schmidt, R. A., Boade, R. R., and Bass, R. C., 1979, A new perspective on well shooting: the behavior of contained explosives and deflagrations: Sandia Laboratories, presented at the Fifth Annual Fall Technical Conference of the Society of Petroleum Engineers, Las Vegas, Nevada, Sept.

Schmidt, R. A., Warpenski, N., and Cooper, P., 1980, In-situ evaluation of several tailored pulse well shooting concepts: Sandia Laboratories, May.

Schmidt, R. A., Warpenski, N., and Cooper, P., 1979, In-situ testing of well shooting concepts: Geotechnical Research Division: Sandia Laboratories, New Mexico, Proceedings of the Third Gas Shale Symposium, Morgantown, West Virginia.

Warpenski, N. L., et al., 1979, High energy gas frac, multiple fracturing in a well bore, Sandia Laboratories, presented at the Twentieth U.S. Symposium on Rock Mechanics, Austin, Texas, June.

22
Progressive-Cavity Pump Applications

J. C. Collum

Introduced in 1936, the progressive cavity pump is of simple design and rugged construction, and its low operating speeds enable the pump to maintain long periods of downhole operation. The pump has only one moving part downhole, with no valves to stick, clog, or wear out. The pump will not gas lock. It can easily handle sandy and abrasive formation fluids and is not plugged by paraffin, gypsum, and scale.

Because there is no recirculation inside the pump, a progressive cavity pump will not emulsify formation fluids. One standard pump handles a wide range of well conditions. The flow can be easily adjusted as fluid volume changes. By simply changing the RPM of the pump, the operator can easily adjust the fluid output. The pumping rate can be changed without pulling the pump.

The basic pumping unit consists of one moving element, a single helical steel rotor, and a stationary element, the double-threaded helical elastomer stator. When the rotor is placed inside the stator, a series of sealed cavities is formed. As the rotor turns, the cavities progress from the bottom of the stator to the top. The effect is similar to a rod pump that is continually on the upstroke. Because the design of a progressive cavity pump is so simple, there are only two basic variables to consider in matching the pump to well conditions. The first is capacity, which is determined by the size of the cavities formed between the rotor and stator. Larger cavities produce higher flow rates at a given well depth and rotation speed. The second is depth capability, which is determined by the number of seal lines. This affects the length of the rotor and stator. A

244 Shallow Oil and Gas Resources

progressive cavity pump will pump from greater depths at higher pressures when equipped with a longer rotor and stator at a given capacity rating.

Current models available have rated depth capabilities of 3,000 ft with flow as great as 500 barrels of fluid per day (BFPD).

In July 1983, STAATSOLIE, The National Oil Company of Suriname, installed two progressive cavity pumps in 15° gravity API crude oil at a depth of 1,600 ft. Well No. 1 increased from 7 to 19 BFPD and Well No. 2 from 26 to 62 BFPD. The company has since installed ten additional progressive cavity pumps.

Driveheads of two sizes support the surface equipment and produce the needed rotation downhole.

1. DH-10 for production of 25 BFPD at 1,500 ft.
2. DH-20 rated at 3,000 ft and 500 BFPD.

There are six different models of rotor and stator.

1. 5 stage 2:32 BFPD at 1,000 ft
2. 5 stage 3:100 BFPD at 1,000 ft
3. 9 stage 2:32 BFPD at 2,000 ft
4. 9 stage 3:100 BFPD at 2,000 ft
5. 9 stage 5:500 BFPD at 2,000 ft
6. 14 stage 3:100 BFPD at 3,000 ft

Either drivehead is interchangeable with any rotor or stator to achieve whatever production capability is desired. Field personnel like the pump because of its simplicity. There are only two points of lubrication and very little upkeep is required.

23
The Economic Advantages of Microdrilling

Josef Macsik, Tomas Dahl, and Lars Oldsberg

This paper describes the development of a new cost-effective drilling system from a previously uneconomic conventional system.

In 1977, on the island of Gotland, Sweden, a conventional 100-ton capacity rig had been operating for some years, but the payback of producing oil wells was not encouraging enough to justify the continuation of the program. In short:

- Costs were too high.
- Rig was too large.
- Impact on nature was too great.
- Production was too small.

The solution to this was obviously to reduce the rig size and increase the number of exploration wells drilled. Different ongoing slimhole operations were studied, but none was found that could drill a 1–2-in. hole to several hundred meters using safe and accepted oil field practice. Consequently a new rig had to be designed in cooperation with the operator of the area, Oljeprospektering AB.

Result: A 2-in. hole could be drilled to approximately 500 m, giving enough information to identify a discovery.

The next step was to develop the technology to turn the slim holes into producing wells. This was achieved, and the first well put into production in 1978 is still producing.

Shallow Oil and Gas Resources

Gradually a completely new system has been developed for drilling shallow wells. The recommended maximum depth at the moment is around 1,500 m, with a final hole diameter of 66 or 86 mm.

The system covers:

- Exploration drilling for oil and gas.
- Production (development) drilling.
- Drilling of stratigraphic wells.
- Drilling of stimulation and injection wells.
- Electric logging service for the preceding.
- Production equipment installation.

Better appreciation of the difference between conventional drilling equipment and this ultraslim hole technique is obtained by looking at some data and drawings.

Approximate comparisons:

	Microdrill	*Conventional*
Drilling depth	5,000 ft	5,000 ft
Hole diameter, TD	2½ in.	6 in.
Rig weight	14 tons	40 tons
Horse power	100 hp	380 hp
Drill string weight	6.5 tons	35 tons
Mud pump	60 hp	300 hp
Mud tank	35 bbls	400 bbls
Circulated mud volume	60 bbls	500 bbls
Mud cost	20%	100%
Casing weight	3 lbs/ft	16 lbs/ft
Drill site	25%	100%

The dimensions of drill strings and casing tubulars are shown in Figure 1, bit sizes in Figure 2, and general rig dimensions in Figure 3.

Time

Due to the lightweight equipment the time to prepare the wellsite, mobilize, rig up, and drill a well is considerably shorter than for a conventional rig.

Casing
Max sizes and depths

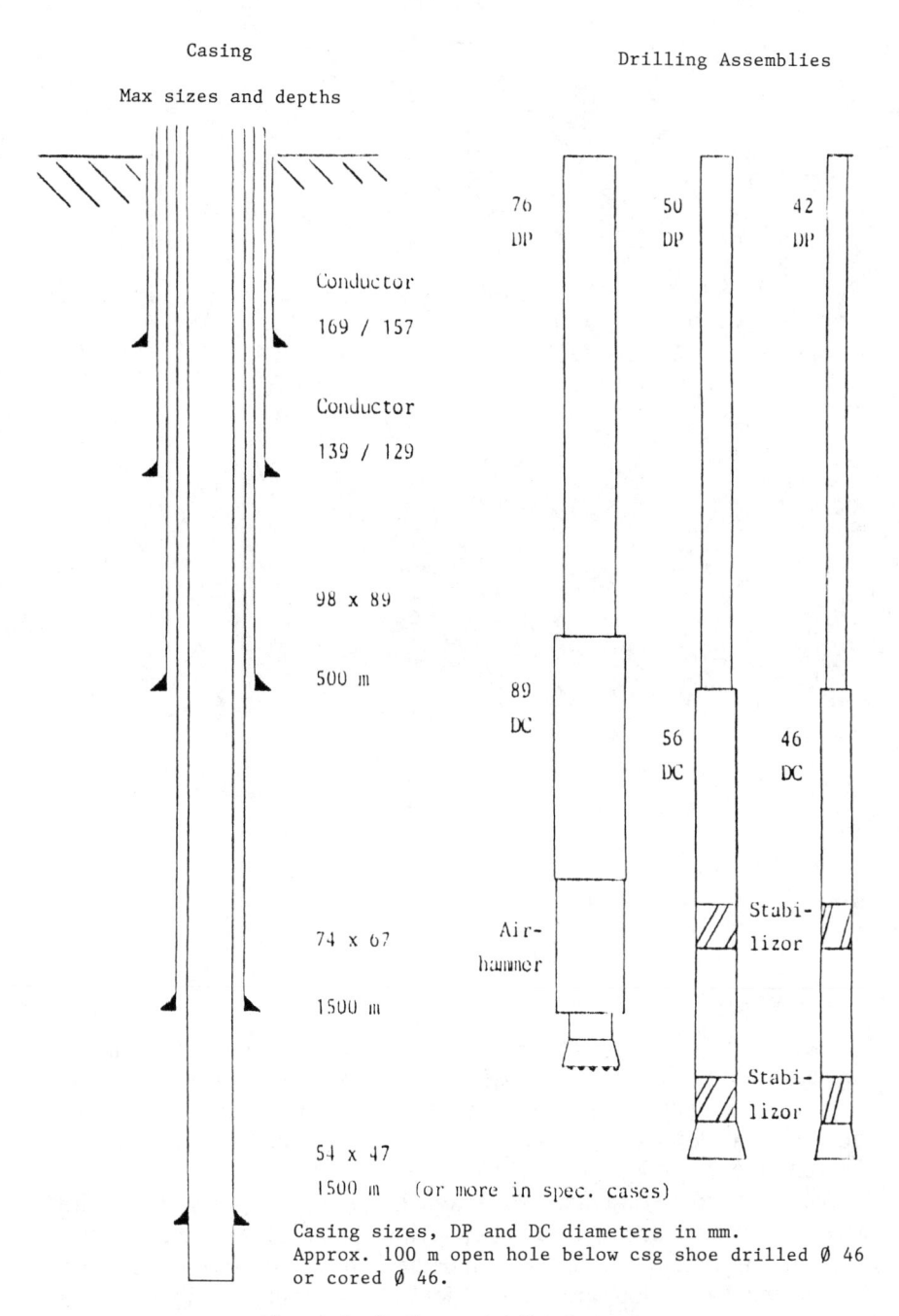

Drilling Assemblies

Conductor
169 / 157

Conductor
139 / 129

98 x 89

500 m

74 x 67

1500 m

54 x 47
1500 m (or more in spec. cases)

76 DP 50 DP 42 DP

89 DC

56 DC 46 DC

Air-hammer

Stabilizor

Stabilizor

Casing sizes, DP and DC diameters in mm.
Approx. 100 m open hole below csg shoe drilled ∅ 46 or cored ∅ 46.

Figure 1. Casing and drill strings.

248 *Shallow Oil and Gas Resources*

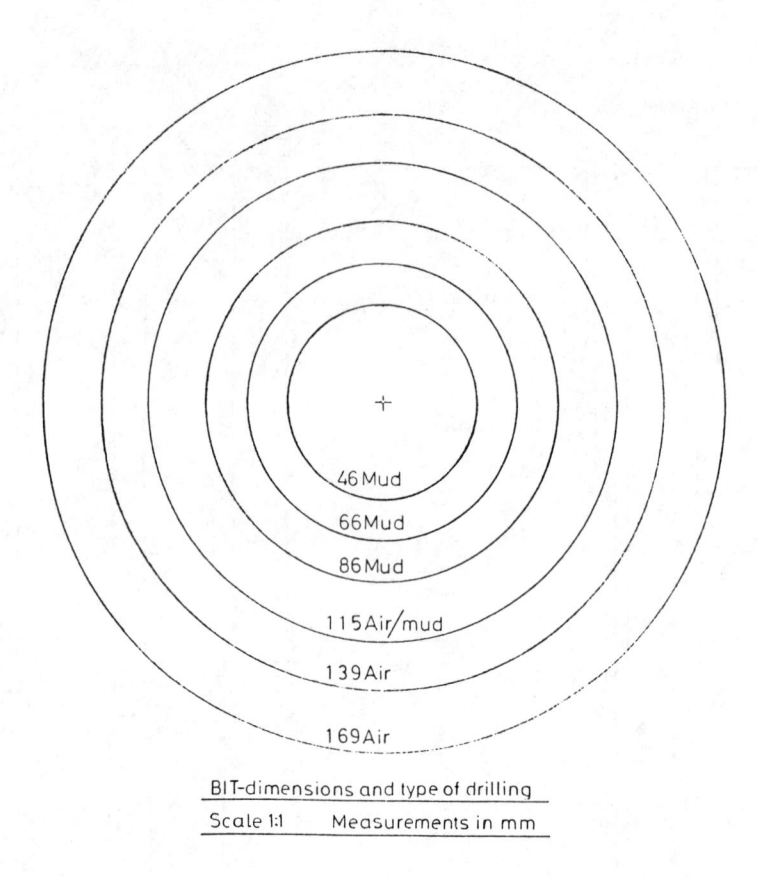

46 Mud
66 Mud
86 Mud
115 Air/mud
139 Air
169 Air

BIT-dimensions and type of drilling
Scale 1:1   Measurements in mm

The wellsite covers an area of only 25 × 32 m, with a reserve mud pit capacity of only about 100 bbls. Only four truckloads are needed to move the complete rig and support equipment (camp not included). Despite the small crew (1 toolpusher, 1 mechanic, 2 drillers, and 2 assistant drillers for continuous operation) rigging up takes only a few hours.

The rig is all hydraulic, semiautomatic, e.g., no power tongs are needed, and all tubulars are light enough to be handled manually. Drill rate and trip time are normally comparable to a conventional rig. Mud mixing and cement jobs are made easier because of the small volumes; a casing job from 500 m to surface requires some 20 sacks of cement to fill the annulus completely. Cement bond logs have shown good bonding for

MD-5 (SC BRITTA'S) DRILLING RIG

Road dimension, mm

A	Length	10.005
B	Width	2.500
C	Heigth	3.050

Rigging up dimension, mm

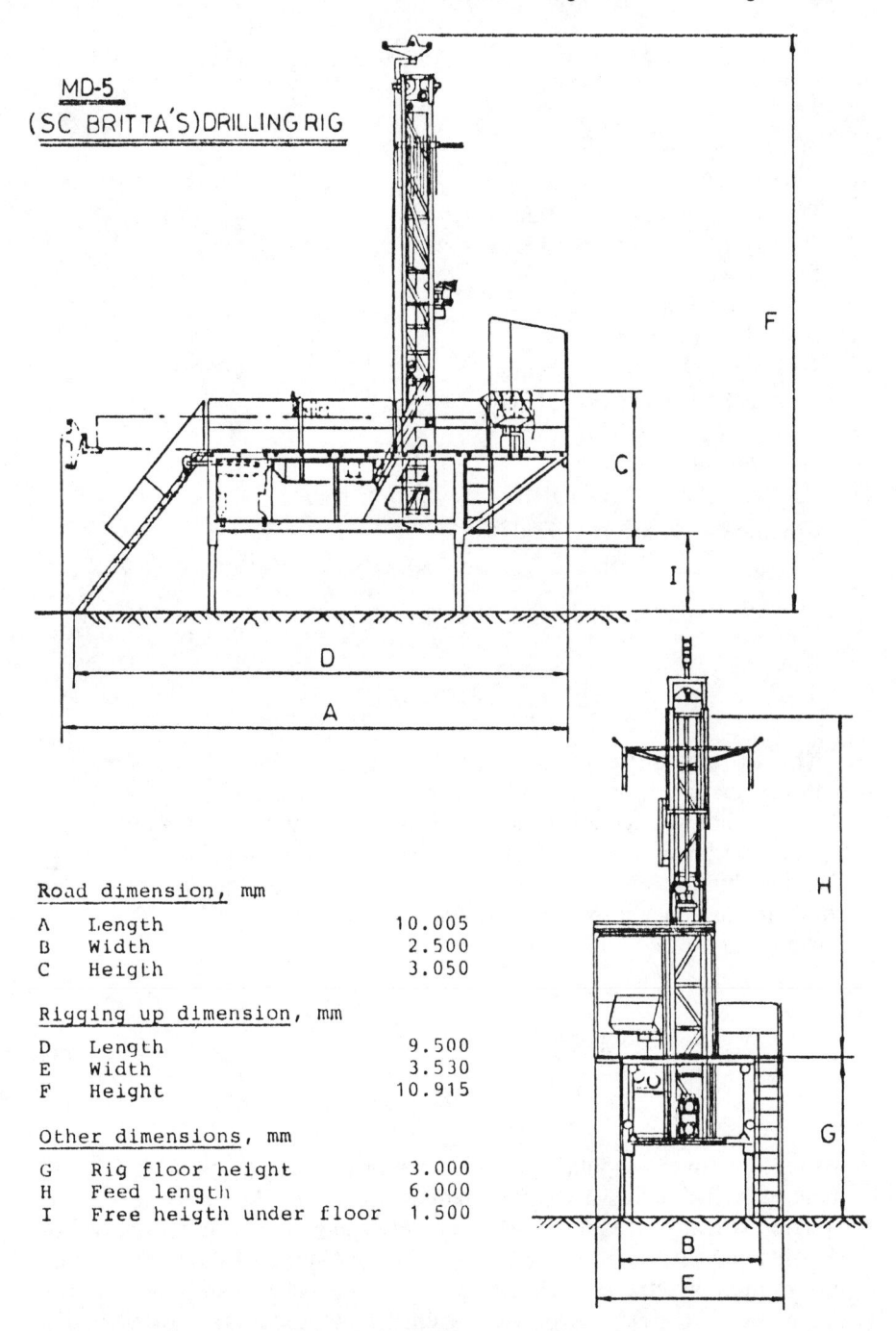

D	Length	9.500
E	Width	3.530
F	Height	10.915

Other dimensions, mm

G	Rig floor height	3.000
H	Feed length	6.000
I	Free heigth under floor	1.500

Figure 3. Rig dimensions.

these small cement volumes. Here follow some examples of duration of operation.

Shallow wells—Baltic area
(incl rig move)
Dry well: 350 m—3 days
Producer: 230 m—5 days

Paris basin
(incl rig move)
Dry well: 630 m—7 days
Producer: 630 m—11 days

Gas storage—Sweden
(incl rig move, 4 casing strings, sandfix, tests)
Completed: 1,028 m—32 days

Services

This drilling method provides a complete system with all necessary services included, a great advantage when operating in remote areas where outside service is difficult to obtain. Cementing, mud control, coring, testing, swabbing, controlled deviation drilling, supply of consumables like bits, casing, wellheads, packers, downhole pumps, or sucker rods, are part of the system and the operator can decide upon quality, quantity, and price within one contract.

Safety

Pressure control is vital when drilling for oil and gas. Standard equipment such as chokes, valves, or remotely controlled BOP's are part of the system. The most frequently used type of casing is external upset, teflon ring-sealed, equivalent to N80 quality. The volumes of the mud system, type of mud, and the mud weight can be controlled as in any conventional operation. The electrical power system (lights, sensors) is of normally accepted, explosion-proof design.

Logging

In order to meet all demands in exploration and production drilling, electric wireline logging service is vital. A complete set of slimhole logging tools fitting these hole sizes has therefore been put together and is available on request.

The logging units are all computerized with the latest in microprocessor technology and logging results are presented in standard format immediately on site. The following tools and services are available:

- Sonic.
- Compensated density.
- Neutron/thermal neutron.
- Computerized neutron.
- Natural gamma ray.
- E-log resistivity.
- Deep-focused resistivity.
- Induction resistivity.
- Caliper.
- Temperature.
- Deviation.
- Perforation.
- Casing cutting.
- Bridge-plug setting.
- Drillstring back off.
- Fishing service.

For control of casing work, a casing collar locator and a cement bond log are available.

Two types of logging units are available, one truck-mounted unit and one skid-mounted unit, in the form of an airplane cargo container.

Environment

The disturbance (noise, earthmoving, pollution, traffic) on nature, farmland, and nearby settlements has always been a problem when oil rigs are operating. In the future, we can probably expect even stricter regulations for these. With this small, lightweight equipment, the impact can be kept at a minimum.

Production

Standard API production equipment can be installed in the slim holes. Completion can be either casing or tubing production. The table below gives some examples of recorded production volumes:

Area	Depth (m)	Casing ID (mm)	Stroke length (m)	Production (bbls/day)
Sweden, Baltic area	230	47	1.0	45
France, Paris basin	1,000	47	3.2	125
Italy, Strangol-lagalli	300	47	1.0	25

Economy

The lightweight equipment, small crew, and low consumption of consumables will of course reflect on the costs of an operation. There are examples of cost savings of 75% compared to conventional drilling.

References

Dahl, T., and Bayless, J. C., 1979, Low cost exploratory drilling system successfully used in Sweden: International Petroleum Times, March 1.
Dahl, T., 1982, Swedish group's small-hole shallow-drilling technique cuts cost: Oil & Gas Journal, April 19.

24
Subsidence and Oil Production from Shallow Reservoirs

Erle C. Donaldson and Saeed N. Mogharabi

Abstract

The production of oil from shallow oil reservoirs frequently leads to surface subsidence, which is the end result of reservoir compaction due to fluid withdrawal. Three notable regions of subsidence due to oil production from shallow reservoirs are Long Beach and the San Joaquin Valley in California, and the Bolivar coastal region of Venezuela. This phenomenon is becoming increasingly important as more land surface is required for human endeavors. Research has recently been intensified in this field to develop a better understanding of the global scope of subsidence, theoretical and engineering evaluation of the problem, and initiation of research to develop means for prediction and abatement of subsidence due to oil production.

The types of compaction (elastic, plastic, etc.) and the circumstances under which each occurs are discussed in relation to the overburden pressure, pore pressure, and compressibilities of the fluids and the geologic framework. The analyses of criteria for identification of reservoirs susceptible to compaction are presented. Oil deposits in shallow young sediments that have undergone monotonic deposition and are loosely consolidated have the greatest potential for subsidence. The specific analysis of a hydrocarbon reservoir for susceptibility to subsidence requires knowledge of the depositional environment and its petrophysical characteristics, which are presented in detail.

The mechanism of compaction in isothermal reservoirs under production by solution gas drive, or limited waterflood, is contrasted to compaction/subsidence occurring in shallow reservoirs under thermal recovery. This analysis is presented with reference to field case histories. Finally, a simplified approach to initial estimation of the anticipated amount of compaction/subsidence is presented together with a discussion of the thorough mathematical description of the phenomena.

Introduction

Subsidence caused by production of fluids from shallow subsurface reservoirs has become more prominent during this century. It is becoming an environmental problem of increasing magnitude. The problem arises because of increased use of land surface for human endeavors, improvements of technology that have resulted in development of efficient high-rate pumps, and the widespread availability of electric energy.

Extraction of large volumes of water for agriculture and production of oil have resulted in damage to buildings, sea inundation, and relocation of river and canal beds. Subsidence has been experienced in many important cities: the Houston-Galveston area of Texas; Long Beach, Los Angeles, and the San Joaquin Valley, California; Mexico City; Venice, Italy; Tokyo, Japan; the Lake Maracaibo region in Venezuela, and many others (Prokopovich, 1983, Gabrysch and Bonnet, 1957, and Allen and Mayuga, 1969).

Subsidence due to oil production from shallow reservoirs is confined to well-defined criteria that can be readily assessed at any time during the development of an oilfield. In general, thick, shallow reservoirs of wide areal extent, composed of sands that are loosely cemented, are susceptible to compaction/subsidence. The specific analysis of a hydrocarbon reservoir for susceptibility to subsidence requires knowledge of its petrophysical characteristics and depositional environment.

Classification of Reservoir Subsidence/Compaction

Subsidence may be classified into two very broad generic groups: *exogenic,* which is related to processes originating near the earth's surface, such as fluid withdrawal and mining, and *endogenic,* which is due to processes that originate deep within the planet, such as faulting, folding, and volcanism (Prokopovich, 1983).

The overburden pressure (P_o) of a surface reservoir is supported by the grain-to-grain pressure of the sand (P_g) and the reservoir fluid pressure (P_p):

$$P_o = P_g + P_p \tag{1}$$

During sedimentary deposition these three stresses attain equilibrium with different degrees of support delegated to the rock-matrix skeletal structure and the fluids occupying the pores. Rapid sedimentation, accompanied by tectonic movement, leads to confined, undercompacted reservoirs. Decrease of the pore pressure of an undercompacted reservoir by removal of interstitial fluids increases the grain-to-grain pressure, which results in one, or a combination of several types of compaction. Two types of recoverable and nonrecoverable deformation may take place in response to the imbalance of the three forces expressed in Equation 1. Increase of the grain-to-grain stress can result in recoverable elastic and viscoelastic (time dependent) deformation of the grains, or the grains may permanently deform by structural yield (crushing under the increased load) and plastic deformation of shape (Dusseault, 1983 and Lofgren, 1976).

In addition to the grain response to increased stress, the subsurface reservoir as a whole undergoes several types of change adjusting to the imbalance of forces caused by fluid withdrawal. Loosely cemented grains undergo permanent rearrangement into pore spaces, causing loss of porosity and permeability. This type of compaction, resulting from grain mobility, is only partially reversible if the fluid pressure is increased once more by water injection.

Cemented sand reservoirs may undergo significant compaction if the reservoir has great thickness (>50 meters) and large areal extent. Shallow reservoirs will undergo viscous creep, deforming horizontally as well as vertically if the overburden stress is supported principally by the grain-to-grain stress over a long period of time. If the conditions of thickness and areal extent do not exist, even a slightly cemented sand will not undergo significant compaction and the problem may consequently be ignored.

Shale and clay beds interspersed within a productive zone can contribute significantly to overall reservoir compaction. In relatively young, shallow oil reservoirs these shale and clay beds are generally undercompacted and exhibit high compressibility. Compaction takes place over a long period of time because the process involves expulsion of water, which is partially controlled by a diffusion mechanism (Dusseault, 1983 and Geertsma, 1973).

Clay and sand layers compact almost to the same extent. The principal difference is the plastic behavior of the clay bodies because of their extremely low effective permeability to the water which must be expelled for compaction to take place. Over a period of several years, however, the slow contribution of shales and clays to compaction (and subsequent subsidence) can be of major importance (Van der Knaap and Van der Vlis, 1967).

Significant compaction, and subsequent subsidence, may take place in productive zones containing interbedded clay and shale even if the productive formation itself has low compressibility. Thus, in the evaluation of a zone for compaction/subsidence, one must consider the total hydraulically-connected zone (Dusseault, 1983, Chilingarian, 1983, and Prokopovich, 1983).

In any of the classifications of subsidence/compaction, however, the amount of potential subsidence decreases exponentially with depth. Thus shallow sedimentary deposits, in which the entire thickness of the reservoir is preserved without uplift, erosion, and re-burial, have the greatest potential for significant surface subsidence if fluids are withdrawn from the subsurface reservoir.

Compressibility of Sedimentary Formations

Several definitions of compressibility apply to geological formations under various circumstances. These are explained with the aid of the three principal stresses acting on a cubical element (σ_z-vertical stress and σ_x-, σ_y-horizontal stresses). When the three stresses acting on the element are equal in magnitude, the resulting force balance is known as hydrostatic loading because this result is obtained when an elemental cube is submerged in a fluid ($\sigma_z = \sigma_x = \sigma_y$). Triaxial loading is attained when the vertical stress is greater than the two horizontal stresses, which are maintained equal in magnitude, but the sides are allowed to expand ($\sigma_z \neq \sigma_x$, ($\sigma_x = \sigma_y$), x and y mobile). The third condition, known as uniaxial loading, results when the vertical stress is maintained greater than the horizontal stresses, but the sides are not allowed to expand ($\sigma_z \neq \sigma_x$, ($\sigma_x = \sigma_y$), x and y immobile).

Hydrostatic loading exists in the early stages of sedimentary deposition, but this changes to triaxial loading as the depth of burial increases the vertical stress. The changing force distributions cause viscous deformation of the reservoir and increase the pore pressure of the fluids.

Compressibility, measured under triaxial loading conditions, is defined as the change in bulk volume per unit volume, as a function of the change of vertical stress (Chilingarian et al., 1983):

$$C_b = -\frac{1}{V_b}\frac{dV_b}{d\sigma_z} \qquad (2)$$

Bulk compressibility includes horizontal deformation as a function of vertical stress; therefore, it is representative of the changes occurring in the subsurface formation as the components of the stress balance change with deepening burial during sedimentation. It also applies to the changes that occur due to reduction of the pore pressure from withdrawal of fluids. The lateral motions (horizontal creep) of the subsurface reservoir during fluid withdrawal, however, are extremely slow. Therefore, uniaxial compressibility can be applied to the analysis of compaction with the assumption that the time-dependent changes in horizontal dimension are insignificant with respect to the vertical change in dimension of the reservoir (compaction).

Uniaxial compressibility is defined as the change in length per unit length, as a function of the change of vertical stress:

$$C_u = -\frac{1}{z}\frac{dz}{d\sigma_z} \qquad (3)$$

Geertsma (1973) proposed a simple means of making a quick evaluation of the magnitude of compaction based on the uniaxial compressibility. By assuming a fixed value of formation compressibility for the pressure range being considered, the change of vertical dimension of the reservoir (Z) can be estimated:

$$\Delta Z = C_u \, \Delta P_p \int_0^h dZ \qquad (4)$$

Numerous authors have presented laboratory data showing a linear-logarithmic relationship between porosity and permeability. Using these data, or similar data obtained from samples of the formation, one can obtain an estimate of the magnitude of changes in porosity and permeability that are anticipated:

$$\phi_2 = \frac{\phi_1 h_1 - \Delta Z}{h_2} \tag{5}$$

$$K = A\, 10^{B\phi} \tag{6}$$

This simple analysis will indicate whether a more detailed computer simulation of the entire reservoir is necessary if fluids are to be withdrawn from an environmentally sensitive area.

Dusseault (1983) lists several other criteria for evaluating potential subsidence, including the geologic history of the specific formation. Compaction of reservoirs due to fluid withdrawal is a phenomenon associated more generally with shallow reservoirs. Furthermore, Dusseault (1983) showed that if a shallow formation has been subjected previously to high stress by deep burial, its compressibility will be reduced. If the formation is then changed to a shallow reservoir by uplift and erosion, it will retain most of its low compressibility. Therefore, in a previously stressed formation, compaction due to fluid withdrawal may be insignificant. Generally only younger (Cenozoic) sediments less than 2,000 meters in depth are important with respect to significant compaction that may create a surface environmental impact.

If a confined aquifer is developed by tectonic movement during deposition of a younger sediment, a larger part of the load of the overburden will be taken by the fluids in the pores of the formation as the depth of burial increases. Thus the zone will have weak grain-to-grain supporting pressure and high pore pressure. A formation in this condition (like those of the geopressure zone of the U.S. Gulf Coast or the fresh water aquifers of the San Joaquin Valley) is undercompacted in relation to its depth of burial (Prokopovich, 1983). Therefore, withdrawal of fluid will increase the effective stress leading to grain rearrangement and compaction. The compressibility of such an undercompacted zone is very high, probably at the maximum for the specific formation.

Heavy Oil Recovery and Subsidence

Heavy oils, less than 20° gravity API, have such high viscosities that production by conventional methods is not possible unless the oil contains a considerable amount of gas in solution, which effectively reduces the viscosity and provides energy for displacement of the oil to the production wells. Therefore, methods of employing heat to reduce oil viscosity and provide energy for displacement have been developed:

1. Fire-flood, in which air is injected in the reservoir to maintain a burning oil zone in situ.
2. Steam-soak, which is the periodic injection of steam alternating with periods of production.
3. Steam-drive, where oil is displaced continuously by injected steam and condensed water, from the injection well to peripheral production wells.
4. Hot-water drive, which is simply the injection of heated water.

The heavy oils frequently occur in shallow, undercompacted, unconsolidated, sands; therefore, the sands are readily susceptible to compaction due to fluid withdrawal, which reduces the pore pressure and begins the chain of events that leads ultimately to surface subsidence.

Compaction of the petroleum reservoir during production serves to maintain the pore pressure that may be a significant part of the energy available for displacement of the oil to the production wells. Thus, if the oilfield is not in an environmentally sensitive location, the subsidence may be tolerated in the interest of maximum recovery of oil. If the oilfield is, however, in a sensitive area, thorough evaluation of the impact of surface subsidence must be made and (if warranted) a water injection-pressure maintenance program must be considered to maintain the pore pressure in the reservoir and thus minimize compaction/subsidence. For example, after subsidence became a severe environmental problem at Long Beach, California, a water injection program for pore pressure maintenance had to be initiated to stabilize the compacting zone (Colazas and Olsen, 1983).

Schenk and Puig (1983) report an analysis of compaction/subsidence in a thermal project area which is part of the Tia Juana oilfield in Venezuela. They report on a 30-meter thick, oil-saturated sand zone containing oil with an average viscosity of 2,000 centipoises, at a temperature of 45°C and initial pressure of 5.86 MPa. Production in three stages (primary, by solution-gas drive; steam soak; and steam drive) resulted in considerable production which can be attributed directly to reservoir pressure maintenance by compaction (Table 1). The greatest amount of subsidence was 2.4 meters at the approximate center of the compacting zone.

Table 1
Production-Subsidence Relations of the M-6 Project Area, Tia Juana oilfield, Venezuela*

Production Mechanism	Oil Produced (m³)	Percent of Original Oil	Subsidence Volume (m³)	Percent of Recovery
Solution gas drive	10.67×10^6	12.8	7.79×10^6	73.0
Steam soak	8.27×10^6	9.9	4.93×10^6	59.6
Steam drive	4.45×10^6	5.3	0.58×10^6	13.0

Original oil in place = 83.48×10^6 m³
Initial reservoir pressure = 5.86 mPa
Final reservoir pressure = 1.03 mPa
* Surface subsidence, measured as cubic meters in the project area, is considered to be equal to the reservoir compaction (Schenk and Puig, 1983).

Approximately 73 percent of the initial production of heavy oil was attributed to pressure maintenance by compaction of the reservoir during the initial phase of production. Compaction, however, decreased in importance with time, ending at 13 percent of the total production. This decrease was probably due to a decrease of compressibility of the sand as compaction proceeded.

Mathematical Evaluation

To make a thorough evaluation of the production-compaction relationship of a virgin field, or one in which compaction is already active, computer simulation of the process is necessary (Raghaven and Miller, 1975).

Evans (1983) presented a mathematical model that is based on the equations of continuity, effective stress, force equilibrium, and energy, for complete mathematical simulation of a productive reservoir which is susceptible to compaction. This model, excluding the energy equation, was used by Donaldson (1984) to prepare a computer program for workshop discussion, publication of the complete listing, and explanation of the program, which can be easily modified for more critical evaluation of a reservoir if supporting data are available. The computer program was developed for isothermal, single-phase flow, and vertical compaction with iterative decrease of porosity and permeability. The program allows for pressure maintenance of water injection, water influx, and reservoir heterogeneity with respect to thickness, porosity, and permeability.

The pressure distribution of the reservoir is calculated implicitly in response to initial reservoir pressure, fluid flow within the zone (production, injection, water influx), and the current values of porosity and permeability, using Equation 7:

$$\frac{\partial}{\partial x}\left(\frac{Kh\partial P}{\mu \partial x}\right) + \frac{\partial}{\partial y}\left(\frac{Kh}{\mu}\frac{\partial P}{\partial y}\right) + Qij = C_b \phi h \left(\frac{\partial P}{\partial t}\right) \qquad (7)$$

Initial values of h, ϕ, and k are entered at each grid node and thereafter the decrease of reservoir thickness as a result of compaction is calculated from Equation 4, the changes in porosity and permeability being derived from laboratory data expressed as Equation 6. After making these calculations for each grid node of the network over the productive zone, a new pressure distribution is calculated implicitly for the reservoir. The process is repeated, with printing or graphical output of production and changes of the reservoir parameters with respect to time or amount of compaction (Figure 1).

Conclusions

Subsidence caused by production of fluids from subsurface reservoirs is becoming an environmental problem of increasing magnitude.

Thick, shallow reservoirs of wide areal extent composed of sands that are loosely cemented are susceptible to compaction/subsidence.

Clay and shale beds interspersed within a productive zone can contribute significantly to the overall compaction of the zone over a long period of time.

Compaction of unconsolidated sand reservoirs leads to permanent loss of porosity and permeability.

The potential for compaction/subsidence decreases in an approximately exponential manner with respect to depth of burial.

Laboratory-measured bulk compressibility is representative of the changes occurring in the subsurface environment during imbalance of stress forces.

The geologic history of the productive zones is a major factor in their behavior with respect to compaction. Cemented reservoirs, or those that have had a previous history of compaction, exhibit low compressibility.

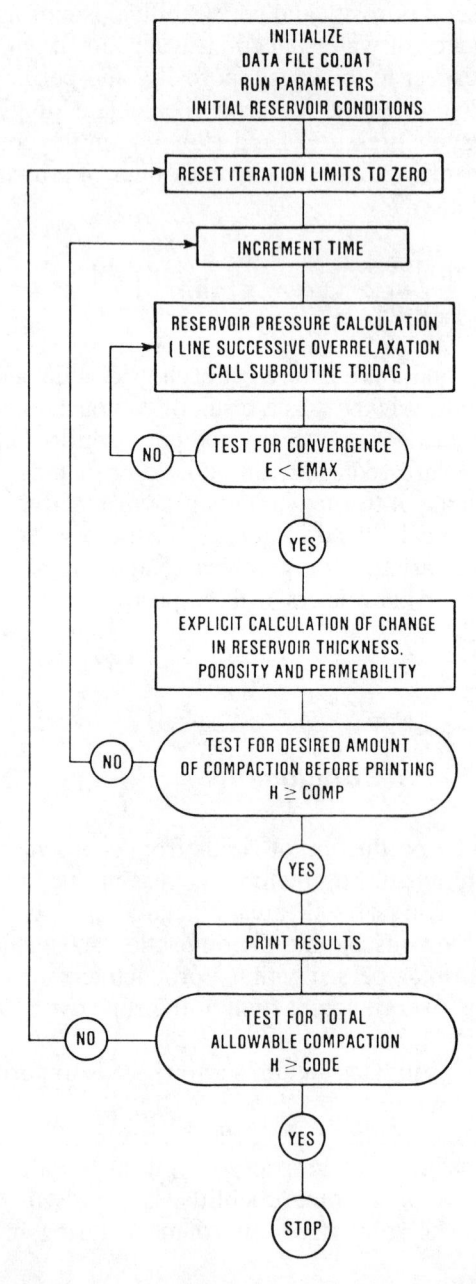

Figure 1. Flow diagram of the calculation procedure of computer program COMPAC.

Compaction of a petroleum reservoir during production may serve as a significant fluid-displacement mechanism by maintaining the pore pressure of the formation.

Detailed computer evaluation of any specific reservoir can be conducted with as much refinement as allowed by the availability of data.

Notation

C_b Bulk compressibility
C_u Uniaxial compressibility
h Reservoir thickness
K Permeability
P_g Grain-to-grain pressure
P_o Overburden pressure
P_p Pore fluid pressure
V_b Bulk volume
X Horizontal dimension
Y Horizontal dimension
Z Vertical dimension
ϕ Porosity
σ Component of stress on an elemental volume of a reservoir

References

Allen, D. R. and Mayuga, M. N., 1969, The mechanics of compaction and rebound: Wilmington Oilfield, Long Beach, California, USA, Land Subsidence, IASH-UNESCO Publication 89, pp. 410–422.

Chilingarian, G. V., et al., 1983, Compressibilities of sands and clays: Proceedings US DOE/Venezuela Forum on Subsidence Due to Fluid Withdrawal, Op. cit., pp. 25–32.

Colazas, X. C. and Olson, L. J., 1983, Subsidence monitoring methods and burch mark elevation response to water injection: Wilmington Oil Field, Long Beach, California. Proceeding US DOE/Venezuela MM Forum on Subsidence Due to Fluid Withdrawal, Op. cit., pp. 121–132.

Donaldson, E. C., 1984, Computer simulation of reservoir compaction due to fluid withdrawal: Proceeding Eng. Foundation Conf. on Compressibility Phenomena in Subsidence, Henniker, New Hampshire, July 29–August 3, 1984.

Dusseault, M. B., 1983, Identifying reservoirs susceptible to subsidence due to fluid withdrawal: Proceeding US DOE/Venezuela Ministry of Mines Forum on Subsidence Due to Fluid Withdrawal. (E. C. Donaldson and H. Van Domselaar, Eds.). NTIS, Springfield, VA 22161, pp. 6–14.

Evans, R. D., 1983, Fluid withdrawal from an anisotropic compressible porous media, US DOE/Venezuela MM Forum, Op. cit., pp. 93–96.

Gabrysch, R. K. and Bonnet, C. W., 1957, Land surface subsidence in the Houston-Galveston region: Texas Water Dev. Board. Report, Austin, TX, pp. 188.

Geertsma, J., 1973, Land subsidence above compacting oil and gas reservoirs: J. Petrol. Tech., v. 25, no. 6, June, pp. 734–744.

Lofgren, B. E., 1976, Land subsidence and aquifer-system compaction in the San Jacinto Valley, California—a progress report: J. Research, U.S. Geol. Survey, v. 4, no. 1, Jan.–Feb., pp. 9–18.

Prokopovich, N. P., 1983, Tectonic framework and detection of aquifers susceptible to subsidence: US DOE/Venezuela MM Forum. Op. cit., pp. 33–44.

Raghaven, R., and Miller, F. G., 1975, Mathematical analysis of sand compaction, *Compaction of Coarse-Grained Sediments,* I: Elsevier Sci. Pub. Co., New York (G. V. Chilingarian and K.H. Wolf, Eds.), pp. 403–524.

Schenk, L., and Puig, F., 1983, Aspects of compaction/subsidence in the Bolivar Coast heavy oil fields: Highlighted by Performance Data of the M-6 Project Area. US DOE/Venezuela MM Forum, Op. cit., pp. 109–120.

Van der Knaap, W., and Van der Vlis, A. C., 1967, On the cause of subsidence in oil-producing area: Seventh World Pet. Cong., Mexico City, Elsevier Publ. Co., v. 3, pp. 85–95.

25
Control of Hazards in Oil Drilling and Well-Servicing Operations

R. J. Murphy

Abstract

Built into the exploration and development of oil/gas resources are explicit hazards in operations associated with sophisticated equipment. Such hazards, when uncontrolled, represent high potential for personal injury and equipment damage. This situation is frequently aggravated by high personnel turnover rates and, field supervisors, though technically skilled, who lack the essential background or training to deal with people-related operational problems. The evidence for this observation is overwhelming when comparing insurance rates for this industry with other comparable high-hazard industries, such as heavy construction. The etiology of equipment damage, operational error, and personal injury have the same common ground, namely, lack of control. The solution to such problems must first begin by defining the cause of the error. This must be done by overcoming the tendency to classify every problem as "carelessness" or negligence on the part of the worker. Rather, a methodology of operations review is demonstrated to show its strength in problem solving. This method is then applied to define the organizational level and function that has failed and contributed to the cause of error.

Introduction

Hazards in petroleum production operations can be characterized as *errors* that arise from interaction of man with equipment or material in performing a mission. Error, in this sense, relates to any interaction that can cause injury to man, damage the system, generate substantial time or material loss, or degrade efficiency in performing the mission. Another consideration related to error is the variation encountered between individuals' perceptions of error. For example, most people would think that there is little need to control leakage of liquid from a pump at the rate of one drop per second. However, the loss associated over a period of time of one year might be a ton of product. The purpose of this paper is to define how such losses occur and a methodology to deal with their causes.

The economic incentive for effecting comprehensive loss control programs in the oil production industry are overwhelming. Consider the following actual accident scenario. A floorhand on a drilling rig near Clinton, Oklahoma, was assisting the crew as pipe was pulled out of the drillhole. The crew was inspecting each joint of drill pipe and was laying down those joints which were damaged. The floorhand was tailing a damaged joint of pipe to the V-door when he lost his balance and fell thirty feet out of the V-door to the matting boards below. The injuries and the direct costs associated with this accident are as follows.

Injuries—Fractured left wrist, fractured left arm, eight fractured ribs, multiple fractures of the pelvis, fractured right ankle and internal injuries.
Hospitalization—Hospitalized for one month.
Disablement—Total temporary disability of seven months. Permanent partial disability as follows:
    40% to right foot.
    45% to body.
    60% to left hand.
    5% to left arm.
Costs—Medical costs as follows:

Doctors:	$ 4,750
Hospital:	16,885
Miscellaneous:	5,245

Costs—Disability payments as follows:

Temporary total disability:	$ 5,250
Permanent partial disability:	39,375
Total Direct Costs	$71,505

Normally, management would view these costs as insured and disregard their significance. However, insurance rates are based on company performance. Consequently, one way or another, the losses come out of company earnings. If we assume that this rig works on a 10% profit margin rate, it will have to make $715,050 in order to pay for the direct costs of this accident. If the day rate for this rig is $7,500 per day, it would translate into needing 95 days of work to recover this loss. If the rig was on a footage rate of $20 per foot, 35,753 feet of drilling is needed to recover the loss.

This cost analysis includes only the direct cost normally insured. There are indirect costs that should also be considered. They include: Lost time of the crew and personnel involved in the accident or in documenting and investigating the accident; cost of hiring and training a replacement for the crew; and, loss of efficiency inherent in changing personnel in the crew. These indirect costs are documented as ranging from 4 to 17 times the direct costs of this accident (Heinrich et al., 1980).

Background

The traditional approach that "things cause accidents" has led to the formulation and promulgation of safety standards in the United States by the Occupational Safety and Health Act (OSHA) of 1970. Thereafter, significant funds were expended by industry to correct condition-related matters to conform with OSHA and promote safety. According to documented studies, approximately 80% of accidents are caused by unsafe acts by people and 20% are caused by unsafe conditions (Heinrich et al., 1980). It would appear that the best that OSHA could achieve in reducing accident rates would be 10% to 20%, and, in fact, there has been a reduction of 13%, based on a comparison of 1970 and 1975 frequency rates. This suggests that the maximum reduction in unsafe conditions has been accomplished.

Probably high accident rates attributed to unsafe acts are explained by the concept of multiple causation, an updated version of the domino sequence of Heinrich et al., (1980). The domino sequence describes the accident event as a chronological series of five factors: Management (lack of control); basic sources of error (personal or operational); immediate cause (unsafe act or condition); contact with an energy source (leads to accident); and loss (damaged property, injury, mission degradation, or material loss). Loss control has traditionally focused on breaking the domino sequence by directing corrective action toward immediate causes. However, unsafe acts or conditions are not the cause of accidents, but are symptoms. The cause more likely lies in failure in the management functions (planning, organizing, directing, and controlling), leading to operational or personal error. The management factor will be discussed later when implementing loss control in organizational functions.

Basic causes of error in the accident sequence have been defined as operational demands or personal factors. An example of the former is the commitment of a worker to a difficult or dangerous task for which he has neither adequate experience nor training. Other examples of operational and personal factors that may contribute to accidents are listed in Table 1.

Operational errors related to non-injury incidents are numerous, such as sending a production rig to the wrong well site. These are unplanned, undesired events. The issue then for an effective loss-control program is for management to recognize and deal with potential losses in the same manner they deal with production errors. Rather, the same tools that lead to successful production, communication, responsibility, authority and accountability, and effective training are inherent to implementing loss control.

Organizational Functions in Loss Control

Having defined the characteristics of loss control, organizational functions must be exercised to implement the necessary elements. The traditional approach in dealing with loss-related programs in an organization

Table 1
Basic Causes of Loss

Job Related	Personal Related
Failure in management functions of planning, directing, and controlling:	*Worker related:*
Skill/experience/training incompatible with job	Aggressive over-commitment of capabilities or resources
Time constraints on job	Marital problems
Resource/equipment limitations	Financial problems
No performance standards	Motivation
No monitoring	Task assignments
Poor example	Physical condition
Failure to create incentive	Reaction time
Poor system design/layout	Training and safety awareness
Lack of communications	System understanding
Lack of production/safety goals	Alcohol/drug influence (on the job) or aftereffects
Poor accountability	

is to appoint a staff safety professional and assign him as an appendage to some organizational element. Frequently, the safety professional is assigned to the personnel, engineering, or operations staff. An alternative is the safety committee. No other daily operational matters are exercised by management through a committee. Consequently, the value of safety committees is limited to recognition of safety interests.

It is more rational to implement loss-control programs through the normal organizational functions of management, supervisors, and workers. These functions are traditional and are shown in Table 2. Foremost is a strong commitment of upper management to the program. It begins with the chief executive officer's commitment to the policy with his personal signature. Policy statements vary from organization to organization and the essential components are that they must *guide* the entire company's thinking. It is not the detailed procedures of *how* it will be done. It is left to the operating elements to develop specific rules or procedures.

If management fails to establish linkage in effectively monitoring and enforcing policy and procedures, the program is for naught. The supervisor will only respond to those matters in which management has demonstrated interest, to the point that a system is developed to measure the supervisor's performance and the performance is consistently evaluated. The rule of recognizing and rewarding improved performance is as important to loss control as to any aspect of productivity.

Loss control within the organization ultimately rests on the workers. There must be complete understanding among crews as to exactly what is expected in terms of compliance with accident-prevention practices. Frequently, supervisors violate procedures themselves. Management must also be evaluated in its capacity to assign tasks and missions that knowingly will require bypassing accepted work practices. Such problems can be anticipated in "hurry-up jobs."

It is especially important in crew operations (where the conduct of one member may jeopardize the rest of the crew) that the workers recognize and correct unsafe practices of fellow crew members. No supervisor can be by the side of all crew members simultaneously. Consequently, there is essential need for all crew members to be interdependent in enforcing safety rules.

Evaluating Losses

The accident, or error, sequence can be characterized as multiple causation, which states that several factors combine in random fashion to cause accidents. To illustrate how multiple causation serves to better de-

Table 2
Organizational Functions

Management	Supervisor	Worker
1. Formulate and communicate policy and procedures (rules).	1. Communicate what is to be done and feedback limitations to management.	1. Follow procedures.
2. Apply rules consistently.	2. Assure resources and capability available to comply.	2. Communicate limitations to performance.
3. Monitor performance.	3. Set example and motivate.	3. Seek training to improve performance.
4. Enforce procedures consistently, recognize and reward performance.	4. Enforce procedures consistently, recognize and reward good performance.	4. Motivate/enforce procedures among coworkers.

fine sources of error, consider the following common accident of a man falling off a step ladder.

- The unsafe act could be use of a defective ladder.
- The unsafe condition could be a defective ladder.
- The correction is riddance of the defective ladder.

Multiple causation investigation generates contributing factors that relate to the incidents. For example:

- Was the worker trained in use of the ladder and evaluating it for defects prior to its use?
- Did he report the deficiency to the supervisor?
- Did the supervisor take action to replace the ladder?
- If he did, was purchasing negligent in ensuring a replacement was issued?
- Why was the ladder not found defective during normal equipment inspections?
- If the supervisor knew the ladder was defective, why did he allow its use?

By addressing these questions, the following corrective actions singularly, or jointly, could be exercised:

- Improving inspection techniques.
- More detailed instruction to workers in their responsibility to evaluate equipment safety and reporting deficiencies.
- Developing follow-up procedures for replacing defective equipment.

An excellent technique for dealing with error is the technique of operations review (TOR). The technique is essentially a tracing system that can also be used as a training technique in safety with crew members. It is applied when an accident or any incident that reflects an error in operation occurs. A cause code (Table 3) is used to focus on problems that could relate to the incident. The TOR steps follow:

1. Describe and clearly state what happened.
2. Select *one number* from the cause code which appears to be the most apparent cause of the incident.
3. Trace and eliminate. Jot down the first number chosen on a paper. Following this number on the cause code form are other numbers immediately across from the chosen number. Jot them down and note their description. Decide whether or not they are contributing

Table 3
TOR Cause Code*

1 Coaching
10 Unusual situation, failure to coach (new man, tool, equipment, process, material, etc.) — 44,24,62
11 No instruction. No instruction available for particular situation — 44,22,24,80
12 Training not formulated or need not foreseen — 24,34,86
13 Correction. Failure to correct or failure to see need to correct — 42,20,30
14 Instruction inadequate. Instruction was attempted but result shows it didn't take — 15,16,42
15 Supervisor failed to tell why — 44,24,83
16 Supervisor failed to listen — 11,81
17
18

2 Responsibility
20 Duties and tasks not clear — 44,34,14,53
21 Conflicting goals — 80
22 Responsibility, not clear or failure to accept — 26,14,54,82
23 Dual responsibility — 47,34,13
24 Pressure of immediate tasks obscures full scope of responsibilities — 36,12,51

5 Disorder (Continued)
53 Property loss. Accidental breakage or damage due to faulty procedure, inspection, supervision, or maintenance — 43,20,80
54 Clutter. Anything unnecessary in the work area. (Excess materials, defective tools and equipment, excess due to faulty work flow, etc.) — 44,36,80
55 Lack. Absence of anything needed. (Proper tools, protective equipment guards, fire equipment, bins, scrap barrels, janitorial service, etc.) — 44,36,80
56 Voluntary compliance. Work group sees no advantage to themselves — 40,15,41
57
58

6 Operational
60 Job procedure. Awkward, unsafe, inefficient, poorly planned — 44,32
61 Work load. Pace too fast, too slow, or erratic — 44,51,63
62 New procedure. New or unusual tasks or hazards not yet understood — 43,44
63 Short handed. High turnover or absenteeism — 80,40,61
64 Unattractive jobs. Job conditions or rewards are not competitive — 81,46

25	Buck passing, responsibility not tied down	80,86	
26	Job descriptions inadequate	80,86	
27			
28			
3	**Authority** (Power To Decide)		
30	Bypassing, conflicting orders, too many bosses	44,13	
31	Decision too far above the problem	36,83,85	
33	Authority inadequate to cope with the situation	81,83	
34	Decision evaded, problem dumped on the boss	36,14,85	
35	Orders failed to produce desired result. Not clear, not understood or not followed	40,46,13,15	
36	Subordinates fail to exercise their power to decide	26,12,83,85	
37			
38			
4	**Supervision**		
40	Morale. Tension, insecurity, lack of faith in the supervisor and the future of the job	15,56,64,80	
41	Conduct. Supervisor sets poor example	13,84	
42	Unsafe Acts. Failure to observe and correct	24,11,52	

65	Job placement. Hasty or improper job selection and placement	80,86	
66	Co-ordination. Departments inadvertently create problems for each other (production, maintenance, purchasing, personnel, sales, etc.)	45,35,13	
67			
68			
7	**Personal traits** (When accident occurs)		
70	Physical condition—strength, agility, poor	44,26,65	
71	Health-sick, tired, taking medicine	44,24,65	
71	Impairment, amputee, vision, hearing, heart, diabetic, epileptic, hernia, etc.	44,24,65	
73	Alcohol—(if definite facts are known)	80	
74	Personality-excitable, lazy, goof-off, unhappy, easily distracted, impulsive, anxious, irritable, complacent, etc.	44,13	
75	Adjustment-aggressive, show off, stubborn, insolent, scorns advice and instruction, defies authority, anti-social, argues, timid, etc.	44,13	
76	Work habits-sloppy. Confusion and disorder in work area. Careless of tools, equipment and procedure	44,13	
77	Work assignment-unsuited for this particular individual	42,65	
78			

table continued

* *Source: Heinrich et al., 1980.*

Table 3 continued

25,36,12,52	43 Rules. Failure to make necessary rules, or to publicize them. Inadequate follow-up and enforcement. Unfair enforcement or weak discipline	79	**8 Management**
		24,81,83	80 Policy. Failure to assert a management will prior to the situation at hand
22,34,30	44 Initiative. Failure to see problems and exert an influence on them	83,86	81 Goals. Not clear, or not projected as an "action image"
10,12,15,81	45 Honest error. Failure to act, or action turned out to be wrong	36	82 Accountability. Failure to measure or appraise results
40,21,56	46 Team spirit. Men are not pulling with the supervisor	12,86	83 Span of attention. Too many irons in the fire. Inadequate delegation. Inadequate development of subordinates
23,25,15,66	47 Cooperation. Poor cooperation. Failure to plan for coordination	20,65	84 Performance appraisals. Inadequate or dwell excessively on short range performance
	48	36	85 Mistakes. Failure to support and encourage subordinates to exercise their power to decide
	49	66	86 Staffing. Assign full or part-time responsibility for related functions
	5 Disorder		87
41,24,31,80	51 Work flow. Inefficient or hazardous layout, scheduling, arrangement, stacking, piling, routing, storing, etc.		88
21,32,14,86	52 Conditions. Inefficient or unsafe due to faulty inspection, supervisor action, or maintenance		

causes to the incident. If not contributing causes, then strike out the number on your list. If they are contributing causes, see what additional numbers are cross listed with them and list them. Continue to trace and eliminate numbers until they run out or the final numbers on the list repeat themselves. Those numbers and descriptors that remain on the list are those you have decided are contributing factors.
4. Now make a list of the contributing factors.
5. Develop alternative solution to factors and select solutions.
6. Remember that some solutions may be beyond your range of authority to implement. These solutions should not be ignored but identified and communicated to your immediate supervisor.

The corrective actions are called the "four by four concept," which states that at each organizational level, four functions must be carried out. Failure to carry out any of the four at each level provides an opportunity for an error or mishap to occur in operations. Each cause code identified in the TOR should be tied to specific organizational functions outlined in Table 2, to define the appropriate level and function within the organization at which to apply corrective action.

Application Example

The application of the operation review technique is best illustrated by example. Specific details follow.

Accident Description: A floorhand, helping out at the mud tanks, was dumping a bag of caustic soda into a mud tank when the caustic splashed back into his face, causing skin burns and chemical burns in both eyes. Personal protective goggles were available for his use, but he did not have them on. The injured man reported that he had watched the derrickman perform the same task and never saw him wear goggles.

Step 1: Select *one* critical primary cause for the accident. Choose 76—careless in procedure.

Step 2: Opposite 76 are the numbers 44 and 13. These should be evaluated to decide if they could be a contributing cause to the accident. *44* is a failure of the supervisor to anticipate this problem ahead of time (probably true in this case). *13* is failure to see need for correction by coaching and training the floorhand properly (again, probably true).

Step 3: Evaluate the numbers now opposite 44 and 13. They are, respectively, 22, 34, and 42, 20, 30. Again we evaluate each, eliminate those that are not applicable in the accident sequence and continue evaluation of other numbers and causes indicated until numbers regenerate in the list or no numbers are generated. A summary of tracing out this accident's contributing causes might look like this.

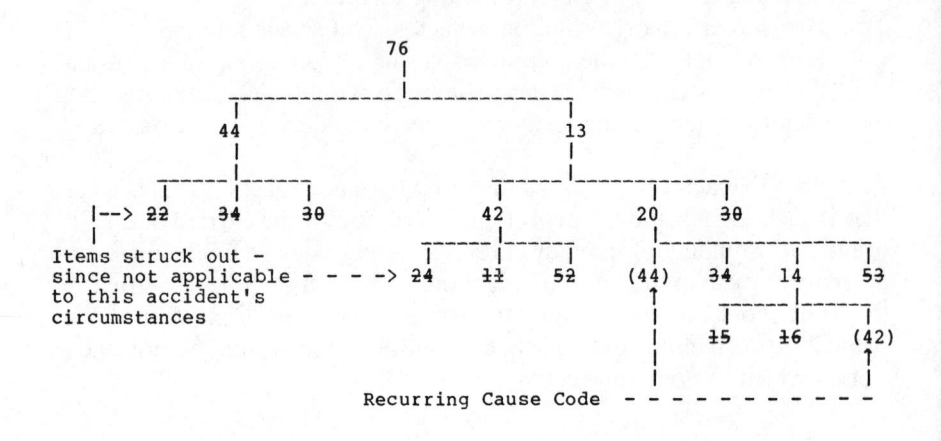

```
                              76
                               |
            _____|_____
            |                                     |
            44                                    13
            |                                     |
       _____|_____                      _____|_____
       |    |    |                      |         |         |
  |--> 22   34   30                     42        20        30
  |                              _____|_____   _____|_____
Items struck out -               |    |    |     |    |    |    |
since not applicable - - - ->   24   11   52   (44)  34   14   53
to this accident's                                    ↑    _____|_____
circumstances                                         |    |    |    |
                                                      |   15   16  (42)
                                                      |                ↑
                                                      |                |
              Recurring Cause Code - - - - - - - - - -
```

Step 4: Final cause code summary and associated level for correction, using Table 2.

Cause Code	Corrective Level/Function
76–Work habits bad	Supervisor/2, 4
13–Failure to correct through coaching	Supervisor/1
20–Duties and tasks not clear	Supervisor or Management/1, 2
42–Failure to observe and correct unsafe acts	Derrickman or Supervisor/2, 4
14–Instruction inadequate	Derrickman/Supervisor/2, 4

Step 5: Analysis of accident cause code points out there is lack of training and follow up to ensure the floorhand knew how to properly handle caustic and ensure he used protective goggles. Plans and procedures to implement this action by a supervisor are required from management, as well as monitoring work performance to ensure compliance.

Summary

Control of hazards and of losses in oil production activities is enhanced if management approaches a solution analytically, rather than subjectively. The causes of error that result in operational losses (productivity, time, job quality, equipment, and personal risk) require use of management tools. Finger pointing or expedient classification of error as work "carelessness" or negligence may ease the responsible manager's mind, but will do little to correct recurring errors. Rather, the operations review method presented provides a disciplined methodology for a manager to scrutinize potential failure in the organization's functions, thereby assuring long-term corrective action and, importantly, economic competitiveness by minimizing such losses.

Reference

Heinrich, et al., *Industrial Accident Prevention:* McGraw-Hill Book Co., New York, N.Y., 1980.

26
Low-Cost Methods for Shallow Oil Exploration and Production

Curtis L. Talbot

Searching for oil at shallow depths in the United States is not as precise a science as is the drilling of deeper wells. The same techniques could be used, but many are costly and time consuming. Most of the time shallow wells are drilled by small companies on limited budgets or by individuals who cannot afford much more than the actual drilling cost. Keeping the cost to a minimum in all phases of the project is a major consideration.

In the exploration phase, costly items such as seismic surveys and expensive field studies are avoided. Oil at shallow depths has not usually gone through as complex a geological history as the more deeply buried horizons. Traps are usually quite simple. Faults that slice up fields at depth commonly die out toward the surface. Periods of folding and refolding often are not reflected in the beds that have been deposited more recently. Where shallow oil occurs above deeper oil accumulations, the structure is usually a subtle reflection of the deeper geology.

The exploration time is best spent reviewing data already available from other sources. The best places to hunt are in the files of the governmental agencies. All data are open to the public and one needs only to find the right field or well and then make liberal use of the copying machines. Well logs are filed with both the state and the Federal government on public land. For state and private land they are available from the state only. These logs can be checked for possible shallow or bypassed oil accumulations. Geological reports are filed with the state and federal government on most wells. These reports often include sample descriptions, drillstem tests, and comments on zones that were not considered worthy of development by a major company. Due to the large overhead of major oil companies, they often set minimum figures on pay thickness or areal extent required before developing a well. A small company often can avoid these requirements.

The government agencies are also excellent sources for maps. When an oil company argues the need for a certain spacing with the state, it supplies a map to strengthen its case. When there is to be any unitization or secondary recovery, maps are provided at the hearings. Some companies even provide a small geological map with the drilling application. These maps are a boon to the explorationist who is studying the shallow zones in the same region. For a small company, drafting is a cost that should be avoided. The maps include a base map with all the wells plotted up to the date of submission. Using a light table, you can duplicate the base, then spot any additional wells. Using the log data, you can add the subsea data points for the horizon you are studying and map the geology.

From the submitted maps you can check to see that your map reflects the deeper zones. You can also extrapolate faults to see if they are present in the shallower horizon. This provides you with a fairly accurate geological structure map without drafting costs.

In reading through old information, as one collects it, one must continuously evaluate what is being said. Your new review can overlook the same oil that was bypassed when the well was drilled. Remember that in the shallower portion of the hole, drilling moves quite rapidly. Evaluation of the upper hole is minimal unless the company is expecting to encounter hydrocarbons they desire to test. A good example of this is the Greybull field. Over one hundred tests were drilled to the Greybull sand between the years 1918 and 1950. Suddenly someone discovered that they were drilling through a sand from 200 to 400 ft that was over 100 ft thick. A small 60 acre block of this sand has now produced about 800,000 barrels of high gravity oil. In comparison, the Greybull sand, the major exploration target, has produced only about 250,000 barrels after sixty years.

Even with a geologist or mud logger on location, showings may be missed. Samples need to be checked for light oil as soon as they reach the surface; a delay of an hour or more may eliminate the sample show. Some excellent wells in late Cretaceous sands of the Powder River basin have no shows at all in the samples by the time they reach the shaker. Also be aware that shallow pays may have low resistivities not apparent to the man evaluating the logs. There are cases of zones with very low resistivities yielding up to 50 BOPD. In an extreme case, a low-resistivity zone yielded only a little gas on a drillstem test. The company chose to skip the zone for there was no gas market and both shows and test suggested gas. When the shallow sand was mapped, no one could explain why gas occurred so far off the top of the structure. The well was later recompleted into this zone and produced 400 BOPD. A good rule of thumb to remember is that the bigger the company, the more mistakes it is likely to make.

Company economics or preconceived ideas may also create bypassed oil zones. A major firm may be playing only a particular formation. Other companies may have done reservoir studies and placed minimum thickness and porosity requirements on the pay. Lastly, history can and does affect the completion decision. Wells drilled when oil was $5.00/barrel or less may not have been profitable to produce but in the present market are economic. The same situation holds true for gas. Right now the market is down; if gas is found the decision may be to abandon the well. Lack of pipelines at the time of discovery might also mean the well is uneconomical.

The other part of the exploration phase is gathering verbal data. Listening is one of the most important tools in this business. You must act as your own scouting service and be keenly aware of what is going on. Comments made by roughnecks, mud engineers, drillers, and other geologists must be screened and checked. Although a man may not be schooled in geology, he may note an event such as oil on the pits or the odor of gas. These clues can be followed up by a careful check of data submitted to government agencies. Some wells drilled strictly on hearsay information given by field hands can be remarkably successful.

It is also possible to go to the oil company that drilled the well and get data. If the prospect is abandoned, companies will share some data or they may allow the wellsite geologist to divulge his personal observations. If the company still has the lease, they may provide data in hopes of acquiring an income from a farmout. It is important to always try the oil company even if it has a history of turning down requests. Sometimes a disgruntled employee may inadvertently provide the clue that you need.

The second phase is the actual drilling of the well. This begins with the acquisition of the land. Usually this can be done, but it often involves a great deal of patience. If the landowner has the minerals, you can deal directly with him. If the land is held by a company, you can negotiate a farmout of the shallow rights. If a farmout places too much burden on the lease, wait for the lease expiration. The expiration process can often be speeded up by getting the mineral owner to request a release. When the land is held by a speculator or a promoter it is best just to file away your knowledge until the lease expires, for the costs of this middleman cut deeply into your pocket. In time the property will change hands and you can make a reasonable deal.

282 Shallow Oil and Gas Resources

Once you have the land, drilling can commence in a short time. Again to cut costs, it is often best to obtain well permits far in advance of the expected spud date. In this way, news of your plans is carried in the trade journals and you can often get bids for services and equipment from sources you may not have considered. Always compare prices and get bids wherever possible. To reduce costs it is best to have a man who is both a geologist and an engineer oversee the well. Such a person can do both the actual drilling and the well evaluation. Shallow wells usually do not require a strong engineering expertise. Most of the problems can be solved by the drilling contractor. In fact, it is best to drill under a footage contract which leaves most of the desisions on the contractor's shoulders. Most contractors have years of experience and will familiarize themselves with the local drilling problems before submitting a bid. The footage contract also gives you a firm price on the job before you begin.

You can do some things to reduce the bid price. First, always check and see if a reentry of an existing well can accomplish your drilling objectives. Where this can be done, you save the cost of surface casing and much of the drilling cost. Where no old holes are available, determine the minimum amount of surface casing needed to prevent surface problems. Most deep holes are overengineered and surface casing length can be shortened considerably. All available data should be studied with an eye to reducing mud weight without increasing the risk. Cutting cost is the objective, but one still does not want to endanger life. Consideration should also be given to air drilling, which is much faster than mud drilling and causes less formation damage. It requires more surface equipment during drilling but eliminates most of the mess that is associated with the mud. At the Lamb field, an air-drilled well encountered two gas horizons at less than 1,500 ft that had been overlooked during the mud drilling of numerous deep production wells in that same field.

Some other suggestions can be offered on drilling. Try to avoid using oil field equipment. Anything in the oil field comes at a premium cost. It is best to use water well drilling rigs or coring rigs, when they are available. They can accomplish the same job as an oil rig on these shallow holes and usually will bid it cheaper. A good evaluation of the hole is necessary. The best tool for good evaluation is a careful examination of the rock cuttings as they come out of the hole. Drillstem tests (DST) and log evaluation are helpful but should be used to confirm other data. One has a tendency to overlog a hole. Log only when needed and remember that cased-hole logs are less expensive, if they will provide the data that you need. In some areas there are slim-hole logging trucks that can provide the same logs as Schlumberger or the other major logging services. On a recent hole, a gamma ray with SP and resistivity log and a gamma-ray-density log with caliper log cost $1,700 for 2,000 feet of hole. Bids from major logging firms were several times this amount.

Assuming that the well is successful, the final phase is the completion. Here again it is best to avoid buying oil field equipment except when it is necessary. On many of the shallow wells it is possible to use a waterwell pump. These pumps are far less expensive than a pumpjack and can be serviced without a pulling unit. In the Newcastle, Wyoming area, waterwell pumps produce far more oil than pumpjacks during their daily operation. The steady pull of the submersible pump does not cause the formation damage that seems to occur with the stroking action of a conventional rod pump. This saves the costs of regular cleanouts of wells that sand up in shallow, loosely-consolidated sands. To prevent burning up the waterwell pump, a timer needs to be installed. A simple timer such as those used to turn on and off lights in a burglar system for a home can be added to the wells' electrical system to prevent burn out and can be set to come on several times a day to keep the well drawn down for optimum production. Total system cost can be under $1,000.

In many farming areas where they haul water, you can often buy tanks designed for water at a savings in cost over oil tanks. The only consideration in buying such a tank is the effect of the produced oil on the plastic in a plastic tank. One can also gain a definite savings by using standard black plastic pipe for flowlines wherever feasible. The original cost is lower than metal pipe and plastic can be repaired without welding. Plastic pipe comes in any length you desire, so you do not have to deal with connections.

After farm supply dealers, the next best source of completion supplies is the used equipment dealer. In many cases an oil company buys equipment for a well that depletes rapidly and the materials have seen only limited service. One can also find surplus material around existing fields; often, this equipment can be purchased directly from another oil company at a considerable savings.

Several techniques can be used to cut completion costs. Instead of using a professional cementer for the casing, hire a local cement company. Establish circulation with water or lightweight mud after running casing. Then dump the cement down the inside of the casing directly from the cement mixer. When the casing is full of cement, or the required volume has been placed in the pipe, put a wiper plug in the casing. Reattach the rig pumps and pump the plug down with water. The plug will be caught at the bottom of the casing, if the last joint has been crimped so that the plug can't go out of the casing. When the surface pressure rises, close your surface valve and hold pressure for at least twelve hours. This method will cut your cementing costs by at least 50% on shallow jobs.

When the chance of formation damage is high, or when there is a risk that the well may be a failure, it is often best not to cement at all. Instead, run an external packer on the casing just above the completion zone. This packer works like a drillstem test packer and prevents materials outside the casing from reaching the producing formation. If the well turns out to be nonproductive, the packer can be released or shot off and the casing brought back out of the hole. In producing zones that are subject to formation damage, this technique can be combined with air drilling to eliminate any possible damage and to allow completion of the well without perforating costs.

One other casing technique is the tubingless completion. This involves the production of the fluid up the casing. It saves one string of pipe and works very well in gas wells. Many low-pressure gas wells can be produced up $2^{3}/_{8}$ in. or $2^{7}/_{8}$ in. tubing. Some wells in the Midwest use tubing for casing in oil wells. In the case of oil wells, the producers use a one-inch pipe string instead of rods, with a slightly different pump on the bottom of the string. This enables the producer to pump the oil up the one-inch string.

Two pumping systems, used in place of the traditional pumpjack, have proven very effective on shallow wells. The best known is the hydraulic pumpjack which hydraulically moves the rods in and out of the hole. The cost is comparable to the standard pumping unit. It provides a much longer than normal stroke and can be repaired easily, without having to hunt for oilfield parts. Older hydraulic units were bulky and costly. The newest generation are lightweight and can be fixed rapidly should they break down. At least one operator has adapted the units so he can pull his own rods and rerun them without need of a pulling unit.

The second pumping method is the progressive cavity pump. It works like a screw. It continuously pulls fluid into the tubing as the screw is rotated. The cost is much less than the traditional pumpjack and can move fluid of virtually any density. It does not seem to be bothered by high sand content in the fluid or by gas locking, it does not have the same number of rod repairs as usual, it is a compact unit that can be set up by one man, and it is easy to install or repair. These units are rapidly gaining acceptance for shallow work and are installed in wells from Montana to Texas.

There are many additional ways to cut costs. I have seen operators perforate wells using charges purchased from regular perforating companies. They run them in on telephone cable or other thin electrical line and set them off with the car battery or whatever electrical source is handy. I've seen dynamite used in the same manner to perforate wells or to cut tubing and casing. There are innumerable varieties of homemade wellheads and tubing hangers. Some fellows even have their own polished rods lathed locally to cut costs. There is no limit to the number of ways that savings can be generated. All one needs is an open mind as to how to do the job and enough imagination to be innovative.

When you put all three phases of the small operator's business together, it is possible to find and produce oil at far less cost than a major producer. Depth and location can vary the cost to some extent. However, it is quite feasible to drill and complete shallow wells for under $20/foot. Dry hole costs, including evaluation, run about $10/foot. Thus a 1,000 foot well might cost $20,000 to put on line. Today, with oil at $30/bbl, this expense could be recovered with production of less than 1,000 bbls of oil. It is impossible to comment for all operators but even wells that start out at 2 BOPD can be profitable.

An extreme example of profitability is the shallow Greybull field, cited earlier. The Peay sand, discovered in the late 1950s, is at depths of 250–350 ft. It has produced over 10,000 barrels per acre on primary recovery. Thus, a well on two acre spacing would cost $6,000 to put on line and would return a gross income of $300,000. These wells started out at 15 to 20 BOPD, so the rate of return on investment was very rapid (payout less than one month). Remarkably, there are still areas on this structure that have not been developed due to leasing problems.

In summary, the shallow operator can do very well. By minimizing his expense he can make a fair profit when a major operator would have a loss. The key is to look for ways to keep the cost low. Never be afraid to try an unconventional method, if it will accomplish the same objective at less expense. Millions of barrels of oil will remain untouched if ways are not found to locate and extract these hydrocarbons at less than oilfield cost.

27
Exploration and Development of Shallow Oil Fields in Thailand

Prakal Oudomugsorn

Abstract

Although oil seeps have been known since 1921 to occur in the Tertiary intermontane Fang Basin, northern Thailand, it was not until 1950 that the presence of a shallow oil field was proved by drilling at Chai Prakarn, Chiangmai Province. In the following years, two additional shallow oil fields, namely, Mae Soon and Prong Nok, were discovered by subsequent drilling. Oil reservoirs in these three fields consist of sandstones of Pliocene age at depths ranging from 500 ft to about 3,000 ft. Production of oil from these fields began in 1954, at rates varying from 500 to about 1,200 barrels per day. Most of the oil produced has a paraffin base, with API gravity varying from 28° to 31°. All crude oil from the Fang Basin is sent to the mini-refinery, set up near the field site, for refining. The mini-refinery has a refining capacity of 1,000 barrels per day.

Several more shallow fields are expected to be found in the other Tertiary intermontane basins in the northern part of Thailand. Additional detailed seismic survey and more intensive drilling programs are required to prove the presence of the new oil fields.

Introduction

Exploration for oil and other mineral fuels deposits in Thailand has been carried out since 1921. Wallace M. Lee, an American geologist, investigated several lignite deposits in the Tertiary basins in the northern part of Thailand, including oil seeps in Fang Basin, Chiangmai Province. In 1950, the first shallow exploration well was drilled near the seepage site. Oil was found in sand reservoirs of Pliocene age at depths ranging from 500 ft to about 3,000 ft. Subsequently, several oil discovery wells were drilled in the nearby areas. Three shallow oil fields were found by a more intensive drilling program, all of which are situated in the so-called Fang Basin. Production of oil from these shallow fields started in 1954 and up to the present, the rate of production has varied from 500 to about 1,200 barrels per day. The oil produced is sent for refining to the minirefinery set up near the field site. A brief geology and development program of three shallow oil fields in the Fang Basin will be discussed in this paper.

The Geology of the Shallow Oil Fields in Northern Thailand

Structurally, northern Thailand comprises several intermontane Tertiary basins, formed by major fault blocks trending to the N-NE. These basins were probably formed during the Himalayan orogeny in late Cretaceous time. Some of the well-known basins are the Chiangmai, Lampang, Prae, Chiangrai, and Fang (Figure 1). To date, shallow oil accumulations of producible amount have been found only in the Fang basin.

The Fang Basin

The Fang Basin is one of the Tertiary intermontane basins of northern Thailand, and is about 50 km in length and 16 km in width. It covers an area of about 275 sq km. Tertiary sequences in the Fang Basin consist of clays, shales, and sandstones interbedded with some thin lignite beds throughout the section. These Tertiary sediments contain small vertebrate bones and teeth, leaves, turtle's back plates, and viviparus gastropods suggesting a lacustrine deposit of fresh or brackish water environment.

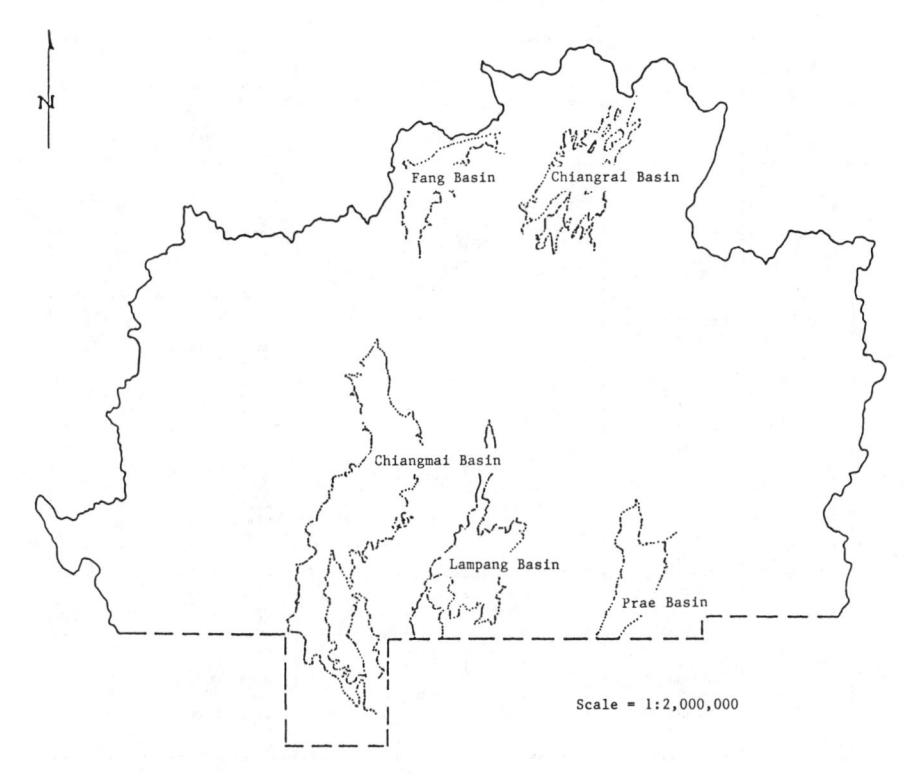

Figure 1. Map of major Tertiary basins in northern Thailand.

Stratigraphy

The stratigraphy of the Fang sediments (Figure 2) is obtained from subsurface data. The lower part of the column consists of sands, clays, and shales overlying basement rocks. The middle section consists of oil sand overlain by gray to brown clays and shales which serve as cap rocks. The upper part lies unconformably on the middle part and is composed mainly of cross-bedded, arkosic sands. Lignite is scattered throughout the entire stratigraphic section; however, it is more concentrated in the cap rock. Cores from the cap rock contain plant leaves, reed stems, fish teeth, viviparid gastropod, and insects. Palynological examinations of rock cores through the section drilled in the Fang Basin proved that the Tertiary rocks at Fang range in age from the Oligocene to the Miocene or possibly Pliocene.

SYSTEM	SERIES	GROUP	FORMATION	AREA Fang, Mae Soon and Chiangmai Basins (Buravas, 1969)
Quaternary	Recent	Chao Phraya	Chao Phraya	Alluvium
	Pleistocene		Mae Fang	Gravel Arkosic sands Clays (1,200 ft)
Tertiary	Pliocene		Mae Sot	Arkosic sands Clay/shale Minor coal Chai Prakarn Oil Sand (1,550 ft)
	Miocene	Krabi	Mae Hom	Calcareous mudstone Shale Thin fossiliferous ls. Minor Coal (240 ft)
	Oligocene		Li	Organic shale Minor coal Mae Soon Oil Sand (360 ft)
	Eocene - Paleocene		Nam Pat	Basal conglomerates Red sands Reddish shale (600 ft)

Figure 2. Stratigraphic column of the Fang Basin, northern Thailand.

The unconformity is probably regional in extent, being contemporaneous with the Pliocene-Pleistocene uplift of the central belt of Burma, to the west of Thailand. The uplift during Pliocene-Pleistocene time probably brought the lacustrine stage of sedimentation in the Fang Basin and other nearby basins to a close and only the fluviatile environment remains up to the present day.

Drilling and Development

After intensive geophysical surveys covering the entire area of the Fang Basin were carried out in 1961-1962, an intensive drilling program was implemented to locate the presence of the oil field. As a result, three shallow oil fields were found, namely, Chai Prakarn, Mae Soon, and Prong Nok (Figure 3).

The Chai Prakarn Oil Field

The Chai Prakarn oil field is located along the eastern rim of the Fang Basin. It is about 5 km south of Fang District. In this field, oil is trapped in an anticlinal structure which is cut by several parallel faults.

Drilling activity in the area started in 1954. Between 1954 and 1960, a total of 108 wells were drilled in the Chai Prakarn structure. Two years later another 8 wells were drilled, making a grand total of 116 wells, 23 of which were completed as producing wells. The Chai Prakarn oil field produced from a Pliocene sand reservoir ranging in depth from 500 to 1,000 ft. The amount of oil produced between 1954 and 1974 totaled 202,725 barrels. The field was abandoned in 1974.

The Mae Soon Field

The Mae Soon oil field is situated on the western rim of the Fang Basin, approximately 5 km west of Chai Prakarn. This field is also on an anticlinal structure which trends in a NNE-SSW direction. The structure is about 1,000 m wide, and about 1,750 m long. The oil discovery well in this field was recorded in 1973; since then, about 40 wells were drilled by the Defense Energy Department of Thailand. There are 18 production wells yielding about 600 to 1,200 barrels of oil per day. The Mae Soon Field has an oil reserve of approximately 1,500,000 barrels, out of which 1,000,000 barrels have already been produced. All of the oil is sent to the Fang refinery, which is situated near the field site.

The Mae Soon crude oil has a paraffin base which solidifies at room temperature but will remain in liquid form in the reservoir; the quality of Mae Soon crude oil is as follows:

API gravity	28.2°–31.2°
Sulfur content	0.26–0.38%
Pour point	85°–96°F

It also contains paraffin wax and lubricating oil of approximately 20% by volume. The rest is light oil and gas.

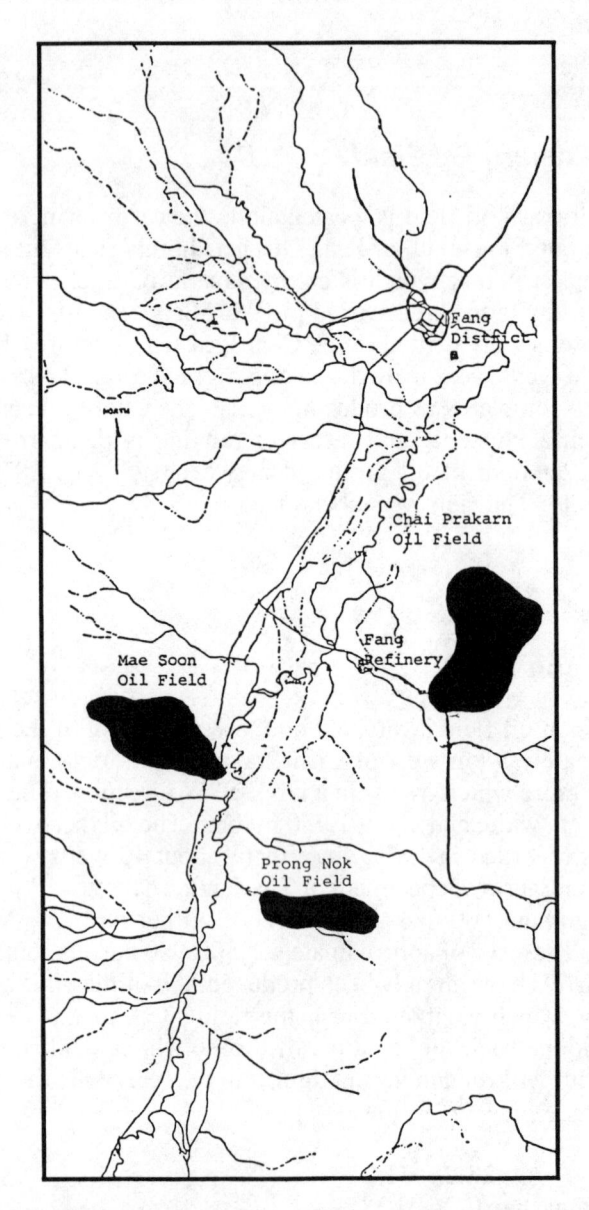

Figure 3. Map showing the location of shallow oil fields, Fang Basin, Chiangmai Province, Thailand.

The Prong Nok Oil Field

The Prong Nok Oil Field is situated south of Prong Nok village. In this field oil is trapped in a structural nose. The oil discovery well was drilled in 1979 with the oil reservoir occurring at about 3,500 ft. The Defense Energy Department drilled four more delineation wells, three of which encountered oil at depths ranging from 1,048 ft to 3,635 ft. These discovery wells were later completed as production wells. Several more wells are planned to be drilled in the Prong Nok area. At present only a few hundred barrels of oil is produced from this field.

Conclusion

It has been proved that at least three shallow oil fields exist in the Tertiary Fang basins in the northern part of Thailand. The Defense Energy Department is responsible for the drilling and production of oil from these shallow fields. Crude oil is produced from sand reservoirs occurring at depths of 500 ft to about 3,000 ft. The rate of production ranges from 500 to 1,200 barrels per day. All of the oil produced is sent for refining at the mini-refinery near the producing site. Additional seismic survey and drilling programs are required in an effort to locate additional oil fields.

References

Buravas, Col. S., 1969, Succession of rocks in Fang and Chiangmai Areas: Defense Energy Depart., Ministry of Defense. (unpublished report).

Endo, S., 1964, Some older Tertiary plant from Thailand: Geology and Paleontology of Southeast Asia, Tokyo University Press, v. 1, pp. 113–115.

Poothai, C., and Chana, A., 1968, Geology of Mineral Fuels in Thailand: Dept. of Mineral Resources, Ministry of Industry. (unpublished report).

Section IV

Shallow Natural Gas Production and Technology

28
Accurate Fluid Property Prediction in Natural Gas Resource Development and Custody Transfer

Jeffrey L. Savidge, K. Hemanth Kumar and Kenneth E. Starling

Introduction

Once considered no more than a nuisance and an unwanted co-product of crude oil production, natural gas has evolved to the point where it now supplies approximately one-fifth of the world's energy needs (Ikoku, 1980). Since its emergence as a valuable resource following World War II, the gas industry has experienced orders of magnitude increases in the value of natural gas. This has stimulated the production and transmission of gases once considered uneconomical. As a result, attention has been focused on the need for more accurate thermodynamic property information than is provided by some of the equations developed in the relative infancy of the gas industry.

In 1969 the American Gas Association (AGA) Transmission Measurement Committee voiced concern regarding accurate volumetric property measurement for custody transfer and helped to establish a program to update AGA Report No. 3 (Bean, 1982), which addresses the use of a standard in volumetric flow calculations for custody transfer situations. This report is a product of that effort. Its objective is to provide a more reliable method for calculating the supercompressibility factor and other gas thermodynamic properties than is possible with currently used methods (AGA, 1963). Though the work has been initiated by the transmission segment of the gas industry, the results have industry-wide value and application. This paper presents a new, extremely accurate equation of state to be used for the calculation of natural gas thermodynamic properties. It has been developed at the University of Oklahoma (OU) under the sponsorship of the Gas Research Institute (GRI). The work has evolved through close collaboration with industry representatives and research cooperation with the European gas research group (Reintsema, 1983). The results provide industry with a state-of-the-art thermodynamic equation of state that is significantly improved in volumetric-prediction accuracy and has expanded computational advantages over current methods.

Accurate thermodynamic-properties information for natural gas and associated liquids are a fundamental concern of the gas industry because of the pervasive role they play in the utilization of gas resources. Properties such as the supercompressibility factor, compressibility factor, density, enthalpy, heat capacity, fugacity, and sound velocity, are important quantities needed in gas reservoir engineering, separation and processing, compressor design and operation, gas transport, storage, and custody transfer calculations. Property data are typically provided from three possible sources:

1. Experimental measurements.
2. Theoretical estimations.
3. Correlations of experimental data.

Direct experimental measurement under most industrial conditions is impractical due to the large number of field and operating conditions encountered, time involved in obtaining accurate data, and high costs. Theoretical estimates provide valuable guideposts but often lack sufficient accuracy to warrant use as a fundamental property value. When applied within specified limitations, a correlation will generally provide the most practical means to obtain a reliable property estimate. It should be kept in mind that a correlation is simply a mathematical representation of a set of data and is obtained by regressing the set to fit a selected functional form or mathematical model. The accuracy of a correlation is dependent on both the quality of data selected for regression and the functional form of the equation. At its very best, a correlation is only as good as the data used in its development. Thermodynamic equations of state are, in a general sense, correlations of thermodynamic property data. If properly developed, they provide a large amount of computational information through the mathematical interrelationships among the thermodynamic properties (e.g., from one basic formulation numerous physically valuable quantities may be obtained). Thus, given a suitable correlation for one of the properties allows evaluation of the other properties, assuming the equation has been developed in a thermodynamically consistent manner. For example, a correlation of compressibility-factor data is capable of providing good enthalpy information, given the preceding restrictions.

The thermodynamic properties of a gas in a pipeline, reservoir, or storage tank are solely dependent upon the conditions of state of the gas (i.e., temperature, pressure or volume, and composition). At low-pressure conditions (P < 1 atm) gases behave almost ideally. The pressure-volume-temperature behavior of an ideal gas follows the familiar ideal gas law:

$$PV = nRT \tag{1}$$

where P is the absolute pressure, V is the volume occupied by the gas, n is the number of moles of gas, T is the absolute temperature, and R is the gas constant.

As pressure increases, a significant variation occurs between the actual experimental volume and that calculated from the ideal gas relation. For moderate pressures, such as those occurring under pipeline conditions or normal reservoir conditions, a gas is compressed more than would be predicted by the ideal gas law and thus occupies less volume or has a higher density than the ideal gas density. For very high pressures, gases tend to compress less than predicted by the ideal gas expression, due to repulsive forces exerted by the molecules. Corrections for the deviations in gas behavior between measured and predicted ideal volume are made using an empirical factor Z, commonly referred to as the gas deviation factor or the compressibility factor. It is defined as:

$$Z = \frac{V_{actual}(T,P)}{V_{ideal}(T,P)} \tag{2}$$

where V_{actual} is the measured volume at temperature T and pressure P, and V_{ideal} is the calculated volume from the ideal gas relation at temperature T and pressure P.

The real gas equation of state accounts for deviations from the ideal and may be expressed as

$$PV = ZnRT \tag{3}$$

The compressibility factor, Z, is a quantity which must be experimentally evaluated. Its value depends on the temperature, pressure, and composition of a gas. For over a century, a great deal of effort has been expended by the scientific and engineering communities to measure as accurately as possible the pressure-volume-temperature (PVT) behavior of gases. This has resulted in a substantial amount of data of varying quality for systems important to development of natural gas correlations. Under the best of conditions, which few experimentalists attain, obtainable experimental uncertainties are on the order of 0.01% to 0.05% in the compressibility factor for the gas phase. As is evident from Equation 3 a small uncertainty in Z has a direct impact on the volume. For the large volumes of gas typically encountered in the gas industry, uncertainties are significant and may amount to millions of dollars daily, worldwide. Thus, there is a strong incentive to use the most accurate prediction method for the compressibility factor available.

Over the years various authors have presented correlations for the compressibility factor. These include Standing and Katz (1942); Benedict, Webb and Rubin (1940); Hall and Yarborough (1974); Peng and Robinson (1976); Soave (1972); Starling (1973); and the present gas industry standard (AGA, 1963), each with differing accuracy limitations based on the selected data chosen for correlation development and the functional form of the equation. The work presented herein is tailored to natural gases and is based upon a thorough analysis of all existing data for natural gas systems. Only the most accurate data have been selected to provide the best possible predictive capabilities currently available.

Equation-of-State Use in Natural Gas Engineering

Equations of state play a central role in reservoir engineering calculations of gas formation factors, gas in place, material balances, reserve and reservoir predictions, and bottom-hole pressure calculations. Equation-of-state use in gas compression calculations include enthalpies, volumetric efficiencies, horsepower, discharge temperatures, and heat loads for interstage coolers. Vapor-liquid-equilibrium predictions are important for both the reservoir engineer and processing engineer, who need accurate fugacity information to evaluate the vapor and liquid split. All these quantities are directly obtainable from a single thermodynamically-consistent equation of state.

By far the most pervasive property used in all phases of gas resource development is the compressibility factor. If the gas were treated as an ideal gas then errors amounting to 40% or more may occur, depending on the gas composition and the temperature-pressure condition. The sensitivity of a calculation to an uncertainty in Z is dependent on the functional form of the equation. It may or may not be proportional. In a general sense, the more accurate the compressibility factor used in a calculation, the more reliable that calculation may be considered to be.

Accurate Fluid Property Prediction

This is important to the gas industry due to the magnitudes of the numbers involved in many of the calculations. Small, seemingly insignificant uncertainties in Z amount to very large quantities of gas worldwide.

The sensitivity of a calculated quantity that uses the compressibility factor, Z, in one form or another, may be obtained by applying the first-order approximation of the standard error propagation formula:

$$\Delta M = \frac{\delta M}{\delta Z} \Delta Z \tag{4}$$

where M is the quantity of interest, ΔZ is the uncertainty in Z, $\delta M/\delta Z$ is the partial derivative of M with respect to Z and ΔM is the resulting uncertainty in M due to an uncertainty in Z. For example, the impact of a 1% uncertainty in Z ($\Delta Z = 1\%$) on any calculation can be directly evaluated from Equation 4.

Custody-transfer calculations are particularly sensitive to Z since the cost of the gas is based on the total amount transferred. Volumetric flow rate calculations require an accurate knowledge of the real gas compressibility factor. Typical flow rate calculations through orifice meters apply the formula:

$$Q_h = C' \sqrt{P_f h_w} \tag{5}$$

where Q_h is the flow rate in cubic feet per hour at standard conditions, P_f is the static pressure in psia, h_w is the differential pressure in inches of water, and C' is the product of a number of factors given by:

$$C' = F_b \times F_{pb} \times F_g \times F_{tf} \times F_r \times Y \times F_{pv} \times F_m \tag{6}$$

Detailed information regarding the definition of each of the factors is presented in AGA Report No. 3. The quantity accounting for deviations from ideal behavior of a gas, due to its condition, is F_{pv}, which is called the supercompressibility factor and is defined as:

$$F_{pv} = \left(\frac{Z_b}{Z}\right)^{1/2} \tag{7}$$

where Z is the compressibility factor at the flowing temperature and pressure and Z_b is the compressibility factor at the standard pressure and the flowing temperature. The U.S. gas industry currently uses supercompressibility-factor tables and equations (PAR Research Project NX-

Shallow Oil and Gas Resources

19) developed in the early 1960s (AGA, 1962). Evaluations of the NX-19 Supercompressibility Factor Correlation (Starling, 1982; Ghannudi, 1982) have illuminated some serious deficiencies and shown the need for an improved supercompressibility factor correlation.

Equation-of-State Applicability Range

Figure 1 presents the temperature-pressure region and targeted uncertainties for the computation of natural gas supercompressibility factors.

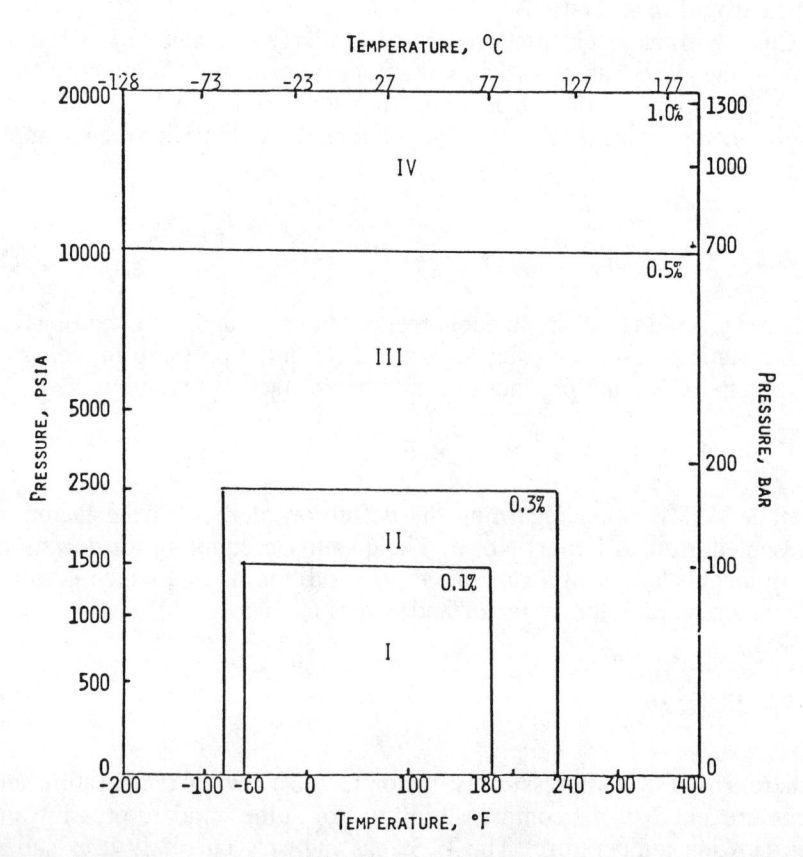

Figure 1. Targeted uncertainty limits for computation of natural gas super-compressibility factor.

The entire surface extends from $-130\,°C$ to $204\,°C$ and pressures to 1,400 bar. Prediction objectives vary within this region, depending on the temperature, pressure and gas characteristics according to gas industry needs. Region I encompasses the majority of custody transfer conditions and ranges from $-50\,°C$ to $80\,°C$ and pressures to 100 bar. The targeted uncertainty within Region I is 0.1% in the supercompressibility factor, F_{pv}, which corresponds to approximately 0.2% in the compressibility factor, Z. Compositional requirements include high-specific-gravity natural gases and diluent levels of N_2 and CO_2 up to 50%. Region II bounds Region I, extending from $-60\,°C$ to $115\,°C$ and pressures to 170 bar and covers the remaining conditions at which custody transfers may occur. The objective for Region II is 0.3% accuracy in supercompressibility factor. Region III bounds the two inner regions and covers typical reservoir and gas processing conditions. It ranges from $-130\,°C$ to $200\,°C$ and pressures to 690 bar. Targeted uncertainties in Region III are 0.5% in F_{pv}, which corresponds roughly to 1.0% in Z. Region IV covers the high-pressure range and extends up to 1,380 bar. Gases with concentration levels of N_2 and CO_2 up to 100% should have accuracies to 1.0% in the supercompressibility factor. The above accuracy levels do not apply to the critical region.

Correlation Methodology

The correlation methodology used in this work is based on a three-parameter corresponding-states approach, using a combination of second virial and conformal solution-mixing rules to describe the effect of gas composition on the behavior of a gas. The corresponding states theorem proposed by J.D. van der Waals in 1873 says essentially that the variation of a real gas from the ideal gas law is basically the same for different gases at the same reduced temperature and reduced pressure conditions. This theory is accurate to within a few percent for simple fluids. It can be expressed by:

$$Z = Z_o (T_r, P_r) \tag{8}$$

where Z is the compressibility factor and $Z_o (T_r, P_r)$ is some function of the reduced temperature, T_r, and reduced pressure, P_r. It is widely known as a two-parameter corresponding-states method. This is the approach applied in many of the early compressibility factor equations, including Standing and Katz (1942) and the NX-19 method (AGA Research Project NX-19, American Gas Association, 1962) for supercompressibility factor calculations. Two-parameter corresponding-states correlations have serious deficiencies, for the components under consideration increase in differences in molecular size and shape. Many natural gas components fall in this category. A three-parameter approach has the capability of correcting for these effects and gives extremely accurate correlation results for natural gases.

The Equation of State for Natural Gases

The equation of state developed for highly accurate prediction of natural gas volumetric and derived properties is a three parameter, nine term, density virial expansion with 43 generalized constants. It was developed from a thorough evaluation of available pure and binary gas data. Only the most accurate data were selected for correlation development, including hydrocarbons ranging from methane through n-octane and diluents including carbon dioxide and nitrogen.

In the three-parameter approach, the compressibility factor can be expressed as:

$$Z = Z_o (T^*, \rho^*) + \gamma Z_\gamma (T^*, \rho^*) \tag{9}$$

where Z_o is the isotropic contribution and Z_γ is the anisotropic contribution. Z_o and Z_γ are both functions of the reduced density ρ^*, and the reduced temperature T^*; γ is the orientation parameter for the particular fluid. The characterization parameters are based on molecular theory, where T and ρ can be reduced using a molecular energy parameter, ϵ, molecular size parameter, σ, and an orientation parameter, γ. The reduced conditions are given by:

$$T^* = kT/\epsilon \tag{10}$$

$$\rho^* = \rho\sigma^3 \tag{11}$$

Values for ϵ and σ are estimated from a knowledge of the critical properties of the pure components.

$$\epsilon = kT_c/1.2593 \tag{12}$$

$$\sigma^3 = 0.3189/\rho_c \tag{13}$$

The OU-GRI equation in reduced form is:

$$Z = 1 + \rho^* B^* + \rho^{*2} C^* + \rho^{*3} D^* + \rho^{*4} E^* + \rho^{*5} F^* \\ + \rho^{*6} G^* + \rho^{*7} H^* + \rho^{*8} I^* + \rho^{*9} J^* \qquad (14)$$

The reduced virial terms B* through J* are given by:

$$B^* = A1 + A2/T^* + A3/T^{*3} + A4/T^{*4} + A5/T^{*5} + A6/T^{*6} \\ + A7/T^{*7} + A8/T^{*8} + \gamma (A30 + A33/T^{*2} + A31/T^{*5} \\ + A34/T^{*7} + A32/T^{*8})$$

$$C^* = A9 + A10/T^{*5} + A11/T^{*6} + A12/T^{*3} + \gamma (A35 \\ + A38/T^{*5} + A37/T^{*6} + A36/T^{*3})$$

$$D^* = A13/T^* + A14/T^{*2} + A15/T^{*4} + \gamma (A39/T^{*4})$$

$$E^* = A16/T^* + A17/T^{*5} + \gamma (A40/T^*)$$

$$F^* = A18/T^{*2} + A19 + A20/T^{*3} + \gamma (A41 + A42/T^{*2} \\ + A43/T^{*3})$$

$$G^* = A21 + A22/T^* + A23/T^{*2}$$

$$H^* = A24 + A25/T^{*2}$$

$$I^* = A26/T^* + A27/T^{*2}$$

$$J^* = A28 + A29/T^{*2}$$

Table 1 is a list of the generalized constants to be used with this equation. Table 2 contains the correlation characterization parameters.

The mixing rules used with the OU-GRI equation of state utilize second virial mixing on the first-order density term and conformal solution mixing on the remaining higher order terms. This approach has the advantage of providing extremely accurate prediction capabilities at low densities, such as those occurring in Region I. This approach avoids problems associated with higher-order virial mixing by using the conformal solution approach for the higher-order density terms. The following semi-empirical, conformal-solution mixing rules are used to obtain the

Table 1
Generalized Constants for Use with Equation of State for Natural Gases

i	Ai	i	Ai
1	1.38666	23	−74.1660
2	−4.11476	24	−87.8341
3	−9.9280	25	72.0647
4	−218.5270	26	91.8779
5	0.901983	27	15.6396
6	269.729	28	66.2220
7	22.1405	29	−44.9259
8	31.0408	30	−32.3963
9	6.40852	31	−53.0080
10	43.9476	32	2.34982
11	20.4458	33	18.8792
12	−36.3165	34	14.8367
13	46.6093	35	−13.0265
14	−40.8286	36	−27.9769
15	28.0044	37	2.41311
16	−6.89997	38	−12.6170
17	−33.1592	39	−32.6122
18	15.0173	40	48.7117
19	19.6545	41	41.5876
20	−52.9974	42	−78.9295
21	−42.2180	43	77.4499
22	−39.1004		

Table 2
Values of Characterization Parameters Used in Correlation

Fluid	Crit. Temp. (K)	Crit. Density (kg-mol/m³)	Orientation Parameter γ
Methane	190.555	10.04999	0.0115
Ethane	305.390	6.7574	0.0918
Propane	369.850	4.9550	0.1520
Carbon Dioxide	304.200	10.6383	0.2106
Nitrogen	126.200	11.1730	0.0427

characterization parameters σ_x, ϵ_x, γ_x for the high-order virial terms C* through J*.

$$\sigma_x^{3.0} = \sum_i \sum_j x_i x_j \, \sigma_{ij}^{3.0} \tag{15}$$

$$\epsilon_x \sigma_x^{4.0} = \sum_i \sum_j x_i x_j \, \epsilon_{ij} \sigma_{ij}^{4.0} \tag{16}$$

$$\gamma_x \epsilon_x^{4.5} = \sum_i \sum_j x_i x_j \, \gamma_{ij} \epsilon_{ij}^{4.5} \tag{17}$$

Second virial mixing is given by:

$$B_{mix} = \sum_i \sum_j x_i x_j \, B_{ij} \tag{18}$$

The pair characterization parameters σ_{ij}, ϵ_{ij}, and γ_{ij} are functions only of the pure-fluid parameters σ, ϵ, and γ of components i and j. To obtain these the following empirical combining rules are used for conformal solution mixing.

$$\sigma_{ij} = (\sigma_i + \sigma_j) \tag{19}$$

$$\epsilon_{ij} = (\epsilon_i + \epsilon_j)^{1/2} \tag{20}$$

$$\gamma_{ij} = {}^{1/2}(\gamma_i + \gamma_j) \tag{21}$$

The combining rules for the second virial mixing are given by:

$$\sigma_{Bij} = (\sigma_i \sigma_j)^{1/2} \tag{22}$$

$$\epsilon_{Bij} = \beta_{ij} \, (\epsilon_i \epsilon_j)^{1/2} \tag{23}$$

$$\gamma_{Bij} = {}^{1/2}(\gamma_i + \gamma_j) \tag{24}$$

The binary-interaction parameter β_{ij} is determined from binary-mixture volumetric data. β_{ij} values to be used are presented in Table 3. A value of unity should be used for β_{ij}, for binary pairs not listed in Table 3.

Equation 18 is solved by calculating the second virial term from the second virial mixing and combining rules; the higher-order terms are calculated using conformal mixing. The density is calculated implicitly by solving Equation 14.

Table 3
Binary Interaction Parameter Values

System	β_{ij}
Methane-nitrogen	0.9707
Methane-carbon dioxide	0.9601
Methane-ethane	1.0000
Methane-propane	1.0000
Methane-n-butane	1.0000
Nitrogen-ethane	0.9634
Nitrogen-carbon dioxide	1.0268
Nitrogen-propane	0.9479
Ethane-carbon dioxide	0.9176

Equation-of-State Results

The details of the correlation capabilities for pure components and important binary systems are given in Starling (1934). Briefly, methane is correlated in Regions I and II to better than 0.05% and to approximately 0.1% in Regions III and IV for the compressibility factor. Good results are also obtained for pure ethane, propane, nitrogen, and carbon dioxide in the gaseous phase. Binary systems are generally correlated better than 0.1% in the compressibility factor.

Figures 2 and 3 illustrate predictive capabilities for actual natural gases, using the NX-19 method and the provisional OU-GRI equation of state respectively. The average absolute percent deviation (AA%D) is defined as

$$AA\%D = \frac{1}{N} \sum_{i=1}^{N} \left| \frac{Exp_i - Calc_i}{Exp_i} \right| \times 100 \qquad (25)$$

The data are from the Gasunie laboratories of the Netherlands (Reintsema, 1983) and are for lean or high diluent-content ($N_2 > 7.5\%$ or $CO_2 > 5\%$) natural gases. The reference line at 0.2% is the targeted uncertainty in the compressibility factor Z. In general the NX-19 method provides fairly good results for lean gases with low diluent content. However as the amount of carbon dioxide increases, predictive accuracies drop significantly. As is evident from Figure 3, the OU-GRI method predicts this class, including high diluent-content gases, well within 0.1%, which approaches the experimental uncertainty in the data. Figures 4 and 5 compare prediction capabilities for rich natural gases (ethane > 5.0%). The data are from Gaz de France. The NX-19 method in general shows unacceptable predictive capabilities for this class of gases. The OU-GRI equation of state provides substantial predictive improvement.

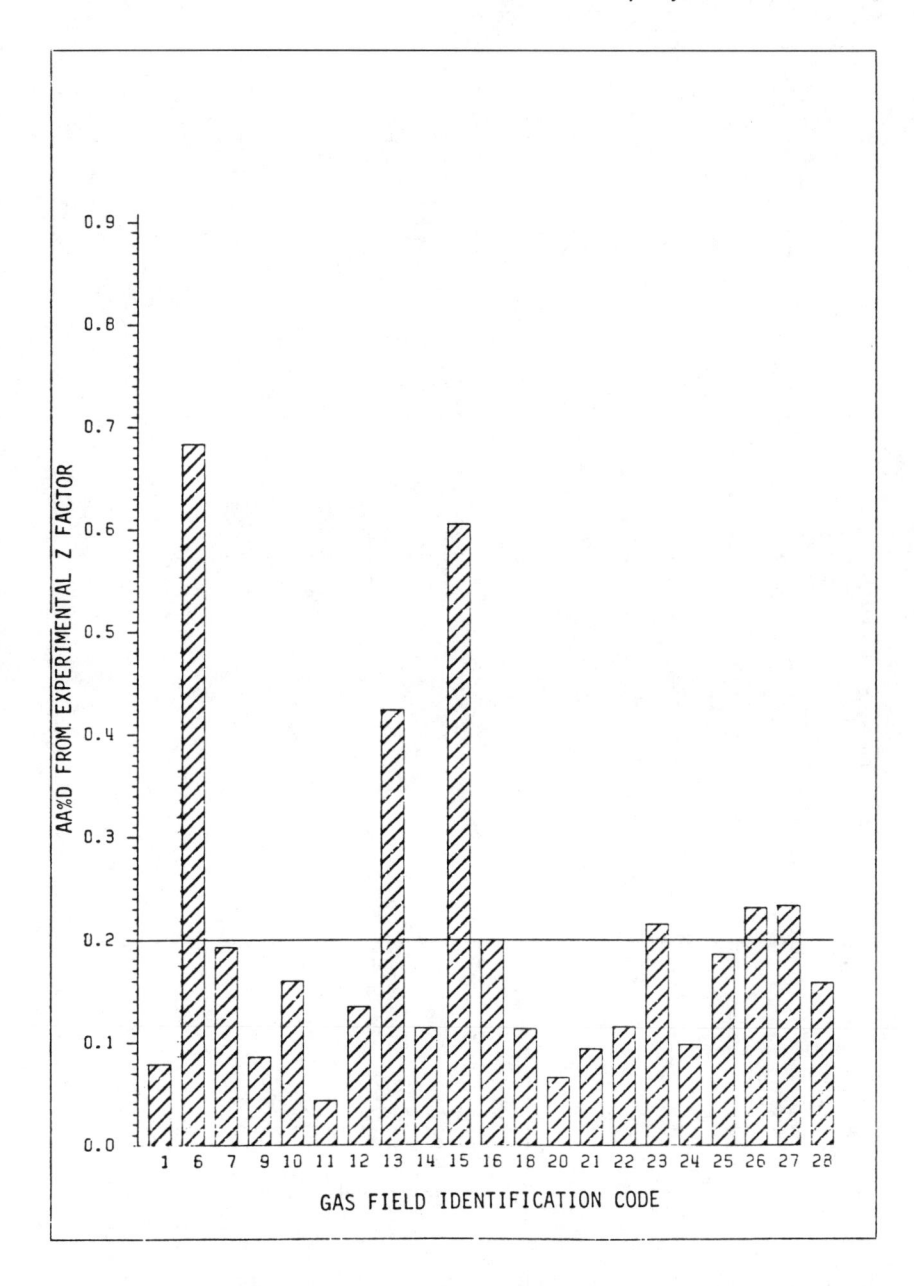

Figure 2. NX-19 prediction results: lean or high diluent natural gas mixtures (from Gasunie).

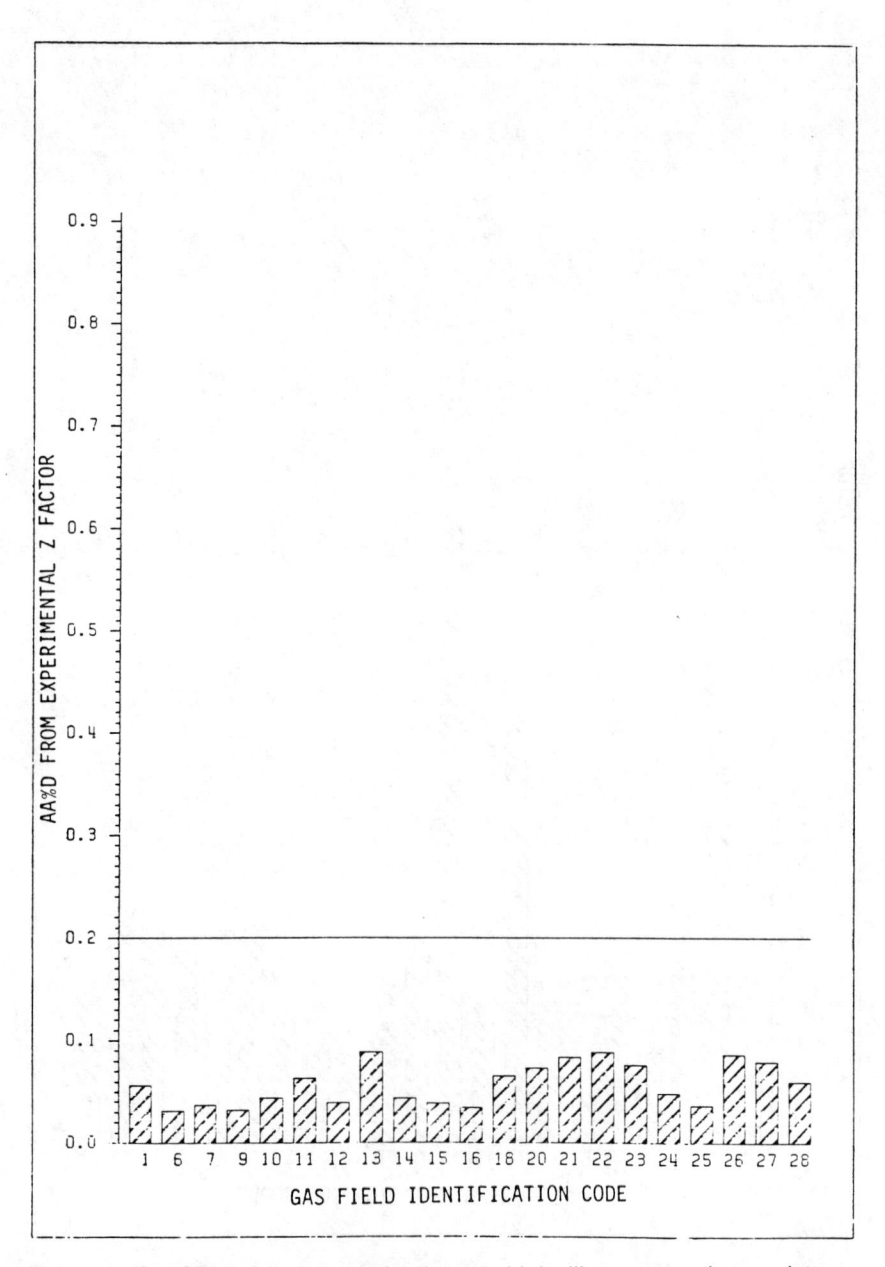

Figure 3. OU-GRI prediction results: lean or high diluent natural gas mixtures (from Gasunie).

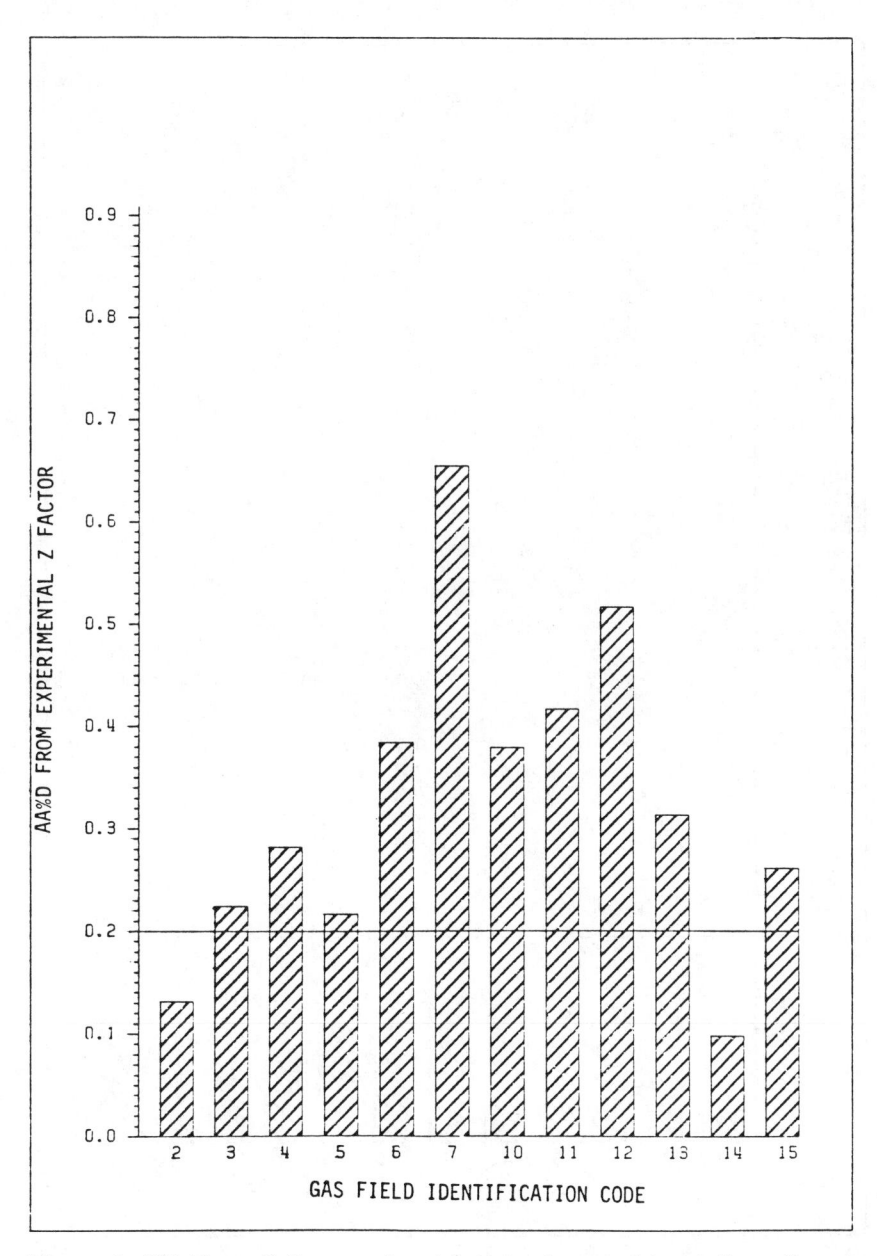

Figure 4. NX-19 prediction results: rich natural gas mixtures (from Gas de France).

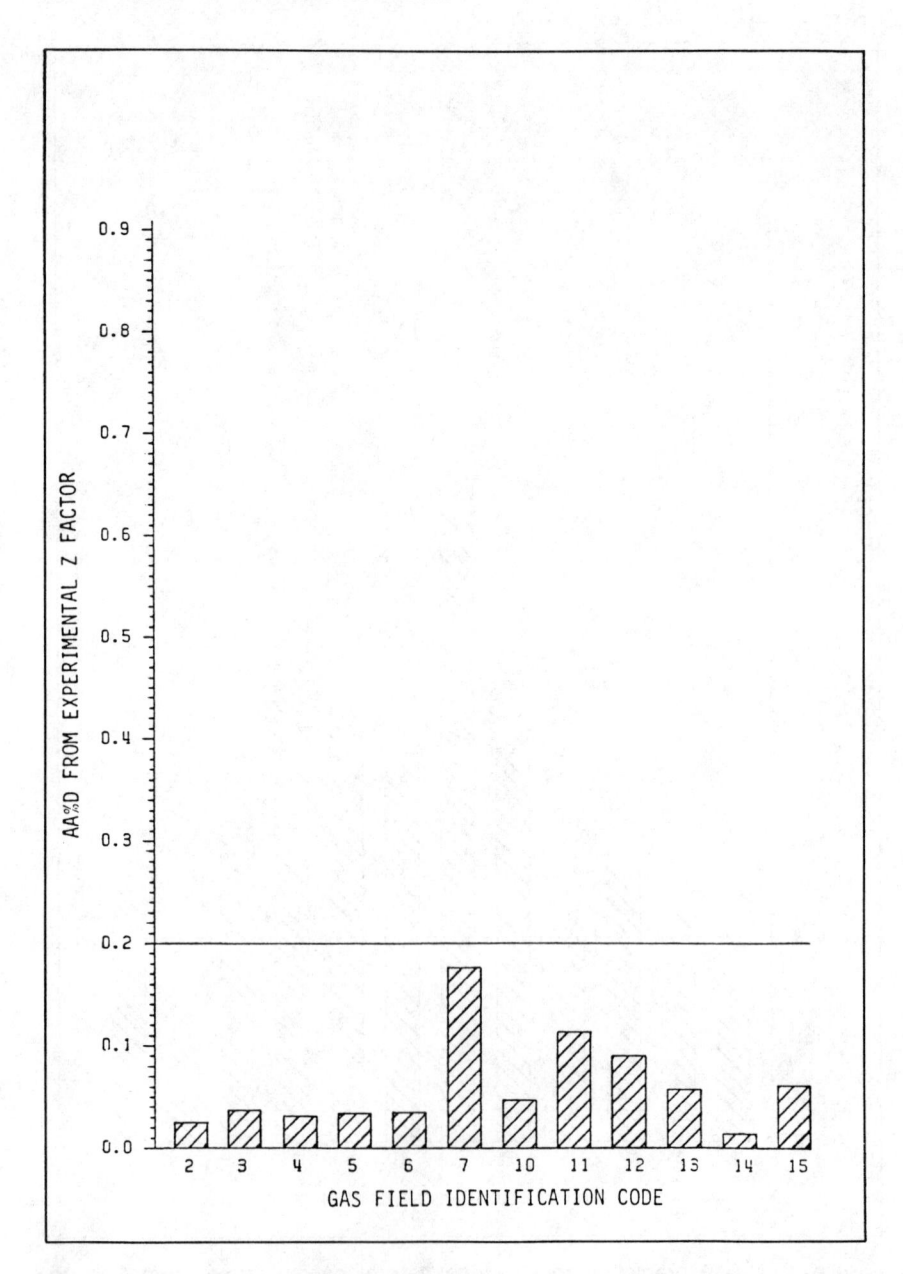

Figure 5. OU-GRI prediction results: rich natural gas mixtures (from Gas de France).

References

American Gas Association, 1963, Manual for the determination of supercompressibility factors for natural gas: AGA Publication, New York.

American Gas Association, 1930, Orifice metering of natural gas: American National Standard, AGA Publication, Arlington, VA.

Bean, H. P., 1972, 61st Gas Processors Association Meeting, Dallas, TX, March 15-17.

Benedict, M., Webb, G. B., and Rubin, L. C., 1940, J. Chem. Phys.: v. 8, p. 334.

Ghannudi, M. A., Kumar, K. H., and Starling, K. E., 1982, Evaluation of present methods of calculation of supercompressibility factors in the U.S. gas industry: Proceedings of the 1981 International Gas Research Conference, Government Institutes Inc. Publication, Rockville, MD.

Ikoku, Chi U., 1980, Natural gas engineering: PennWell Publishing Company, Tulsa, OK.

Lee, T. J., Lee, L. L., and Starling, K. E., 1979, Conformal solution theory parameter mixing rules: ACS Adv. Chem. Ser., v. 182, pp. 125-141.

Peng, D. Y., and Robinson, D. B., 1976, Ind. Eng. Chem. Fundam.:, v. 15, p. 59.

Reintsema, S. R., 1983, Gasunie Laboratories, Groningen, The Netherlands, private communication.

Reintsema, S. R., et al., 1983, Measurements and improved prediction method of compressibility factors in custody transfer situations: paper E-07 presented at the 1983 International Gas Research Conference, London, June 13-16.

Soave, G., 1972, Chem. Eng. Sci.: v. 27, p. 1197.

Standing, M. B., and Katz, D. L., 1942, Trans AIME: v. 146, p. 140.

Starling, K. E., 1973, Fluid thermodynamic properties for light petroleum systems: Gulf Publishing Co., Houston, TX.

Starling, K. E., et al., 1982, Research on supercompressibility of natural gas: presented at the American Gas Association Transmission Conference, Chicago, IL, May 17-19.

Starling, K. E., et al., 1984, American Gas Association Distribution/Transmission Conference, San Francisco, May 7-9.

van der Waals, J. D., 1873, thesis, Leiden.

Yarborough, L., and Hall, K. R., 1974, Oil and Gas Journal; Feb. 18, p. 86-88.

29
A Techno-Economic Model for Shallow Gas Reservoirs under Partial Water Drive

Ashis K. Das and Djebbar Tiab

Abstract

A mathematical model is presented that is capable of analyzing reservoir and production performance of a shallow gas field under partial water drive conditions and of assisting management in achieving the best economic indices in its production development.

The model works in the semi-steady-state flow regime, and for non-uniform well spacing; but holds only for single-phase gas flow in the uninvaded region. Two-dimensional numerical calculations of the model would help design an optimum development drilling sequence that minimizes the present-day cost of drilling and production operations, locate gas-water contact at any time, calculate optimum production rate of each well, identify effects of alternate well spacing, examine lease-line drainage, determine degree of depletion in specific zones in the reservoir, and choose between blowdown or gas cycling, depending on the strength of the driving aquifer.

A system analysis for the surface gas-gathering scheme is also presented to investigate the vital role of compression cost and to help the practicing engineer derive a criterion for an optimum management decision.

Introduction

For any oil and gas field, new or old, it is not feasible to continue to profit from its development indefinitely; there is always an operational limit. This limit is defined in the literature on several grounds. Typically, such definitions are based on both economic and technical bounds. Regardless of what the defined bounds are, it is the purpose of the producing company to maximize the present value of all future incomes, and minimize the present value of all future costs. Since such ventures are always constrained by physical, technical, operational, or other limitations, the problem of oil or gas field development is one of constrained optimization. In other words, the problem is (Rowan and Warren, 1967) "What drilling and/or production policy must be adopted to maximize the return from a given operation when certain practical limitations are present?" Ideally, given a reservoir, a designated production demand schedule (which may be either deterministic or stochastic), and a number of possible sites for new wells, the problem is to choose the optimum well locations and the sequence of drilling at those locations to maximize the net present worth of future earnings.

The most accurate and versatile approach to attack such a problem would be dynamic programming, where each successive step of pressure depletion constitutes a dynamic stage. However, such an approach appears to be rather complex and prohibitive in computation cost. A more approximate approach, accurate enough for most gas-reservoir cases and comparatively simple, is a trial-and-error method based on an optimized mathematical model. Henderson et al. (1968) used such an approach to study gas-storage-field development problems. Coats (1969) suggested a multiple-trial method for determining optimum well locations in a depletion-drive gas reservoir. Similar avenues were reported by Burns (1969) and Hessing (1966). Other workers, however, have presented more complicated but exact linear (Lee and Aronofsky, 1958; Aronofsky and Williams, 1962; Wattengarger, 1970; Coats et al., 1970; Green et al., 1971; Nandan, 1970) and integer (Rosenwald, 1972) programming approaches. A truly dynamic programming approach to such problems does not appear to have been used much in the petroleum literature.

The present work is basically a modification of Coats' (1969) original work, generalized to include the effect of an aquifer partially or totally enclosing the gas reservoir. The concept of the "f" function, introduced by Kumar (1977), has been used to represent an aquifer of any strength. Thus, a value of f equal to zero would generate the case of a volumetric or closed reservoir (Coats' original case), whereas a value of f equal to one would represent the case of an active aquifer (full strength). Any value of f lying between zero and one would represent various partial water drives. It is to be noted at this point that the so-called constant-pressure condition at the gas-water-contact is true only for f equal to one. For a more typical partial water drive, pressure is generally a monotonically decreasing function of time, at the gas-water contact.

The previously mentioned generalization of Coats' model is necessary in particular for shallow gas reservoirs adjacent to aquifers, because aquifers at moderate depths are most likely to be semi-active or weak. However, the model presented would be equally applicable to any gas reservoir, shallow or deep, depletion type or water-drive, as long as the assumption of single-phase flow in both the invaded and the uninvaded regions is valid, as will be discussed later.

One critical aspect of shallow gas field development is the necessary compression cost associated with the low-pressure gas produced. To meet the sales demand, the compressor load is expected to be rather high and hence the surface production system, of which the compressor station forms a component, plays a vital role. The present model therefore extends the Coats model further to include this vital aspect of production system analysis.

In what follows, we present a mathematical model, and a production system analysis model. The former represents the reservoir flow and the latter represents flow through the production system up to the gas delivery line. The two are combined to obtain an optimized development drilling schedule, optimum well production rates, and optimum equipment sizes that will minimize the compression cost.

The Problem Statement: Solution Approach

The original Coats (1969) model attempts to optimize one single aspect of production development, the drilling sequence. Given a maximum number of allowable well locations, two-dimensional pressure calculations are performed, each time with a new well (out of the remaining allowables) included, until the resulting calculated remaining gas in place (G_r) turns out to be minimum. Thus an effort is made to keep the plot between cumulative number of wells drilled (n_w) and G_r as low as possible (Figure 1). In Figure 1, curve 3 is the preferred and curve 1 the least desirable.

Coats' approach attempts to seek the smallest number of wells at any time during the field development. However, this may not be equivalent to maximizing the present value of net profit when the two-dimensional pressure calculations are coupled with the production system analysis

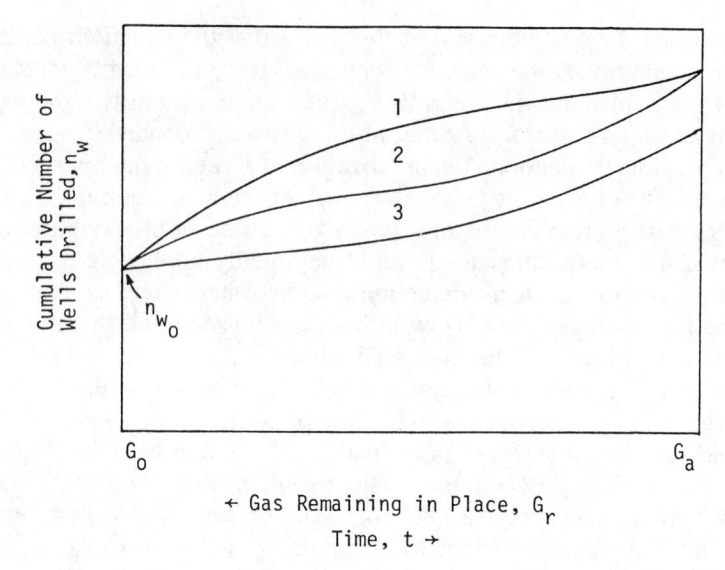

Figure 1. Cumulative wells vs. time for a producing gas field.

model. This is primarily because the gas compression and drilling cost play dominant roles and the total optimization should consider minimizing the compressor horsepower and drilling cost.

The effect of an adjacent aquifer and hence the resulting water influx further complicates the problem. It appears obvious that prediction of a drilling location now should consider the advance of the gas-water contact. If the location suggested for a new well happens to have already been invaded by an aquifer then the location would not be drilled and another would be sought. The model would be required to trace the advancing gas-water contact with production time.

The problem at hand, therefore, can be stated as follows. We assume the specified field production rate, Q_F (which is expected to decline after some time); the reservoir heterogeneity, that is, distribution of kh and ϕh, obtained from actual or simulated well tests and past production history; existing gas in place; existing number of wells and their locations; the present location of the gas-water contact; and a set of admissible well locations. We seek an optimum drilling sequence for some or all of these locations, the timing of such drilling, optimum well rates at any time, the compressor horsepower at any time, and the size of equipment in the production system that will minimize the compressor load.

In addition to the answers obtainable, as just described, the present model would also help analyze the following problems, which will be discussed later in more detail:

1. Choice between blowdown or gas cycling.
2. Lease-line drainage.
3. Depletion of specific zones of the reservoir.
4. Life of a new well and its economic justification.
5. Effect of an imposed minimum flowing bottomhole-pressure condition on the producing wells.

The Mathematical Model

Underlying Assumptions

The major assumptions of Coats' (1969) original model are:

1. Single-phase gas flow.
2. Quick establishment of pseudo-steady-state flow.
3. No condensation or turbulence near the wellbore.

To extend Coats' model to the case of a partial water-drive gas reservoir, the preceding assumptions, particularly 1 and 2, need to be removed to allow the two-phase flow of gas and water, and to incorporate a now-significantly long transient period before the pseudo-steady-state. The length of the transient period is expected to be significant, particularly during two-phase flow of gas and water. However, to keep the model simple, let us assume that only single-phase gas flow persists in the uninvaded region; whereas, gas is trapped completely, if at all, in the water-invaded region. This has been demonstrated experimentally by Geffen et al. (1952) and Chierici et al. (1963). The possibility of gas percolation from the water-invaded zone to the uninvaded zone cannot, however, be ruled out, and needs to be explored in the future. Such percolation may result in two-phase flow near the gas-water contact.

Such assumptions would probably be accurate for shallow gas reservoirs, since, due to low-pressure gas, the amount trapped would be insignificant ($S_g \propto P$), and mobile-gas saturation would not be achieved in the invaded region. It is to be noted at this point that assumption 1 essentially means that for any well's history, the production stream abruptly changes from all gas to all water, with no two-phase flow. Given the assumption of single phase flow, assumption 2 of Coats' model is well justified. Thus, some of the original equations of Coats' model may be retained, whereas, others need to be modified or introduced to consider partial water drive. The present model has been designed to incorporate non-Darcy flow near the wellbore, but the assumption of no liquid condensation still applies, as in Coats' original model.

Gas Flow in the Uninvaded Zone

The equation for compressible transient gas flow through porous media is given as:

$$\nabla \cdot \left(\frac{kh}{\mu_g} \rho_g \nabla P\right) + hq_{inj}^m - h \frac{TP_s}{T_s} q^m = h\phi \frac{\partial p}{\partial t} \tag{1}$$

Using the definition of gas density and the real gas pseudo-pressure (Al-Hussainy et al., 1966):

$$\rho_g = \frac{PM}{zRT} \tag{2}$$

and

$$m(P) = \int^P \frac{2P}{\mu_g(P)z(P)} dP \tag{3}$$

Equation 1 can be reduced to

$$\frac{1}{2} \nabla \cdot \{kh \nabla m(P)\} + h \left\{q_{inj} - \frac{TP_s}{T_s} q\right\} = \frac{h\phi}{2} \frac{\partial \left(\frac{P}{z}\right)}{\partial t} \tag{4}$$

$$= \frac{\phi hC}{2} \frac{\partial m(P)}{\partial t} \tag{5}$$

where $C = \dfrac{\partial\left(\frac{P}{z}\right)}{\partial m(P)}$ is a unique function of $m(P)$.

For a partial water-drive gas reservoir, the semi-steady-state flow equations in terms of Kumar's (1977) f index and real-gas pseudo-pressure are:

$$q = \frac{2\pi \, kh\{m(P) - m(P_w)\}}{T\{\ln\left(\frac{r_e}{r_w}\right) - 0.5(1-f) + s'\}} \tag{6}$$

$$q = \frac{2\pi \, kh\{m(\bar{P}) - m(P_w)\}}{T\{\ln\left(\frac{r_e}{r_w}\right) - 0.25(3-f) + s'\}} \tag{7}$$

where $s' = s + Dq$

In shallow gas reservoirs, pressures and temperatures are low. In addition, reservoir permeabilities are relatively high and unsteady-state effects are small. Therefore turbulence effects are likely to be negligible.

Equations 6 and 7 are in terms of point pseudo-pressure and average pseudo-pressure, respectively. Using Equation 6 in Equation 4 we obtain:

$$\nabla \cdot \{kh \, \nabla m(P)\} + h\left\{q_{inj} - \frac{P_s}{T_s} \frac{2\pi \, kh\{m(P) - m(P_w)\}}{\left\{\ln\frac{r_e}{r_w} - 0.5(1-f) + s'\right\}}\right\}$$

$$= h \, \phi \, C \, \frac{\partial m(P)}{\partial t} \tag{8}$$

$$= h \, \phi \, \frac{\partial\left(\frac{P}{z}\right)}{\partial t} \tag{9}$$

Semi-Steady State Approximation

Under the stated semi-steady-state approximation, Coats gives the following equation

$$\frac{\partial\left(\frac{P}{z}\right)}{\partial t} = -\frac{P_s T}{T_s} \frac{Q_F}{V_p} \tag{10}$$

Substituting this in Equation 9, we get, setting $q_{inj} = 0$,

$$\left(\frac{T_s}{P_s T}\right) \nabla \cdot \{kh \nabla m(P)\} - \frac{2\pi kh^2}{T}$$
$$\left\{\frac{m(P) - m(P_w)}{\ln\left(\frac{r_e}{r_w}\right) - 0.5(1-f) + s'}\right\} = -h\phi \frac{Q_F}{V_p} \quad (11)$$

A finite-difference scheme for Equation 11 can be formulated either in r-z or in x-y (i.e., 2-dimensional) coordinates to seek a solution for m(P), for a specified Q_F and P_w, and hence P at any location in the uninvaded region, subject to the boundary conditions to be discussed next. Note that Equation 11 is good for point-to-point calculation. If, however, m(P) for the grid block is chosen as the block-average m(P) then the calculations would be more correct for smaller block sizes.

Boundary Conditions

Equation 11 can be solved for each block in the grid subject to a generalized mixed-boundary condition, which imposes a Neumann type (i.e., closed or no-flow) condition over the sealed periphery of the reservoir, and a pseudo-Dirichlet-type boundary condition (i.e., pressure falling with time and not constant, as typical of partial water drive) on the rest of the periphery (Figure 2). Two simple cases appear easy to solve. One is the case of a circular reservoir completely enclosed by an aquifer (an r-z model would be helpful); the other is the case of an elongated reservoir with edge-water drive (a one-dimensional model, basically). Whatever the case, the preceding discussion appears to indicate that the solution process would be required to trace the moving gas-water-contact with time and hence be able to calculate water influx as well.

Locating Gas Water Contact

For a bottom-water drive, an r-z model is particularly helpful to calculate the shrinking thickness between the gas-water contact and the top of the gas cap (Zakirov et al., 1966). However, for a two-dimensional areal calculation, as in Coats' paper, one is required to follow areal movement

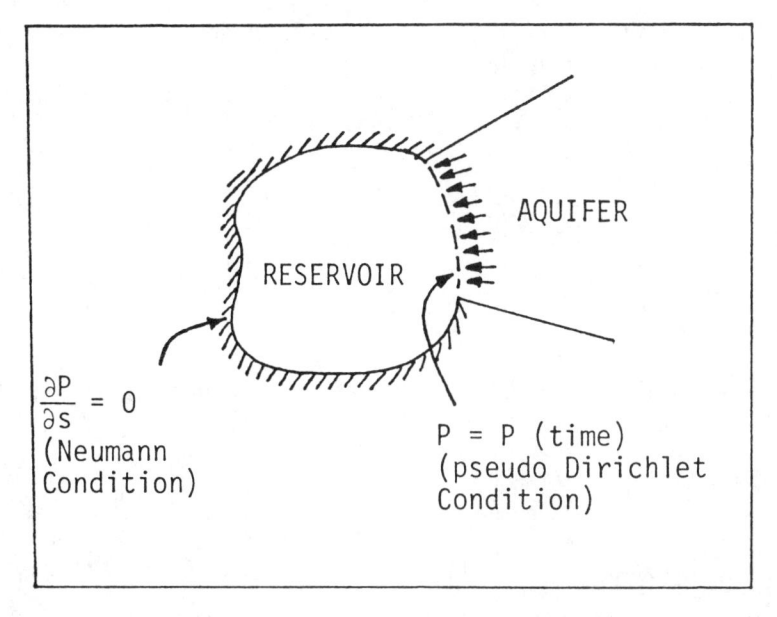

Figure 2. Boundary conditions for gas reservoir under partial water drive.

of the gas-water contact. For this purpose, a finite number of points, located on the initial gas-water contact, are first selected. These points are then moved according to the calculated fluid-flow velocity at their location. The fluid-flow velocities in x and y directions are calculated from the pressure distribution and Darcy's law (Field, 1973).

$$(v_x)_{i,j} = \frac{0.006329 \, (P_{i,j} - P_{i+1,j}) \, (k_x h)_{i,j}}{\mu_g \, \Delta y \, \phi \, (1 - S_{wi}) \, \{0.5(h_{i,j} + h_{i,j+1})\}} \tag{12}$$

$$(v_y)_{i,j} = \frac{0.006329 \, (P_{i,j} - P_{i,j+1}) \, (k_y h)_{i,j}}{\mu_g \, \Delta x \, \phi \, (1 - S_{wi}) \, \{0.5(h_{i,j} + h_{i+1,j})\}} \tag{13}$$

The velocity of the selected points then is easily computed as interpolated values from the v_x and v_y matrices. Field et al. (1973) have indicated the need to limit the time-step size in such calculations to maintain accuracy. The time step is limited by:

$$\Delta t \leq \frac{(\Delta x)(\Delta y)}{v_{max}} \qquad (14)$$

where

$$v_{max} = \sqrt{(v_x^2 + v_y^2)} \qquad (15)$$

The new position of a point P on the gas-water contact is then given by

$$(x_p)_{n+1} = (x_p)_n + (v_x)_p \left(\frac{\Delta t}{\Delta x}\right) \qquad (16a)$$

$$(y_p)_{n+1} = (y_p)_n + (v_y)_p \left(\frac{\Delta t}{\Delta y}\right) \qquad (16b)$$

where $(x_p)_n$ and $(y_p)_n$ are coordinate locations of the point P at time-step n. Such front-tracking calculations as described may also be conveniently obtained as a two-dimensional plot overlaid on the field-well locations.

Gas-in-Place Calculation

Gas in place in the entire field is related to pressure, production rate, and water-influx data, using a material balance equation for each grid block (represented by node i,j) and summing over the entire field. Thus, for each grid block i,j

$$\left(\frac{P}{z}\right)_{i,j} = \left(\frac{P}{z}\right)_i \left\{ \frac{1 - \dfrac{G_{P_{i,j}}}{G}}{\dfrac{G_{P_{i,j}}}{G} \dfrac{We_{i,j} \, E_i}{G}} \right\} \qquad (17)$$

where

$$E_i = \left(\frac{P}{z}\right)_i \left(\frac{T_{sc}}{P_{sc} \, T}\right) \qquad (18)$$

and for summation over the entire field

$$G = We\, E_i + \left(\frac{P}{z}\right)_i \sum_{i,j} \left\{\frac{We_{i,j}\, E_i - G_{P_{i,j}}}{\left(\frac{P}{z}\right)_{i,j} - \left(\frac{P}{z}\right)_i}\right\} \qquad (19)$$

Note that Equation 19 requires calculation of cumulative water influx and cumulative gas produced from each grid block (i,j). Cumulative water-influx calculation will be discussed next. Cumulative gas produced from each grid block can be obtained from the block's production rate history.

Water-Influx Calculation

Equation 19 requires calculation of field water-influx, We, and grid block water influx, $We_{i,j}$, as functions of production time. Field water-influx can be obtained using either van Everdingen-Hurst (1979) or Carter-Tracy (1958) model. Using the van Everdingen-Hurst model, which is suited for computer calculation

$$We = B \sum_{j=1}^{n} \Delta P_j\, Q_D\, \{t_{D(n)} - t_{D(j-1)}\} \qquad (20)$$

where

$$B = 2\pi h \phi_a c_{f+w} r_R^2\, f_R \qquad (21)$$

$$\Delta P_j = \begin{cases} \dfrac{P_{j-2} - P_j}{2} & \text{for } j > 1 \qquad (22a) \\ \dfrac{P_o - P_j}{2} & \text{for } j = 1 \qquad (22b) \end{cases}$$

where ΔP_j represents the reservoir pressure change for the j-th time step, in psi. Using the cumulative water influx calculated for the previous time step, the net water influx for the total field in any time step is determined. The net water influx is distributed among the invaded grid blocks (known from the initial and current position of the gas-water contact) according to their pressure and flow capacity, using the following equation:

$$\Delta We_{i,j} = \frac{(\Delta W_e) N_f \left\{1 - \left(\frac{Z_n}{Z_{n+1}}\right)\left(\frac{P_{n+1}}{P_n}\right)\right\}_{i,j} (kh)_{i,j}}{\sum_i \sum_j \left\{1 - \left(\frac{Z_n}{Z_{n+1}}\right)\left(\frac{P_{n+1}}{P_n}\right)\right\}_{i,j} (kh)_{i,j}} \quad (23)$$

From the knowledge of $\Delta We_{i,j}$, the incremental time-step water influx, one can obtain the cumulative water flux in the grid block, $We_{i,j}$, at any time.

The System Analysis Model

The mathematical model predicts the semi-steady state pressure distribution and the well production rates at any time. However, when the well is produced with technical equipment, the natural producing potential of the formation is not fully realized. The purpose of the system analysis model is to maximize the technical capacities of the producing system to minimize the loss in natural production potential.

For a gas reservoir producing only dry gas, the major elements (or components) in the system analysis scheme are vertical flow string, compressors, surface flow lines and accessories, and the gas delivery lines. The additional problems of dehydration, sweetening, and storage would require inclusion of additional elements in the production system analysis. In this paper, only the major elements have been considered. A typical production system installed on a gas well is shown in Figure 3. (The well will be produced only up to the time of total water invasion, and will remain abandoned thereafter or may be considered for injection, if necessary.) The equations of gas flow in three major segments, i.e., reservoir-to-bottomhole, bottomhole-to-wellhead, and wellhead-to-delivery exit lines are required for system analysis. Solutions to these equations would yield pressures at various locations in the entire flow net. Based on the pressures, and the optimum flow rates obtained from the solution to the mathematical model, an engineering estimate of equipment sizing can be obtained.

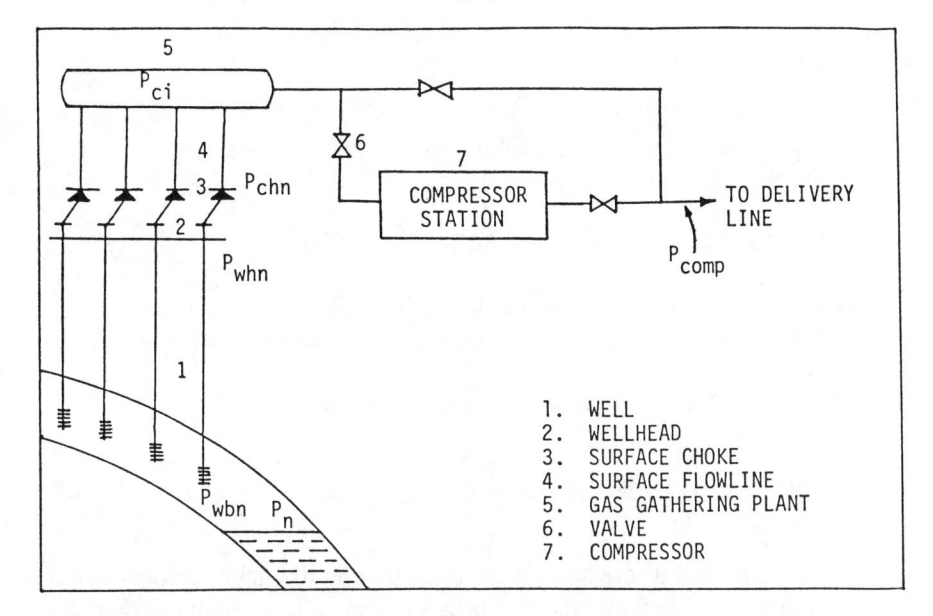

Figure 3. Typical production system for a dry gas field (modified from Zakirov et al., 1982).

Flow Equations

All equations below pertain to any well n in the field. These equations are given in terms of real pressures rather than pseudo-pressures (Zakirov et al. 1982).

1. Flow from reservoir-to-bottomhole

$$P_n^2 - P_{wbn}^2 = aq_n + bq_n^2 \tag{24}$$

where a and b are deliverability constants obtainable from well tests; q_n is obtained from solution of the mathematical model.

2. Flow from bottomhole-to-wellhead

$$P_{whn}^2 = P_{wbn}^2 - \frac{q_n^2 \, G \, \overline{T} \, \overline{z} \, f_1 \, L_n}{40{,}000 \, d_n^5} \left(\frac{e^{M_n} - 1}{M_n} \right) e^{-M_n} \tag{25}$$

328 Shallow Oil and Gas Resources

where $M_n = 0.0375 \dfrac{GL_n}{\overline{T}\,\overline{z}}$ (26)

and d_n and L_n are tubing diameter and length respectively.

3. Flow through wellhead choke

$$P^2_{chn} = P^2_{whn} - \psi_n q_n^2 \tag{27}$$

where ψ_n is a constant incorporating flow geometry (i.e., choke and surface-pipe diameters) and average physical properties of the flowing gas.

4. Flow from wellhead choke to common compressor

$$P^2_{ci} = P^2_{chn} - 94\,\lambda_{p_n}\,\overline{\rho}\,\overline{z}\,\overline{T}\,L_{p_n}\,q_n^2\,d_{p_n}^{-5} \tag{28}$$

where L_p and d_p are surface-pipe length and diameter, respectively, between the well and the compressor inlet. $\overline{\rho}$ is average gas density in the surface lines to the compressor.

All the preceding equations can be combined to obtain a single equation and then solved, or they may be solved simultaneously using an efficient iterative scheme, to obtain the pressures at various important locations in the entire flow system.

Compressor Capacity

Particular emphasis is required on compressor capacity while developing shallow gas reservoirs, hence, in the entire optimization scheme the compressor load is expected to play a vital role. The factors on which compressor capacity depends, the production rates of different wells, pressure losses in flow strings, and decline rate of reservoir pressure, are all functions of time. Hence, compressor load will vary with time and an optimum may be sought at all times. The capacity N may be calculated from

$$N = A\,Q_F(t) \left\{ \left\{ \dfrac{P_{comp}\,e^\gamma}{\sqrt{\dfrac{P^2_{ci}(t) - f_1\,Q_F(t)}{n(t)}} - f_2\sqrt{\dfrac{Q_F(t)}{n(t)}}} \right\}^{\frac{K-1}{K}} - 1 \right\} \tag{29}$$

where $\quad A = 870 \dfrac{\overline{P}}{\eta} \dfrac{K}{K-1}$ (30)

η = the compressor efficiency
n(t) = the number of wells as a function of time
f_1 = the friction factor in the vertical flow string
f_2 = the friction factor in the surface flow lines

In a practical case the optimization would be considered under the case of limited available compressor horsepower. Note that the optimum production rates suggested by the mathematical model may not correspond to the optimum production rates for minimum compressor horsepower, for a given equipment sizing and therefore a combined optimum may have to be established.

Economical Considerations

Once the mathematical model, coupled with the system analysis model, is able to predict an optimized production-development scheme, the resulting optimum parameters, production rate, number of wells drilled, compressor horsepower utilized, etc., can be used together with the data on gas price, compression cost, and interest and inflation rates, to forecast the total present worth of all future earnings and expenditures. Management decisions will then be based on these forecasted parameters.

Solution Algorithm

1. For the current number of producing wells, specified Q_F and P_w, perform a two-dimensional pressure calculation to obtain current remaining gas in place. Plot this as (n_{wo}, G_o) on an $n_w - G_r$ plot as shown in Figure 1.
2. a. Locate the current position of gas-water contact (GWC).
   b. For W admissible wells, perform two-dimensional pressure calculations W times, each time considering all the current wells plus one of the W admissible wells, thus obtaining G_r for all the W cases.
   c. Drill that wellsite in Step 2b that resulted in the lowest value of G, if the corresponding new GWC does not imply invasion of that location. If this location is invaded then ignore this from fur-

ther consideration and repeat Steps 2a through 2c. Continue until a new wellsite has been chosen. Plot the results on the n_w vs. G_r plot. Note that it is also necessary to check, during each pressure calculation, the number of already-drilled wells which have been invaded by water, and to ignore them in further calculation.
3. Perform two-dimensional pressure calculations for the existing wells, plus the well drilled in Step 2, and for one of the remaining admissible wells (accounting for the ones rejected due to water influx), as many times as the number of remaining wells, each time considering a different new well. Obtain a point $n_w - G_r$.
4. Repeat Step 3 until all the admissible wellsites are drilled or G_r calculated is less than or equal to the gas in place at abandonment.
5. From the auxiliary information available from Steps 1 through 4 (i.e., well production rates, number of wells producing, etc.), make a table of producing wells, wells watered out, wells drilled, drilling locations, individual well production rates, and reservoir average pressure against production time.
6. Using the table obtained in Step 5, do a system analysis for the entire field (note that only producing wells are considered in system analysis) at various selected times, chosen as in the preceding table. This obtains variation of compressor capacity with time for a fixed field rate Q_F and fixed bottomhole flowing pressure Pw.
7. Whenever the compressor load exceeds the available capacity, it is indicated that the specified field rate cannot be maintained. Set a new Q_F, if the economics justify, and repeat Steps 1 through 7. Note the time when Q_F is re-specified. Continue until the last Q_F specified differs from the Q_F at the economic limit, by a set tolerance level, or falls below it; or the cumulative gas produced equals or exceeds the gas at abandonment, earlier defined; or a new specification of Q_F causes immediate need for more than available compressor capacity.

Discussion

The techno-economic model discussed so far achieves the major objective of establishing an optimum drilling sequence to maintain a desired field production schedule under the available technical limitations, for a gas reservoir under partial water drive. In addition to this, several other management problems can be handled by this model.

Two alternative choices often faced when developing a gas reservoir under water drive are gas cycling or blow-down. Gas cycling is aimed at preventing loss of hydrocarbon by condensation to immobile saturation in the reservoir, or maintaining bottomhole flowing pressure above the sales-line pressure, thus reducing compression cost. Blow-down is adopted primarily to prevent loss of gas by trapping due to an active water influx. Since shallow-gas reservoirs are expected to be associated with low-pressure aquifers, blow-down of a shallow gas reservoir appears to be a remote possibility. Thus the presented model can also be used with an injection term in the mathematical model (Equation 8, for example, set to zero in the presented discussion) to explore the possibilities of better economics by gas cycling rather than incurring compression cost.

Our model, and Coats' original model, allows us to investigate the effect of reservoir heterogeneity on development decisions; or working in the reverse direction, allows us to establish reservoir heterogeneity distribution from field production and pressure data. The model also allows us to investigate the effect of different well spacing, and of decreasing well spacing with advancing GWC to compensate for lost production due to watering out of some wells.

Since the mathematical model keeps track of the cumulative gas production from each grid block, the degree of depletion of specific parts of the field (for example, different leases in the field) can be explored. Application of this model to a field example will be included in the final report of this on-going study.

Conclusions

1. A techno-economic model has been presented that performs a two-dimensional, semi-steady-state pressure calculation to obtain an optimum development drilling sequence for any shallow gas reservoirs under partial water drive.
2. A system analysis model also developed enables one to obtain required compression capacity against time for the well production rates given by the mathematical model of gas flow in the reservoir.
3. The system analysis model can also be used to obtain optimum equipment sizing for a minimized compression requirement, for given well production rates.
4. In addition to obtaining an optimum development program for a gas field under partial water drive, the techno-economic model also allows calculation of specific zone depletion, tracking of GWC, decisions regarding gas cycling or blow-down, and estimation of reservoir heterogeneity distribution.

Notation

A	constant in compressor capacity equation
a	deliverability constant
b	deliverability constant
B	water influx constant
C	a unique function of m(P)
c_{f+w}	compressibility of the aquifer
D	turbulence coefficient
d_n	diameter of vertical string in n-th well
d_{p_n}	diameter of surface pipe between wellhead and inlet of compressor for n-th well
e	exponential
E_i	initial expansion factor
f	Kumar's partial water drive index
f_1	friction factor for vertical tubing
f_2	friction factor for surface pipe
f_r	fractional angle of water encroachment
G	gas in place
G_o	gas initially in place
G_r	gas-in-place remaining
G_p	cumulative gas produced
G_a	gas in place at abandonment
h	thickness of pay
i	grid block node index
j	grid block node index
K	ratio of specific heats
k	permeability
k_x	permeability in x direction
k_y	permeability in y direction
L_n	length of tubing in n-th well
L_{p_n}	length of surface pipe from n-th well to compressor inlet
ln	natural logarithm
M_n	constant in (Equation 25) for n-th well
m(P)	real gas pseudo pressure
n	number of well; summation counter
N	compressor capacity
N_f	number of grid blocks flooded
n_w	cumulative number of wells drilled
n_{w_o}	cumulative number of wells drilled initially
P	pressure
P_{comp}	compressor outlet pressure, i.e., delivery pressure

P_{ci} pressure at compressor inlet
$\overline{P}$ average reservoir pressure
P_{wbn} pressure of n-th well bottom
P_{whn} pressure at n-th wellhead
P_{chn} pressure on the downstream side of choke for n-th well
P_w well-bottom flowing pressure
P_s standard pressure at surface
P_o initial pressure in the water influx equation
ΔP_j pressure drop in j-th time interval in the water influx equation
q_n well production rate of the n-th well
Q_D dimensionless water influx rate
q_{inj} well injection rate, volume per unit time
q well production rate, volume per unit time
q_{inj} well injection rate, mass per unit time
q^m well production rate, mass per unit time
Q_F total field production rate
R gas constant
r_R effective reservoir radius
r_e drainage radius; for a grid block: $r_e = \sqrt{(\Delta x)(\Delta y)/\pi}$
r_w wellbore radius
s skin factor
S_{wi} connate water saturation
T temperature
$\overline{T}$ average wellbore temperature
T_s standard temperature at surface
t time
Δt time step size
$t_{D(n)}$ dimension time corresponding to end of a period in water influx calculation
v_x velocity in x direction
v_y velocity in y direction
v_{max} resultant or maximum velocity
V_P field's total pore volume
We field water influx (cumulative)
ΔW_e incremental water influx for the field
$We_{i,j}$ cumulative water influx in the grid block i,j
$\Delta We_{i,j}$ incremental water influx in the grid block i,j
Δx grid size in x direction
Δy grid size in y direction
x_p location of a point P on GWC along x coordinate
y_p location of a point P on GWC along y coordinate
z gas deviation factor
$\overline{z}$ average gas deviation factor in wellbore

Greek Symbols

μ_g gas viscosity
$\bar{\rho}$ average gas density in surface pipes
e^γ coefficient accounting for weight of gas in compressor capacity equation
η compressor efficiency
∇ gradient operator
Δ operator to denote increment in quantity
∂ del or differential operator
ϕ porosity
$\int$ integral sign
π 3.14159 . . . .
ψ_n constant in (Equation 27) for well n
λ_{p_n} constant in (Equation 28) for well n

References

Al-Hussainy, R., et al., 1966, The flow of real gases through porous media: JPT, May, p. 624.

Aronofsky, J. S., and Williams, A. C., 1962, The use of linear programming and mathematical models in underground oil production: Management Science, July, Vol. 8, p. 374.

Burns, P. G., 1969, Gas field development: automatic selection of location for new producing wells: M.S. Thesis, University of Texas at Austin.

Carter, R. D., and Tracy, G. W., 1958, An improved method for calculating water influx: Trans., AIME, Vol. 219, p. 415.

Chierici, G. L., et al., 1963, Experimental research on gas saturation behind the water front in gas reservoirs subjected to water drive: paper 17, proc., 6th World Petroleum Congress, Section II, p. 483.

Coats, K. H., 1969, An approach to locating new wells in heterogeneous gas producing fields: JPT, May, p. 549.

Coats, K. H., et al., 1970, A new technique for determining reservoir description from field performance data: SPEJ, March, p. 66, Trans., AIME, Vol. 249.

van Everdingen, A. F., and Hurst, W., 1979, Application of the Laplace transformation to flow problems in reservoirs: Trans., AIME, Vol. 186, p. 305.

Field, M. S., et al., 1973, Kaybob south-reservoir simulation of a gas cycling project with bottom water drive: SPE Reprint Series No. 11 on Numerical Simulation, SPE of AIME, Dallas, Texas.

Geffen, T. M., et al., 1952, Efficiency of gas displacement from porous media by liquid flooding: Trans., AIME, 1952, Vol. 195, p. 37.

Green, D. W., et al., 1971, Some recent advances in methods of oil and gas reserves estimates: paper presented at Eighth World Petroleum Congress, Moscow, June 13-19.

Henderson, J. H., et al., 1968, Use of numerical reservoir models in the development and operation of gas storage reservoirs: *JPT,* Nov., p. 1239.

Hessing, R. C., 1966, How to evaluate well locations in a gas storage reservoir, OGJ, March 7, Vol. 64, p. 105.

Kumar, Anil, 1977, Steady flow equations for wells in partial water drive reservoirs: *JPT,* December, p. 1654.

Lee, A. S., and Aronofsky, J. S., A linear programming model for scheduling crude oil production: *JPT,* July, p. 51, Trans. AIME, Vol. 213.

Nandan, Braj, 1970, An automatic history matching procedure for use in mathematical modeling of hydrocarbon reservoirs: M.S. Thesis, University of Kansas, Lawrence.

Rosenwald, G. W., 1972, A method for determining the optimum location of wells in an underground reservoir: Ph.D. Dissertation, University of Kansas, Lawrence.

Rowan, G., and Warren, J. E., 1967, A system approach to reservoir engineering: optimum development planning: *JCPT,* July-Sept., p. 84.

Wattengarger, R. A., 1970, Maximizing seasonal withdrawals from gas storage reservoirs: *JPT,* Aug., p. 994.

Zakirov, S. N., et al., 1982, Mathematical modeling of gas field development and production optimization: paper SPE 11181 presented at the 57th Annual Fall Technical Conference and Exhibition of SPE of AIME, Sept. 26-29, New Orleans, Louisiana.

30
Natural Gas-Fired Co-generation Topping Plant

Thomas J. George

Introduction

In 1977 a crippling shortage of gas, real or contrived, was experienced in the eastern half of the United States. As a result, the Federal Government formed the Department of Energy; new gas price regulations were formulated; and gas exploration and production drilling increased vastly.

In 1981-82 the 1977 situation began to reverse. Large amounts of available gas added during an economic recession caused a glut of gas. Public gas utility companies were forced to curtail purchases in 1982 and predictions of shut-ins are projected through at least 1988.

It is becoming obvious that until economic conditions improve, many gas wells in Appalachia will either have to remain shut-in for part of the year or alternative industrial uses for the gas will have to be found. One logical alternative use for natural gas is the generation of electric power.

Niagara-Mohawk Power & Light Co. of New York and Penelec Co., two major electrical companies in Western Pennsylvania and Western New York States have both evidenced an interest in purchasing electric power under certain circumstances. A group of five public utility companies, including Penelec, built a nuclear generating station at Three Mile

Island, near Harrisburg, Penn.; this operation required a massive capital investment. The plant, now closed, may never reopen. This loss of generation is one reason for the utility company's interest in purchasing power from outside facilities.

The availability of large amounts of shut-in gas in Appalachia, coupled with the demand for electric power, suggests the desirability of establishing gas-fired electric-power-generating facilities. Generating electric power with natural gas is not new; however, as will be explained later, achieving high efficiency and profitability demands finding a use for the heat generated by the engine. Determining the feasibility of such cogeneration of heat and power necessitates:

1. Locating available shut-in gas wells having sufficient reserves to justify investment in an electric generating facility.
2. An acceptable point of sale near a utility, industry, or industrial user.
3. A market for the heat, steam, or exhaust emissions.
4. Ability to raise necessary capital, through joint ventures, to own and operate the generating facilities with or without participation of the gas producer.

In most cases, if a supply of natural gas is present, power lines are sufficiently close to justify a generating plant. The problem lies in attaining full efficiency, a way to utilize the engine heat. Unfortunately, industrial users are seldom adjacent to the generating plants. In-field use for the heat must be found. Among the possibilities are greenhouses or peat moss driers, but these uses may not always be practical.

This writer has developed an in-field use, whereby gas from wells is used to fuel engines to run generators. Heat from the engines is used to distill crude oil from the wells. The distillates are then used to dissolve the paraffin which builds up in the well or on the equipment.

The natural gas-fired, cogeneration topping plant (GFGTP) is designed to be a comprehensive unit that will utilize the maximum energy contained in a Btu of field natural gas. This not only provides an alternate use for field natural gas, other than sale to a utility company, but also solves a problem faced by producers of hot natural gas produced in the Appalachian region: namely, how does a producer capitalize on the additional Btu's present in a cubic foot of gas, for which the utility companies do not compensate him?

In defense of the utility companies in this part of the country (northeastern U.S.A.), they are not compensated for additional Btu's. Utility companies, in fact, are penalized for taking high Btu gas from the producer, because the utility customer then needs fewer cubic feet of gas to obtain equivalent heat, but pays only for standard gas of 1,000 Btu's per cubic foot.

Unless there is sufficient volume to justify a stripping plant, the utility companies must take the high Btu gas into their systems and, in various ways, must adjust the total Btu value of the gas supply to meet the standards called for in their tariffs.

Using high Btu natural gas directly, as a fuel source for a power generator or a drive unit, eliminates the stripping operation which, of course, is energy intensive. The GFGTP is designed to utilize the total heat contained in a cubic foot of natural gas in a profitable way.

The GFGTP Co-Generation Unit A is a rather simple and relatively inexpensive cogeneration application. It is designed to utilize the natural gas from the well directly, to drive an electric-power generator. The exhaust and manifold heat are then utilized to heat crude oil to a point sufficient to vaporize the light ends. The vapors are distilled into a paraffin solvent which we shall call "naphthalene." The residual crude is then collected and sent to market at a price at least as high as that for the crude on a direct sale, undistilled.

Figure 1 shows the process. Gas is piped from the well (A) to the drive unit of the power generator (B). The exhaust from the power generator is piped through a heat tube into vessel C, which contains crude oil. Manifold heat also passes through a water or steam system to coils within the heating vessel. Vapors from crude oil heating vessel (C) are piped to a coil immersed in water in the distilling tank (D). The vapors are distilled and piped to the naphthalene collection tank (E). In a batch operation a measured amount of crude oil is piped to the heating vessel, where at 400°F, the distillation is completed. When the residual is sufficiently cooled, it is transferred to crude collection tanks for transportation to market.

The produced naphthalene has many uses, in this example as a paraffin solvent. Paraffin accumulation in the equipment and on the sand face is a major problem with Penn-grade crude oil and in many other producing areas. Since naphthalene goes into solution with the crude oil, it is an excellent paraffin solvent. The naphthalene is completely recoverable and can go directly to market as part of the bulk. The end product is completely compatible with the refining process, whereas other paraffin sol-

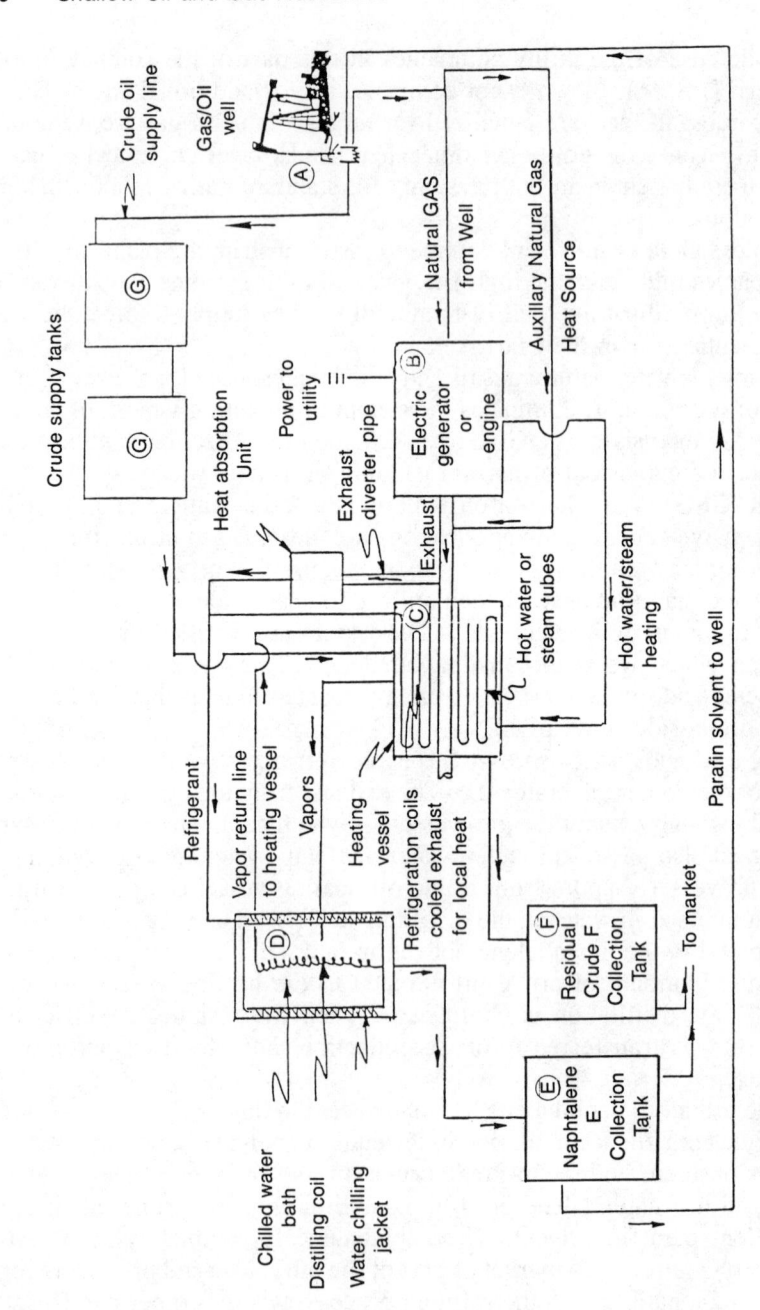

Figure 1. Cogeneration unit for generating electric power and distilling crude oil.

vents have chloride or other bases that are rejected by refineries because they are corrosive.

This entire process can be looked at as a self-greasing axle. That is, natural gas is piped to the generator, heat from the generator to the heating vessel, vapors from the heating vessel to the distilling tank, distillate to the solvent collection tank, and residual crude to a collection tank and to market. Subsequently, the naphthalene is pumped back to the well for paraffin removal and to market, and the oil is pumped from the well to the crude supply tanks (G) and thence, to the heating chamber. The only items "lost" are the low-grade heat from the cool exhaust, which can be used for local heating in nearby shops or buildings, and waste heat from the heating vessel. The idea, of course, is to maximize the economics of the system by utilizing all energy.

Natural gas is used not only to run the generator drive unit, but also is piped into the exhaust tube, where the gas is burned to heat the oil in the event the generator is down during the distillation batch. An exhaust diverter pipe is designed to divert the exhaust, when the batch is completed, to another batch tank or into the air during cooling of the batch. In addition, a heat absorption unit, using exhaust and manifold heat, can be used to cool the heating vessel after the batch is completed, as well as the water in the distillation tank. Otherwise, cool replacement water is required to maintain the necessary temperature for the distillation process. Since some vapors are not distilled in the process, a return line brings undistilled vapors back to the heat tank for recycling.

FERC regulations state: "When natural gas is used as a prime source to drive a generating unit, there must be an efficiency factor of at least 42%." In the case of Niagara-Mohawk Power & Electric Co. of NY, if one does not qualify as a cogeneration unit, Niagara-Mohawk (N-M) can pay only 3.9% per KWH. If the unit does qualify, N-M must pay its "avoided cost," currently 6¢ per KWH. This 21% additional payment can be the deciding factor as to whether it is feasible to utilize natural gas and pay for the cost of the cogeneration unit.

If a congeneration unit is properly designed and the economics are favorable, it will pay to install a cogeneration unit, based on a known price per cubic foot of gas. With escalators, the unit can be satisfactorily amortized and the economics of the well justified. Table 1 shows an economic analysis of an electrical power-generating facility, with comparative returns on investment utilizing gas with 905 and 1,200 Btu's per cubic foot.

Table 1
Analysis of Electrical Power-Generating Facility Utilizing Natural Gas-Fired Internal Combustion Engine One Caterpillar Model G 399NA
(Based on 7-year amortization of equipment)

Gas engine specifications:

  5,560 cu ft/hr yields 460 kW
  Note: Gas having LHV of 905 Btu/cu ft

Gas costs:

$$\frac{5{,}560 \text{ cu ft/hr}}{460 \text{ kW}} \times 460 \text{ kW} \times \$3.00/1{,}000 \text{ cu ft} = \$16.68/\text{hr gas cost per hr of operation}$$

Equipment costs:

  G 399NA $ 200,000.00
  Installation   + 25,000.00
                   $225,000.00 ÷ 60,000
                   (Est. life of equip.) = $3.75/hr cost per hr to own equipment

Operating and administrative costs:

  Cost per hour to operate equipment $2.50/hr
  Administrative costs (insurance)   + .25
                                        $2.75/hr

  Total generating cost   +2.75/hr Total generating costs
  (less cost of fuel) $ 6.50

Natural Gas-Fired Cogeneration Topping Plant

$16.68/hr
Fuel costs @ $3.00/mcf gas @ 905 Btu = $23.18/hr

Total cost to own equipment and produce power = $23.18

$16.68/hr
Fuel Costs @ $3.00/mcf gas @ 1,200 Btu = $13.00/hr

Sales:

460 kW × $.051/kW-hr = $23.46 (Assets)
                        −23.18 (Liabilities)
             Net profit = $00.28

Investment return

@ $.28/$3.75 × 100 = 6.133%
@ 1,200 Btu/cu ft = 118.6%

905 Btu Gas

$ 225,000.00	Int. investment	
150,000.00	Oper. costs (over 7 yrs)	
1,000,800.00	Fuel costs	
1,375,800.00		

Sales @ $23.46 × 60,000 hrs = $ 1,407,600.00
                              −1,375,800.00
                    Return = $ 31,800.00

1,200 Btu Gas

$ 225,000.00	Int. investment	
150,000.00	Oper. costs (over 7 yrs)	
780,000.00	Fuel costs	
$1,155,000.00		

Sales @ $23.46 × 60,000 hrs. = $ 1,407,600.00
                               −1,155,000.00
                     Return = $ 252,600.00

31
Natural Gas—The Politically and Environmentally Benign Least-Cost Energy for Successful 21st Century Economics, the Energy Path to a Better World

Robert A. Hefner III

Introduction

The motor of our economic systems is driven by energy: the agricultural system energizes people, and people, utilizing all additional forms of energy, energize the other productive systems of civilization. No product can be made nor any service provided without the expenditure of energy. Therefore, most fundamental in the successful development of an economy is access to an abundant, acceptably priced source of energy.

There are 3.6 billion people living in the lesser-developed nations. To close the widening gap between the standard of living in the developed industrial economies and the developing economies (often almost hopelessly burdened by debt), a goal so vital to world peace and increased global quality of life, the per capita consumption of primary energy must significantly increase in the developing economies over the near- and mid-term. I believe the only abundant, timely energy source that can economically meet that demand is natural gas, which has the added advantage of meeting that demand without pollution or proliferation.

Without increased per capita energy consumption in the developing countries, permitting real growth in standards of living and reduction of debt, there will be further loss of hope and additional economic erosion. This may aggravate the existing global financial strain, which undoubtedly will lead to extended periods of high inflation and high unemployment, frequent recessions, and possible political chaos.

I therefore believe that an efficient, least cost, intelligently integrated energy system is fundamental to the success of all economies, developed or undeveloped, whether struggling to maintain present high standards of living or to raise lagging standards of living, in meeting the challenge of economic efficiency, growth, and global competition as we move toward and into the 21st century. Efficient integrated energy systems fueled by methane may become the energy bridge by which successful nations will meet this challenge and achieve goals of development and economic security as the world moves ahead.

Differentiation Between the Oil and Methane Industries

The understanding of natural gas and methane technology has been buried within, and blinded by the focus on crude oil. Natural gas is geologically, technologically, economically, and politically different from oil. Until these fundamental differences are clearly and fully understood, we will not be able to competently assess our energy options.

- A significant geologic difference between oil and natural gas is that natural gas is literally pervasive throughout the world's sediments and, therefore, not limited in commercial quantity by the unique trapping circumstances necessary for the commercial production of oil. How many countries now, or in the recent past, believed that they possessed little or no natural gas because their *oil* exploration results over the past decades have been negative? To name just a few: Pakistan, India, Japan, Taiwan, Brazil, Argentina, New Zealand, Malaysia, West Germany, and China.
- A significant technological difference between oil and gas is that natural gas is compressible and oil is not (Boyle's law). Therefore, gas can be produced commercially from rocks through which oil cannot even move. Additionally, a barrel of oil produced from 30,000 feet is still a barrel of oil at the surface; whereas a barrel of gas at 30,000 feet can be as many as 500 barrels of methane at the surface. Thus, as one drills deeper, (1) the quantity of gas recoverable from any given reservoir is increased by each additional increment of pressure,* and (2) reservoirs

---

* Between 0.5 and 1.0 psi per foot of depth

of lesser quality become more and more capable of commercial gas production.
- A significant political and economic difference between oil and gas is that in the United States, United Kingdom, and many other nations, nearly all of the gas consumed in the past three decades has been found either during the search for oil or produced in association with oil, with the gas subject to stringently low price controls. Those controls maintained the end use price for natural gas at a fraction of the cost of competing energies, thus distorting supply and demand and creating the perception of shortages amongst plenty.

The setting for our recent misconception that we were running out of natural gas began with our journey to the moon. Our astronauts by their descriptions and photographs provided us images of our "spaceship" earth, a tiny and fragile-appearing "jewel" floating in a sea of infinity. It was, I believe, this image of our planet which led to the perception of resource shortages embodied in the Club of Rome's Malthusian statement to mankind in 1972 (Meadows et al., 1972). The finite resource conception embodied in that report, combined with the belief held by most economists and experts that the consumption of energy in an economic system was largely, if not totally, inelastic (a "fact" thoroughly "proven" by the statistics of the past), led to the "understanding" by most experts and policy makers that natural resource commodities, particularly oil and gas, were finite and nearing exhaustion at their current rates of consumption. They were thinking, of course, more of oil, which in fact may be reaching a global state of mature development. Grounded upon these misperceptions was the birth of OPEC, the "oil shocks" of the 1970s, and the "certainty" of $100 per barrel oil; this, if nothing else, led us back to the principles of supply and demand and economic elasticity, even in energy consumption.

In hindsight, it is easy to understand the fundamental fallacy in the Club of Rome's concern over finite resources, for the rate of consumption of the world's resources is inversely proportional to the full economic per-capita cost of primary energy. Therefore, I would say: "to whatever extent mankind is eternal, resources are infinite." When has civilization ever "run out" of a natural resource? Never. We simply stop using a resource in short supply and substitute something else. Man has great ability to adapt and to create, thus allowing civilization to move from an increasingly costly technology to a better, less costly one.

Oil mentality has not only dominated and distorted the search for natural gas, but oil-based distortions have influenced statistics, and therefore statistically based econometric projections regarding the methane resource base. At best the most prestigious statistically based econometric projections of the natural gas resource base were merely conservative projections of the amount of gas that may in the future be found and produced along with oil and at worst were the basis for tragic and costly errors made by economic planners and political leaders. It is vital to understand how previous misconceptions regarding the abundance of the methane resource base have led to seriously misdirected energy policies. In the United States, I believe the misconceptions of methane shortage will be recorded in history as one of the most economically, geopolitically, and geostrategically costly of the post-World War II era.

Had policy makers in both industry and government, dominated by an oil mentality, been aware of the domestic abundance of economic natural gas, the United States economy would have benefited by savings of hundreds of billions of dollars and hundreds of thousands of man years which were wasted on nuclear plants, new coal-fired electric-generating facilities, LNG projects, shale oil, the entire synfuel program, and the husbanding of Alaska gas.

The developing nations of the world must seize the opportunity of hindsight and benefit through an understanding of the costly errors committed by the developed economies as we reacted to the oil shocks of the past decade. All nations can benefit by understanding thoroughly the oil-related reasons for the unrealistically conservative statistical projections of the remaining natural gas resource base in the United States. All statistical approaches to determining the natural gas resource base are so distorted by their inseparable economic association with oil as to be totally flawed and their results to be unrealistically conservative. Estimates such as those used by the U.S. Office of Technology Assessment in its publication entitled *U.S. Natural Gas Availability* (Plotkin, 1983) appear in Figure 1, which shows dramatically that "sound" statistical projections of natural gas supply by U.S. experts have, to my knowledge, always been so conservative as to be useless. The reason for these errors lies in want of understanding of the distinction between oil and gas and of the necessity of separating consideration of gas not associated with oil from gas that has been a byproduct of oil exploration and exploitation.

Methane Is Pervasive and in Abundant Supply

The important point is that I believe methane to be pervasive in sediments throughout the world. My focus is not on either shallow or deep methane, both of which are abundant, but on "economic" methane. I believe the world contains many more times the number of Btu's of energy in the form of commercially available nonassociated methane than in the form of commercially available oil.

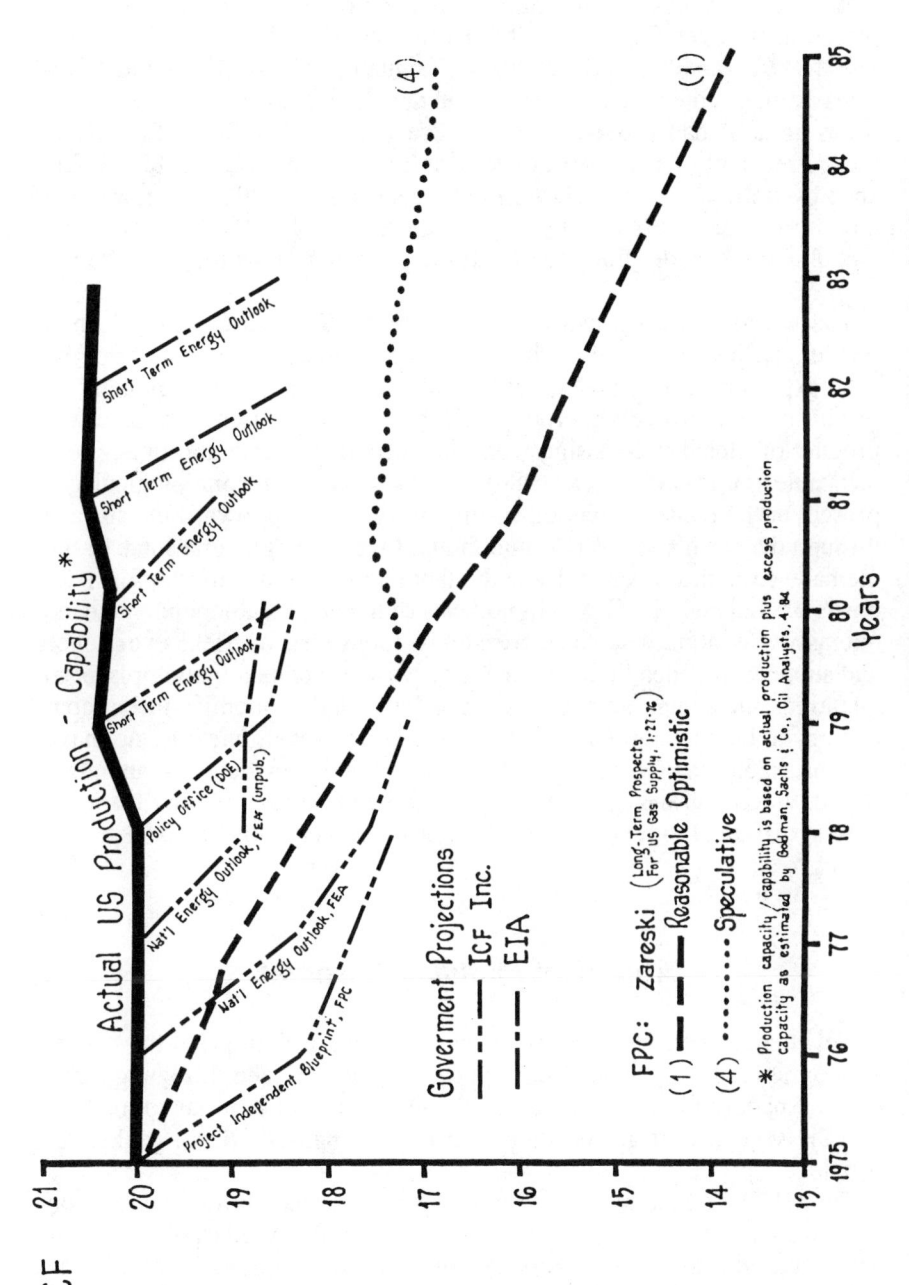

Natural Gas—The Energy Path to a Better World 349

Figure 1. Projected production capability vs. actual production capability of conventional U.S. gas.

Authoritative estimations of worldwide proved reserves of gas are placed at between 30,000–40,000 trillion cubic feet (TCF) *(World Oil,* 1983; *Petroleum Information,* 1984; Trigueiro, 1984). The United Nations estimates the potential reserves in developing countries to be 4 to 24 times greater than proved reserves (Brewster et al., 1981). Eric Giorgios, President of the World Gas Conference, opened the 1982 conference by stating, "The world will have consumed by 2020 no more than one-third of its total natural gas. It is, therefore, obvious that there is no risk of a world-wide shortage of natural gas either in the medium or long term."

The world-renowned astronomer, Professor Thomas Gold of Cornell University, has advanced the hypothesis, now increasingly supported by field experience, that most quantities of natural gas may be abiogenic in origin, not a hydrocarbon product of fossil decomposition, but rather the product of global out-gassing from the core of the earth and thus, inexhaustable (Gold and Soter, 1980). He theorizes that nonbiological gas present at the planet's creation continuously seeps up toward the surface through the earth's major rifts and faults. Over geologic time some of the methane combined chemically with plant and animal material to form deposits of coal and oil. Gold's hypothesis of the earth's abiogenic methane outgassing continues to be increasingly reinforced by field experience and supports the idea that methane is pervasive, abundant, and present in quantities more than sufficient to support global economies throughout and beyond the next great energy transition. However, one does not have to subscribe to the Gold theory to conclude that the evidence of abundant world-wide natural gas reserves is overwhelming and that the resource base is sufficient to fuel the world's economies well into the next century.

Abundant Methane Reserves

New major methane discoveries are occurring in all areas of the world. A geographic list of major finds appears in Table 1. The following maps of the Anadarko Basin (Figures 2–7), which is most familiar to me, depict the sequence of gas development in the Anadarko Basin geological province.

The GHK Companies have 25 years of experience in the exploration and exploitation of high-pressure natural gas in the Anadarko Basin. We have focused, during our company's history, on natural gas, not oil. Although the Anadarko Basin has always been a very gassy province, statistics until recently have underestimated the potential of unassociated gas

Table 1
Worldwide New Major Methane Discoveries

Africa:

    Egypt—Badr al-Din, Timsah field, Tinah field
    Cameroon—North Matenda
    Ivory Coast—Espoir
    Tunisia—Bir Aouine

Australia/New Guinea:

    Northwest Shelf (Gorgon), Barrow Island, Cooper/Eromanga basin, Amadeus Basin, Gippsland Basin, Papua

Europe:

    United Kingdom—Southern North Sea, Central North Sea, Celtic Sea Basin
    Norway—Troll Field, Tromso-Hammerfest Basin, Traena-Helgeland Basin, Sleipner Field, Askeladden Field
    Netherlands—Dutch North Sea
    Ireland—Kinsale Field
    France/UK—Anglo-Parisian Basin
    Austria—Vienna Basin, Molasse Basin
    Germany—Kinsau
    Italy—Cantabrian Basin, Campobasso area, Po Basin (deep)
    Spain—Gaviota Field, El Serrablo Field
    Switzerland—Finsterwald Field
    Yugoslavia—Adriatic Sea

Far East:

    China—Hainan Island
    India—Bombay High Field, South Bassein Field
    Malaysia—Sabah, Trengganu
    Pakistan—Northern Potwar, Badin Block, Baluchistan Province, Pirkoh Field, Kandhkot Field, Mari Field, Sui Field
    Burma—South Irrawaddy River

Middle East:

    Saudi Arabia—Permian Khuff formation (Ghawar, Uthmanigah, Haradh, Abu Safah)
    United Arab Emirates
        Abu Dhabi—Sheif Field, Marghan Field
        Sharjah—Saaja Field
        Dubai—Margham
        Ras al Khaimah—offshore
    Oman—Russaya, Gulf of Oman
    Iran—Kangan Field, Khangiram
    Bahrain—Khuff Field
    Israel—Negev Desert
    Qatar—Khuff formation

table continued

**Table 1
continued**

North America:

United States—Texas Gulf Coast/Texas Tabasco Field
　　　　　　　Alaska—Beaufort Sea (Seal Island), Taktovaktak Peninsula, North Slope Point Thomson area, Bering Sea (Navarin Basin, Norton Sound, St. George Basin)
Canada—Newfoundland (Hibernia), Sable Island (Venture)
Mexico—Monclova area, Huimanguillo Field

South America:

Venezuela—Monagas, Anzoatagui, Guarico, Zulia, Gulf of Paria
Brazil—Upper Amazon, Potiquar Basin, Reconcavo, Marajo
Argentina—Neuquen Basin, Isla Grande
Colombia—Llanos Basin
Trinidad and Tobago—Teak Field, Poui Field
Bolivia—Vuelta Grande, Porvenir Field
Chile—Magellan
Peru—Camisea River

U.S.S.R:

West Siberian Basin (Kamal, Yamburg, Belyy Island, Kara Sea), Golitsyn, Turkmenia (Dauletabad, Donmez), Yamburgskoye, Urengoy, Pechora Basin (Vuktyl Field, Vozey-Usa), Karachaganak, East Turkmenistan, Southern Uzbekistan

---

wells. The maps clearly indicate that the Anadarko province, for all practical purposes, is one giant gas field.

Proven ultimate reserves of natural gas in the Anadarko Basin stand today at about 140 TCF, equaled in North America only by the entire U.S. Gulf Coast. The much more thoroughly explored shallow horizons, less than 15,000 ft (4,573 m), account for about 80% of the proven reserves and will probably account for 20% to 30% of the yet undeveloped reserves. We confidently predict with 90% certainty that approximately 60 TCF of gas remains to be developed from all depths. We believe there is at least a 30% chance of equaling the already developed 140 TCF of ultimate proven reserves.

These estimates obviously cannot be reconciled with other recent estimates, by such experts as those at Exxon. They estimate that the total undiscovered gas potential of the United States is only about 300 TCF (Wheeler, 1984). However, we believe our 25 years of active experience provides us greater ability to predict with significantly higher reliability

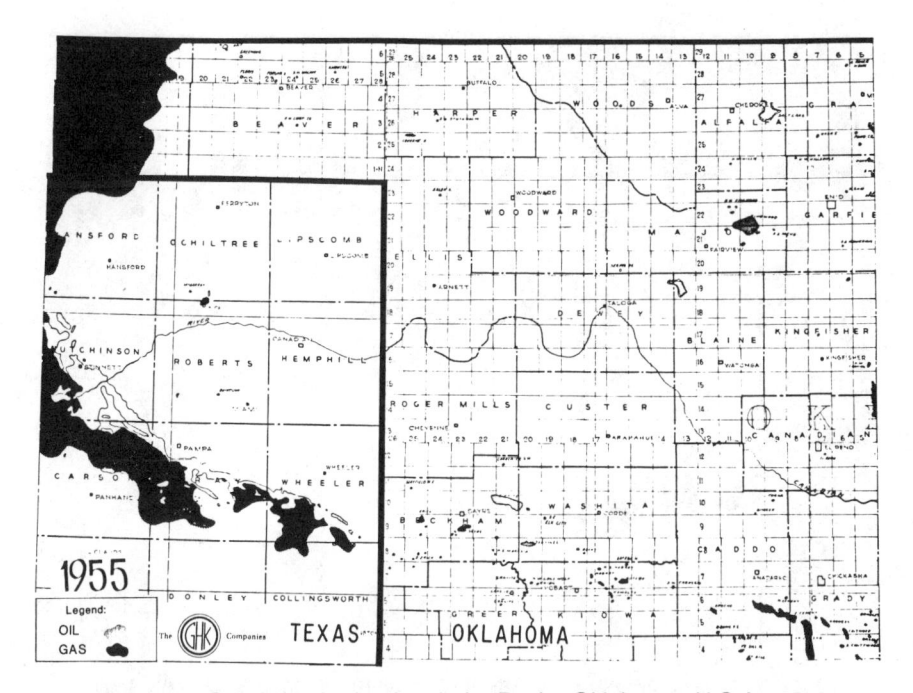

Figure 2. Gas fields in the Anadarko Basin, Oklahoma, U.S.A., 1955.

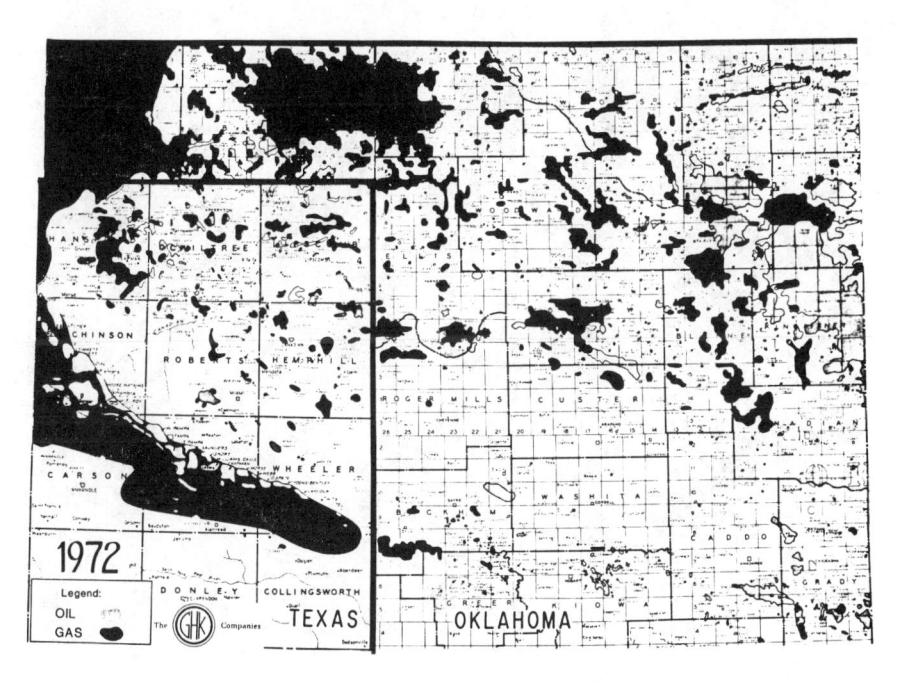

Figure 3. Gas fields in the Anadarko Basin, Oklahoma, U.S.A., 1972.

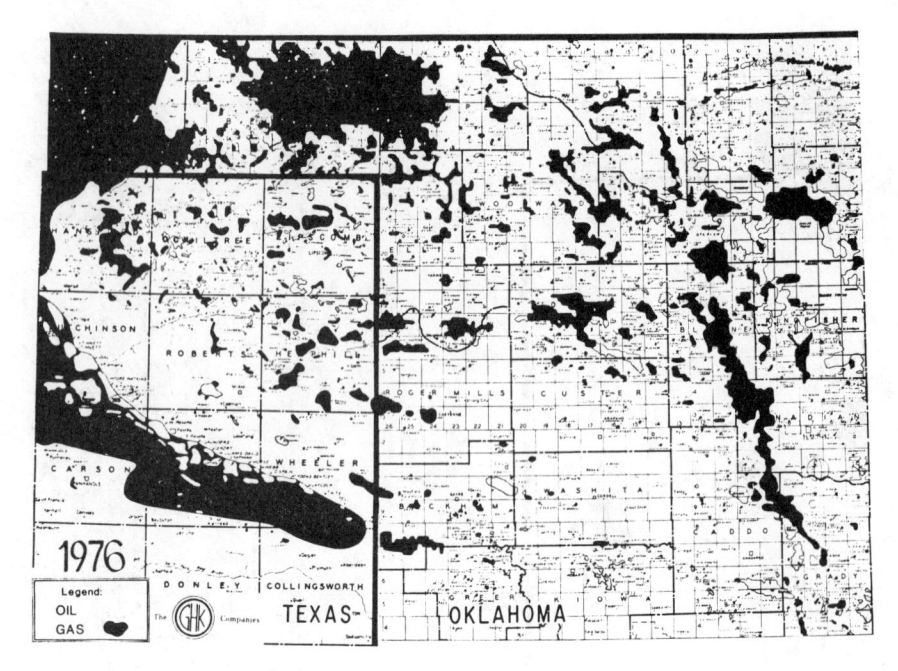

Figure 4. Gas fields in the Anadarko Basin, Oklahoma, U.S.A., 1976.

Figure 5. Gas fields in the Anadarko Basin, Oklahoma, U.S.A., 1979.

Figure 6. Gas fields in the Anadarko Basin, Oklahoma, U.S.A., 1982.

the gas resources of the Anadarko Basin than those who are focused on oil and who have been comparatively inactive in the basin for most of its history. In assessing the probability for significant remaining natural gas in the Anadarko Basin, it is important to realize that less than 10% of its volume of deep sediments has been developed, yet this volume yields over 6 TCF of proven natural gas reserves.

Therefore, having experienced for over two decades unrealistically conservative natural gas resource projections for one of the world's largest gas provinces, the Anadarko Basin, made by the same experts who today remain equally pessimistic for all of North America, I have come to the conclusion that the "Anadarko error" is a result of oil-dominated thinking and is applicable to all of North America and for that matter the entire world. The size of the resource base for natural gas must no longer be an issue for debate.

356 Shallow Oil and Gas Resources

Figure 7. Gas fields in the Anadarko Basin, Oklahoma, U.S.A., 1983.

Primary Energy Market Penetration and the Future for Methane

Methane has an abundant economically recoverable resource base. Given that fact and our need for a national energy policy, it is useful to examine the history of primary energy use, during which one fuel source has succeeded another in market dominance.

The International Institute for Applied Systems Analysis (IIASA)* in Vienna is a world-recognized agency supported by the United States (40% approximately), the USSR (40%) and other nations, including several OPEC countries (20%). At the Institute, Dr. Cesare Marchetti analyzed the rates at which new forms of energy displaced existing energy

---

* IIASA members are: Canada, Czechoslovakia, France, E. Germany, West Germany, Soviet Union, Japan, Bulgaria, United States, Italy, Poland, United Kingdom, Austria, Hungary, Sweden, Finland, and The Netherlands.

forms (Marchetti, 1979). From energy market-share data, beginning with the 19th century Industrial Revolution, Dr. Marchetti discerned a mathematically consistent pattern of bell-shaped curves. Beginning in about 1840, the world was fueled by wood; wood then slowly was displaced by coal, coal by oil, and now oil by methane (Figures 8-10).

About half-way through the work, Dr. Marchetti and his co-workers observed that their "Logistic Substitution Analysis" had much broader application than just to energy. The broader concept to be examined became "technology substitution," the rate at which new technologies displace old technologies. The energy analysis (coal displacing wood, oil displacing coal, methane displacing oil) was reexamined on the basis of technology substitution rather than energy substitution. This analysis was observed to be sound, in that world energy systems like coal, oil, and methane are complex and require complex infrastructures. For a new technology to replace an old technology requires the emplacement of a new infrastructure and the retraining of millions of people.

Coal technology is relatively simple. Lumps of coal can be thrown together under a boiler, the burners ignited, and the coal burned easily. The technology of oil is much more complex. First, one has to invent chemical engineering and then the chemical engineers have to invent the refin-

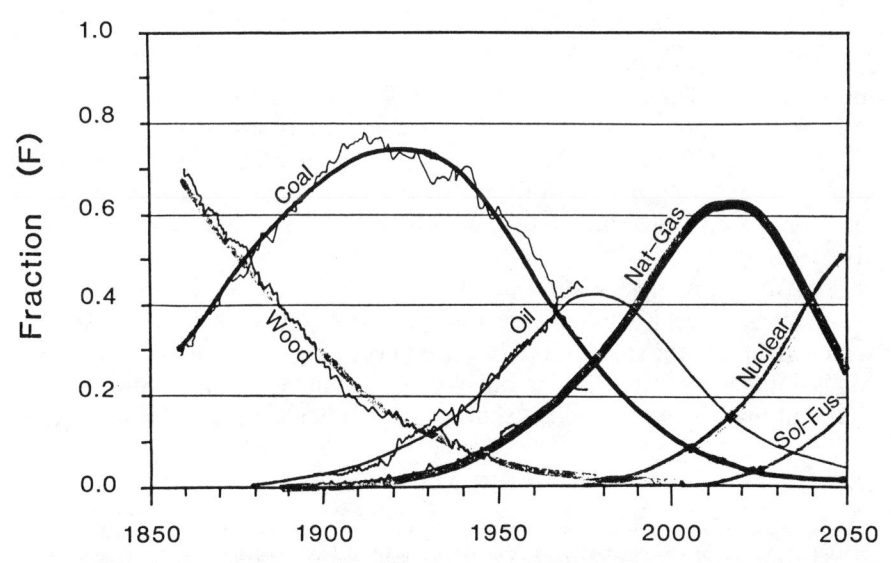

Figure 8. World primary energy substitution (from Marchetti, 1979).

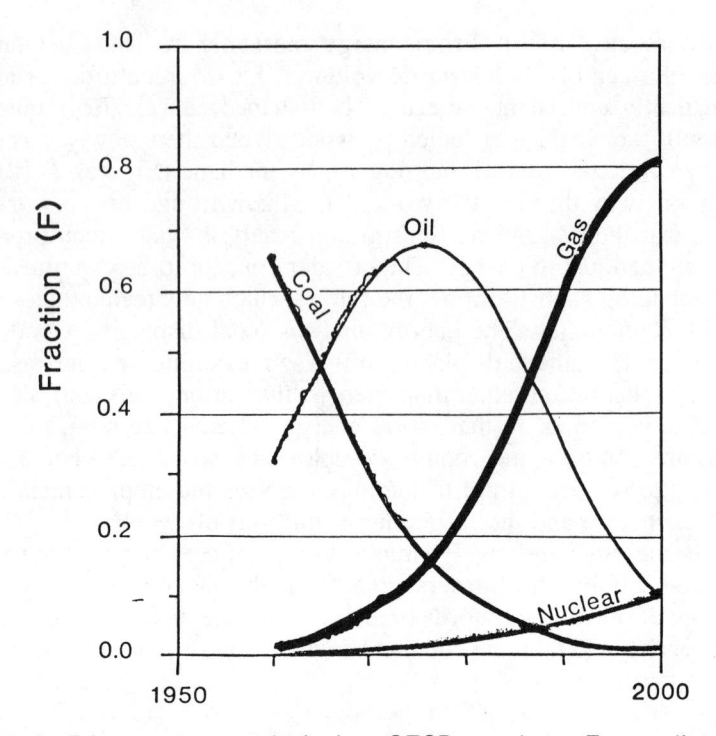

Figure 9. Primary energy substitution, OECD members, Europe (from Marchetti, 1979).

ing operation. Then, the Detroits of the world have to invent the mass production of motor cars, which are the primary customers of the refined oil products. This process involves the development of a complex infrastructure involving tens of millions of people around the world and is, therefore, a slow process which can only proceed at a relatively measurable fixed rate.

The World Energy Substitution chart (Figure 8) shows that on a market-share basis coal increased steadily against wood, peaking about 1920 with a 70% market-share. Coal has been declining ever since.† The reason is not that we were running out of coal or running out of people to dig coal, but that a new, least-cost technology, oil, began to increase its mar-

---

† The exception is the U.S. where omnibus energy legislation was passed in 1978 prohibiting the use of natural gas and thereby forcing increased consumption of coal for generating electricity, thus temporarily distorting the natural energy substitution pattern.

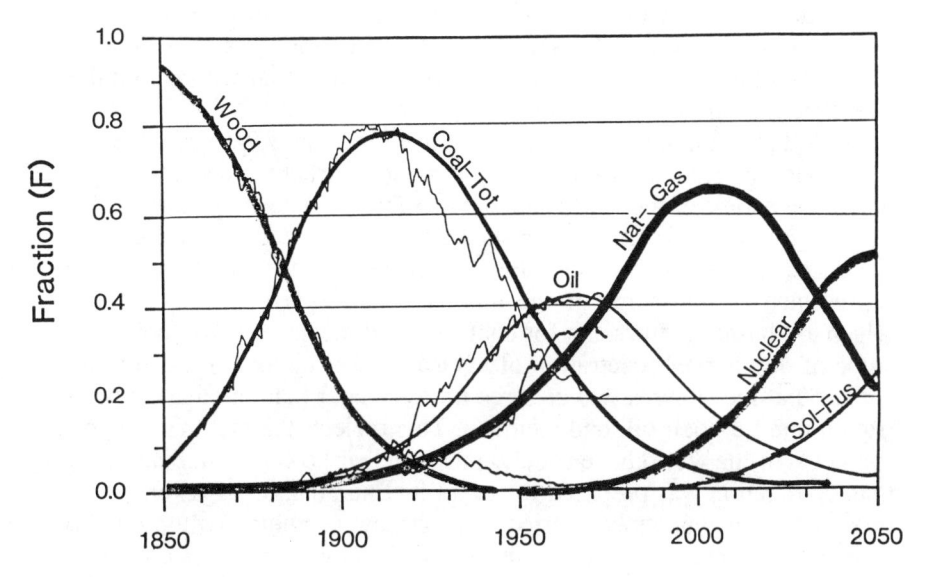

Figure 10. Primary energy substitution in the United States (from Marchetti, 1979).

ket share against coal. Oil gained market share steadily until the middle 1970s when it, in turn, began to lose ground to a better, least-cost technology, methane. Methane continues to gain market share around the world, with the exception of the United States. Based on the misconception of resource shortages, particularly natural gas, the U.S. passed laws which prohibited certain uses of methane and discouraged all industrial use. These laws have temporarily distorted methane's growth and caused many large, inefficient capital expenditures such as for the production of electricity.

As compared to oil, which is commercially produced from relatively small unique traps, methane is a more geographically pervasive and globally abundant resource. Methane is much cleaner burning, requires no refining operations, is more efficiently consumed, and has a far higher hydrogen/carbon ratio. Methane is also easy on engine metallurgy, for gas turbines fueled by methane will run 30–40,000 hours without overhaul.

As in the case of wood and coal, the world is not running out of a finite resource, oil; our civilization is simply reacting to both dramatically higher commodity prices distributed through a mature and slightly obsolete infrastructure and the true and very real geopolitical, geostrategic, and geoeconomic risks and costs of today's global oil production and consumption patterns. Methane is simply a superior, least-cost energy technology that is, as the Marchetti analysis demonstrates, forcing older, less efficient oil and coal technologies out of the market.

The substitution analysis reflects and, within limits, projects the rate the economic system can develop a new infrastructure and the rate at which the people of the world can learn to adapt to a new technology. The rate seems to stay relatively constant decade after decade. If there is a fixed rate at which a new technology can penetrate the market, that rate is represented by the slope of the market penetration lines in Figure 8, which are basically the same for coal, oil, and methane. If we look at the slope of the market penetration of nuclear power on an expanded scale, we see that the slope is two or three times steeper than the market penetration line for coal, oil, and methane. This reflects the fact that the governments of the world have tried to push nuclear power harder and faster than civilization was prepared to accept it. Examination of several cases in which governments have tried to push new technologies too quickly demonstrates that each time this happens, the new technology collapses, lies dormant for five or more years and then recovers elastically after this hiatus to the "normal" market penetration rate. This is what is happening in nuclear power now. We do not expect to see any significant number of new orders, especially in the United States, for the next ten or so years. We also expect eventually to see nuclear power recover to its normal market penetration rate. When this happens, we can expect to see nuclear power reintroduced not primarily as a source of energy for electric generation but rather as a high-temperature energy source for processing methane to be distributed by the existing infrastructure.

Although I have addressed only the energy technologies, the Marchetti Market Penetration Analysis was applied to 401 cases, many of which were not energy technologies. Market penetration analysis held in each of the 401 cases examined. They have never found a case in which a logistic substitution analysis failed to describe the market conditions that actually existed.

This analytical concept is important in future planning. In the 401 cases examined, no case has been found in which a technology recovered a long-term lost market share. This indicates that no amount of money can be spent over the long term to bring back coal or oil technology. It also indicates that no amount of money can be spent to accelerate nuclear technology. Methane, not nuclear, is the technology which will replace depleting oil and become the bridge fuel to the 21st century.

Costly Consequences of U.S. Energy Policies Based on the Misperceptions of Natural Gas Shortages

Failure to recognize the abundance of the natural gas resource base and its natural role as the oncoming primary energy source, and failure to understand the lessons of the Marchetti analysis have led the United States to experience the most expensive energy fiascos of its history.

Nuclear Power

The nuclear industry would have the public believe that electricity generated by nuclear energy is inexpensive and efficient. In fact, tax subsidies now pay 100 percent of the financing costs of a reactor and underwrite nearly half of the price of delivered nuclear electricity (Totten, 1984). In addition, almost half of the nuclear capacity which has been ordered since the inception of commercial nuclear power has been cancelled, and the costs will be met primarily by the consumer and taxpayer. Plant cancellations and enormous cost overruns are too numerous to cite, and statistics on nuclear energy change so often that maintaining current data is difficult.

The Seabrook, New Hampshire, nuclear plant at which construction has been suspended, was originally budgeted at less than $1 billion. The completion cost is now estimated at $6 billion and New England utility customers will have to pay for the 800 percent cost overrun through higher rates (Wald, 1984). If Seabrook 2 ever goes on line, rates to utility customers will have to be increased 40% to 50% to recover investment; if it does not, they may still have to pay. These new electric rates will be more expensive than the deepest of deep natural gas.

Seabrook is only one in a long line of nuclear fiascos: the Washington Public Power Supply System's (WPPSS) default of $2.25 billion was the largest municipal bond failure in history. The Marble Hill, Indiana, plant was cancelled last summer at a cost of $2.5 billion. The Zimmer plant in Ohio is to be converted, at a cost of $1.7 billion, into a coal-burning plant. The Zimmer plant will trade one out-moded technology for another. In Oklahoma, there has been a nuclear failure at Black Fox, at a cost of $250 million for the first 1% of construction. Increased electric rates cover the waste. The Oklahoma ratepayer's share of the cancellation cost is a perfect example of distorted government thinking and policy in the midst of natural gas abundance.

The one figure that tells the woebegone nuclear tale most accurately is the more than $11 billion in current cancellation costs. This figure will climb higher, for many projects are up to 12 years behind schedule and many times over budget. Long Island's Shoreham, for example, was supposed to cost $261 million but will cost about $4 billion. Bill Perrin, First Boston Corporation, says that "In the current climate—political, social and financial—you cannot rule out the possibility of utility bankruptcy."

Besides the cost cancellations and cost overruns, other problems with nuclear exist which are even more ominous:

1. How does a nuclear nation rid itself of its used nuclear fuel rods? In the United States rods are being stored "temporarily" in on-site "swimming pools." With states having the veto right to any site selection, it will be very difficult to locate a permanent repository. It has been 27 years since the first commercial reactor began operating in Shippingport, Pennsylvania, but no permanent disposal system has been adopted and the government continues to spend over $300 million a year to study the problem. England has a storage depot and recycling plant at Seascale, for waste from around the world. According to a recent article in the *Wall Street Journal* (Hudson, 1984), during its 32 years of operation, the plant has deposited over half a ton of waste plutonium into the sea. Testing of the local sea waters showed that they contained up to 250 times the normal amount of radioactive cesium. Contamination of fish in the area, and above normal doses of radiation among the local inhabitants, is prevalent. Many accidents have happened at the plant, and during one of those, an enormous dose of radiation was spewed over England—40 to 400 times as much as was released in the accident at Three Mile Island. The consideration of siting and environmental issues in England is clearly separated from the consideration of safety issues concerning nuclear. There is an option to hold a public inquiry only at the earliest site-approval stage, and yet the British Department of Energy says Britain is "strongly committed to safe nuclear power." How can England be considered a nuclear "success story"?
2. There is a great deal of criticism in the United States among scientists and experts concerning industry regulation and safety. The public, aware of that criticism and of such mishaps as the near meltdown at Three Mile Island, is fearful of a nuclear accident with severe consequences. The likelihood of a catastrophic accident is perceived as greater than that estimated by safety analysts in industry and government, creating a significant credibility gap with its many inherent political difficulties (Office of Technology Assessment, 1984).
3. A more universal fear is of the violence of nuclear proliferation. That this potentially devastating technology is being foisted largely

by private industry and the governments of the developed countries upon weaker economies and less stable political systems should be a concern to all (Akhund, 1983). It is a real fear that "an underdeveloped country with a nominal nuclear power program might obtain enough technology and equipment on the world market to build facilities to produce weapons-useable materials; or that a rapidly industrializing state with a nuclear power base in a troubled part of the world might be tempted to use the civilian program as a base for developing nuclear weapons" (Office of Technology Assessment, 1984).

Nuclear was never a necessity in the United States but was developed perhaps out of guilt for dropping the atomic bomb, as expressed by our "Atoms for Peace" program and justified by the misperception of natural gas shortages in the United States. The LDC's must not duplicate the tragic nuclear mistake which has been such a large-scale waste of both capital and human talent in the United States as to seriously strain our energy and economic systems.

Coal Power

Coal, as indicated on the preceding Marchetti charts, is an outmoded form of primary energy. This dangerous to mine, awkward to handle, and dirty to burn, source of energy is a technology of the past. Once civilization moves to a better technology, it never goes back! The recent increase in coal use for the production of electricity is simply a very costly and temporary distortion of the natural transition to methane caused by the enactment of laws prohibiting use and limiting demand for natural gas.

Studies link the major cause of acid rain to sulfur dioxide emissions from coal-fired power plants. Reputable, scientific research, as conducted by the National Academy of Science's National Research Council and even the government's own Environmental Protection Agency's study at Research Triangle Park (Peterson, 1984), maintains conclusively that the evidence linking acid rain to coal-fired utilities is indisputable.

The $3.5 billion cost of adding scrubbers to the 50 dirtiest utilities, and annual acid-rain damage of $5 billion (Marcy, 1983), added to the already high costs of new coal-fired electric utilities, would price coal out of the market should legislation be passed requiring that the full cost of the environmental burden be borne by the coal industry and its consumers.

Synthetic Fuels

Synthetic fuels, although not the social and environmental issue of nuclear and coal, are simply uneconomic and a tremendous burden on United States taxpayers. The Synthetic Fuels Corporation, a U.S. Government entity, was authorized to obligate $20 billion for synthetic fuels development, of which $7.5 billion in executed letters of intent has already been authorized. This government program of subsidies would support the production of synthetic fuels at two to three times the current cost of oil or gas and have no hope of being either economic or needed in the foreseeable future. The SFC is looking at price guarantees of up to $67 per barrel for the production of shale oil and $11 per MCF for high-Btu gas, which is currently selling below $3.00 at the wellhead (March, 1984).

The Rapid Acceleration of Productivity in the Methane Industry

History clearly tells us that human ingenuity proliferates during the emergence of a new industry and multiplies the efficiency of "in hand" technology by orders of magnitude in its early decades. As a better technology arises within the economic system as a result of the emplacement of the infrastructure and the training of the people, new industrial units organize to respond to changing market forces and opportunity.

Methane technology is roughly where the aviation industry was in the 1930s. In its first 30 years the aviation industry accomplished, without revolutionary scientific breakthroughs, a productivity gain of roughly 100 times, comparing the DC3 to the 747. It is within the first few decades that an emerging new industry rapidly increases the productivity of "in hand" technology and brings forth enormous new markets and unforeseen horizons just as we have experienced in the global development and proliferation of the aviation industry. We have no doubt that the methane industry is set to launch itself into equivalent productivity gains and begin to feel the effects of evolutionary improvements in methane technology as the industry expands into new horizons.

The methane industry may experience the multiplying effect of the combination of technological improvements and productivity gains in exploration, drilling and production, pipelining and distribution, ocean transport, conversion to electricity (high efficiency combined-cycle gas turbines), and direct conversion of natural gas to competitively priced, clean liquid fuels with enormous implications for the future of the world's transportation sector. Simply stated, as methane technology improves, the amount of methane available as reserves and resources will increase exponentially (Osborne, 1984).

Conclusion

Although natural gas fuel has been in use for a long time, past political policies and resource-base misconceptions have restrained and impeded the development of the methane industry. However as we progress toward the 21st century, economic forces of change and necessity are as always overcoming unwarranted conservatism and political resistance, leading in the case of energy, to the global development of an integrated, effectively-competitive methane industry. Methane technology is now in its early decades of market expansion as a least-cost, politically and environmentally benign energy source for successful 21st century economies. Such a course will improve the quality of life on earth.

References

Akhand, Iqbal, 1983, On revitalizing the international order: Aspen Institute for Humanistic Studies.
Brewster, W. H., et al., 1981, Summary of the World Bank—BEICIP report on natural gas: May.
Gold, Thomas, and Soter, Steven, 1980, The deep earth gas hypothesis: Center for Radiophysics and Space Research, Cornell University, June.
Hudson, Richard L., 1984, Atomic-age dump: Wall Street Journal, April 11.
Marchetti, C., and Nakicenovic, N., 1979, The dynamics of energy systems and the logistic substitution model: International Institute for Applied Systems Analysis, December.
Marcy, Steve, 1983, Scrubbers are at the heart of the latest acid rain bill: Oil Daily, June 28.
Meadows, Donella, et al., 1972, Limits to growth: First Report to the Club of Rome.
Office of Technology Assessment, 1984, Nuclear Power in an Age of Uncertainty, February.
Osborne, David, 1984, America's plentiful energy resource: The Atlantic Monthly, March, pp. 86–102.

Peterson, I., 1984, Written on the wind: tracing acid rain's elemental signature: Science News, January 21, p. 39.

Petroleum Information International, January 9, 1984.

Plotkin, Steven, et al., 1983, U.S. natural gas availability: Office of Technology Assessment, September.

Steele, William, 1984, The deep-earth gas hypothesis: Funk and Wagnall's Science Yearbook.

Totten, Michael, 1984, The road to trillion-dollar energy savings: New York Times, May 24.

Trigueiro, David, 1984, Is gas fuel of the future?: AAPG Explorer, April, p. 7.

Wald, Matthew L., 1984, Seabrook cost may burden New Englanders: New York Times, April 4.

World Oil, World trends, August 15, 1983, p. 27.

Wheeler, C. B., 1984, Statement before the Subcommittee on Energy Regulation, Committee on Energy and Natural Resources: United States Senate, April 26.

32
Completion of Shallow "Barefoot" Wells in Appalachia

Thomas J. George

Introduction

The first oil well completed in modern times is generally accepted to be the Drake well, drilled in Titusville, Pennsylvania in 1859. During the next fifty years the oil fever spread to other areas in western Pennsylvania, southwestern New York State, eastern Ohio, and northern West Virginia. This area is known as Appalachia, a term derived from the Appalachian mountain range.

In the early years of oil exploration and production, the drive was to get the "easy" oil, that is, the "gushers." The oil flowed under natural gas drive from very permeable and porous sand formations. When the easy oil formations became scarce, prospectors generally pulled the pipe and found other, similar areas to drill.

In the mid-1920s, water-flooding was discovered. Fresh water was injected under pressure into a series of intake wells, flushing oil into producer wells. The oil and some water flowed from the producing well into tanks, where the oil and water were separated. The oil was sent to market and the fresh water and salt water from the formation were disposed in the streams. Water-flooding is still being widely used in Appalachia and, in recent years, tertiary methods of well enhancement came into use. Streams are no longer used legally for the disposition of the water.

Most secondary and tertiary projects are being conducted by the oil refining companies, namely Witco-Kendall, Quaker State, Pennzoil, and a few major independent operators.

In the past ten years, due to the increased price of crude oil, there has been a sharp upsurge in drilling in Appalachia. Unfortunately, as stated before, the "easy" oil is gone. Today most of the drilling is done into tight formations that previously were bypassed. Because of low permeability and porosity, new methods of stimulation of the formations were discovered to break up the oil- and gas-producing sands and expose more surface drainage area. This new method of stimulation was pioneered by a man named Earl Halliburton. The Halliburton Company is still the largest in the field of "hydrofracturing" in the world.

Current Methods of Well Treatment

When Colonel Drake drilled the first well and for some years thereafter, when the bit penetrated the sand, natural pressure forced the oil and gas to the surface. The surface area of the bore hole provided the only drainage area; basically, the circumference of the hole times the sand thickness equalled the drainage area in square feet. In later years, a man by the name of Robertson discovered that by lowering nitroglycerin into the sand formation and detonating it, the subsequent fracturing dramatically increased the drainage-surface area.

In the late 1940s a new method of artificially fracturing the sand came into use. In this process a sand formation is isolated by means of packers. Fluids are introduced into the isolated area under extreme pressure. Because the fluids are prevented from moving up or down hole by the packers, the pressures are exerted laterally, splitting or "fracing" the sands. It was soon found that the method was ineffective, however, for once the pressures were relieved the fractures would heal.

This was corrected by including proppants, such as hard sand or glass beads, with the fluid. After the pressure was released, the fluid was forced back into the drill hole but the proppant remained behind in the fracture, propping it open and vastly increasing the drainage area.

Over the years various fluids were used in fracturing: kerosene, crude oil, salt water, but mostly fresh water due to its low cost, safety, and availability. In recent years so-called "foam" fracs, using a combination of nitrogen, surfactants, and fresh water have been used. CO_2 is also coming into wide use. One of the main reasons for the effectiveness of nitrogen is its ability to expand 300 times its volume in changing from liquid to gas. This factor further increases the drainage area. The surfactants aid in the flow of the fracturing medium. Refinements in the types of proppants have also increased the effectiveness of fracturing. Very round silica sands (plain and resin coated), zirconia, sintered bauxite, and glass and ceramic beads are some of the many new proppants.

The fracturing process has greatly enhanced the economies of drilling and producing the "hard", low-permeability sands that were noneconomical in the past.

Economics of Appalachian Wells

In Appalachia there are a number of oil and gas producing horizons of different geological ages. These sands generally bear the names of the areas in which they were discovered. The name of the driller or the mineral-rights owner also is given to the names of sands. Often, sands of the same geologic age are known by different names in different areas. The same sands in different areas may contain greater or lesser amounts of oil, gas, or salt water. Appalachian sands may thin and thicken locally but are generally quite uniform over a large area, often being referred to as blanket sands.

In some areas there may be as many as six or eight sands of varying thickness and productivity. In years past, several sands might be penetrated, with only the richest completed. Currently, drillers are going back to the same areas and completing the formerly noneconomic sands. However, because of the high cost of drilling and equipment today, and the relatively low productivity of hard sand wells, very close attention must be given to economic factors. Hydrofracturing costs normally can run from one-third to one-half the total cost of a well; therefore, the method used must be as economical as possible, the most important variable being the "frac" cost.

Barefoot Hydrofracturing Method

In most parts of the world, after a well is drilled, steel pipe is cemented in the hole. The cement prevents gas and fluids from escaping between the pipe and the bore hole, keeps fresh water from the oil and gas formations, and holds the pipe in place.

Once the pipe is cemented in place, the formations to be treated are determined. The pipe is then perforated at the desired points by explosive charges, which perforate the steel and cement, and penetrate the formation a short distance, rarely more than an inch. The shot holes provide a conduit for the frac medium and give the fracture a beginning; this process is called "notching."

The perforating process and the pipe are costly. In a marginal well, the cost is prohibitive. Therefore an alternative has been developed in other parts of Appalachia called "The Barefoot Method."

In this process, after the hole is drilled to its desired depth, a "water string" is introduced below the fresh water-bearing sands. In some areas, at a depth of approximately 450 ft in a 1,800 ft well, a cement shoe is placed at the bottom of the water string and cement is pumped under pipe to surface. This prevents fresh water from entering the bore hole and hydrocarbons from entering and polluting the fresh water zones. In these low-pressure wells, this provides sufficient protection to keep the pipe from "blowing out" after completion. The lower part of the hole has no pipe and is then called "barefoot." The obvious question is, then, "How do you contain fracturing stages in a rough and ragged open bore hole?"

This is accomplished in the following way. Sand or pea gravel, known as "plug-back," is carefully dropped into the bore hole to a point at the bottom of the uppermost fracturing stages (there could be as many as twenty stages in a well). If the plug-back is not slowly introduced, it will "bridge" in the hole or cause an obstruction and not allow the plug-back to fill the bore hole. A rubber "hook-wall" packer, one that will grip the sides of the rough bore hole securely, is lowered by means of heavy-duty steel pipes (3 or 4 in.) to the top of the uppermost stage.

The medium, the fracturing fluid, normally fresh water, is introduced into the isolated areas under extreme pressure (1,500 to 5,000 psi). When "breakdown" is experienced, pumping rate of the fluid is set to accomplish an even rate of flow as the fracture increases and expands. Normally the flow rate ranges from 15 to 25 bbls per minute. Once breakdown, or the first point of overcoming the resistance of the sand body occurs, proppants, surfactants, and other chemicals are introduced with the fracturing medium; this continues until the desired frac is obtained.

The hydraulic pressure is then relieved and the medium is allowed to flow back to the well hole. This is known as "blow-back." When the introduced pressure has been completely relieved, the packer is mechanically unsealed and water is forced down the annulus of the drill pipe, flushing plug-back up through the frac pipe and onto the ground. The frac pipe is slowly lowered during the washdown, staying just inches from the plug-back.

When the frac pipe has been lowered to the bottom of the next stage, it is then pulled back up hole to the top of that stage. The packer is then reset and the fracturing process is repeated. The plug-back and the packer isolates the stage to be fractured.

The process is repeated until the bottommost stage is completed. The flushing of the plug-back is continued for approximately another 50 ft to the bottom of the hole. This lower part of the bore hole is referred to as the "pocket." The pocket serves as a receptacle for sludge, pieces of rock, and other foreign objects and provides, in effect, a sludge pit.

The well is then "washed down" to remove debris and dislodge loose rocks, leaving the hole clean. The frac pipe and packer are then removed from the bore hole.

A production string of pipe (normally 2 in.) is then inserted into the hole and a perforated nipple attached to the string below the nipple with a 20 ft anchor. The purpose of the anchor is to indicate total depth. The tubing is then raised from the bottom to the desired depth.

Finally, a pump is lowered inside the tubing by means of $1/2$ in. or $5/8$ in. sucker rods. At the surface, a pump jack is attached to the rods and production pipe is installed, completing the well.

Conclusion

This method of fracturing a shallow well is viewed with amazement by those from outside the Appalachian area. It is believed that this is the only part of the world where this technique is used. Those not familiar with the procedure cannot understand how one can treat a well that is not under pipe. If the cost of running casing to total depth, cementing, and perforating were added to current well costs, it is probable that the number of shallow, marginal oil and gas wells drilled in Appalachia would be cut in half.

Many of the alternate and new methods being used in other parts of the world are still uneconomical here. If similar conditions exist in other parts of the world, perhaps the "barefoot" technique could make the difference between success and failure.

33
The Natural Gas Resources Development of Kathmandu Valley

G. S. Thapa

Abstract

Natural gas dissolved in water occurs in the fluvio-lacustrine sediments of Quaternary age in Kathmandu Valley. The sediments are blanketed on the basement of Precambrian-Middle Paleozoic rocks of the Kathmandu complex. The basement rocks are intruded with granite in the south and augen gneiss in the north of valley.

The Quaternary sediments are mainly composed of mud, silt, sand, and gravel. The methane natural gas (CH_4) occurs in the coarse sediments below a depth of 180 m in the southern half of the valley.

Introduction

In view of the acute scarcity of domestic fuel in Kathmandu and the existence of a natural gas resource within the valley, the Department of Mines and Geology initiated systematic exploration work for natural gas deposits. Surface and subsurface exploration work disclosed a natural gas reserve of 47.6 million m^3 within a 4 km^2 area and to a depth of 300 m. A prospective area of 26 km^2 for natural gas has been classified on the basis of surface geology, bore-hole data, resistivity and gravity data, and gas showings from water wells. The observation of a model gas plant, constructed for a study of stable supply and economic potentiality of natural gas, indicates that the gas can be used domestically as a source of fuel.

Distribution of natural gas deposits in the Kathmandu Valley on a commercial basis, together with their further exploration, is under consideration.

History and Previous Work

The natural gas occurrences in Kathmandu Valley have been known since time immemorial. At many places, such as the Coronation Hotel (now Leo Hotel), Crystal Hotel, and Bir hospital, the gas was used as domestic fuel. The occurrence of gas was first known from tube wells drilled for ground water. Commercialization of the natural gas was not considered due to the lack of sufficient information about the size and reserve of the deposit.

Many geologists have worked in Kathmandu Valley in connection with surface geology and geophysical studies and ground water investigations (Nantiyal and Sharma, 1961; Sharma and Singh, 1966; Tandukar and Pandey, 1969; Stocklin and Bhattarai, 1978). No systematic exploration for natural gas has been carried out in the past. Misra and Srivastav (1964) and a Chinese petroleum investigation team (CPIT, 1973) made surface observations of natural gas occurrences in Kathmandu Valley; it was marsh gas formed in a discontinuous lenticular sand at different levels. They concluded that the amount of production is not large enough to be of industrial value.

Considering the acute scarcity of domestic fuel in Kathmandu City and existence of a natural gas resource within the valley, the Department of Mines and Geology, in collaboration with JICA, initiated systematic exploration work in 1978. The surface geological and geochemical work disclosed that a natural gas deposit exists in Kathmandu Valley and that the gas is of "dissolved-in-water type" originating from the Quaternary fluvio-lacustrine sediments. A tentative natural gas reserve of 42 million m^3 was inferred within the surveyed area of 4 km^2 south of Kathmandu City.

As a follow-up to the surface exploration, 3 test wells were drilled in 1980 and 1982 within a 4 km^2 area, each to a depth of 300 m and, a model gas plant was constructed. The test wells are connected with the gas holder at the model gas plant site through underground polyethylene pipe lines and distributed to few nearby places on an experimental basis to study the reliability of the gas in terms of production, quality, and supply.

Topography and Geology

Topography

Kathmandu Valley is a small, semicircular, intramountain basin situated in the lesser Himalayas of central Nepal and covering an area of about 400 km^2. The lowest elevation is observed in the Bagmati gorge in the southern part, about 1,250 m, and the highest elevation within the valley is about 1,400 m. The valley is surrounded by high mountains, such as Sheopuri (2,732 m) in the north and Phulchoki (2,762 m) in the south. The mountains have undulating topography with steep slopes reflecting the general nature of the geological structure. Piedmont fans are deposited along the valley margin, forming a prominent topographic feature in the foot hills.

The rivers traversing the Kathmandu Valley are generally broad and flat, except the upstream and downstream of the valley, where they are deeply incised. Well-developed erosional terraces are common in the valley. The main rivers within the valley are the Bisnumati, flowing from north to south, the Manhara, flowing from east to west, and the Bagmati, flowing south from the confluence of the two rivers at Teku Dobhan. Since the valley is semicircular in shape, the main drainage system is towards the center of the basin, where it joins the Bagmati river system; this in turn flows through Chobar gorge in the south.

Geology

The basement of Kathmandu Valley consists of the rocks belonging to Phulchauki and partly the Bhimphedi group of Kathmandu complex (Precambrian to Middle Paleozoic). The rocks are mainly metasedimentary consisting of quartzites, phyllites, schists, slates, limestones, and marbles with igneous and basic intrusives. The rocks are highly folded, faulted, and fractured, the general trend being WNW-ESE. They are overlain unconformably by the Quaternary fluvio-lacustrine deposits (Dhoundiyal, 1961, 1966; Gandotra, 1968a, b; Biunie and others, 1973; Nadgir, 1973; Sharma, 1977; Stocklin and Bhattarai, 1978; and JICA 1980).

The geological succession of Kathmandu Valley is shown tentatively in Table 1. The geology of the Kathmandu complex in and around Kathmandu Valley can be broadly described as follows:

Quaternary—Fluvio-lacustrine sediment
Lower Paleozoic—Phulchauki Group⎫
Precambrian   —Bhimphedi Group⎬ – Kathmandu complex

Precambrian

Bhimphedi Group: The rocks of the Bhimphedi group are exposed at the distal, peripheral margin of Kathmandu Valley and consist of quartzite, schists, and carbonates. They are intruded with granites, pegmatites, and basic rocks south of the Kathmandu Valley; in the north, well-developed Sheopuri gneiss is common.

Lower Paleozoic

Phulchauki Group: The rocks of this group are widely exposed around the hills, particularly those encircling the valley from east, west, and south. The rocks in general are metasandstone, phyllites, slate, dolomite, and limestone. They occur in a syncline, the core being the youngest unit of the group, the Godavari limestone, trending from Godavari to Phurping in WNW-ESE direction along the southern part of the valley. The Silurian Chitland formation of this group has been found to contain trilobites in ferrugenous concretions. The lowest formation, the Tistung, is unconformable on the underlying Bhimphedi Group.

Quaternary Fluvio-Lacustrine Sediments

Kathmandu Valley consists of Quaternary fluvio-lacustrine sediments within an area of about 250 km^2. The area is low and flat, with elevations varying from 1,250 to 1,450 m.

The sediments are composed of clay, silt, sand, and gravels, thus presenting various sedimentary environments, such as lakes, swamps,

Table 1
Geological Succession of Kathmandu Valley

Cenozoic Quaternary		
Holocene		Fan gravels
		Talus
Pleistocene		Lake deposits (gravel, sand, silt, clay, peat, lignite, and diatomaceous earth)
		Fluvial deposits (gravel, sand, silt, clay and lignite)
Lower Paleozoic	Phulchauki Group	
Devonian		Godavari Limestone—limestone, dolomite
Silurian		Chitland Formation—slate
Cambro-Ordovician		Chandragiri Limestone—limestone
Cambrian		Sopyang Formation—slate, calcphyllite
Early Cambrian		Tistung Formation—metasandstone, phyllite
	Unconformity	
Precambrian	Bhimphedi Group	
Late Precambrian		Markhu Formation—marble, schist
Precambrian		Kulekhani Formation—quartzite, schist

Metamorphics—Sheopuri gneiss
Igneous rocks—pegmatite, granite, basic intrusive

marshes, rivers, estuarine deltas, and piedmont fans. The maximum thickness of the sediment is estimated to be 650 m. The northern half of Kathmandu Valley consists of granular sediments derived from Sheopuri gneisses, whereas in the southern half, the sediments are finer, the source being the metasedimentaries of the Phulchoki Group.

Structure (Pre-Quaternary)

The Bhimphedi and Phulchauki group of rocks constituting Kathmandu complex form a major Mahabharat synclinorium with WNW-ESE trend. The center of the synclinorium is occupied by the Kathmandu Basin and the hills south of it (Stocklin and Bhattarai, 1978).

The lake which produced the lacustrine deposits formed in late Pleistocene to Quaternary time by the rapid development of the Mahabharat range in the south of the valley; this preceded the downward erosion of the antecedent Bagmati River (Sharma, 1977).

Facies Classification of Quaternary Sediments

The fluvio-lacustrine sediments are composed of clays, silts, sands, gravels, and diatomaceous earth. These sediments can be grouped into the following three facies (Table 2):

Facies I—Fluvial deposits
Facies II—Lake delta deposits
Facies III—Proper lake deposits

Beside these three facies of fluvio-lacustrine sediments, piedmont fan deposits are observed along the valley margin. All facies show a horizontal change in a north-south direction. The Quaternary sediments in general are flat-bedded, but at some places a low angle of dip (about 5°) towards south is observed in the northern part of the valley, where sediments of Facies I prevail. The basement depth of Quaternary sediments varies from place to place. The maximum depth is expected to be about 650 m from the surface.

Facies I (Fluvial deposits)

The sediments of this facies are well developed in the northern half of Kathmandu Valley. It is composed of coarse-grained sand with gravels and clays and silts with sand. This facies presents a small-scale sedimentary cycle of alluvial flood plain, in which grain size shows upward fin-

Table 2
Facies Classification of Quaternary Sediments
(After Jica, 1980)

Sedimentary Environment	Lithofacies
Facies I Fluviatile (Alluvial flood plain):	Fining upward cycle
Ia Stream channel fill	Coarse-grained sand with gravel
Ib Over-bank (included marsh)	Clay and silt with sand
Facies II Lake-delta:	
IIa Delta-front and delta-Slope	Fine to medium grained sand
IIb On-delta marsh	Carbonaceous clay and silt
Facies III Lake (proper):	
IIIa Proximal lake	Silt-sand alteration
IIIb Proximal lake	Laminated clay and fine sand
IIIc Distal lake (included marsh)	Clay occasionally diatomaceous or carbonaceous.

ing (granular sand to fine sand). This sequence has several repetitions in the northern part of the valley, as revealed by the bore-hole logs. The muddy upper half of the cycle consists mainly of grey clay and silt with occasional thin sand beds. The black-colored carbonaceous clay locally known as "Kalimati" occurs at the top and contains 4% to 5% organic carbon.

Facies II (Lake delta deposits)

The distribution of this facies is observed west of Tribhuwan Airport in the central part of the valley and occurs as transition facies between Facies I and III. It consists of bedded sand, mainly medium and fine-grained (Facies II-a) and laminated clay and silt (Facies II-b). The cross bedding and parallel laminations are common in sand beds of Facies II-a. The sand grains are fairly well sorted. Facies II shows a sedimentary cycle of upward fining sequence and appears to have originated from lake-deltas in an estuarine environment. Facies II-b on the other hand is considered to be derived from swamps or marshes.

Facies III (Proper lake deposits)

This facies is widely distributed in the southern half of Kathmandu Valley and consists mainly of unconsolidated to semi-consolidated mud beds and thin beds of fine sand and diatomaceous earth. It represents the sediments on the bottom of the ancient lake and is classified into three subfacies.

Facies III-a and III-b consist mainly of sandy mud with alternating silt and sand, and silty clay with frequent intercalations of very thin layers of fine sand and silt, respectively. These facies are distributed from east to west in the central part of the valley.

Facies III-c occupies a large area south of Patan city to Bhaktapur in the east and has an E-W trend. It is composed of gray to dark-gray clay, occasionally with light-yellow diatomaceous earth. This subfacies includes carbonaceous clay (Kalimati) in certain places and is exposed extensively in the southern part of the valley. It is considered that this subfacies was formed in the distal lake environment where the flow of the lake was suspended.

The bore-hole logs show that the deposits of Facies III are distributed uniformly to a depth of 180 m south of Kathmandu City. Along Dhapa Khel Harisidhi area, the clay of this facies is developed to a depth more than 300 m. Below these depths the sediments of Facies I are likely to be present as extensions of the northern half.

Natural Gas Deposit

Natural gas dissolved in water occurs in the fluvio-lacustrine deposits of Kathmandu Valley. The composition is CH_4 (75% −80%), CO_2 (14% −23%), and N_2 (1.5% −6%). The muddy sediments rich in organic matter are found in the marsh deposits of Facies I and the lacustrine deposits of Facies III and are considered to be the source rock of natural gas. The coarse sediments in any of the three facies could be the reservoirs, the gravels and sand layers of Facies I being considered the most important gas reservoirs in the valley. The potential gas reservoir occurs at a depth of more than 180 m in the southern half of the valley, where the sediments of lake facies (III-a and III-b) are distributed on the surface. Probably due to the absence of adequate source material, the northern half of the valley is devoid of natural gas, though it has excellent reservoirs.

Gas-water ratio, as observed at the casinghead of the three test wells drilled to a depth of 300 m each within 4 km^2 of the prospective area, and another existing water well, is about 1:2.5. The average thickness of res-

ervoirs in the test wells at 300 m depth, is 85 m. Assuming a porosity of 35%, the reserve of natural gas within 4 km² is as follows:
Area = 2 km × 2 km (4,000,000 m²)
Reservoir thickness = 85 m
Porosity = 35% (0.35)
Gas-water ratio = 1:2.5 (0.4)
Reserve of natural gas (m³) = 47,600,000 m³ (47.6 million m³ = 4,000,000 × 85 × 0.35 × 0.4

Since the thickness of Quaternary sediments varies from place to place in Kathmandu Valley, the maximum being around Baneswar (650 m) in the central part, it is quite likely that gas reservoirs extend over the entire prospective area. In gas dissolved in water, the saturation of methane increases with depth as a result of hydrostatic pressure.

Development of Natural Gas Deposit

Surface Exploration

Surface exploration consisted mainly of geological and geochemical exploration. The geochemical work included measurement at the casinghead of gas-water production per day, gas-water ratio, ground-water temperature, and pH and RpH value of water. The gas and water samples collected from various water wells were analyzed in the laboratory to determine the chemical composition of gas and the water.

Where the log data of the water wells drilled by private individuals were not available, the depth of the aquifers was calculated on the basis of ground-water temperature at the casingheads, using the following formula.

$$D = (Tw - 20.1) \times 30$$

where D = depth (m) of the aquifers
Tw = ground water temperature (°C) at the casing head
20.1 = average annual temperature (°C) in Kathmandu in 1970
30 = geothermal gradient (1°C/30 m)

The result of surface exploration indicated the existence of natural gas reservoirs at a depth of about 200 m within the surveyed 4 km² area.

Subsurface Exploration

As follow-up to the surface exploration a test-well drilling to a depth of 300 m was undertaken in 1981 within the 4 km² area in Tripureshwar. Several aquifers containing natural gas dissolved in water were observed below 180 m in a thick clay-silt sequence, often with carbonaceous clay (Kalimati). The clay-silt sequence occurs as alternate layers with aqueous reservoirs. The total thickness of the reservoirs was found to be 83.25 m to 300 m depth. Two more test wells were drilled in 1983 in the same area at an interval of about 600 m (at Pachali and Teku Doban), which revealed similar characteristics. The production and chemical data of the test wells are shown in Tables 3 and 4.

Prospective Area for Further Exploration

The geological and geochemical data, and three test wells have proved a reserve of 47.6 million m³ of natural gas within an area of 4 km² and to a depth 300 m in Teku-Tripureswar area of Kathmandu Valley. It is con-

Table 3
Production Data of the Test Wells

Test Well No.	Location	Res. Thickness (m)	Free Flowing Gas m³/d	Free Flowing Water kl/d	Free Flowing Gas/Water ratio	Gas Lift Gas m³/d	Gas Lift Water kl/d	Gas Lift Gas/Water ratio
DMG-1	Tripureswar	83.25	98.3	541	0.18	400	1045	0.38
DMG-2	Pachali	84.40	28	77	0.36	–	–	–
DMG-3	Teku Doban	102.38	382	600	0.64	1638	1045	1.56
	Total		508	1218	0.41	2038	2090	0.97

Table 4
Chemical Data of the Test Wells

Test Well No.	Location	CH_4%	CO_2%	N_2%	O_2%	Calorific Value Kcal/Nm³	Wobbe Index (WI)	Combustion Potential (C.P.)
DMG-1	Tripureswar	78.36	13.55	7.17	0.89	7473.19	8862.89	27.88
DMG-2	Pachali	77.16	13.55	6.33	0.7	7358.75	8818.15	27.73
DMG-3	Teku Doban	75.93	17.42	5.49	1.16	7241.44	8418.32	26.48

sidered that the southern half of the valley is more prospective for natural gas. The maximum depth of sedimentation in the northern half is deduced from the following data.

1. The deepest well, drilled to a depth of 457 m at Hari Sidhi in the southeast, has not reached the basement rock.
2. The basement topography of Kathmandu Valley, prepared from the gravity survey by Moribayashi and Maruo (1980), shows the maximum depth of sedimentation to be about 650 m around Baneswar area, trending in a NW-SE direction.
3. The basement contours prepared from the resistivity survey by WHO project Nepal 0025 in 1973 also shows the maximum depth of basement in the same area.

Cross sections projected on the basis of bore-hole logs, resistivity and gravity data, and the nature of the sedimentation process, indicate that the deepest part of the basin is about 650 m in a NW-SW direction from Baneswar.

Considering these facts and the gas showings from the water wells, a total area of 26 km^2 has been found prospective for natural gas. Out of the total prospective area, only a 4 km^2 area has been explored to a depth as great as 300 m.

It is therefore planned to explore systematically the remaining 22 km^2 area for natural gas. For the purpose of exploration, drilling 600-m and 300-m exploratory wells is proposed. If these exploratory wells find potential gas sources, then they can be converted into production wells at a later stage, together with additional production wells required to produce and supply gas on a commercial scale.

Model Gas Plant

In order to study the consistency and economic potentiality of the natural gas deposit a model gas plant with a capacity of 500 Nm^3 per day was constructed by the end of fiscal 1982/83. The model plant has a gas holder of 500 Nm^3 capacity, an odorization plant governer unit and main and branch supply pipe lines of 1.9 km length. The gas holder is connected with three test wells through underground pipe lines (gathering lines) and gas is supplied to the holder. From the holder the gas is distributed at low pressure to some government-owned cottage industries, offices, hospitals, and laboratories, on an experimental basis. The results obtained from the model plant indicate that the gas can be used domestically as a source of fuel. The specification of supply gas standard of Kathmandu gas is as follows:

1. Standard calorific value: 7,200 Kcal/Nm^3
   Minimum calorific value: 7,000 Kcal/Nm^3
   Maximum calorific value: 7,400 Kcal/Nm^3
2. Supply pressure: 100–250 mm H$_2$O
3. Wobbe index (WI): 8,000–8,600
4. Combustion potential (C.P.): 25–30
5. Hydrogen sulphide (H$_2$S): 0.02 g/Nm^3 (13 ppm)
   Ammonia (NH$_3$): 0.2 g/Nm^3
   Total sulphur: 0.5 g/Nm^3

The Cost and Utilization of Gas

The economic analysis the natural gas deposit of Kathmandu Valley is tentatively made for the commercial distribution of 2,000 Nm^3 per day which can be produced with gas lift from the existing three test wells of the model plant. This 2,000 Nm^3 of gas is sufficient for 3,000 families as domestic fuel. The commercial distribution of gas through an independent organization or a project has been envisaged.

The total cost for the commercialization of gas is estimated at Rs 22,072,000 (US$ 1,471,467). The annual cost of production for 2,000 m^3 of natural gas comes to Rs 3,816,674. The unit cost of production is Rs 5.23 per m^3. The sales price is fixed at Rs 7.50/m^3. The project will generate sales revenue amounting to Rs 5,475,000 annually.

Market Prospect

The market survey carried out for 1,000 families shows good response from the public. They are willing to use it if available. Further, the acute shortage of firewood due to deforestation and the hard currency required to import kerosene and LPG requires alternate energy in Kathmandu.

Project Justification

Natural gas as an energy source will not only save foreign currency spent to import kerosene but will also save the country's forest resources from depletion. At present kerosene worth Rs 71,050,000 per year, firewood worth Rs 15,958,800 per year, and LPG worth Rs 6,804,000 per year are consumed in Kathmandu Valley. The savings in the initial stage will not be substantial but will increase in later stages, when the production capacity of the project will be further expanded and people will be more accustomed to using it.

The cash flow statement and internal rate of return (IRR) show positive balance throughout the period and the cumulative cash flow at the end of 20 years will be Rs 78,663,895. The calculated IRR is 14.3% and the cost-benefit ratio at 14% discount is 1:1.02. For the present study the life of the project is taken as 20 years while the actual life of the project is expected to be more than 50 years.

The promotion of this project will provide a direct and regular source of revenue to HMG/Nepal in the form of royalty and surface rental. The total revenue to HMG/Nepal comes to Rs 689,371 per year. Besides, the project will provide direct employment to 80 persons. The financial highlights of the project are shown in Table 5.

Table 5
Profitability Indicators

Items		Rs
1. Sales revenue		5,475,000
2. Total cost of production		3,816,674
3. Gross profit (1–2)		3,230,509
Total investment		
1. Fixed investment	21,253,000	
2. Working capital	819,000	22,072,000
Equity investment		
1. Equity assets	4,426,330	
2. Additional investment	8,822,709	13,249,039
Profitability indicators		
A. Percentage of net profit to sales revenue (%)		59.0
B. Percentage of net profit to equity (govt. finance) (%)		24.4
C. Percentage of net profit to loan (%)		36.6
D. Return on investment (%)		24.8
E. Break-even point (%)		58.12
F. Internal rate of return (%)		14.3
G. Cost-benefit ratio at 14% discount		1: 1.02
H. Value added %		68.24

Conclusion

The surface and subsurface exploration of natural gas deposits in Kathmandu Valley revealed the existence of gas reservoirs at depths from 180 m to 300 m in the southern half of the valley. The gas deposits have originated from fluvio-lacustrine deposits of Quaternary age. According to the gravity survey the sediments have a maximum thickness of 650 m, overlying Precambrian and Paleozoic bed rocks.

The gas is composed of CH_4 (75%-80%), CO_2 (14%-23%), and N_2 1.5%-6%). A tentative gas reserve of 47.6 million m^3 has been estimated within a 4 km^2 area. A total prospective area of 26 km^2 is classified on the basis of bore-hole logs, gravity and resistivity surveys, and gas showings from the water wells.

The model gas plant constructed for the study of gas consistency and the results of its observation during the last ten months show that the gas can be used as domestic fuel. The tentative economic feasibility study of the deposit for 2,000 m^3/d production and supply has given positive figures. Further exploration for natural gas in the prospective area of Kathmandu Valley is necessary. It may include exploratory wells of 300 m and 600 m depth.

References

Binnie and Partners, 1973, Master plan for the water supply and sewerage of Greater Kathmandu and Bhaktapur: Report, World Health Organisation Project Nepal 0025, Vol II C Water Supply.

Chinese Petroleum Investigation team (CPIT), 1973, Report on the investigation of petroleum geology in the kingdom of Nepal: Chinese Petroleum Investigation Team Peking.

Dhoundial, D. P., 1961, Report on the geology of Paleozoic fossileferous rocks of Phulchouki range and detailed investigations of the Godavari marble, Kathmandu valley Nepal: Geological Survey of India, unpublished report.

Dhoundial, D. P., 1966, Lower Paleozoic fossileferous rocks south of Kathmandu Valley, (with map): Geological Survey of India unpublished report.

Gandotra, V. M., 1968-a, Geological mapping of Chitland Formations in the area south east of Kathmandu, Nepal (Progress report No. 1 for the field season 1965-66): unpublished report, Geological Survey of India.

Gandotra V. M. 1968-b, Geological mapping in parts of Mahabharat and Churia Ranges south of Kathmandu Valley, Nepal: Progress report for the field season, 1966-1967.

Gansser, A., 1964, The Geology of the Himalayas: Interscience Publishers, London.

(JICA) Japan International Co-operation Agency, 1980, Natural Gas Resources in Kathmandu Valley.

Moribayashi, S., and Maruo, Y., 1980, Basement topography of Kathmandu Valley, Nepal—an application of gravitational method to the survey of a tectonic basin in the Himalayas: Jour. Japan Soc. Engineering Geol., Vol. 21, No. 2.

Nantiyal, S. B., and Sharma, P. N., 1961, A geological report on the ground water investigation of Kathmandu Valley: Geological Survey of India, unpublished report.

Sharma, C. K., 1977, Geology of Nepal, Navana Printing Works Calcutta.

Sharma, P. N., and Singh, O. R., 1966, Groundwater Resources of Kathmandu Valley, Supplementary Report. Geological Survey of India, unpublished report.

Stocklin, J., and Bhattarai, K. D., 1978, Geology of Kathmandu area and Central Mahabharat Range, Nepal Himalaya: unpublished report, HMG/UNDP Mineral Exploration Project Kathmandu.

Tandukar, R. P., and Pandey, M. R., 1969, Report of the magnetic survey of Kathmandu Valley: Nepal Geological Survey, unpublished report.

Water and Energy Commission, HMG/Nepal, 1978, Report on the investigation of natural gas occurrences in Kathmandu Valley, unpublished.

34
Prospects of Shallow Gas Resources in Bangladesh

M. A. Ghafur

Abstract

The principal sedimentary basin of Bangladesh, Bengal Basin, originated in early Cretaceous time and may be divided into a stable continental shelf area and a geosynclinal area separated by the transition zone known as the "hinge-zone." The hydrocarbon potential of Bangladesh has so far been tested by only 69 wells, including GIB wells, which resulted in the discovery of 13 gas fields with total estimated recoverable gas resources 10.52 trillion cubic feet (tcf). Gas shows at shallow depth and reserves proved in some of the wells indicate prospective shallow gas resources in Bangladesh. Further appraisal of existing fields and continued search for hydrocarbons in the country may result in more commercial gas fields at shallow depths.

Introduction

The hydrocarbon potential of Bangladesh so far has been tested by only 69 wells, which resulted in the discovery of 13 gas fields with estimated recoverable gas reserves of 10.52 tcf. In some of the wells the reserves are at shallow depth. "Hinge-zone" and other promising geological structures and areas are as yet unexplored or else only scanty exploration has been carried out. From drilling and geophysical data, it is possible to identify some areas which are prospective for gas exploration at shallow depth. Further appraisal of existing fields and continued search may result in additional commercial shallow gas fields.

Geology

Bangladesh is constituted mainly of the sedimentary Bengal Basin, which originated in early Cretaceous time with the separation of India from Australia and Antarctica. The Bengal Basin is bounded on the west by the Indian shield, on the east by the Arakan Yoma-Naga geanticlinorium, and on the north by the Shillong Plateau and the Mikir Hills; and on the south it is open for an unknown distance through the Bay of Bengal into the Indian Ocean (Figure 1). The geophysical results and drilling data lead to the concept of a subdivision of the Bengal Basin into a stable continental shelf area to the west and northwest, with a comparatively reduced sedimentary succession, and a geosynclinal area to the south and south-east with very thick sedimentary fill. The zone of transition between the two regions is known as the hinge zone and is characterized by a relatively rapid increase in thickness of individual formations and concomitant facies changes from shallow shelf to deep basin sediments.

Tectonically, Bangladesh may be divided into three main regions: the Indian Platform slope, comprising the Dinajpur Slope, Rangpur Saddle, and Bogra Slope; the Platform slope of the Bengal Foredeep, comprising the Faridpur Trough, Sylhet Trough, Madhupur High, Barisal High, and Hatiya Trough; and the folded belt of the Bengal Foredeep, comprising the West and East zones.

Stratigraphy

The stratigraphy of Bangladesh is not yet fully established. However, on the basis of the available data, an attempt is made to establish the stratigraphic succession of the country, shown in Figure 2.

Geophysical Investigations

Figure 3, from the aeromagnetic survey of 1980, shows the estimated depth to magnetic basement. The basement depth in Dinajpur platform

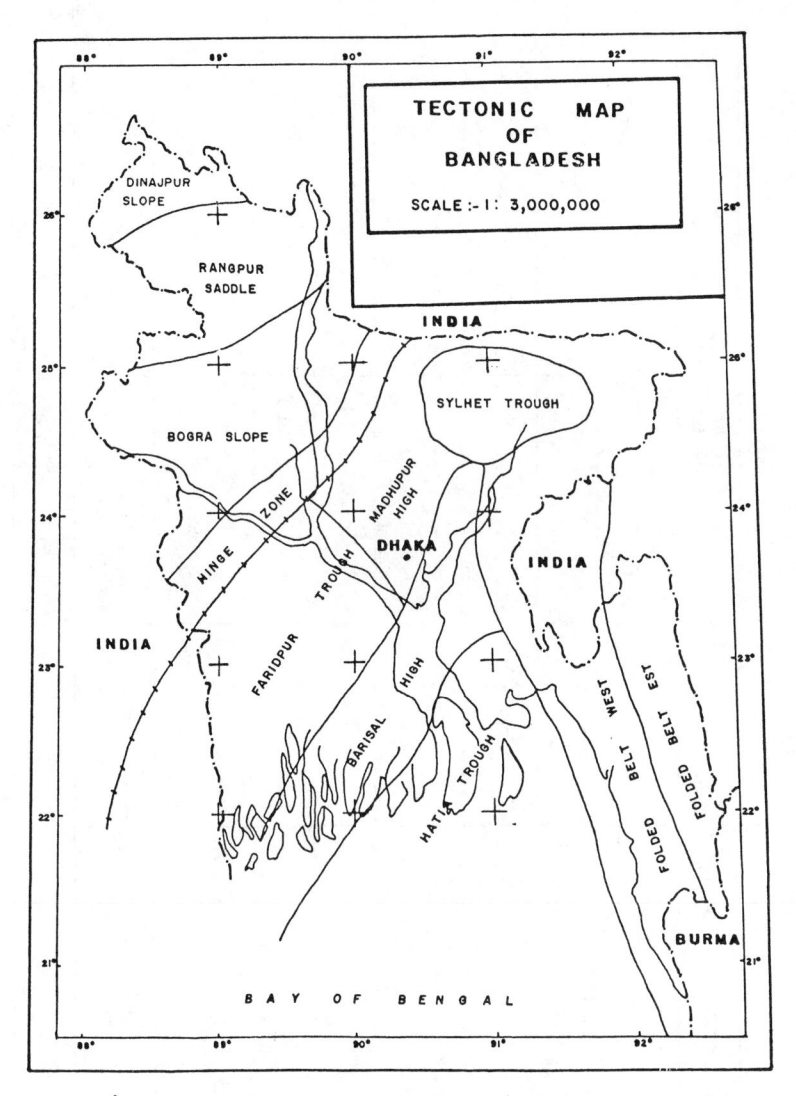

Prospects of Shallow Gas Resources in Bangladesh 391

Figure 1. Tectonic map of Bangladesh (scale: 1:3,000,000).

Shallow Oil and Gas Resources

ERA	SYSTEM	SERIES	STAGE	SHELF GROUP	SHELF FORMATION	BASIN GROUP	BASIN FORMATION
	QUATERNARY	RECENT			Alluvium (270')		Alluvium
		PLEISTOCENE			Dihing (650'+)		Madhupur Clay (151'+)
		PLIOCENE	UPPER		Dupi Tila (905')		Dupi Tila (6000 ft+)
			LOWER			Tipam (7450')	Girojan Clay (3500')
			UPPER				Tipam ss (3950')
		MIOCENE	MIDDLE	Surma (Undiff-1355')		Surma (19000')	Bokabil (6000')
			LOWER				Bhuban (13000')
		OLIGOCENE	UPPER	Barail (535')	Bogra (535')	Barail (2120'+)	Jenam (2120'+)
			MIDDLE				
			LOWER				
		EOCENE	UPPER	Jaintia	Kopili (780')		
			MIDDLE		Sylhet ls (800')		
			LOWER		Tura ss (790'+)		
		PALEOCENE			Sibganj Trapwash (750')		
	CRETACEOUS	UPPER			Rajmahal Trap (2000')		
		LOWER					
	JURASSIC	UPPER					
		MIDDLE					
		LOWER					
	TRIASSIC	UPPER					
		MIDDLE					
		LOWER					
	PERMIAN	UPPER		Ranigang (1443'+)	Pahrpur (1520'+)		
		LOWER		Barakar (160'+)	Kuchma (160'+)		
PRECAMBRIAN					Basement Complex		

Figure 2. Tentative stratigraphic column for Bangladesh.

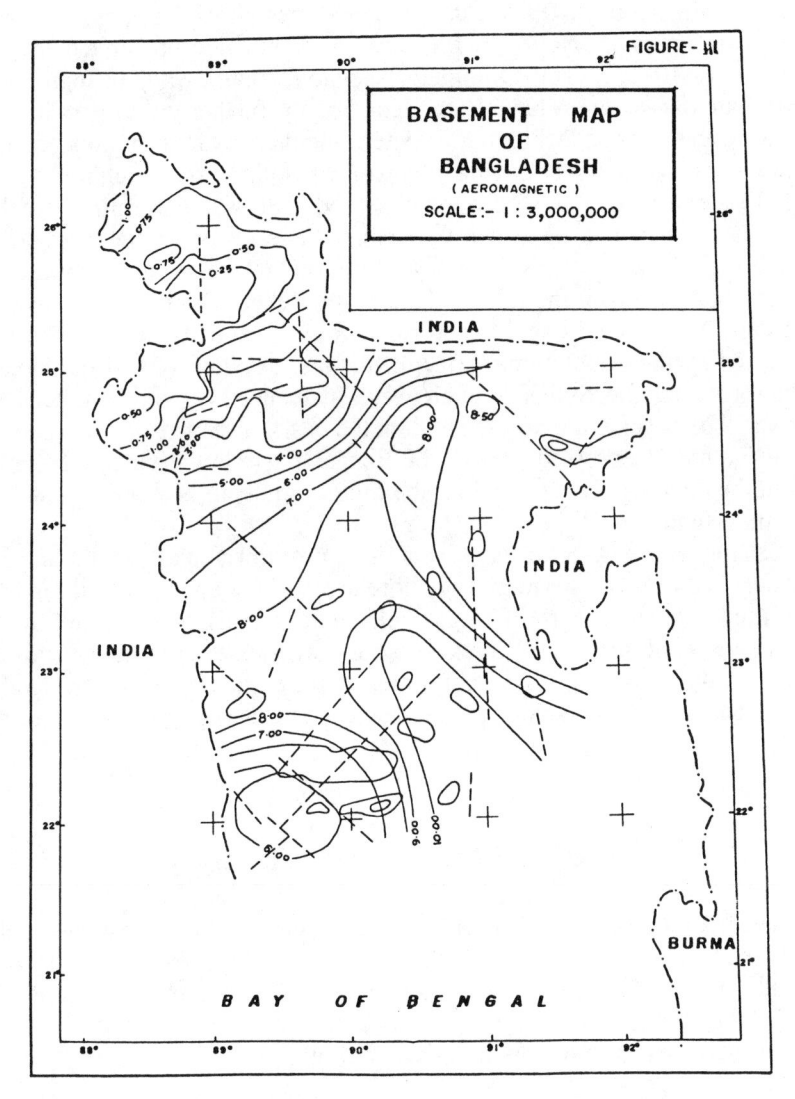

Figure 3. Aeromagnetic map of Bangladesh (scale: 1:3,000,000).

slope of the north-western part of the country is as much as 1 km. A basement high with depths of less than 6 km has been interpreted in the southwest deltaic zone. The greatest thickness of sediments in Bangladesh occurs south of Dhaka, in a basement trough which opens seaward to the S-SW. The maximum thickness of sediments interpreted in this trough is 10 km. The Dinajpur, Naogaon, Bogra-Mymensingh, and Faridpur-Barisal areas have been identified for further investigation with gravity and seismic techniques to determine their hydrocarbon potential.

Figure 4 is a Bouguer anomaly map of the country, on which it is possible to identify regions where the observed field was determined by the prevalent influence of either the sedimentary cover or the basement. Such regions are the western slope of the Bengal Basin, the central part of the Bengal Basin, the folded flank of the Bengal Basin, and the Sylhet Trough. Within the Bengal Basin and its folded flank the field was highly affected by the gravitational effect of the mantle; the slope of the Indian Platform and the Sylhet Trough were influenced by the sedimentary cover. The basement map prepared on the basis of the gravity data shows that the most depressed part of the Bengal Basin is the Hatiya Trough, with the depth exceeding 18 km; the minimum depth is about 10–11 km on the Madhupur High.

Seismic surveys so far have been carried out principally in the northeastern, central-southern, and offshore areas of the country. Drilling has resulted in the discovery of 13 gas fields. Other areas, such as the Chittagong and Chittagong Hill tracts and the hinge-zone, are under exploration. In the next few years exploration will be undertaken in the southwestern part of the country; thus, seismic surveying intensifies day by day.

Wells Drilled and Gas Reserves

The hydrocarbon potential of Bangladesh has so far been tested by only 69 wells, including the GIB wells (Table 1; Fig. 5), resulting in 13 gas fields (Figure 6) with total estimated recoverable gas reserves of 10.52 tcf (Table 2). Potential natural gas resources may be as much as 20 tcf. A number of reasons support this observation:

1. Many of the gas fields have not been delineated and the true extent of many gas reservoirs and reserves is not known precisely.
2. Promising geological structures have not yet been drilled.

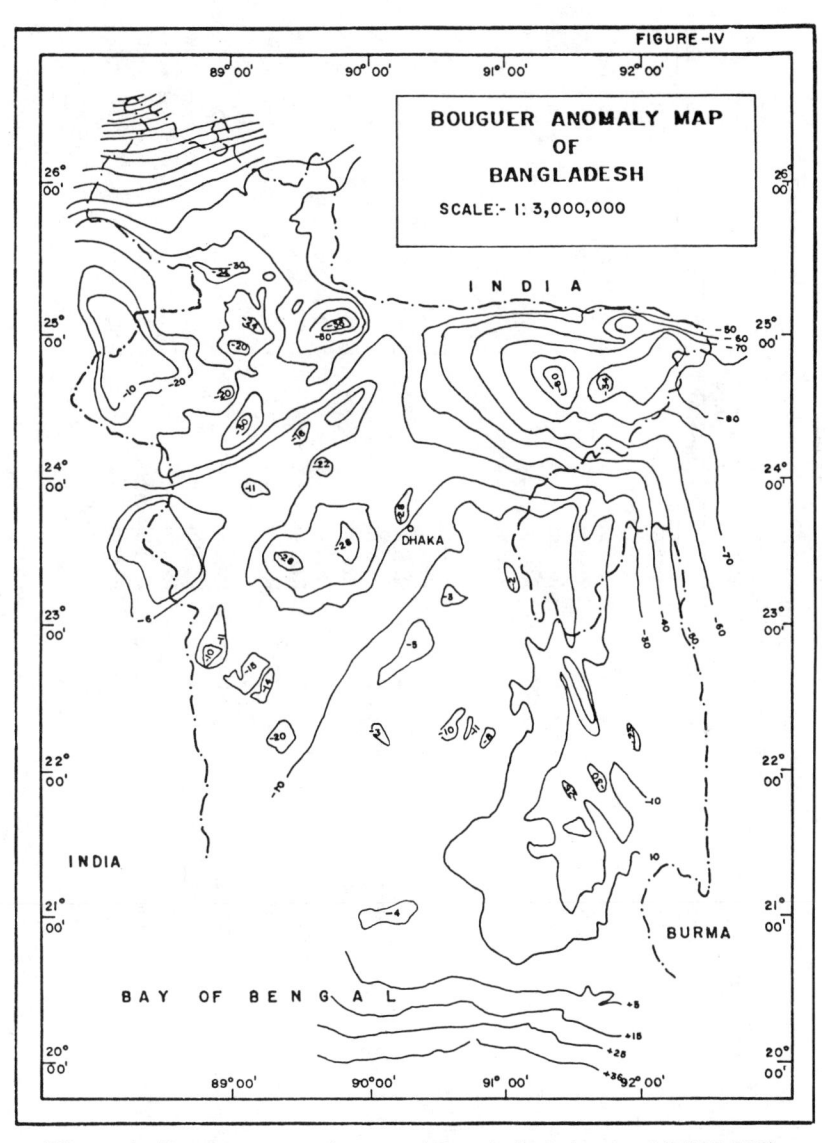

Prospects of Shallow Gas Resources in Bangladesh 395

Figure 4. Bouguer anomaly map of Bangladesh (scale: 1:3,000,000).

Table 1
Wells Drilled in Bangladesh

Name of the well	Year of Completion	Total Depth (ft)	Age At Surface	Age At Total Depth	Status/Result
1. A R C O—1	1976	12,804	Recent	Mid. Miocene	Dry offshore
2. Atgram—1	1982	16,276	Pleistocene	L. Miocene	Gas show
3. Bakhrabad—2	1969	9,311	Recent	L. Miocene	Gas well
4. Bakhrabad—2	1981	8,550	Recent	L. Miocene	Dev. well
5. Bakhrabad—3	1981	9,339	Recent	L. Miocene	Dev. well
6. Bakhrabad—4	1981	9,430	Recent	L. Miocene	Dev. well
7. Bakhrabad—5	1981	9,675	Recent	L. Miocene	Dev. well
8. Beani Bazar—1	1980	13,480	Recent	L. Miocene	Gas well
9. Begumganj—1	1977	11,992	Recent	L. Miocene	Gas well
10. Begumganj—2	1977	11,736	Recent	L. Miocene	Dry
11. B I N A—1	1976	13,436	Recent	L. Miocene	Dry offshore
12. B I N A—2	1977	14,088	Recent	L. Miocene	Dry offshore
13. B O D C—1	1976	15,086	Recent	L. Miocene	Dry offshore
14. B O D C—2	1976	14,551	Recent	L. Miocene	Dry offshore
15. B O D C—3	1978	14,725	Recent	L. Miocene	Dry offshore
16. B O G R A—1	1960	7,175	Recent	Pre-Cambrian	Gas show
17. Chhatak—1	1959	7,000	U. Pliocene	L. Miocene	Gas well
18. Cox Bazar—1	1969	12,133	Recent	L. Miocene	Dry offshore
19. Fenchuganj—1	1960	8,000	U. Pliocene	L. Miocene	Gas show
20. F E N I—1	1980	10,496	Recent	L. Miocene	Gas well
21. Habiganj—1	1963	11,500	U. Pliocene	L. Miocene	Gas well
22. Hazipur—1	1960	12,520	Recent	Oligocene	Oil show
23. Habiganj—2	1963	5,100	U. Pliocene	L. Miocene	Gas well
24. J A L D I—1	1965	7,546	Plio-Miocene	L. Miocene	Dry

Prospects of Shallow Gas Resources in Bangladesh   397

25.	J A L D I—2	1966	11,024	Plio-Miocene	L. Miocene	Dry
26.	J A L D I—3	1970	14,764	Plio-Miocene	L. Miocene	Dry
27.	Kailas Tila—1	1968	13,576	U. Miocene	L. Miocene	Gas well
28.	K A M T A—1	1982	11,857	Pleistocene	L. Miocene	Gas well
29.	Kutchma—1	1960	9,433	Recent	Cretaceous	Dry
30.	Kutubdia—1	1977	11,500	Recent	L. Miocene	Offshore gas
31.	Lalmai—1	1958	9,813	U. Pliocene	L. Miocene	Dry
32.	Lalmai—2	1960	13,506	U. Pliocene	L. Miocene	Dry
33.	Muladi—1	1976	15,520	Recent	L. Miocene	Dry
34.	Muladi—2	1979	14,944	Recent	L. Miocene	Dry
35.	Patiya—1	1953	10,176	Mid. Miocene	L. Miocene	Oil show
36.	Patharia—1	1933	2,871	L. Miocene	L. Miocene	Oil show
37.	Patharia—2	1933	3,436	L. Miocene	L. Miocene	Gas show
38.	Patharia—3	1951	5,411	L. Miocene	L. Miocene	Dry
39.	Patharia—4	1953	2,723	L. Miocene	L. Miocene	Gas well
40.	Rashidpur—1	1960	12,663	U. Miocene	L. Miocene	Gas well
41.	Rashidpur—2	1961	15,071	U. Miocene	L. Miocene	Gas well
42.	Semutang—1	1969	13,412	Plio-Miocene	Mid. Miocene	Appraisal well
43.	Semutang—2	1970	5,036	Plio-Miocene	Mid. Miocene	Dry
44.	Semutang—3	1970	5,092	Plio-Miocene	Mid. Miocene	Dry
45.	Semutang—4	1971	4,803	Plio-Miocene	Cretaceous	Dry
46.	Singra—1	1980	13,452	Recent	L. Miocene	Gas show
47.	Sitakund—1	1910	2,500	L. Miocene	L. Miocene	Gas show
48.	Sitakund—2	1910	2,500	L. Miocene	L. Miocene	Gas show
49.	Sitakund—3	1914	2,500	L. Miocene	L. Miocene	Gas show
50.	Sitakund—4	1914	3,358	L. Miocene	L. Miocene	Gas show

Table 1 continued

Name of the well	Year of Completion	Total Depth (ft)	Age At Surface	Age At Total Depth	Status/Result
51. Sitakund—X	—	—	U. Miocene	—	Drilling in progress
52. Sylhet—1	1955	7,800	U. Pliocene	L. Miocene	Gas show blowout
53. Sylhet—2	1956	9,245	U. Pliocene	L. Miocene	Gas show
54. Sylhet—3	1957	5,497	U. Pliocene	Mid. Miocene	Gas well
55. Sylhet—4	1962	1,035	U. Pliocene	Plio-Miocene	Dry
56. Sylhet—5	1965	1,885	U. Pliocene	Plio-Miocene	Gas show
57. Sylhet—6	1964	4,610	U. Pliocene	Mid. Miocene	Gas well
58. Titas—1	1962	12,323	Recent	L. Miocene	Gas well
59. Titas—2	1963	10,574	Recent	L. Miocene	Dev. well
60. Titas—3	1969	9,315	Recent	L. Miocene	Dev. well
61. Titas—4	1969	9,350	Recent	L. Miocene	Dev. well
62. Titas—5	1981	10,805	Recent	L. Miocene	Dev. well
63. Titas—6	1983	10,070	Recent	L. Miocene	Dev. well
64. Union—76—1	1976	765	Recent	—	Dry offshore

Gib

Name of the well	Year of Completion	Total Depth (ft)	Age At Surface	Age At Total Depth	Status/Result
1. Sitakund—1 to 25 Core holes	1939	100–600	L. Miocene	—	2 major thrusts on the western flank
2. Sylhet Gib—1	1951	—	U. Pliocene	—	—
3. Sylhet Gib—2	1952	—	U. Pliocene	—	—
4. Jaldi Gib—1	1965	4,429	Plio-Miocene	Mid. Miocene	Gas shows
5. Jaldi Gib—2	1966	4,488	Plio-Miocene	Mid. Miocene	Gas shows
6. Semutang Gib—1	1966	1,109	Plio-Miocene	Mid. Miocene	Gas shows & blown out.

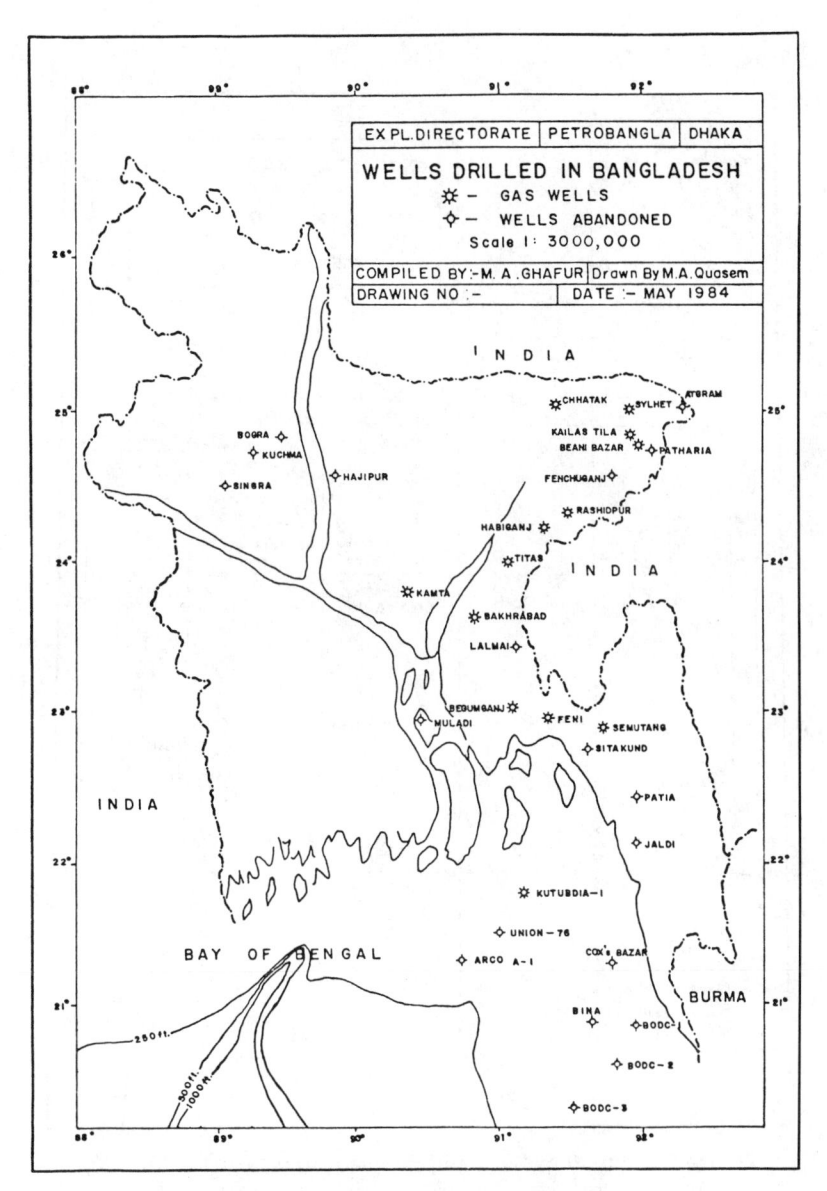

Prospects of Shallow Gas Resources in Bangladesh 399

Figure 5. Wells drilled in Bangladesh.

400 Shallow Oil and Gas Resources

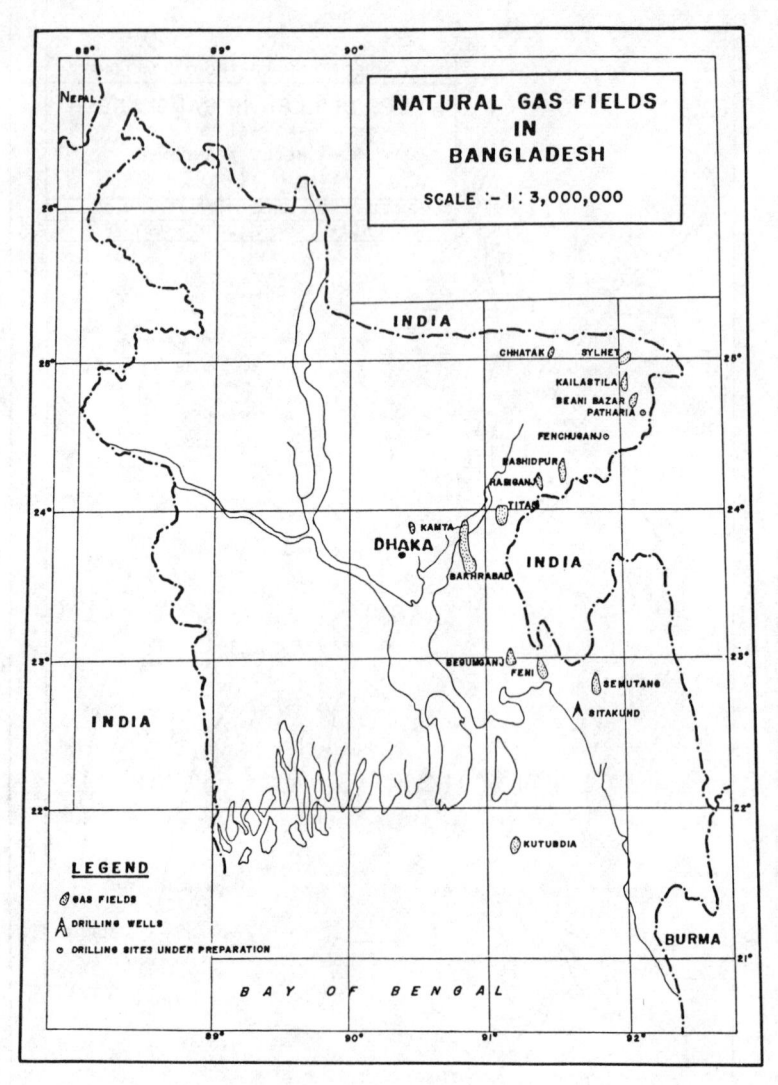

Figure 6. Natural gas fields in Bangladesh (scale: 1:3,000,000).

Table 2
Estimated Recoverable Gas Reserves, with Probable Reserves Discounted 50% and Possible Reserves Discounted 25% (Tcf)

Sl. No.	Field	Year of Discovery	Proven	Probable	Possible	Total	Condensate Recovery (Bbls/mmcft)
1.	Sylhet*	1955	0.44	0.24	—	0.68	3.40
2.	Chhatak*	1959	0.02	0.02	0.02	0.06	Trace
3.	Rashidpur	1960	0.49	0.38	0.19	1.06	0.03
4.	Kailas Tila	1962	0.38	0.15	0.07	0.60	10.00
5.	Titas*	1962	0.95	0.90	0.40	2.25	1.50
6.	Habiganj*	1963	1.00	0.19	0.09	1.28	0.03
7.	Bakhrabad*	1968	0.74	0.74	1.30	2.78	2.00
8.	Semutang	1968	0.04	0.03	0.03	0.10	—
9.	Kutubdia	1977	—	1.00(?)	—	1.00(?)	Trace
10.	Begumganj	1980	0.08	0.03	—	0.11	0.29
11.	Feni	1980	0.07	0.18	—	0.25	3.82
12.	Beani Bazar	1982	0.098	0.076	0.069	0.243	18.0–20.0
13.	Kamta	1982	0.0096	0.0938	—	0.1034	0.16
	Total		4.3176	4.0298	2.169	10.5164	

* Producing wells.

3. The "hinge-zone" is one of the areas deserving priority attention.
4. The single offshore discovery at Kutubdia promises other offshore discoveries in the Bay of Bengal.
5. About 40% of the total area of Bangladesh is unexplored or else only scanty exploration has been carried out.

Prospective Areas

The depths of the wells vary from 1,109 ft (338 m) to 15,086 ft (4,598 m). Gas shows at shallow depths and proved reserves found in such wells as Semutang, Feni, Beani Bazar, Sylhet, Chatak, Kailas Tila, Bakhrabad, and Kamta indicate that the prospectivity of shallow gas in Bangladesh cannot be ruled out (Figure 7).

From the data, it appears that there is a greater possibility of finding gas at shallow depth in the gas belt area; other areas, such as the hinge-zone lower shelf, and platform slope of the sub-Himalayan foredeep, have not been studied fully.

402 Shallow Oil and Gas Resources

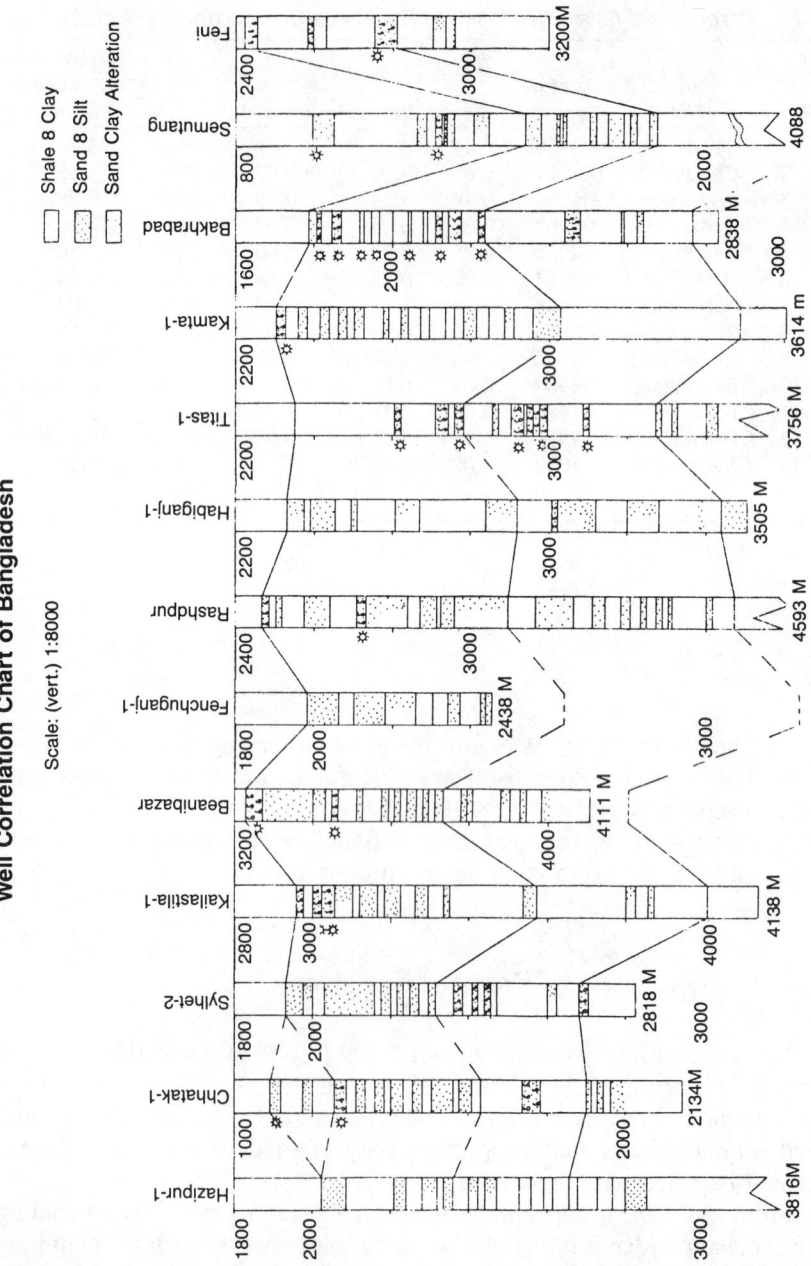

Figure 7. Well correlation chart of Bangladesh (scale: (vert.) 1:8,000).

As evidenced by seismic data, the western and northwestern parts of Bangladesh offer mainly stratigraphic traps, fault traps, reefs, and unconformities. The presence of suitable source rocks downdip and reservoir rocks updip makes the area prospective. The hinge-zone area has been tested only by the nearby Hazipur well. The Singra, Bogra, and Kuchma wells in the lower shelf area were drilled too far updip and were dry. The Dinajpur platform slope, covering the northwestern part of the country, has not been explored seismically or tested by drilling. The basement depth in this area is about 1 km and the area may be prospective for shallow gas resources if traps are found.

Conclusion

The prospects for shallow gas in Bangladesh have yet to be evaluated but are believed to be good.

35
The Possibility of Exploring and Producing Shallow Oil and Gas Reserves for Rural Communities and Small-Scale Industries in Indonesia

M. A. Warga Dalem and July Usman

Introduction

Development plans conducted since 1969 under "Five Year Development Plan" (now entering the fourth), projected a national energy requirement to steadily increase at a rate of 8% per year. Better economic life, prosperity, and industrialization contribute most to such increasing energy demand. While energy, in the form of liquid fuel, gas, electricity, LPG, coal, etc. are readily available in the cities and industrialized areas, the situation is a little bit different in the rural areas, where most of the population lives as farmers.

Distances and relatively low requirements are two major factors which prevent energy from being provided to rural communities at a reasonable price. This explains why most villages still have no electricity.

The possible solution is to supply the rural communities and small-scale industries with oil and gas produced from shallow oil and gas reserves. Based on the economic model developed and the underlying assumptions, we believe that energy at a price lower than conventional liquid fuel could be supplied to these places at a dependable, stable rate.

The availability of cheaper energy will encourage and accelerate further development in rural communities and small scale industries. This in turn will open the possibility of constructing small power plants for local consumption.

Petroleum Industry in Indonesia

Geography

Indonesia is an archipelagic state, strategically situated between two continents, Asia and Australia, and between two oceans, the Indian and Pacific. It consists of 13,677 big and small islands. Figure 1 is a general map of Indonesia.

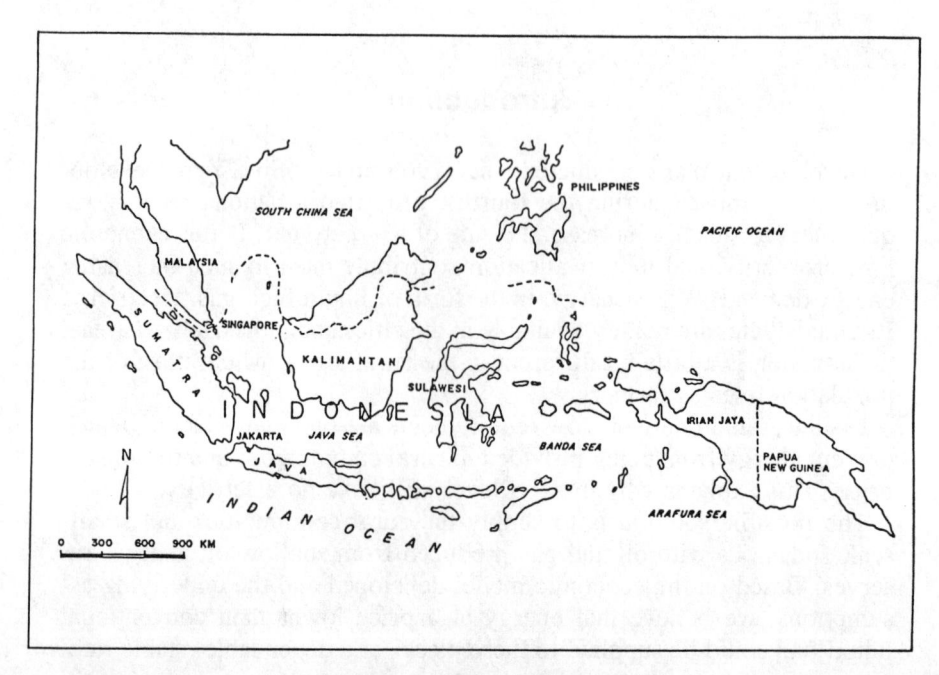

Figure 1. Map of Indonesia.

Petroleum industry in Indonesia

The 1945 Constitution of the Republic of Indonesia stipulates that all minerals, including oil and gas, existing within the statutory mining territory of Indonesia are *national riches controlled by the State* for the benefit and prosperity of all the Indonesian people; all activities of the petroleum industry may, therefore, be *executed by the State* only. *National riches* means that in-kind rights to the oil and gas are in the hands of the people. *Controlled by the State* means that only a State-owned company is allowed to conduct the activities of the petroleum industry. Law No. 8/1971 states that the only such State-owned company is Pertamina.

The petroleum industry is a highly risky business which requires large capital investment and high technical knowhow and skills. Therefore, in addition to the activities conducted by the State through Pertamina, Indonesia also invites foreign oil companies to cooperate and work in Indonesia as *contractors* to Pertamina under the so-called "production sharing contract." Under this contract, Pertamina is the management, while the foreign oil companies are the contractors who are responsible to Pertamina for conducting the petroleum operations; they should provide the required funds and take risks. Under this arrangement, Pertamina is currently producing around 100,000 bopd, 460 million scfgd, and is managing the foreign oil companies' production of 1,550,000 bopd and 3,500 million scfgd. Currently, there are 40 foreign contractors working in Indonesia, operating in 80 contract areas; among these areas, 21 are producing or will start production soon, while the rest are still in the exploration state.

Shallow Oil and Gas Accumulations

For the purpose of this discussion shallow oil and gas accumulations are defined as those occurring at relatively shallow depth—at less than 750 m. Sufficient study and attention had been paid in the past to the occurrence of oil accumulations, both shallow and deep. Therefore, our discussion will be focused on shallow gas only. Shallow gas may occur as:

- *Biogenic gas* is composed mostly of methane. It is generated at low temperatures through decomposition of organic matter by anaerobic micro-organisms.
- *Thermogenic gas* is composed mostly of methane and heavier hydrocarbon components, either in association with oil and condensate or occurring as free gas. The gas is generated at higher temperatures related to geothermal gradients in the subsurface.

Our experience in the past, while drilling to deeper horizons, indicates the existence of such shallow gas. Frequently these gas accumulations caused costly and hazardous problems during the drilling phase for the surface and intermediate strings, especially if in combination with thief zones. Our records indicate that several rigs were lost in the past due to shallow gas blowouts. Little or no attention was paid to this kind of gas accumulation, other than for drilling safety and for supplying field gas to support oil production. Another reason for producers to ignore the shallow gas accumulation, especially in areas close to oil-producing fields, is that it could not compete with the associated gas produced at practically no extra cost.

With increasing energy demand and prices at present, the situation has changed toward greater appreciation of such gas accumulations, especially in those areas located at a distance from existing oil and gas fields and gas pipeline systems.

Figure 2 shows the location of geological basins in which hydrocarbon accumulations, including shallow oil and gas, may potentially be discovered. Taking economic factors into consideration, such as location of potential consumers relative to the oil and gas accumulations and the capital investment required to explore and develop the accumulations, potential areas for shallow oil and gas utilization are the following:

- Onshore portion of the geological basins located along the eastern coast of Sumatra, namely the North Sumatra, Central Sumatra, and South Sumatra basins.
- Onshore portion of the geological basins located along the northern coast of Java, namely the Sunda, North-West Java, and East Java basins.

According to the geological data available from these areas the biogenic gas occurs mostly in relatively young rock sequences of Pleistocene, Pliocene, and late Miocene ages. The sands are dominantly uncompacted and are both marine and nonmarine in origin.

The thermogenic gas and oil although normally occurring in a relatively deeper formation of Miocene age, may also be found at shallow depths; the rock is dominantly sandstone and limestone.

Although production of shallow gas is relatively neglected in Indonesia and interest is more concentrated on onshore operations, it is surprising that it is the shallow offshore L-Parigi gas field, at 550 m subsea, which has become the backbone to the gas-supply system in West Java. This field is 21 km × 4.5 km in size (about 23,000 acres) and is located in relatively shallow water (to 35 m); the producing formation is a very permeable reef limestone 35 m thick on average. Containing about 1 trillion

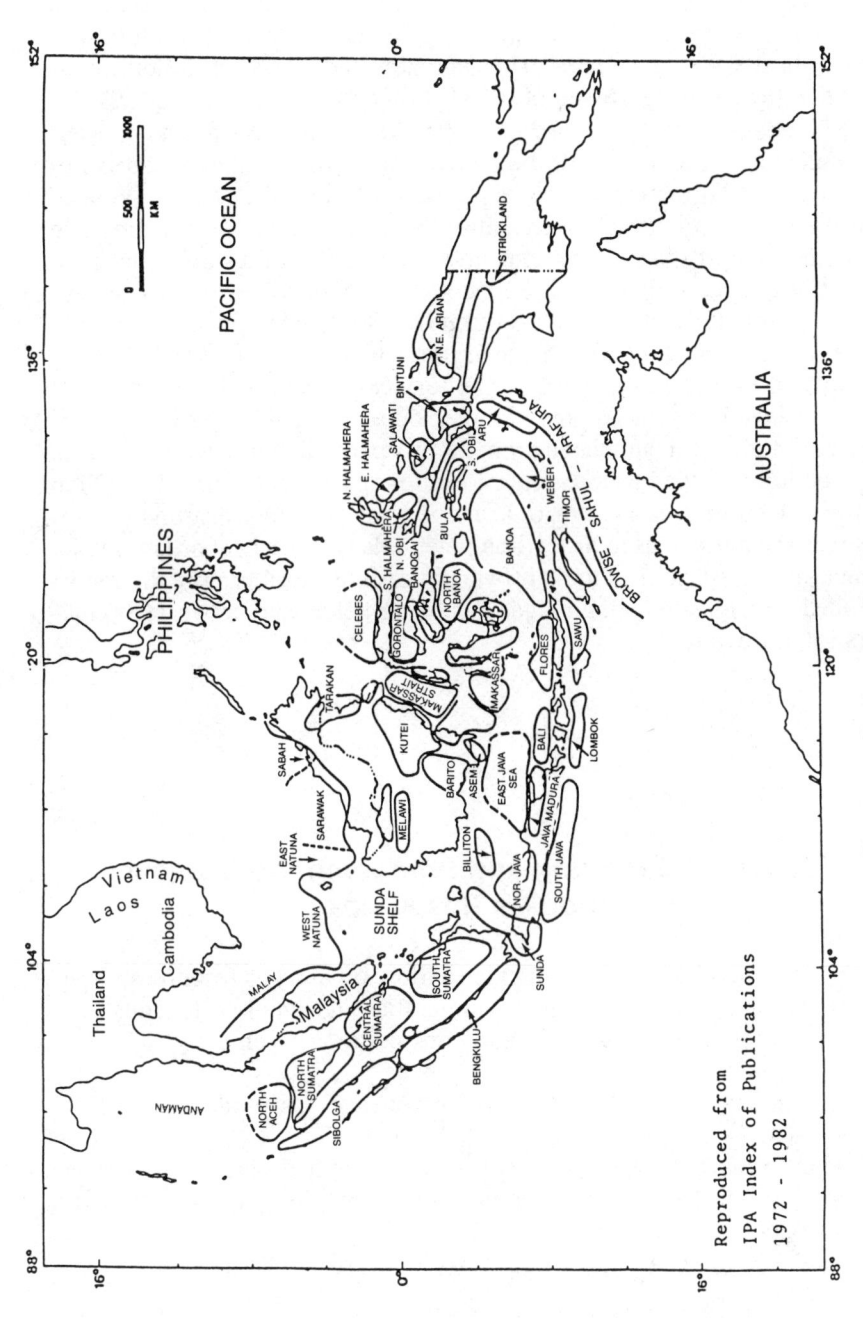

Figure 2. Basin map of Indonesia.

scf of recoverable gas, the field is projected to produce a constant gas supply of 137 milion scfd for 20 years, or about 60% of the total gas supply in the West Java area. The separator and wellhead samples indicate that the gas is composed of 0.6–1.4% N_2; 0.1–0.7% of CO_2; 97.7 – 98.8% methane; 0.4% ethane; and traces (0.1%) of propane and heavier.

Similar gas reserves might be discovered in many other prospects of reefal Parigi Formation. The seismic surveys carried out in the area of North-West Java both offshore and onshore have indicated the presence of numerous prospects. Unfortunately many of them have not been assessed yet because they are not concordantly sitting over the deeper objectives. In a study titled "Java Natural Gas Study—Supply and Demand-Utilization of Shallow Gas Beneficiation to the Small Size Industry" completed by Pertamina and PT. Geoservice in 1984, it is reported that onshore Java island has a potential of 2.60 trillion scf of shallow gas resources in place, accumulated from 55 prospects already identified. This is in addition to the onshore proved shallow gas reserves of 0.77 trillion scf and deep gas reserves of 0.22 trillion scf. The hydrocarbon gas resources are defined as hydrocarbon potentials with vertical or lateral distribution but still subject to be proven by drilling; furthermore the report defined "deep" and "shallow gas" as being below and above 1,500 meters subsea depth.

Feasibility of Exploring and Producing the Shallow Oil and Gas Resources

To be successful in developing the shallow oil and gas resources, especially for the benefit of the rural communities and small-scale industries, several factors should be taken into consideration, such as:

- The population, to whom the shallow oil and gas production will be dedicated.
- The location of the geological basins, in which hydrocarbon accumulations, including shallow oil and gas, may potentially be discovered and developed.
- The Government's energy policy with regard to shallow oil and gas pricing.
- The economics of supplying the shallow oil and gas to the rural communities and small-scale industries.

Population

The size of the population, its distribution, its density, the degree of prosperity as well as the level of their economic activities, will determine the number of potential consumers and their capability for absorbing the shallow oil and gas production.

According to the 1980 data published by the Central Bureau of Statistics, Indonesia covers a total land area of 1,919,443 square kilometers and is populated by 147.5 million people, an average population density of 77 people per square kilometer. Among those, 91.3 million people, or 62%, are living in the small Java island, which contributes a mere 7%, or 132,187 square kilometers, to the Indonesian land area; with an average population density of 690 people per square kilometer, this island is the most densely populated area in Indonesia. The balance, or 56.2 million, lives on the other islands, with an average population density of only 31 people per square kilometer; however, there are 5 provinces in those islands which have a population density of more than 100 inhabitants per square kilometer. In total, therefore, 10 provinces in Indonesia have a population density of more than 100 people per square kilometer (Table 1). Figure 3 shows the population, as well as the population density, of the 27 provinces in Indonesia.

Although the development of shallow oil and gas resources is governed more by economics than by the population, for the time being we believe that the population and its distribution as well as the population density are the key factors to be considered first for the success of a shallow oil and gas development project. Therefore, our consideration will be temporarily limited to those 10 provinces, although we realize that there is no reasonable basis for taking 100 people per square kilometer as the cutoff; in-depth studies will be required to determine the best figure.

Location of the Geological Basins to Support the Development of Shallow Oil and Gas Resources

As mentioned earlier, Figure 2 shows the location of geological basins in Indonesia. Superimposing the population map (Figure 3), in order to secure the market and minimize the capital investment for the pipeline, it is shown clearly in Figure 4 that the most prospective basins to support shallow oil and gas production are the ones mentioned earlier:

- The North Sumatra, Central Sumatra, and South Sumatra basins, which cover the whole eastern part of Sumatra
- The Sunda, North-West Java, and East Java basins, which cover the northern parts of West, Middle, and East Java provinces.

Table 1
**Population, Area, and Population Density
of the Ten Most Densely Populated Provinces in Indonesia
(1980)**

Province	Population	Percent of Total Indonesia Population	Area (km^2)	Percent of Total Indonesia Area	Population Density/ km^2
1. North Sumatra	8,360,894	5.67	70,787	3.69	118
2. Lampung	4,624,785	3.14	33,307	1.74	139
3. Dki Jakarta	6,503,449	4.41	590	0.03	11,023
4. West Java	27,453,525	18.61	46,300	2.41	593
5. Middle Java	25,372,889	17.20	34,206	1.78	742
6. Yogyakarta	2,750,813	1.87	3,169	0.17	868
7. East Java	29,188,852	19.79	47,922	2.50	609
8. Bali	2,469,930	1.67	5,561	0.29	444
9. West Nusa Tenggara	2,724,664	1.85	20,177	1.05	135
10. North Sulawesi	2,115,384	1.43	19,023	0.99	111
11. Other 17 provinces	35,925,113	24.36	1,638,401	85.35	22
Indonesia	147,490,298	100.00	1,919,443	100.00	77
– Java Island (3,4,5,6,7)	91,269,528	61.88	132,187	6.89	690
– Other Indonesia	56,220,770	38.12	1,787,256	93.11	31
Indonesia	147,490,298	100.00	1,919,443	100.00	77

The other four densely populated provinces (with population density of more than 100 people per square kilometer), namely Yogyakarta, Bali, West Nusa Tenggara, and North Sulawesi, will therefore be temporarily put aside, since based on current available data, they are located outside known geological basins.

Shallow Oil Production

Crude oil production started in Indonesia at the end of the 19th century. As happened in other parts of the world in those days, the oil was found at relatively shallow depths. When shallow oil became more and more difficult to discover, people started to explore for and exploit deeper oil accumulations onshore as well as offshore in deeper and deeper water. As a result, the oil (and gas) reserves become more and more expensive to produce.

While most of the earlier oil fields were closed a long time ago for economic reasons, some of them are still producing oil economically to-

Producing Shallow Oil and Gas Reserves for Indonesia   413

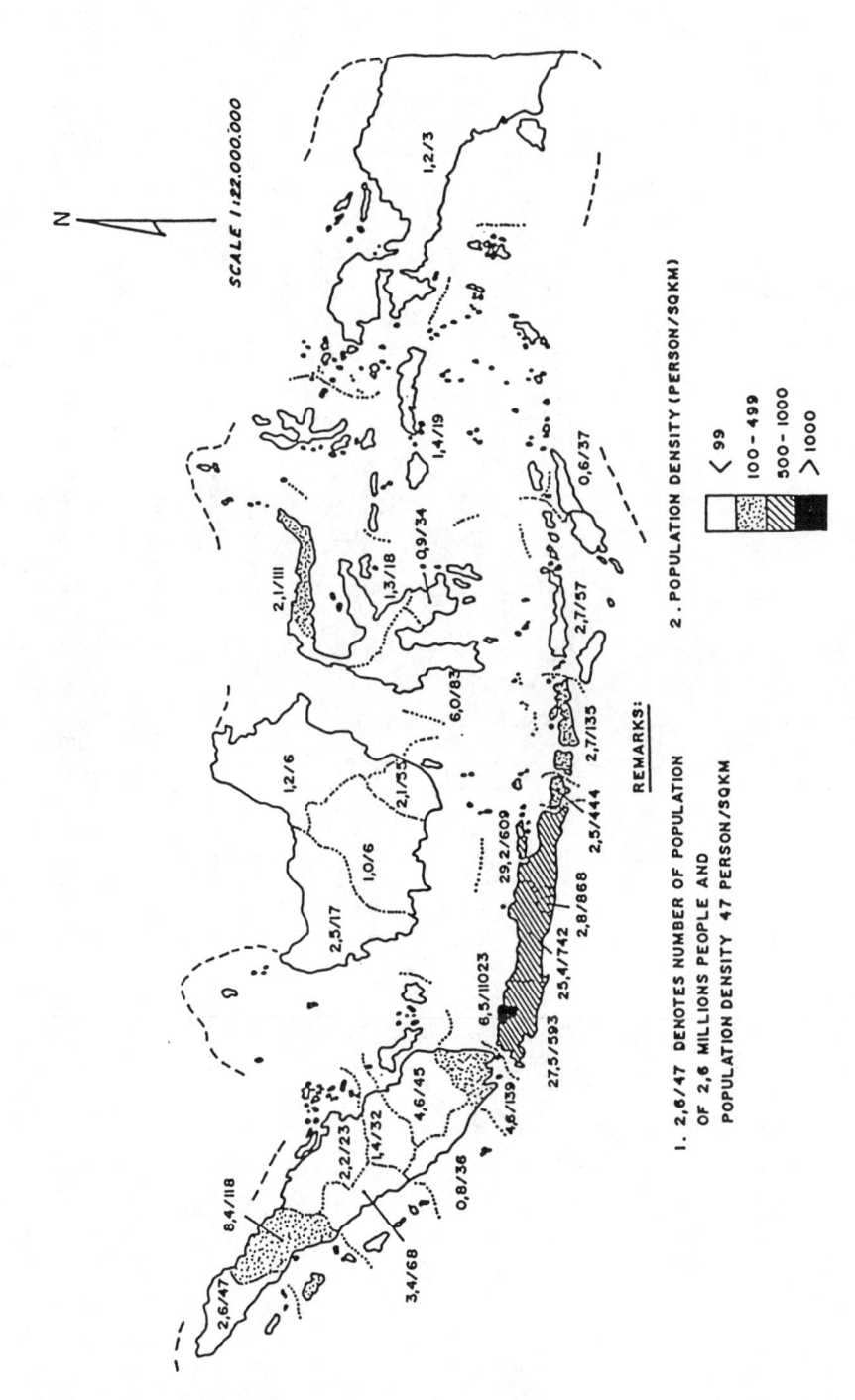

Figure 3. Population and population density by province in Indonesia, 1980.

414   Shallow Oil and Gas Resources

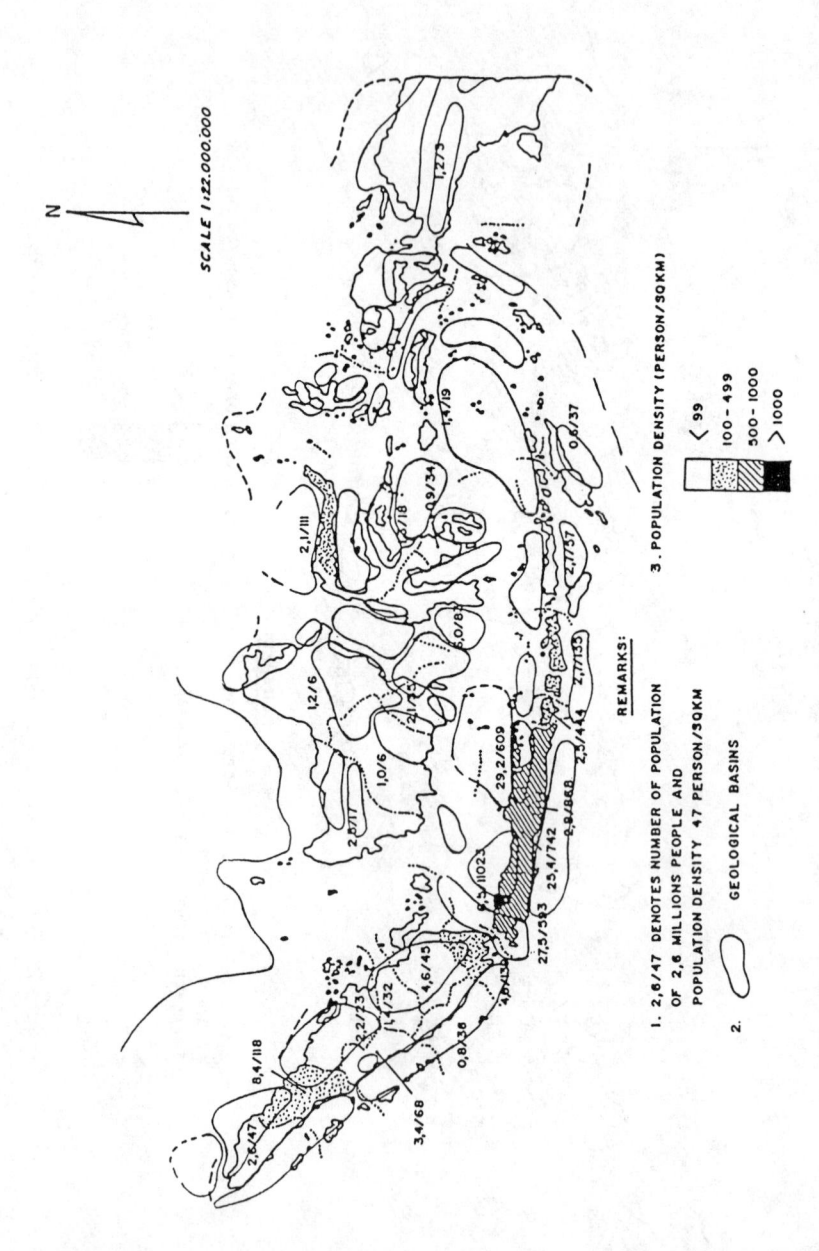

Figure 4. Overlay of Indonesia basin map on population map.

day, using conventional methods. Unfortunately those oil fields which had been closed are located far away from the densely-populated areas, except for one, the Wonocolo field in East Java, which was closed for a short while and then put into production again by the local population, which produces and refines the crude oil in a simple manner for local consumption.

Oil is usually supplied to the consumers in the form of refined products which should meet certain specifications for consumer-protection purposes. Combined with the availability of a good domestic liquid-fuels distribution system throughout, and a subsidized kerosene and diesel oil price, this makes the production of shallow oil for direct use by rural communities and small-scale industries look unattractive. Another factor which supports this opinion is the Government's policy to diversify the development of domestic energy resources by encouraging more and more utilization of gas and liquefied gas, coal, and water energy instead of liquid fuel. This is intended to save crude oil for export purposes only, to generate foreign currencies required to fund national development. The situation may change if Pertamina, the national oil company, discovers shallow, small oil reserves which otherwise cannot be produced economically with conventional methods. Local conditions, for example the existence of nearby small-scale industries such as lime-kilns which could absorb the unrefined crude oil directly, could force the Government to offer lower crude prices to the local consumers than the current subsidized domestic liquid-fuel price.

Shallow Gas Production and Pricing

The utilization of gas at present is confined to major industries involving large capital investment and gas resources. Due to their limited gas requirements, small-scale industries and rural communities are not suitable for a conventional gas-production system requiring a relatively large gas consumption in order to be commercial. Consideration should therefore be focused on exploration and production of shallow gas accumulations requiring lower capital investment.

This production may serve as:

- One of the promising energy sources to meet future increase in energy demand.
- An alternate energy source to partially replace the current liquid-fuel requirement.
- A catalyst to encourage the further development of rural communities and small-scale industries.

Savings in liquid fuel which will otherwise be consumed domestically could therefore be expected and more crude oil could be made available for export. The shallow gas production and its utilization fit very well with the Indonesian Government's current policy to diversify the energy sources.

More and more gas utilization is being encouraged by the Government, among others, by offering an attractive gas price which is significantly lower than the liquid-fuel price (up to US $ 3.00/million Btu's for gas versus an equivalent domestic liquid-fuel price of US $ 5.00–5.67/million Btu's). Another advantage for the gas consumer is lower expenditures, which in turn will result in improved economics of their operations, since liquid-fuel transportation costs and the capital investment required to install the corresponding liquid-fuel stock tanks could be avoided. In addition, the gas is clean and causes less pollution than liquid fuel.

The Economics of Exploring for and Producing Shallow Gas Accumulations

We have discussed previously the prospecive areas which are favorable for the shallow oil and gas production. The economics of exploring and producing shallow gas accumulations are analyzed in a simple manner by using West Java as a model. This province is chosen due to its dense population, progressive small-scale industries, and location close to the existing West Java producing gas and oil fields and gas pipeline. In addition, quite intensive exploration activities, including seismic surveys and drilling of exploratory wells, have been conducted over the last two decades to explore for hydrocarbons in deeper horizons. The current producing fields in this area are the result of such exploration activities. With exploration data becoming available, it is expected that exploration costs for shallow hydrocarbon accumulations could be made significantly lower.

The underlying assumptions for such economic analysis are as follows:

- Reinterpretation of all seismic data (4,000 kilometers) to identify shallow prospects, estimated at US $ 37.50/kilometer.
- A maximum of 40 shallow prospects.
- Geological study resulting in 20 shallow prospects which require further seismic survey. Assume 20 kms/prospect at US $ 5,000/km.
- Shallow seismic interpretation resulting in 10 shallow drillable prospects.
- Ten exploratory wells (1 well/prospect) required.

- Average drilling costs of US $ 500,000/well, typical for a well of 750 m depth.
- Successful exploratory wells produced directly, with no further development wells required.
- Length of gas pipeline: 30 km.
- Project life: 20 years.
- Simple gas processing method.
- Government tax rate: 60%.
- Gas price along the pipeline: US $ 3.00/thousand scf; 1,000 Btus/scf.
- Operating cost: US $ 0.75/million Btu's.

The analysis is made using drilling success ratios of 50%, 30%, 20%, and 10%. Figure 5 shows the conceptual steps required to explore and produce the shallow oil and gas accumulation in West Java, while Table 2 shows the effective estimated expenditures for the shallow gas for the preceding success ratios. The results of the economics exercise are presented in Figures 6 and 7. Assuming that a DCF rate of return of 20% p.a. is appropriate for Pertamina as the producer (although this is not the only economic indicator to be considered), then it may be shown, based on the previous assumptions that the minimum recoverable reserves and consumers' total energy requirements in order for Pertamina to explore and produce the shallow gas reserves economically are relatively low:

Drilling Success Ratio (%)	Minimum Total Consumer's Requirement, thousand scfd	Minimum Recoverable Reserves (20 years supply) (billion scf)
50	1,400	10.22
30	1,830	13.36
20	2,370	17.30
10	3,980	29.05

Although the capability of any single consumer (rural communities or small-scale industry) to absorb the shallow gas production is relatively low, it is expected that there will be a sufficient number of consumers along the pipeline route so that the preceding minimum gas requirement could be exceeded. While the success ratio of the shallow gas wildcat drilling is still unknown, our statistical data indicate that the success ratio of wildcat drillings in onshore West Java for hydrocarbons in deeper horizons is relatively high, about 30%. Such information suggests that shallow hydrocarbons potentially may be discovered in larger quantities in the onshore part of the northern West Java province. The fact that the

418 *Shallow Oil and Gas Resources*

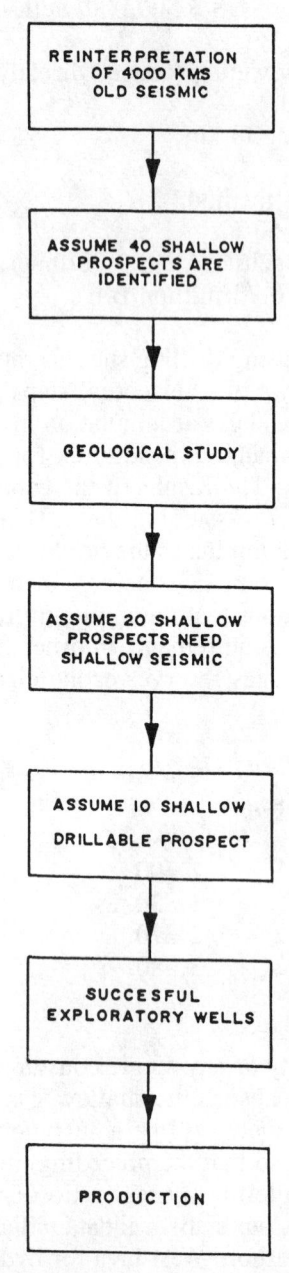

```
┌─────────────────────┐
│   REINTERPRETATION  │
│      OF 4000 KMS    │
│      OLD SEISMIC    │
└──────────┬──────────┘
           ▼
┌─────────────────────┐
│  ASSUME 40 SHALLOW  │
│    PROSPECTS ARE    │
│      IDENTIFIED     │
└──────────┬──────────┘
           ▼
┌─────────────────────┐
│   GEOLOGICAL STUDY  │
└──────────┬──────────┘
           ▼
┌─────────────────────┐
│  ASSUME 20 SHALLOW  │
│   PROSPECTS NEED    │
│    SHALLOW SEISMIC  │
└──────────┬──────────┘
           ▼
┌─────────────────────┐
│  ASSUME 10 SHALLOW  │
│  DRILLABLE PROSPECT │
└──────────┬──────────┘
           ▼
┌─────────────────────┐
│      SUCCESFUL      │
│  EXPLORATORY WELLS  │
└──────────┬──────────┘
           ▼
┌─────────────────────┐
│     PRODUCTION      │
└─────────────────────┘
```

Figure 5. Conceptual steps required to explore and produce the shallow oil and gas accumulation in West Java.

Table 2
Estimated Expenditures Exploring and Producing Shallow Gas in West Java

Description	\multicolumn{4}{c}{Expenditures in US $ at Drilling Success Ratio}			
	50%	30%	20%	10%
1. Assume:				
A. Number of identified shallow prospects	40	40	40	40
B. Number of shallow prospects for shallow seismic survey	20	20	20	20
C. Number of drillable shallow prospects	10	10	10	10
D. Number of exploratory wells	10	10	10	10
E. Number of successful wells	5	3	2	1
2. Reinterpretation of all seismic data 4,000 kms at US $ 37.50/km				
A. Total	150,000	150,000	150,000	150,000
B. Per successful well	30,000	50,000	75,000	150,000
3. Shallow seismic survey for 20 prospects, 20 kms/prospects at US $ 5,000/km				
A. Total	2,000,000	2,000,000	2,000,000	2,000,000
B. Per successful well	400,000	666,667	1,000,000	2,000,000
4. Drilling cost				
A. Per well	500,000	500,000	500,000	500,000
B. Total 10 wells	5,000,000	5,000,000	5,000,000	5,000,000
C. Per successful well	1,000,000	1,666,666	2,500,000	5,000,000
5. Surface facilities, including 1 scrubber and 30 kms of 6 in. pipeline	1,820,000	1,820,000	1,820,000	1,820,000
6. Operating costs $/million Btu	0.75	0.75	0.75	0.75

420 Shallow Oil and Gas Resources

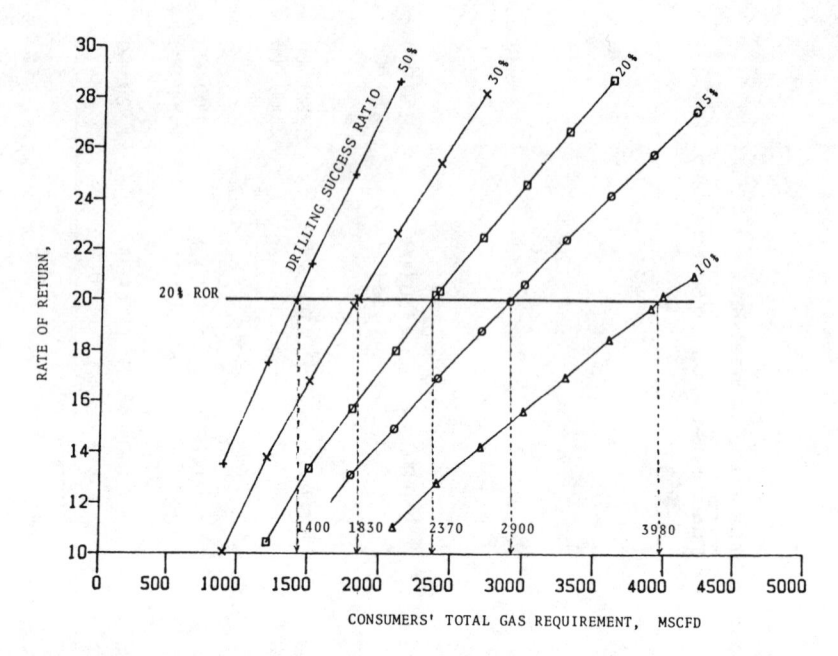

Figure 6. The economics of exploring and producing shallow gas in West Java, based on the economic model under consideration.

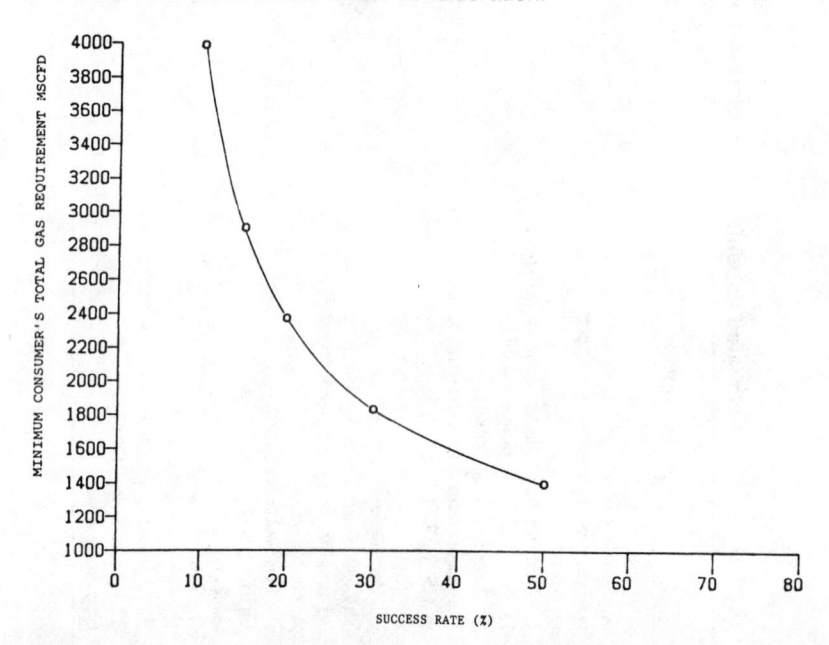

Figure 7. Minimum consumer's total gas requirement at various drilling success ratios.

offshore L-Parigi gas field described previously is located in the same area, strongly supports such expectation.

Another point of interest is with regard to the size of the prospect itself. A daily rate of 1,830 thousand scf production for a success ratio of 30% will require at least 13.36 billion scf of recoverable reserves for 20 years of continuous supply. By assuming:

- Reservoir depth: 750 m
- Sand thickness: 20 m
- Porosity: 25%
- Water saturation: 30%
- Gas formation volume factor: 75 vol/vol
- Recovery factor: 60%

then the required gas accumulation size is around 2.4 square kilometers (equivalent to 593 acres). Current information indicates that such reservoir size is not too difficult to discover.

The same economic model and analysis could similarly be prepared for other provinces or areas. In addition, it should be pointed out that the shallow gas production is very attractive to the Government, since this could partly eliminate the need to subsidize liquid-fuel consumption. It should therefore be encouraged and started as soon as possible.

Lesson Learned

Producing shallow oil and gas reserves is an attractive and promising undertaking. It will serve as one of the promising energy sources to meet future increases in energy demand, as an alternate energy to conserve liquid fuel and as a catalyst to encourage rural communities and small-scale industries for further development. The economic exercise gives indications that energy could be provided to the rural community and small-scale industry at a price which is significantly lower than the current liquid-fuel price. Shallow oil and gas exploration and production should therefore be encouraged.

Taking the economic factors and limited experience into consideration for the time being, attention should be focused on those portions of the geological basins located in densely populated areas, where further detailed studies should be conducted. However, with more experience obtained at a later stage on the shallow oil and gas exploration, production and distribution, including their economics, expansion of similar activities to other, less attractive areas or provinces may be considered. Our experience with utilization of the shallow L-Parigi gas and other shallow and deep gas production to back up the oil production, e.g., for gas lift or for gas supply to fuel the power plants, indicates that gas could be produced in a simple manner requiring no special treatment if the gas CO_2 H_2S composition is normal (low CO_2 content, no H_2S).

Acknowledgments

The authors wish to thank the Management of Pertamina, Jakarta, Indonesia for their permission to publish and present this paper. The authors are also grateful for the cooperation and assistance of Mr. F. X. Sujanto of the Exploration Department, Pertamina for his helpful advice, discussion, and assistance, especially on the geological aspects.

References

Pertamina, and PT. Geoservice, 1984, Java natural gas study—supply and demand—utilization of shallow gas beneficiation to the small size industry.

Pertamina's internal reports.

Rice, Dudley D., and Claypool, George. E., 1981, Generation, accumulation and resource potential of biogenic gas: AAPG Bulletin, Vol. 65-1, January.

Sujanto, F. X., and Purnama, A., 1983, An overview of geologic hazards in Indonesia related to offshore construction with sedimentation and blow-out as examples of cases: presented at Ascope Workshop on Technical Aspects of Offshore Geological and Natural Environment Hazards, held in Jakarta, Indonesia, March 29-30.

Section V

Economics and Financing

36
Drilling Statistics and Costs: An International Comparison by Target Depth

Emil D. Attanasi

Abstract

In the United States, 50% of the exploratory wells drilled by the petroleum industry are drilled to depths of less than 5,000 ft. Data for South American and African wells, however, show only a relatively small proportion of wildcats there drilled to shallow depths. In this paper the substantial differences that exist among drilling depths are demonstrated with international drilling statistics. Cost data are also presented to help explain why such differences exist. The final section of the paper describes the economic conditions peculiar to the United States that serve as incentives for development of shallow oil and gas resources.

Introduction

During the last 25 years, over 60% of the wells drilled in the United States by the petroleum industry were drilled to depths of less than 5,000 ft. Most of the successful exploratory wells drilled at these shallow depths found small fields containing probably less than 1 million barrels of oil equivalent. The generally lower cost for field development at shallow depths reduces the minimum commercial field size and allows development of rather small fields. Furthermore, the lower initial investments and risks involved in developing and producing from shallow fields (at depths of 5,000 ft or less) provide an opportunity for many smaller independent firms to enter the oil and gas industry. This may explain why, in the United States, independent operators have consistently drilled over 80% of the wildcat wells and a majority of the development wells. In most non-Communist areas outside North America, the number of wells drilled to shallow depths is a much smaller percentage of total wells than in the United States.

This paper presents statistics on historical drilling and drilling costs to demonstrate and explain why there are such differences in international drilling patterns. The incentives and economic conditions required for the development of shallow oil and gas resources are also discussed.

Historical Drilling Statistics—United States and Other Areas

In 1959, over 70% of all the wells drilled in the United States were drilled to depths of less than 5,000 ft. Even in 1982, nearly two-thirds of all U.S. wells were still drilled to depths of less than 5,000 ft. Figure 1 shows the relative proportion by depth interval of all wells drilled in the United States from 1959 and 1982. The figure shows that, even though total drilling had declined and increased beyond its previous peak, the proportion of wells drilled to depths of less than 5,000 ft never fell below 50% of the total. Even with the significant oil and gas price increases since the late 1960s, the percentage of total onshore exploratory wells drilled to shallow depths has hovered around 50 percent. Figure 2 illustrates the relative proportion of U.S. onshore exploratory wells drilled to depths of 5,000 ft, 5,000 to 10,000 ft, and deeper than 10,000 ft from 1968 to 1982. Data used as a basis for Figure 2 indicate there was typically a slightly smaller proportion of gas wells (perhaps two-fifths) at shallow depths than there were oil wells or dry holes.

The number of U.S. offshore wildcat wells* usually represents less than 10 percent of the total U.S. wildcat drilling. The number of wildcat wells drilled in the United States is at least six or seven times the total number of wildcat wells drilled in all non-Communist areas outside the United States and Canada. However, wildcat drilling in most non-Communist areas outside the United States and Canada has shifted dramati-

---

* The term wildcat well when used here refers to new field wildcats, that is, wells drilled to find new and distinct fields. These wells were typically drilled some distance from known fields. Exploratory wells include new field wildcats and also wells drilled to define and find additional reservoirs that are above or below a contiguous or existing field.

Figure 1. Total U.S. exploratory and development wells drilled from 1959 to 1982 by depth ranges (1 foot = .3048 meters). (Drilling data from *Twentieth Century,* DeGolyer and MacNaughton, 1983.)

cally to offshore areas and to hostile natural environments. During 1960, less than 3% of the wildcat wells drilled in non-Communist areas outside the U.S. and Canada were in offshore waters, whereas by 1980, 53% were offshore. During these same years for South America, offshore wildcat drilling went from 7% to 41% of total South American wildcat drilling; for Africa, it went from .5% to 53% of African wildcat drilling.

Tables 1 and 2 show depths of the cumulative (through 1983) onshore and offshore wildcat wells drilled in various countries in Central America, South America, and Africa. These tables are based on data from Petroconsultants S.A. This data source included little wildcat drilling prior to 1950 and is probably missing some South American wildcat drilling data during the 1950s. However, the drilling data for Africa are reasonably complete and represent nearly the entire exploration history of that continent. Modern petroleum exploration in Africa did not begin on a large scale until the 1950s and nearly all of Africa's known oil has been discovered since 1950. In contrast, only 40% of South America's known oil has been discovered since 1950.

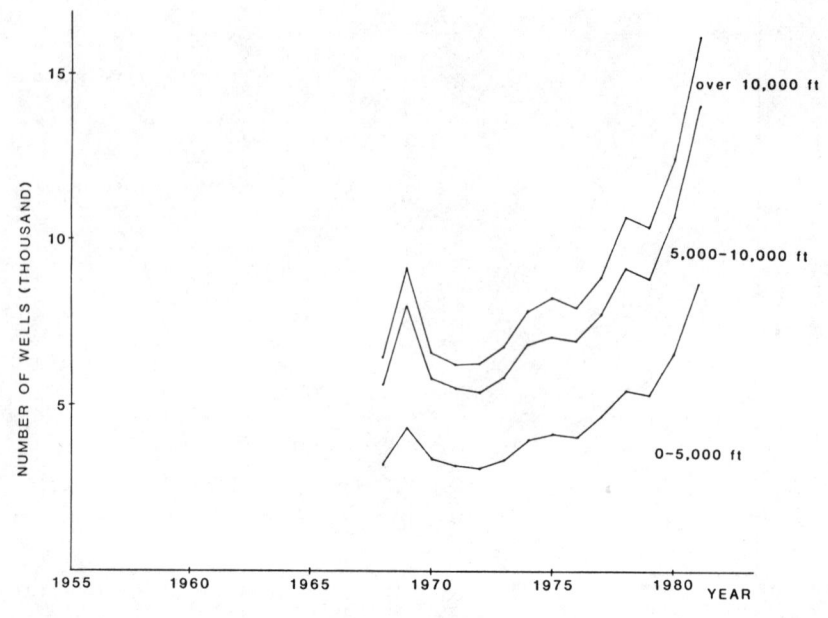

Figure 2. Total U.S. exploratory wells drilled from 1968 to 1981 shown by depth range (1 foot = .3048 meters). (Source: Drilling Statistics for 1960 to 1981 and Drilling and Production Costs for Nonassociated Gas Wells, 1976 to 1981, EIA, 1982, Dallas Field Office, original data compiled from American Association of Petroleum Geologists data.)

Table 1 shows that for all of South America only 29% of the onshore wildcat wells and 26% of the offshore wildcat wells were drilled to depths of less than 5,000 ft. The table shows that nearly 44% of the onshore wells in Central America were drilled to less than 5,000 ft. However, the total onshore wildcat drilling in Central America is only about 4% of the South American onshore wildcat drilling. The table shows wildcat drilling, both onshore and offshore, for nearly all individual countries to be concentrated at depths beyond 5,000 ft.

Cumulative wildcat drilling statistics for Africa, presented in Table 2, show that only 20% of all the onshore wildcat wells had depths of less than 5,000 ft. Only one in ten offshore wildcat wells did not go beyond 5,000 ft in depth. Areas with the largest amounts of oil, such as Nigeria and most of North Africa, have only a relatively small fraction of their wildcat wells drilled to depths of less than 5,000 ft. Although the area encompassing Swaziland, Namibia, Lesotho, and South Africa has a rel-

Table 1
Fraction of Cumulative Central American and South American Onshore and Offshore Wildcat Wells by Drilling Depth Range (through 1982)

Area	Onshore Drilling Depth (ft) < 5,000	Onshore 5,000–10,000	Onshore > 10,000	Offshore Drilling Depth (ft) < 5,000	Offshore 5,000–10,000	Offshore > 10,000
Total Central America	.44	.40	.16	.07	.54	.39
Chile	.19	.70	.11	.25	.63	.13
Argentina, Uruguay, and Paraguay	.24	.59	.17	.32	.54	.14
Ecuador	.21	.48	.31	.33	.17	.50
Venezuela and Trinidad	.34	.42	.24	.13	.39	.48
Bolivia	.30	.36	.34			
Brazil	.40	.48	.12	.31	.23	.46
Guyana, Surinam, and French Guinea	.68	.32	0	.05	.47	.47
Combined Total South America	.29	.49	.22	.26	.33	.41

Note: Totals of depth intervals may not add to 100% because of rounding.

Table 2
Fraction of Cumulative African Onshore and Offshore Wildcat Wells by Drilling Depth Range (through 1982)

Area	Onshore < 5,000	Onshore 5,000–10,000	Onshore > 10,000	Offshore < 5,000	Offshore 5,000–10,000	Offshore > 10,000
Algeria and Morocco	.26	.47	.27	.0	.46	.54
Libya and Tunisia	.13	.58	.29	.09	.36	.55
Egypt	.30	.35	.35	.09	.36	.55
W. Sahara, Mauritania, Niger, Sudan; Upper Volta, and Chad	.22	.36	.42			
Ethiopia, Djibouti, Uganda, Kenya, and Somali Republic	.16	.26	.58	.0	.54	.46
Senegal, Gambia, Guinea-Bissau, Guinea, Sierra Leone, Liberia, Ivory Coast, Ghana, Toga, and Benin	.22	.47	.31	.08	.23	.69
Nigeria	.01	.21	.78	.14	.42	.44
Angola	.35	.51	.13	.04	.51	.45
Cameroon, Central African Republic, Equitorial Guinea, Gabon, Zaire Congo, and Sao Tome Principe	.34	.58	.08	.15	.38	.47
Swaziland, Namibia, Lesotho, and South Africa	.55	.28	.17	.13	.69	.18
Tanzania, Malawi, and Mozambique	.06	.48	.46	.11	.56	.34
Combined Total Africa	.20	.46	.34	.19	.06	.75
				.10	.50	.40

Note: Totals of depth intervals may not add to 100% because of rounding.

atively large proportion of its onshore wildcat wells at shallow depths, it accounts for only about 3% of the onshore wildcat wells in Africa.

Figure 3 shows the onshore wildcat wells drilled in South America for each year from 1950 through 1982. This figure indicates that the shallow wells as a fraction of total wildcats declined in the 1960s and then increased in 1981 and 1982. Whether this trend is the result of missing data is uncertain at this time. Onshore wildcat wells drilled in Africa for each year from 1950 through 1982 by depth ranges are shown in Figure 4. These data show that the proportion of onshore wells drilled to shallow depths has declined since the 1950s. By 1981 and 1982 less than 9% of the onshore African wildcat wells were drilled to shallow depths.

Wildcat-drilling statistics for Africa and South America probably are not representative of development well drilling depths in these areas. Many oil and gas fields discovered in South America and Africa lie at shallow depths. International development-well data (by depth) are presently too scanty to provide meaningful statistics.

According to a survey of international and local drilling contractors, over 90% of the onshore rigs in South America and Africa had drilling

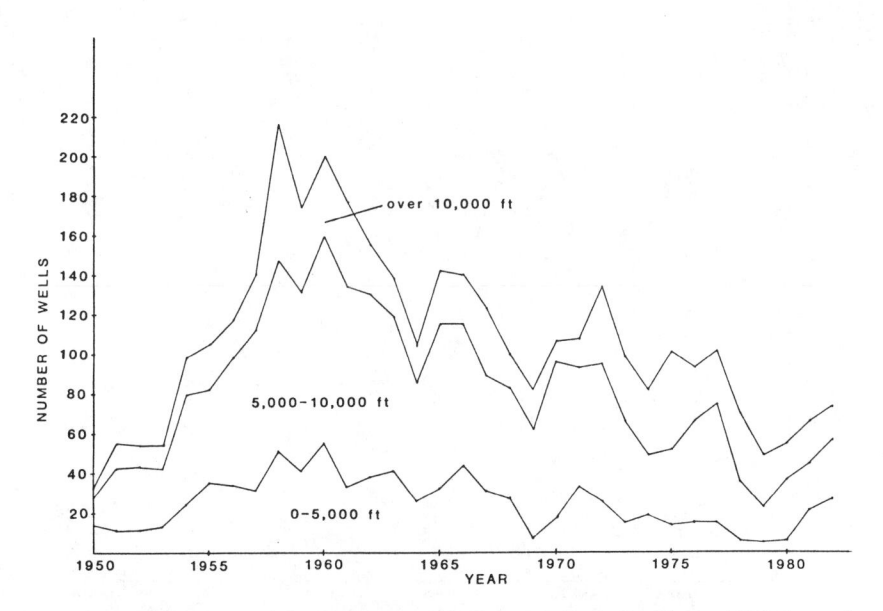

Figure 3. Total onshore South American wildcat wells drilled from 1950 through 1982 shown by depth ranges. (Data from Petroconsultants S.A.)

432 Shallow Oil and Gas Resources

capabilities beyond 5,000 ft (Petroleum Engineer International, 1982). This survey may have missed rigs owned by government oil companies. It seems apparent, though, that with the significant expenses associated with concession negotiations and rig mobilization and demobilization charges, operators desire and do use the capabilities of the deeper-rated rigs to test deeper formations.

Drilling Depth and Cost Relationships

International drilling statistics show that 83% of all wildcat wells drilled in non-Communist areas in 1980 were drilled in the United States. In fact, the United States and Canada together accounted for 95% of all wildcat wells drilled in non-Communist areas. Therefore, the drilling statistics by depth cited earlier represent a relatively small number of wells when compared to the total number of U.S. wells. There are about

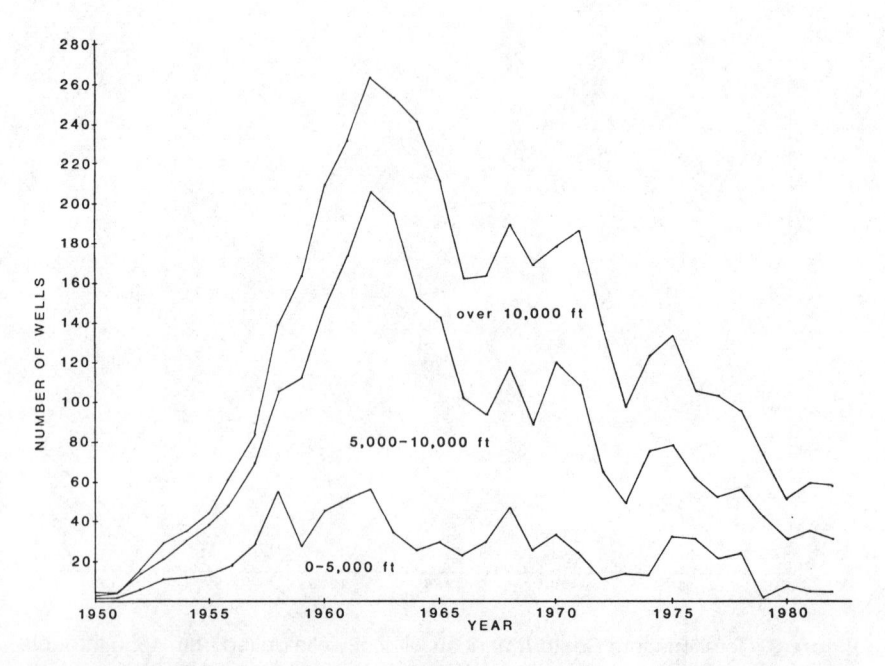

Figure 4. Total onshore African wildcat wells drilled from 1950 through 1982 by depth ranges. (Data are from Petroconsultants S.A.)

five or six times as many onshore rigs operating in the U.S. as there are rigs operating in all other non-Communist countries outside of Canada.

No single source compiles and publishes costs associated with drilling and equipping these wells. The rather wide geographical dispersion of these wells makes assembling a statistically reliable drilling cost series difficult. For the United States, the Joint Association Survey (JAS) publishes historical drilling costs and the U.S. Bureau of Census in the "Annual Survey of Oil and Gas" publishes statistics on exploration and development expenditures.

For the purpose of planning and estimating costs for a drilling program for major areas outside the United States, the known components of drilling costs and the JAS data might be used as a reference point. Drilling costs attributable to labor and materials should reflect local costs. Contract drilling costs should be adjusted to incorporate the prevailing day rates along with expected mobilization and demobilization charges. Although estimation of international drilling costs using such a method is possible, it is beyond the scope of this paper.

According to aggregate statistics published in the 1980 "Annual Survey of Oil and Gas" (U.S. Bureau of Census, 1982), drilling and equipping costs of exploratory wells represented just over two-thirds of U.S. exploration expenditures—exclusive of lease acquisition costs. These aggregate statistics also show that 63% of field development costs—exclusive of lease acquisition costs—are attributable to drilling and equipping production wells. On an individual field basis, the cost of production wells probably accounts for an even larger proportion (up to 80%) of field development costs as depth increases. Generally, at a given depth, the larger the field the more productive the individual wells will be. Thus, the share of total field-development costs accounted for by development-well drilling increases as field size decreases.

At shallow depths, little difference exists between the cost of drilling and equipping an oil or gas well. Figure 5 shows the estimated cost of drilling and equipping an oil and a gas development well in west Texas. Below 9,000 ft, the cost of a gas well increases much more rapidly than the cost of an oil well. This is because high pressure zones are much more likely to be encountered in gas fields at such depths. The figure also shows the rapid increase in costs of both types of wells as depth increases beyond 5,000 ft.

Although lease (production) equipment cost increases with depth, the increase is much smaller than the drilling-cost increase. For example, going from 3,400 ft to 7,200 ft, drilling costs increase by about 266%, whereas lease equipment cost increases by about 31%. From 7,200 ft to 11,400 ft, drilling costs increase 218% and lease equipment costs increase only 27%. (See Attanasi and others, 1981, for detailed costs of field development at alternative depths.)

434 Shallow Oil and Gas Resources

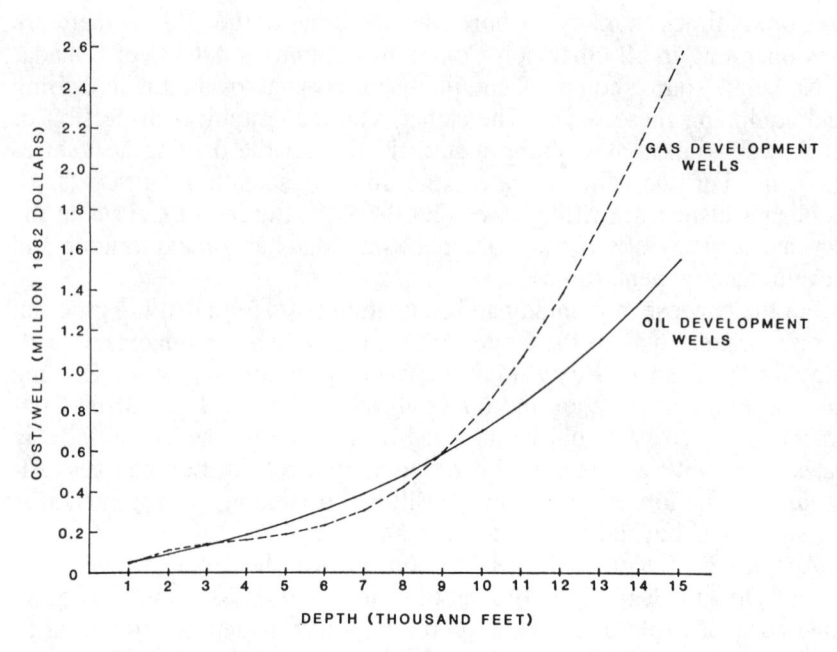

Figure 5. Cost of drilling and equipping oil and gas wells at various depths in West Texas (in 1982 dollars) (1 foot = .3048 meters). (Cost data were updated from Attanasi and others, 1981.)

Economic Incentives for Shallow Field Development

In the United States, drilling-rig day rates can fluctuate over wide ranges depending on depth capabilities of available rigs and total rig utilization. Figure 6 presents data from the Reed Annual Rotary Rig Census for the United States (American Petroleum Institute, 1984). The 1983 total of U.S. onshore rigs represents a 200% increase in the number of onshore rigs since 1973. In 1967, nearly 30% of the onshore rotary rigs were rated for 5,000 ft or less. Although the number of these rigs increased, by 1983 they represented only 15% of the total onshore rigs. Because total required rig time increases more than proportionately (nonlinearly) with drilling depth, many more shallow wells can be drilled per rig per year than deep wells (Adelman and Ward, 1980).

U.S. drilling activity typically follows expected or actual changes in the domestic oil and gas prices. When drilling demand slackens, rig-day

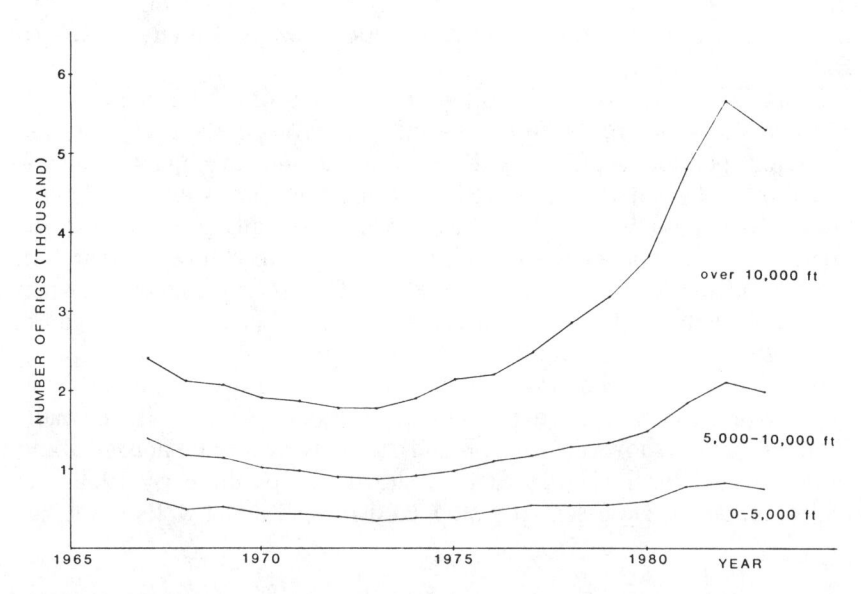

Figure 6. Total number of onshore rotary rigs in the United States from 1967 to 1983 shown by drilling depth capabilities. (Data from American Petroleum Institute; *Basic Petroleum Data Book,* V. IV, No. 1.)

rates can decline steeply. During periods of slack demand, rigs with deeper capabilities will compete with the smaller rigs to drill the shallow wells. This causes day rates for shallow drilling to decline even further.

When drilling demand in countries *outside* the United States, Canada, and Western Europe declines, international contract drillers simply move rigs elsewhere, and day rates typically do not fall as sharply as they do in the United States. Customers are charged for mobilization and demobilization and rigs are moved in response to geographical changes in drilling demand.

The relatively low costs of shallow drilling in the United States allow commercial development of small oil and gas fields. However, profitable development of small fields requires that product transport costs be a small proportion of total costs. Roads or pipeline systems should already be in place, and the local markets for both oil and gas should exist.

Western Europe seems to meet these conditions reasonably well. During the last 30 years, over 40% of the onshore wildcat wells drilled in

Western Europe were less than 5,000 ft in depth. Figure 7, which shows the annual onshore wildcat wells, indicates that shallow wells still continue to represent a large share of all wildcat wells drilled in Western Europe.

Small fields are profitable to find and develop in the United States. Most small fields are developed by independent operators. Exploration by major operators is driven by the need to find and develop large quantities of oil and gas that are required for refinery feedstock and retail operations. Even though small fields may be commercially profitable, major operators are compelled to use their scarce exploration funds to search in areas expected to have very large fields. This is true especially when they explore in areas far from markets and where there is no transportation system to get the product to market.

In fact, the major operators will even reduce their exploration effort in productive areas after it is apparent that future discoveries will be small. Figure 8, which shows the time profile of discoveries and wildcat drilling in the Denver basin (United States), shows that by the early 1960s the major operators had essentially stopped drilling wildcat wells in the ba-

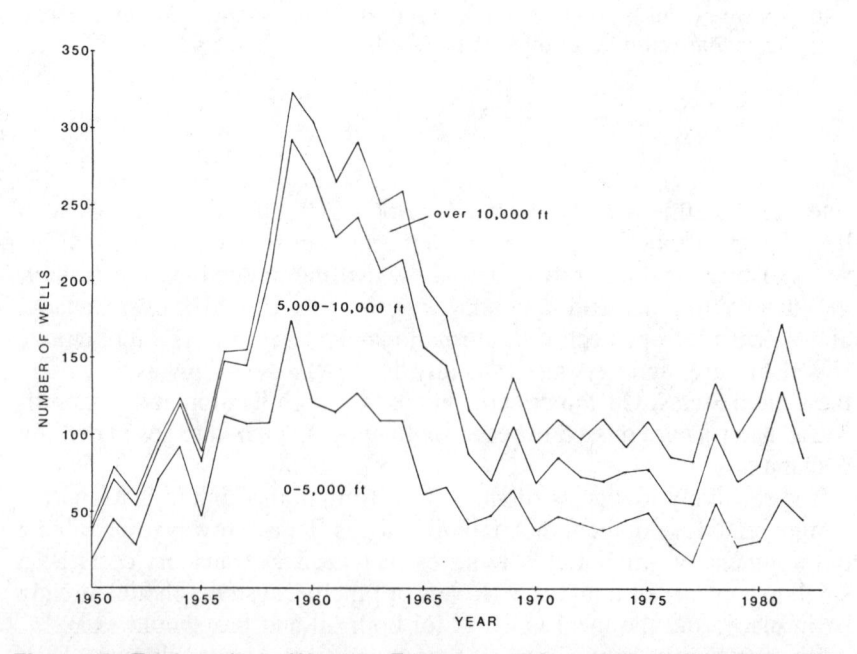

Figure 7. Total onshore Western European wildcat wells drilled from 1951 through 1982 shown by depth ranges.

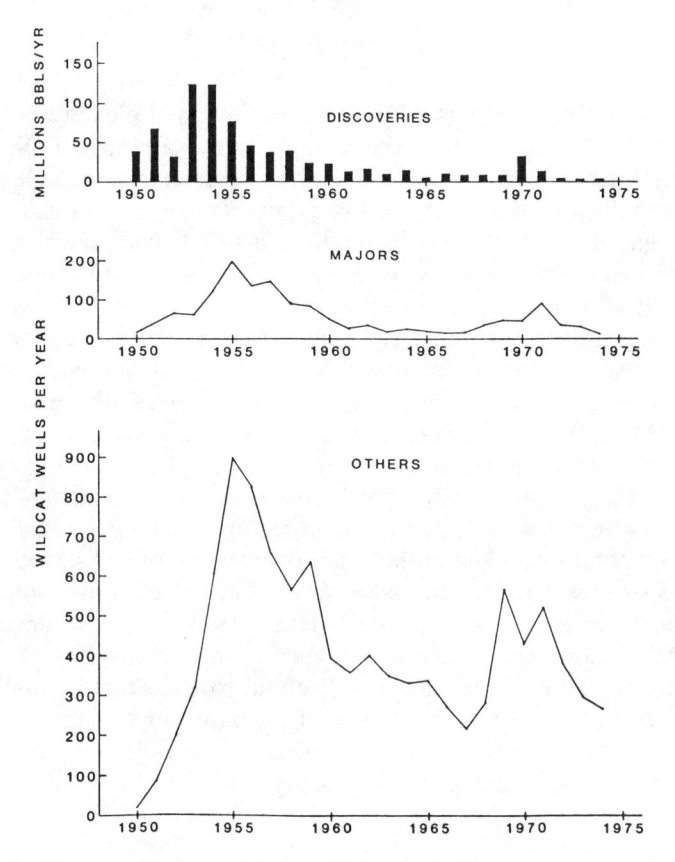

Figure 8. Time profile from 1950 to 1974 of oil discoveries and wildcat wells drilled in the Denver basin for the major and independent operators. (Data are from Drew, 1980.)

sin. Similarly, in the state waters in the Gulf of Mexico, the major operators' share of wildcats dropped from 85% of total wildcats for the 1951-55 period to 16% for the 1976-78 period (Attanasi and Drew, 1984).

In the United States, independent operators drill nearly 80% of the wildcat wells and the majority of development wells. They frequently raise exploration funds by selling drilling programs on a subscription basis to individuals subject to high marginal income tax rates. The objective of some of these operators is to drill as many profitable wells as possible. A significant portion of this drilling is possible only because of their unique sources and ways of raising exploration funds.

Implications

There are several reasons why small oil and gas fields are profitably produced at shallow depths in the United States. The large U.S. inventory of drilling rigs and a very competitive contract drilling industry assure that shallow drilling costs will be relatively low. In most parts of the world a gas discovery is no better than a dry hole. However, in the United States even small gas fields can be commercially produced because markets and product transportation systems exist. The tax treatment of investment in oil and gas exploration by individuals and small operators also aids the commercial viability of small field development.

Major operators will probably continue to preferentially search in areas where large fields can be expected to be found. Developing countries must try to attract the independent operators or to develop their own indigenous industry. Actions by developing country governments that increase risks and costs will generally affect smaller operators more than the major operators. Whether the small operator is foreign or local, governments cannot expect exploration to continue if the governments recover all the economic rent (through taxes) associated with developing successful discoveries. Exploration expenditures are usually paid from equity capital and retained earnings. Without participation by foreign independent operators or development of an indigenous industry to perform the entrepreneurial function, these marginally commercial resources will not be identified and produced.

References

Adelman, M. A., and Ward, G. L., 1980, Worldwide costs for oil and gas, *in* P. S. Pindyck, Advances in the Economics of Energy and Resources, v. 3: Greenwich, Connecticut, JAI Press Inc., p. 123.

American Petroleum Institute, 1984, Basic petroleum data book, v. IV, no. 1, Washington, D.C.

American Petroleum Institute, Joint Association Survey on Drilling Costs, American Petroleum Institute, Washington, D.C. (annual report).

Attanasi, E. D., and Drew, L. J., 1984, Offshore exploration performance and industry change: the case of the Gulf of Mexico, Journal of petroleum technology, March, p. 437–442.

Attanasi, E. D., et al., 1981, Economics and resource appraisal: the case of the Permian basin, Journal of petroleum technology, April, p. 603–616

DeGolyer and MacNaughton, 1983, Twentieth century petroleum statistics, Dallas, Texas.

Drew, L. J., 1980, Firm size in the search for petroleum, MS Thesis, Virginia Polytechnic Institute and State University, Blacksburg.

Energy Information Administration, 1982, Drilling statistics for 1960–1981 and drilling and production costs for nonassociated gas wells, 1976 to 1981, EIA, Dallas Office, U.S. Department of Energy.

Petroleum Engineer International, 1982, Land rig locator: Petroleum engineer international, March supplement, p. 1–75.

Petroleum engineer international, 1984, Free world drilling rigs off by 158 rigs: March, p. 86.

U.S. Bureau of the Census, 1982, Annual survey of oil and gas, 1980, U.S. Bureau of the Census, Current Industrial Reports, Series MA-13K (80)-1, 74 pp.

37
Attracting Capital for Oil and Gas Exploration in a Competitive Environment

David Berberian and William J. Hibbitt

Introduction

The ever-shifting world petroleum picture during the past few years has had an adverse impact on oil-producing countries and developing countries with potential petroleum reserves. The decline in the demand for petroleum due to slowed economic growth, energy conservation and fuel switching, and the reduction in the price of crude oil has forced foreign oil companies to slash exploration and development programs because of reduced income.

The developing countries must take steps to create a positive business and political climate which will attract foreign oil companies (FOC) to invest in exploration and development of their petroleum resources. There already appears to be a trend on the part of developing countries to provide incentives, and improve contract terms. No doubt there is a close relationship between petroleum activity and available incentives and developing countries have to focus on means to attract exploration capital.

Any package to attract exploration capital from FOCs must be based on the principle that the FOCs and the developing host country (DHC) will be partners and each will share in the rewards of success.

From the point of view of an FOC, the following factors are considered in evaluating an investment in petroleum operations outside its home country:

- Geological prospects
- Political and operational environment
- Potential for profits and viability of economic arrangement
- Taxation

Geological Prospects

The DHC must endeavor to develop adequate geologic data and make such data available to FOCs at a minimum charge. The development of the geologic data by a reputable consulting company would add credibility to the geological prospects of the DHC.

Political and Operational Environment

The political climate in the host country plays an important role in the decision-making process. The DHC must strive to portray a positive atmosphere and provide assurance that a contract negotiated with the DHC will be honored throughout its term without unilateral changes. The contract should provide for neutral arbitration by recognized international agencies in the event the terms of the contract are violated or serious disputes arise.

From an operational point of view, the following should be considered:

- The FOC should be allowed to retain title to all equipment utilized in connection with petroleum operations during the term of the contract. This avoids the unnecessary and incorrect perception that the host country is preparing for early nationalization.
- The FOC should be required to maintain its operations in the host country in accordance with reasonable petroleum industry standards and guarantee the DHC that, at termination of a successful contract, the equipment necessary to carry forward the operations would be in place. Title to such equipment should vest in the DHC government, or state oil company, as the case may be, at the termination of the contract. The FOC should have the right to remove all movable facilities, machinery, and equipment at the termination of an unsuccessful contract—that is, one under which the FOC did not recover all its costs from the exploitation area, or areas, covered by such contract.

- The FOC should be given the right to export a reasonable portion of its share of hydrocarbons produced and retain the proceeds abroad. The value determined by reference to world prices of such exported hydrocarbons should be deemed to be the gross income from such exports, realized at the time of export, for DHC income tax and cost recovery purposes.
- The DHC government should continue to maintain foreign exchange regulations and investment laws which allow free movement of funds into and out of the country. The DHC government also should continue to permit the free convertability of foreign currency into and out of local currency at the free market exchange rate. The FOC should be allowed to repatriate its capital and earnings freely and without withholding of additional income taxes.
- The DHC should continue to minimize the burden of import duties and licensing requirements on imported equipment and supplies.
- The contract should provide maximum flexibility in negotiating the size of the exploration blocks in order to encourage exploration of smaller potential reservoirs. The signature bonus due on execution of a contract should be minimal to allow more capital to be invested in actual exploration.
- The contract should allow the FOC maximum flexibility in controlling and managing operations and scheduling times for exploration.

Potential for Profit

FOCs are largely driven by expected cash flow, profit, and access to oil. The contract between the DHC and FOC should provide that the FOC will operate pursuant to a petroleum operations contract. This contract will cover all items, including royalty rates, the cost recovery procedure, and all other aspects of the relationship between the parties as to exploration, development, and production in the host country during the term of the contract.

Such contract should provide that the DHC retains ownership of all hydrocarbon reservoirs, but that the FOC performing exploration, development, and production activities is entitled to, as its sole economic interest in hydrocarbons, a share of production from the applicable exploitation area equal to its total costs allocable to such exploitation area, plus a reasonable percentage of shareable hydrocarbons from such area. The FOC would market the DHC's share of hydrocarbons, unless otherwise agreed.

The value of hydrocarbons from a particular exploitation area would be shared under the petroleum operations contract as follows:

DHC	FOC
1. A reasonable royalty percentage of gross production (without reduction for any expenses) during the entire period of production from the applicable exploitation area.	1. Production equal to all costs allocable to the applicable exploitation area.
2. A reasonable percentage of shareable hydrocarbons (gross production less royalty and FOC costs) generated after recovery by the FOC of costs allocable to the applicable exploitation area.	2. A reasonable percentage of shareable hydrocarbons (gross production less royalty and FOC costs) generated after recovery by the FOC of costs allocable to the applicable exploitation area, reduced by DHC income tax payable thereon.

Under the foregoing, the FOC would be able to evaluate a potential investment in the host country by using a clearly-defined economic formula and its estimates of geological potential for production during the fixed term of the contract.

As a first step toward creating economic attractiveness, the DHC should dedicate the value of all hydrocarbon production from an exploitation area, after deduction of its royalty, to allow the FOC to recover all of its costs allocable to such exploitation area.

After full cost recovery, the remaining shareable hydrocarbons would be divided between the DHC and the FOC. The exact percentages allocable to each party would be agreed upon in the petroleum operations contract based upon production levels which might result from an exploitation area. The higher the production level, the greater the percentage of shareable hydrocarbons the DHC would receive and the lower the percentage the FOC would receive.

As another step toward creating economic attractiveness, the DHC royalty should be fixed at 12.5 percent of the value of hydrocarbons sold or deemed sold until the calendar year subsequent to the calendar year in which the FOC has recovered its costs. All production in such subsequent calendar year and all succeeding years should be subject to a higher royalty. Under this concept, the reduced royalty of 12.5 percent would apply until the FOC recovered its costs, thereby allowing a faster return of capital to the FOC and making the economic arrangement more attractive.

Taxation

In determining the income taxes payable by the FOC, the following should be considered:

- The income taxes payable to DHC must be creditable as a foreign tax for limited state income tax purposes. (See following discussion.)
- FOC's gross income should be equal to the value of the share of production attributed to it for cost-recovery purposes, plus the value of its percentage of shareable hydrocarbons.
- All ordinary and necessary costs and expenses attributable to petroleum operations paid or incurred during the taxable year, including proper home office costs and interest costs subject to certain limitations which the FOC can prove are directly allocable to DHC petroleum operations must be deductible in computing taxable income.
- Any excess of deductions in any taxable year over gross income for such year could be carried forward to the following taxable year and be treated as a deduction incurred in such following year for DHC income tax purposes. Excess deductions could be carried forward indefinitely until utilized against taxable income resulting from all DHC operations.

U.S. Federal Income Tax Considerations

When a United States citizen engages in oil and gas exploration and exploitation within a foreign country, it is of critical importance that (a) the person acquires an economic interest in the minerals in place within the foreign country, and (b) any taxes paid by that person to the foreign country qualify as creditable foreign income taxes for United States Federal income tax purposes.

Economic Interest

The acquisition by the U.S. person of an economic interest in the mineral in place is a prerequisite for claiming deductions for intangible drilling and development costs and cost depletion, and claiming foreign tax credit for certain foreign tax payments. The U.S. person need not acquire title to the minerals in order to have an economic interest, but he must have an interest in mineral in place. In Revenue Ruling 73-470, 1973-2 C.B. 88, the Internal Revenue Service ruled that a domestic corporation has an operating economic interest when it holds contractual rights to explore for oil and gas in a foreign country even though it does not have legal title to the minerals.

Ownership of an economic interest in minerals in place allows a U.S. person to claim a deduction for cost depletion in calculating income subject to Federal income tax. This cost depletion deduction will enable the U.S. person to obtain a tax deduction for capitalized leasehold costs. In addition to the deduction for cost depletion, ownership of an economic interest will enable the U.S. person to obtain a current deduction for intangible drilling and development costs which are incurred in connection with the oil and gas exploration. The current deduction of intangible drilling and development costs is a significant benefit which is available to U.S. taxpayers engaged in U.S. oil and gas exploration. If a U.S. person does not possess an economic interest in the foreign oil and gas property, intangible drilling and development costs would need to be capitalized as part of the leasehold cost. Generally, leasehold costs are depleted over the life of the property, however, since ownership of an economic interest is a prerequisite for claiming a depletion deduction, capitalized lease hold costs could not be depleted, and would be recovered only on sale, disposition, or abandonment of the lease.

Foreign Tax Credit

U.S. Tax Rules in General

A U.S. person is subject to United States Federal income tax on taxable income earned from sources within and outside the United States. Where income which is earned from sources outside the United States is subject to income tax in the source jurisdiction, the United States avoids double taxation of that income by allowing a U.S. person to claim credit (a dollar-for-dollar reduction) against his U.S. tax for the income taxes paid to the foreign jurisdiction. The maximum amount of credit allowable in any taxable year is an amount equal to the U.S. income tax attributable to the U.S. person's foreign source income. Taxes paid by a U.S. person which do not qualify as creditable income taxes may be deducted in calculating income subject to U.S. tax. However, since the tax benefit of a deduction is significantly less than the tax benefit derived from a credit, in order for a U.S. person to minimize his total tax liability, it is extremely important that payments to a foreign government qualify as creditable income taxes.

Code Section 901(f) provides that an amount paid or accrued to a foreign country in a tax year ending after December 31, 1974 in connection with the purchase and sale of oil or gas extracted in that country is not considered under Sections 901 or 275 as a tax, and as such, does not qualify as a creditable income tax, if (a) the taxpayer has no economic interest in the oil and gas, and (b) either the purchase or sale is made at a price other than fair market value. It is not unusual for production-sharing contracts to allow the government to purchase the U.S. person's share of oil at a price to be determined by the government under some agreed-upon formula. In order for taxes paid to the foreign government by the U.S. person to qualify for credit under Section 901(f), the price at which oil and gas is sold to foreign government must represent a fair market value price.

Code Section 901(b)(1) provides that U.S. citizens and domestic corporations may, subject to certain limitations, claim credit for the amount of any income, war profits, or excess profits taxes paid or accrued during a taxable year to a foreign country or to any possession of the United States. Code Section 903 provides that an income tax includes an amount paid in lieu of an income, war profits, or excess profits tax. The Code does not explain which foreign taxes qualify as income, war profits, or excess profits taxes. That determination has been left to case law, revenue rulings, U.S. income tax treaties, and regulations. In addition, the Internal Revenue Service has issued numerous letter rulings explaining the creditability of the taxes of various foreign countries. Table 1 is a summary of the private letter rulings.

In determining whether a foreign tax is within this concept, the courts have considered such factors as the name of the levy, its place in the country's tax system, and whether it is imposed on only one type of business or industry. Much weight is placed on whether it is imposed on net income. If it is levied on the value of the mineral extracted, without allowance for expenses, and the mineral is owned by the government, it is likely to be characterized as a privilege tax.

Since 1976, the Internal Revenue Service and the Regulations branch of the U.S. Treasury Department have provided numerous sets of guidelines on the characteristics of creditable income, war profits, and excess profits taxes and taxes in lieu of income taxes. These guidelines have taken the form of revenue rulings and income tax regulations. The most recent guidelines are final regulations under Sections 901 and 903 which were published in the Federal Register on October 12, 1983 (48 FR 46272). The regulations are effective for taxable years beginning after

Table 1
Status of Creditability of Various Foreign Taxes Under Private Letter Rulings

Country	Tax	Creditable	Noncreditable
Angola			
	Taxes on Oil and Gas Extraction (Ltr. Rul. Ref. No. 8042023)		X
Canada			
1.	British Columbia Mining Tax Act (Ltr. Rul. Ref. No. 7926028)	X	
2.	Saskatchewan Income Tax Act (Ltr. Rul. Ref. No. 8140047)	X	
3.	Alberta Income Tax Act (Ltr. Rul. Ref. No. 8234003)	X	
4.	Ontario's Corporate Tax Act of 1972 (Ltr. Rul. Ref. No. 8224023)	X	
5.	Manitoba Income Tax (Ltr. Rul. Ref. No. 8230311)	X	
6.	Prince Edward Island's Income Tax (Ltr. Rul. Ref. No. 8236027)	X	
7.	Yukon Territory's Income Tax (Ltr. Rul. Ref. No. 8236026)	X	
8.	Northwest Territories' Income Tax (Ltr. Rul. Ref. No. 8234009)	X	
9.	New Brunswick's Income Tax (Ltr. Rul. Ref. No. 8234001)	X	
Ecuador			
	Taxes on Oil and Gas Extraction (Ltr. Rul. Ref. No. 8042023)		X
Haiti			
	Haitian Income Tax Law (Ltr. Rul. Ref. No. 8127094)	X	
Indonesia			
	Indonesia Corporate Tax Law (Ltr. Rul. Ref. No. 8038083)	X	
Ivory Coast			
	Production Shariney Contract Tax (Ltr. Rul. Ref. No. 8302017)	X	
Kuwait			
	Law No. 34/1970 (Ltr. Rul. Ref. No. 8028080)		X
Netherlands			
	Netherlands Company Income Tax Act and Netherlands Dividend Tax Act (Ltr. Rul. Ref. No. 8227039)	X	
Philippines			
	Philippines Income Tax Law (Ltr. Rul. Ref. No. 8106040)	X	

Table 1
continued

Country	Tax	Creditable	Noncreditable
Saudi Arabia	Royal Decree No. 17/2/28/7634, December 27, 1950 (Ltr. Rul. Ref. No. 8104049)		X
Thailand	Thailand Branch Profits Tax (Ltr. Rul. Ref. No. 8104049)	X	
Trinidad	Petroleum Profit Tax and Income Tax Act (Ltr. Rul. Ref. No. 8222069)		X
United Kingdom	U.K. Corporation Tax (Ltr. Rul. Ref. No. 8207052)	X	
Venezuela	Venezuela Income Tax Law (Ltr. Rul. Ref. No. 8116003)		X

November 14, 1983. In addition, a taxpayer may elect to apply the regulations retroactively to earlier open taxable years.

Under the final regulations, a foreign levy is a creditable tax only if it is a tax and its predominant character is that of an income tax in the U.S. sense. This rule is applied to each separate levy. A foreign levy is a tax if it requires a compulsory payment pursuant to the foreign country's taxing authority. However, a payment for a specific economic benefit is not a tax. The predominant character of a foreign tax is that of an income tax if the foreign tax is likely to reach net gain in the normal circumstances in which it applies. Three tests are applied in determining whether a foreign levy is likely to reach net gain: the realization test, the gross receipts test, and the net income test.

The realization test is satisfied if the tax is imposed:

1. Upon or after an event that would result in realization of income under the Code, or
2. Before a realization event if the result is the recapture of a tax deduction, tax credit, or other tax allowance previously claimed, or
3. Upon the occurrence of a prerealization event (excluding deemed distributions) if the tax is based on either (a) the difference in the

value of property at the beginning and end of a period or (b) there is a physical transfer, processing or export of readily marketable property.

The regulations state that the foreign country must not impose a second tax upon the occurrence of a later event with respect to such income from a prerealization event, unless it provides for a credit or other comparable relief. The regulations also include an example where a branch profits tax on a deemed distribution of income meets the realization test.

The gross receipts test is met if the tax is imposed on gross receipts, or gross receipts computed under a method that is likely to produce an amount that is not greater than fair market value.

The net income test is satisfied if, judged on the basis of its predominant character, the tax is imposed on the basis of gross receipts reduced by significant costs and expenses. This reduction can take the form of a deferred recovery of the significant costs in computing the foreign tax as long as such deferral is not an effective denial of such costs. The nondeductibility of another tax, which is itself an income tax, would satisfy the net income test. The regulations also require that losses incurred in one activity in a trade or business be allowed to offset profits earned by the same person in another activity within that same trade or business. The regulations state that it is generally immaterial whether such offset is allowed in the taxable period or in a different taxable period. The regulations do not require that losses from one trade or business be allowed to offset income from another trade or business. In the case of oil exploration, a separate contract area (lease) is a separate activity while marketing, refining, and production are separate trades or businesses.

Once a foreign income, war profits, or excess profits tax is found to qualify as creditable income tax, a taxpayer must prove that the tax has either been paid or accrued by or on behalf of the taxpayer. Amounts of tax paid or accrued to a foreign country do not include amounts that are reasonably likely to be refunded, credited, rebated, abated, or forgiven, or amounts that are used to subsidize the taxpayer either directly or indirectly. Under the regulations, an amount is not treated as paid if the amount is a noncompulsory payment. The taxpayer's decision to treat a payment as compulsory must be based on a reasonable interpretation and application of foreign law involved. A taxpayer is also required to make reasonable efforts to minimize its foreign tax liability over time.

The regulations have introduced a new rule with respect to multiple levies. If, under foreign law, the initial amount of one levy is or can be reduced by the amount of a second levy, the amount of the first levy that is treated as paid or accrued is the excess of the amount paid over the amount of the second levy. If the taxpayer's liability is the greater or lesser of two amounts, the taxpayer is considered to pay or accrue only the levy for which he is liable during the period involved. Thus, if the taxpayer is liable for the greater of an income tax or excess tax, and the income tax is larger, the taxpayer is considered to be liable only for the income tax and not for the excise tax.

Application of U.S. Tax Rules to Foreign Tax Provisions

The determination of whether a foreign levy qualifies as a creditable income tax is made by applying the regulatory guidelines already outlined to the provisions of the foreign law. If the foreign law is drafted with these regulatory guidelines in mind, payments made pursuant to the foreign law should qualify as creditable foreign income taxes. It is important to note, however, that the manner in which the foreign law is applied is equally as important as the stated provisions of the foreign law and that U.S. taxing principles generally are applied in determining whether the provisions of the foreign law satisfy the regulatory requirements.

The first requirement of the regulations is that the foreign levy qualify as a tax. A tax is defined as a compulsory payment made pursuant to the tax authority of the foreign jurisdiction. As a general rule, a tax payment made by U.S. oil companies with respect to income derived from the sale of oil and gas will not qualify in its entirety as a payment pursuant to the foreign jurisdictions taxing authority because the U.S. oil company will have received a specific economic benefit from the foreign jurisdiction. The specific economic benefit received by the oil company is the right to explore for and exploit natural resources within the country. Since this is a privilege that is not generally available to taxpayers, it will be viewed as a specific economic benefit and will require that payments made to the foreign country be allocated between tax payments and payments for the specific economic benefit.

The second regulatory requirement is that the predominant character of the foreign tax be that of an income tax. This requirement is satisfied if the foreign tax satisfies the realization test, the gross receipts test, and the net income test. The provisions of the foreign law in the manner in which the law is applied will determine whether these three tests are satisfied. As a general rule, in order for the realization test to be satisfied, the U.S. person must not be subject to income tax until he either sells or exports the production. If the person is deemed to have realized income when the mineral is extracted, the realization requirement will not be satisfied.

The gross receipts test requires that the tax be imposed on the basis of gross receipts or gross receipts computed under a method which is likely to produce an amount that is not greater than fair market value. In order to satisfy this requirement, it is imperative that the foreign law provide that taxable income is based on actual gross receipts or a realistic estimate of fair market value. Where the contractor sells oil and gas to unrelated third parties, gross receipts from such a sale should, as a general rule, equal the sales price. Where a contractor sells oil and gas to the government, Code Section 901(f) requires that the contractor be paid an amount equal to the fair market value and this same amount should be the gross receipts on which taxable income is based. Regulation Section 1.901-2(b)(3)(i) Example 3, provides an example of a country which imposes tax on income from extraction of petroleum where gross receipts from extraction are deemed to equal 105 percent of the fair market value of the petroleum extracted. This tax does not satisfy the gross receipts requirement because it is designed to produce an amount that is greater than the fair market value. Conversely, Internal Revenue Service has acknowledged that a foreign government has the right to examine the price at which oil and gas is sold to determine whether that price represents fair market value and has indicated that a system (such as Norway's norm price) which bases gross receipts on an amount which is designed to and does in fact approximate fair market value will satisfy the gross receipts test, as such a system is an acceptable exercise of the source country's right to protect its tax revenue by requiring that petroleum sales be made at fair market value.

In calculating the fair market value of oil and gas for purposes of determining a purchase price or export value, the pricing mechanism utilized by the government should consider:

1. The fair market value of the various grades of crude as determined by unrelated third party oil industry publications, if available.
2. The fair market value of similar grades of crude as determined by unrelated third party industry publications and adjusted to account for differences in product yield, quality of crude, shipping, costs and other relevant factors.
3. The prices at which the extracted crude and similar crudes are sold under long-term contracts and on the spot market.

The final test for determining whether the predominant character of a foreign levy is that of an income tax is whether the base of the tax is computed by reducing gross receipts by either significant costs and expenses attributable to such gross receipts or an amount computed under a method that is likely to produce an amount that approximates or is greater than significant costs and expenses attributable to such gross receipts. Generally, a contractor should be allowed a deduction either as an expense or through depreciation or amortization of capitalized expenditures, for all expenditures relating to its oil and gas activity. Deductible expenditures should include reasonable head-office expenses, interest on funds borrowed within and outside the foreign country, all preparatory expenses including startup expenses, travel and transportation costs, geological and geophysical costs and, of course, all drilling, operating, and local general and administrative expenses.

It is generally desirable for the foreign country to combine all activities of a contractor within the foreign country in determining, on an annual basis, the income subject to tax. Under this concept, oil and gas revenue from all contract areas within a country would be combined with non-oil and gas activity in calculating taxable income. It is not uncommon for foreign laws to prohibit the deduction of losses in one exploitation area from profits of another exploitation area. Additionally, it is very common for countries to tax oil and gas activity separately from non-oil and gas activity. These separate net income calculations are acceptable under the regulations, provided the foreign law contains reasonable net operating loss carryover provisions which allow losses in one year to be carried forward to offset profits in a subsequent year.

In addition to satisfying the realization gross receipts and net income tests, in order for the predominant character of a foreign levy to be that of an income tax, the foreign levy must not constitute a soak-up tax. A soak-up tax is defined generally as a tax which is imposed only if credit for that tax will be granted by that country.

Once a determination is made that a foreign levy qualifies as an income tax, credit can be claimed only for the amount of such tax which is paid or accrued by the taxpayer. The person by whom a tax is considered paid is the person on whom foreign law imposes legal liability of such tax even though another person may remit the tax. Therefore, it is important that provisions of the foreign law clearly indicate that a tax on income derived from the sale or export of the U.S. person's share of production is imposed on the U.S. person.

Other U.S. Tax Considerations

As has been previously indicated, U.S. tax law allows a taxpayer to deduct all intangible drilling costs in the year paid or accrued. Costs incurred in the acquisition of leaseholds and tangible property must be capitalized and amortized or depreciated over the life of the asset. From a U.S. tax standpoint, it is desirable for the foreign law to allow the taxpayer considerable flexibility in the timing of deductions under the foreign law. Such flexibility allows the taxpayer to match U.S. income and foreign income, thereby maximizing his available U.S. foreign-tax-credit benefits. Additionally, from an economic standpoint, a provision in a foreign tax law which allows a taxpayer sufficient flexibility in the timing of deductions to obtain an accelerated deduction for expenditures when U.S. tax considerations are not significant or to postpone deductions when his foreign tax credit position requires that foreign taxes be paid is viewed as a very attractive provision.

Frequently, foreign petroleum laws provide that, at some point in time, ownership of all equipment utilized in the exploration and exploitation activities transfers from the U.S. person to an agency of the foreign government. From a U.S. tax standpoint, it is preferable for ownership to transfer when the production-sharing contract terminates, thereby enabling the U.S. person to depreciate the equipment during the exploration and exploitation period. If the U.S. person does not own the equipment during the period of use, depreciation deductions will not be allowed.

38
Problem of Financing of Investment in Exploration and Production of Petroleum in Third World Countries

Gerald Jolin

This paper addresses the problem of financing exploration and production of petroleum in third world countries. The total financing consists of the following elements:

1. Geology and geophysics—feasibility studies done by governments before proposals for granting concessions are made or requested.
2. The concessions.
3. After the concessions are granted, exploratory drilling takes place.
4. Developmental drilling of a discovered field.
5. Completion of the first development prospects and attendant testing for porosity, viscosity, and other important factors.
6. The processing of the product into its parts or the separation of, for example, oil from gas and/or water.
7. Marketing of the product in its raw or processed state.

In relation to the marketing and processing of the product, some ingenuity and creativity is possible and almost necessary in many of the third world countries, wherein the distribution system infrastructure is not highly developed. For example, it may be that if we are drilling gas wells and there are no pipelines, it will be necessary to market the gas by liquefying it, or by stripping the liquids out of the gas, since the liquids can be easily transported and shipped to the nearest dock. The dry gas can be pumped back into the formation and thus maintain pressure on the field. Conversion of the energy in the gas to kilowatts by installing a gas turbine and boiler would also allow marketing.

The present systems of financing independents in United States are as follows:

1. Inside-industry deals, wherein a stronger company can assist a smaller company financially for a piece of the production.
2. Limited partnerships sold directly to private investors by the producer. Normally these are set-up so that they offer a one-quarter working interest in return for investment of one-third of the cost. This method is age old, and requires the producer to serve the investor-relations function.
3. Limited partnerships placed through broker-dealers. In this method the broker-dealer normally qualifies a limited partnership or joint venture by filing the necesary requirements with the Securities and Exchange Commission, and/or with the state regulatory agency governing blue-sky laws. In this situation the broker-dealer, for a fee of 8%-10%, assumes the responsibility for complying with the regulations, and for servicing the investor-relations functions.
4. Reinvesting internally-generated gross profit from prior production.
5. Domestic bank loans.

Among the emerging methods of financing for independents in the United States we find the following:

1. Hybrid loan-equity structures by institutions have become popular. These are efforts by pension funds, insurance companies, and foreign bank subsidiaries to enter the relatively safe field of monetizing oil reserves. Within charter, policy, and regulatory restrictions, a rapid movement is now taking place in this activity. The economist from Claremont, Peter Drucker, reports that over $700 billion is now managed by pension funds. These funds increase annually at approximately 20% (10% from yield and 10% from new pension checks arriving). Some of these funds are already being invested. Damson Oil claims that it has 1,100 institutional customers at this time, according to *Petroleum Investment News*.
2. A combination of completion funds for capitalizing the accelerated cost-recovery-system items and/or other depreciable factors. This,

coupled with production borrowing from equity investment and other sources rounds out the total needs for the producer.
3. Combinations of geology-geophysics-lease funds with completion funds, with driller short-term credit, or other short-term credit. This combination is becoming more common as cash becomes more scarce.
4. Minimum annual royalty programs are a sophisticated tax-shelter program. Critics call them the ultimate tax shelter, but they have been treated favorably in the tax courts so far.

Any or all of the preceding methods now possible and popular in the United States will probably be adapted with modifications for use in third world nations.

It is my belief that items such as gas-stripping and electric generating plants will be more prominent in third world countries by reason of the difference in industry infrastructure. The use of electric-generating and gas-liquid stripping plants does not require the highly developed road and pipeline systems that are common in the United States. Also, the manufacturers of such units are normally in a position to finance operations internally, or by affiliated finance companies. Westinghouse, General Electric, Ingersoll-Rand, Siemens-Allis, and their competitors all have affiliated finance companies to handle the transactions.

Each of the developing countries, more than 100 or them, has its own rules, regulations, and local customs for conversion of local currencies. These variations result from the differences among the countries and the foreign trade balance of each. There are no general rules that apply to all. Normally, either a government agency or designated banks are involved in issuing permits in relation to expatriation of funds.

The 1983 annual report of the International Monetary Fund, entitled "Exchange Arrangements and Exchange Restrictions," carries detailed instructions regarding the regulations in each nation. For example, in Somalia the Ministry of Finance, upon the recommendation of other ministries, may extend total or partial exemption from import and export taxes, excise duty, income tax, and municipal tax to any enterprise for a period not exceeding five years. Enterprises registered on or after March 1968 must employ nationals to the greatest extent possible, and each enterprise must submit to the Committee on Foreign Investments a program for the substitution of resident employees for nonresidents.

Broad, catch-all provisions such as the foregoing appear to be common in most countries, and they have a devastating effect on other specific regulations. A company going into a third world country should first retain competent help to gain understanding of the regulations, and requirements for compliance with the regulations. Then, in view of the catch-all provisions just cited, the management of the third world operation must keep in constant contact with the local governing bodies, as is the case in the United States.

It is well known that the international agencies, such as International Monetary Fund, World Bank, and other international and regional funding agencies, with the assistance of OPIC, have been active in assisting producers to operate in the third world countries. Their staffs have experience in these areas of activity and have knowledge of local situations which is invaluable to potential producers. There also are consulting firms which specialize in establishing industrial activities in third world countries, many of whom have nationals from the various countries upon whom they may call for detailed assistance. It is well known that the smaller third world countries need industrial development generally, and energy development specifically, to increase their standards of living.

39
Developing Country/Independent Oil Company: Considerations for Financing Exploration and Development of Shallow Oil and Gas Deposits

Dorothy Mercer

Introduction

This conference brings together representatives of interested independent oil companies and of countries seeking to develop their small shallow oil deposits. The purpose of this meeting is to explore mutual economic interests and assess the possibilities for mutually rewarding interactions. We are evaluating an uncommon venture by other than the traditional combination of participants in international oil and gas projects. Consequently, the traditional procedures and conventional rules-of-thumb may have to be adapted to fit these special arrangements.

I will explore some economic aspects of such ventures and examine some of the financing techniques that have been used in the traditional operations of the major international oil companies and larger indepen-

---

* U.S. Department of the Treasury. Views expressed in this paper are solely those of the author and do not necessarily represent the official policies or positions of the U.S. Department of the Treasury.

dents in foreign export-oriented projects and of U.S. independents in domestic shallow oil projects. It is hoped some inferences can be drawn for application or adaptation of these well-developed patterns to the special cases under review here.

Historical Perspective

The recent history of oil markets and industry activity provides the setting for the issues we are discussing today. During the oil crisis of the 1970s, the oil-importing developing countries suffered severe economic buffeting due to oil supply-and-price instability. These countries, like all oil importers, have sought to reduce their dependence on imported oil to insulate their economies from this external destabilizing force. One obvious response has been to explore for and develop indigenous oil and gas resources.

The oil price scourge served as a strong incentive for some countries to develop policies and the institutional frameworks to encourage and accommodate oil exploration and production. It also prompted companies in the business of finding and developing oil and gas to diversify their sources and to look again at the geology of parts of the world they had dismissed as marginal or uneconomic at lower prices and in a more stable market. Oil exploration and development expanded in the oil-importing developing countries as well as in the industrial countries and those developing countries which had already been substantial producers and exporters. Some former oil-importing countries became exporters, and others were able to substantially expand their domestic supplies and reduce their level of imports and their vulnerability to reliance on fuel from external sources.

However, the emergence of these additional supplies along with structural adjustments in demand in the form of conservation practices and increased efficiency in the use of hydrocarbon fuels have significantly and permanently altered the market's structure and brought a reversal in oil price trends, at least for the time being. As a result, the expected economic returns from successful exploratory activity have fallen and with them the companies' interest in investing resources in new exploration.

As is economically appropriate, the areas regarded as most geologically promising and those which offered the most rewarding terms have received priority for development. However there may remain small shallow deposits which did not attract activity before the exploration and development boom receded. There are several reasons why such deposits might not have attracted investment. They may have been too small to develop economically. They may have been more costly or less convenient to develop, due to geologic conditions or geographic location relative to markets, which lowered their expected return relative to alternative resources, or they may not have been able to compete for exploration and development resources due to institutional restrictions imposed by the controlling government. There may be such deposits which, although too small for large-scale exports, have the potential to provide some relief from the foreign-exchange drain of a nation's oil imports if they are economic to develop under current market conditions. Our concern here is to explore the financial mechanisms which may be employed to facilitate development of deposits which have economic merit.

New Players on the Scene

It is appropriate that this meeting is being convened with representatives of the independent oil and gas companies for, while most independents have limited their operations to domestic projects, there have always been firms ready to venture abroad, and independents as a group have characteristics which may make them particularly well qualified as potential partners in the solution to this specialized situation. Independent oil companies have typically operated at the frontiers, opening new areas to exploration worldwide, and they have been responsible for the initial exploratory work in a number of developing countries which are among the world's most active oil producers today. In his 1983 study of private oil companies actively drilling for oil in developing countries, Wollstadt (1983) identified 117 independents exploring in a sample of 97 countries. Independents participated in some manner in 6 of the 7 countries the World Bank identified as having become oil exporters between 1979 and 1982.

While most of the cases identified by Wollstadt reflect some form of partnership participation in a major's project for export, they clearly reflect inclination and initiative to work abroad, as well as some degree of foreign experience on the part of at least 117 of the 15,000 U.S. independents. Many other smaller independents may be interested but simply have not figured out how to go about it or located the right opportunity. Insights these 117 companies have gained through their experience as junior partners in foreign projects have improved their capability to initiate and manage projects on their own or as project leaders with other independents. In this sense, developing countries should regard them as potential candidates for undertaking their shallow oil and gas exploration and development projects, and independents lacking experience and expertise in foreign ventures may regard these companies as possible lead partners with whom they might be able to expand their operations outside the United States.

In the U.S. industry, oil companies originate as independents, and the industry is constantly evolving with companies in all degrees of maturity seeking opportunities to employ their resources to generate additional profits. Independents come in all sizes, all degrees of experience and expertise, and in the full range of project interests and capabilities. They will not necessarily be looking for export oil, if they can make money finding and producing oil for the host country's domestic market under conditions which assure the safety of their investments.

As a group the independents have historically tended to be flexible and innovative and more willing than the majors to take risks on the chance of obtaining compensating rewards. Among them, parties can be found who are interested in taking on almost any type project that might be available—so long as it will produce a competitive return on investment. Consequently, if a project can be structured in such a way as to be economic, a country has a good chance of finding a taker among the independents.

Furthermore, the independent sector of the oil industry has suffered its own setbacks as a result of market vacillations. Many new entrants were attracted when oil prices were escalating sharply, and many of that number have been left with their resources less than fully employed as the oil market glut developed and prices ceased to rise. It is likely to be a long time before their full productive potential will again be fully employed in U.S. fields. Consequently, there seems to exist today something close to a textbook case of a mutual coincidence of wants between the oil-importing developing countries seeking to have small shallow deposits developed and the independent oil companies looking for work. The time appears very right for the developing countries to talk terms with some of these hungry "Yankee traders."

Talking Terms

However, I would not lead either party to believe that striking a deal will be easy or quickly accomplished. It is likely to involve a substantial learning experience by both parties, and both parties will probably need to engage in careful cultivation of the prospective relationship. The "terms" to be discussed should not be taken lightly. The conditions imposed by a country on the developer of its oil and gas resources are absolutely crucial to its ability to attract the resources—both real and financial—required to carry out the task.

Broadman's [1983] preliminary research on the factors constraining investment in oil exploration and development in the East Asian developing countries suggested that variations in patterns of foreign investment from one country to the next where there are geologic similarities may be due to country-specific factors outside the realm of geology, such as political risk or contractual and fiscal arrangements. His more recent econometric analysis of the experience of 47 developing countries in attracting exploratory oil drilling during the period 1980-1982 reveals a large discrepancy between resource prospectiveness and the level of exploratory activity. The data indicate that low levels of exploratory activity in geologically promising areas are closely related to the existence of restrictive contractual and taxation schemes (Broadman, 1984).

Analysis of data by the MIT energy group [1984] indicates that oil and gas exploration and development have been skewed away from the developing countries in favor of the industrialized countries—contrary to both geologic prospectiveness and technical efficiency. They blame institutional and political factors for the inability of the parties to reach contractual agreements regarding the distribution of the projects' risks and potential benefits between the foreign oil companies and host governments. The problem has been compounded further for the companies by the added uncertainty about whether previously agreed contract terms will be unilaterally altered at a later date. The same economic and institutional elements that deter an integrated multinational will be doubly formidable to the more vulnerable independent company which typically cannot diversify its operations and spread its risks geographically to the degree that the established multinationals can and lacks the resources to engage in a prolonged legal defense if its investment is under threat.

An independent oil company undertakes an exploration or development project for the same reasons an integrated international company does—to earn profits. However, aside from the few largest independent companies whose behavior is in most ways indistinguishable from that of the multinational majors, the independents (particularly those who would be best suited and most likely to be interested in developing limited shallow deposits for the local market) are a very different breed with different managerial objectives, company strategies, and decision-making styles. Because independents are small, with compact decision-making entities, they can be flexible and responsive to specific conditions. Since the independent's goal is to expand market shares rather than guard an existing dominant share, the independents have historically tended to be more likely than the majors to respond favorably to high-risk projects in areas of unproven resource potential. These characteristics are in their favor for successful participation in shallow oil projects in developing countries. However, the same leanness that yields these characteristics contributes other hazards.

The smallest independent oil companies are typically individual or family-owned, -managed, and -operated. Company assets directly represent the owners' personal wealth, the accumulated fruits of their past labor and management, and their families' present and future financial security. When the independent oil company enters a foreign country to explore for and develop oil resources, it is betting that its financial wellbeing will be improved more by investing in this project than in any of the other alternative projects available, including domestic ones. The country can rest assured that the company is making a major commitment, even in a small project, and is going to take a *vital* and *personal* interest in every aspect of the project and its successful and profitable completion.

The first surprise to the uninitiated will be the disparity in the accustomed division of the proceeds from a project; whereas, developing countries dealing with majors on exploration and development of very large deposits for export are accustomed to receiving 70%-90% of the oil and additionally taxing profits on the company's oil at 50% or more, the independent company, developing shallow U.S. oil deposits, is accustomed to awarding the royalty owner one-eighth to three-sixteenths of production and paying corporate tax rates of 46% after significant tax write-offs for depletion, etc. This gap must be bridged, and both parties will have to carefully examine what trade-offs can be made to structure a project that will make participation economically rational for both parties.

The gap perceived by the oil company will be further widened by the costs of going international. For example, the independent venturing into a foreign project will face high information costs in deciphering and adapting operations to comply with a whole new set of legal, institutional, and cultural patterns. This cost may be alleviated if the independent is able to find a local partner who is familiar with local laws, customs, and procedures and is geographically well located to handle local arrangements. Institutional arrangements and an investment environment which facilitates the development of indigenous private oil companies, can provide double benefits by helping the country develop its own industry and attract the foreign independents with the technical knowledge and skills required to initiate development of local oil and gas resources.

The service and support systems taken for granted in the independent's domestic setting are probably nonexistent and will have to be developed in the host country. The host country that encourages and facilitates the development of local private enterprises to organize and provide such services, has made the environment more attractive to the independent oil company. However, even then, these systems will initially probably function in a less timely and efficient manner than was true in the independent's established and experienced domestic setting. Equipment transporting costs and logistics will add another new and costly dimension to the firm's operations. On entering the international environment, the firm will be exposing itself to currency conversion risks and fluctuations not encountered in moving from one state to another within the domestic setting. This will influence how the firm will want to be paid for its services. All of these items add costs which erode the return on the firm's invested resources. It is the absence of these costs which has accounted for a large share of the apparent cost advantage of the independents shown in the aggregate industry data.

The independent must recover the costs of going international in order to make a foreign project financially competitive with a domestic one. Consequently, a country improves its chances of attracting an independent oil company and improves the terms it must agree to, and the price it will have to pay, by the extent to which it avoids imposing additional local costs on the company. A country improves its chances by refraining from interference with the company's ability to operate and conduct business efficiently, i.e., to contract for services, bring in equipment, bring in personnel, and make management and operating decisions. A country can lighten costs by providing the infrastructure required for the project and creating an environment which encourages development of a healthy product market; the multilateral development banks, such as the World Bank group and the regional multilateral development banks, lend funds for infrastructure projects to support development-related projects of this type.

In addition to these direct costs, a company with a captive operation in a foreign country must be concerned about the sanctity and interpretation of contracts, about later-imposed levies and operating conditions, about an unhampered market that will return a fair price, and about ability to convert its earnings into dollars and take them out of the country. All of these things affect its return from the project and its ability to secure the financial support required.

Gas finds are a special problem with respect to market development and price issues. A number of major international companies have been deterred from development of gas discoveries, and others have rejected opportunities to explore for oil in gas-prone areas, because they lacked assurance that the product would be priced at a level that would yield a competitive return on investment. Companies have particular reason for concern about gas projects because they involve very long pay-out periods during which the product is captive to a single price-administered market. Anything the country can do to hasten the pay-out period and guarantee security of the investment will improve its chances of securing help.

In "talking terms" the chances for negotiating an agreement will be improved if both parties approach the discussions with an attitude of flexibility and willingness to explore trade-offs. This is not to imply that either party must sacrifice all the benefits from the project to the other party—such an arrangement will not produce a contract. It simply means there are many possible combinations of terms for arriving at any given revenue-producing result. Both sides need to be free to be good "horse traders" in order to custom tailor an efficient agreement that addresses all parties' needs and interests.

Generic model contracts can serve as check lists for points to be discussed in the negotiations, but they should be limited to that. A better model may be the contract signed by a successful neighboring country with similar resource potential; but even that cannot be regarded as grounds for a rigid negotiating position, because it does not take account of the different financial and market positions, business styles, and objectives of different companies. It does not allow for differing country economic and political stability, reliability, and geologic prospectiveness or different companies' independent assessments and weighting of these factors. And it ignores the effects of changing market conditions over even a short period of time.

Furthermore, the terms that will close a deal and produce a contract for any given project are project-specific and reflect the relative valuations of the various interests of the two parties. If a country strictly imposes the same terms on all projects in all areas, it automatically excludes the areas that companies find least desirable for development and excludes a number of companies from being interested in *any* involvement in the country. Companies, too, need to approach the talks with an open mind about arrangements that can serve their interests as well as meet the special concerns of the country. Every country/company combination is unique and will produce its own unique arrangements, satisfactory for each specific project.

This does not mean that the government must necessarily make concessions on things near and dear to the hearts of its people to get their resources developed, and it does not mean that either party should fail to negotiate the best deal it can make from its own point of view. However, if those items which the citizens regard as nonnegotiable impose costs and consequently decrease the earnings of the developer, the price of attracting the company to do the job will have to rise to cover that cost. This must be fully explained to the population and appreciated and agreed to by those who are going to live with the consequences of the decision over the long run if there is to be a smooth-functioning, long-term relationship. It is equally important that the general population understand the companies' profit requirements to maintain resources in the industry; these requirements are greater than is true for some other industries. For example, oil exploration is a risky business with many dry holes, the costs of which the company must offset with large returns from successful strikes. This is without regard to the country in which the dry holes and the strikes occur. Unless these matters are understood and appreciated by the population, they hold considerable potential for souring the relationship, and both the company and the government must view the situation with caution.

The decision to undertake a small oil or gas project in an oil-importing developing country is, in principle, no different from the decision to undertake any other oil or gas project. Changing the players does not substantially change the questions that must be answered, the economic criteria that apply, or the institutional accommodations that must be in place. Arriving at answers may be a bit more complex where large companies and large deposits for export are involved, and the negotiating parties may be a bit less experienced and more economically vulnerable where independents and small deposits are involved.

Should The Resource Be Developed?

The first question that must be asked by the country that sets out to have small shallow oil reserves developed is whether the project will be economic to develop. This depends upon the cost of developing the project *vis a vis* the cost of energy supplies from alternative sources. Shallow deposits are typically technically efficient and inexpensive to produce compared to deep or offshore deposits. However, if the deposit is in a location which is difficult to reach or from which it is difficult to transport the oil to locations where it can be refined and used, or if the geological structure is a particularly difficult and costly one to drill or produce, resources expended in producing small deposits may exceed those required for an equivalent level of imports. Developing countries should examine with great caution the argument that exchange savings alone justify a project even if it is not otherwise economic. A very large share of the resources of oil exploration and development must be imported—including specialized and experienced labor. Therefore, a very large part of the cost of developing oil resources is foreign exchange. If local resources are developed at too high a cost, there will be little or no net foreign exchange earnings and hence no net contributions to the balance of payments. The question then becomes whether the country can afford to pay that premium to assure stability of supplies, i.e., an even higher outflow of foreign exchange for a given quantity of oil.

Sharing the Risks and Rewards

If the government decides to proceed with development of the project, a number of options are available to it for organizing the task, depending upon the country's expertise and the availability of technical and financial resources. The options extend along a continuum from the country managing the project itself and securing service contracts for the various tasks, through various joint-venture agreements under which the risks and responsibility for raising the financing are shared as well as the returns, to having an oil company organize and manage all aspects of the project in exchange for a share of production to compensate for the added risk and services performed. Financing, both the source and the mode, is implicit in this decision.

Other things equal, countries generally prefer to minimize the amount of services hired in order to retain the largest possible share of the return. On the other hand, other things equal, oil companies generally prefer to assume full responsibility, including securing the financing for the project. This is a role the companies are accustomed to playing in domestic projects and they have their own sources with which they have an established working relationship. Furthermore, performing this service enables them to increase their return from a project. However, it behooves both parties to examine the risks and incentives inherent in the various contractual arrangements and evaluate how appropriate they are to the project under consideration given the resources, abilities, and interests of each party.

Blitzer and others [1984] emphasize "contract efficiency" as a critical element in whether an agreement can be successfully negotiated in the first place, whether it can be maintained in effect over the life of a project, and whether an optimal level of resources will flow into the development of the country's resources. Efficiency refers to the negotiated point after which no further change in terms can make either party better off without making the other worse off. Blitzer and others [1984] concentrate on the sources of risk at various stages of the oil exploration and development project and the adaptation of the various contract forms to the comparative ability of each party to bear geological, investment-cost, and market-price risks, the influence of contract structures on managerial incentives, and contract nonperformance risks to both parties.

A country's specifications for an efficient contract will depend upon its degree of exposure and ability to diversify risk, its own capacity to manage exploration and development, its ability to monitor contractors' performance, and its knowledge of its own geologic prospects. In negotiating a contract, the party with a greater comparative advantage in bearing a particular type of risk will place higher value on that share than the other party and thus be willing to pay more for it. The degree of risk is determined by the extent of exposure of each party as well as ability to diversify the risk.

Under a service contract, the country arranges for all financing. The country bears all the geological and oil price risk, and the company's incentive to produce a successful result is the value of its reputation and consequent ability to get more contracts. If the project is successful, the company is paid its agreed price, i.e., direct cost plus a rate of return, and goes its own way; all remaining proceeds—great or small—accrue to the country. If the project is *not* successful, the company is paid its full price and goes its own way, leaving the country to draw down its scarce foreign exchange reserves to pay interest on the costs of drilling a dry hole.

In the other exclusive case, the company is responsible for all costs; it bears all exploration and development risks. If the project is successful, the company's costs are repaid at a mutually agreed rate with oil produced from the project until all costs of exploration and development have been repaid. Thereafter, the company continues to receive a share of the production as long as production continues or for some agreed length of time. If the project is not successful, the country pays nothing for the array of services performed by the company; the company assumes all costs as losses and goes its own way. Even under this agreement, the form of payment—physical quantities which the company must market or cash payment of an agreed price per barrel—will shift the market-price risk from one party to the other. Generally, the company has compelling incentives to evaluate accurately and respond appropriately to maximize production and hence its own return. However, Lessard, et al. caution that if the rate of cost recovery is unsatisfactorily low or the company feels its interests are threatened the company may feel a need to produce at an above-optimal level, lowering the ultimate productivity of the field. A different problem may arise if a larger than expected discovery is made or oil prices increase sharply, yielding a windfall for the company. A country suffering fiscal reversals from oil price increases which yielded a windfall to the company might have an incentive to seek to recapture some of the money through new or increased taxes or other changes in terms.

Clearly, there is much room for tailoring the terms of the project to serve the interests of both participants, by the contract form that is chosen and within the chosen contract form. The outcome will depend upon the resourcefulness of the parties in seeking to reach an agreement that will be to the advantage of both to abide by over the long run.

Seeking Financing

Since development of small projects abroad for host country consumption is a departure from the common practice with which the traditional lenders have experience, and the parties are essentially breaking new ground, extra effort may be required to sell the idea to parties from which the funds will be obtained to finance the undertaking. Lenders will be concerned not only with geological prospectiveness but also with viability of the project and the contract. Essentially the same financing sources, methods, and instruments should be available to either a local government with control over oil and gas royalty or the private company, foreign or domestic, seeking to finance a project. Either would initially seek financing from the same sources they have used before and with which they have established credit credentials and a working relationship. The oil company may have some advantage over a local government as a result of having established a performance record in oil and gas work.

Any party participating in the financing must examine its own ability to absorb losses or sustain delays in the project. Unless the party seeking financing already has other significant cash flows or reserves, financing methods which generate fixed charges can be dangerous. Even in a development project where a fair amount of geologic certainty has been established, revenues may not be generated for 2-5 years. It is important that the financing plan be able to accommodate such contingencies.

In the domestic setting, the U.S. independent typically leases a prospective area and then seeks financing for drilling and production. Funds may come from internal or external sources. In 1975, independents as a group, including all sizes, drew 78% of their financing from external sources, according to Chase Manhattan Bank surveys (Energy Finance, 1983). That share fell steadily to 68% in 1980 and is expected to decline to 56% by 1990 as internally-generated funds become increasingly available, due to growing earnings from new production coming on stream and cash flow from depreciation.

The size of the project will determine the approach to external financing. Financing for small-scale projects may be available from a single source whereas a large project or a series of small projects may exceed the degree to which a single lender is willing to concentrate its risk. In the latter case syndicated financing may be sought, with one institution coordinating the assembly of funds from a variety of sources.

Commercial banks have typically provided the bulk of external financing for the independents. In 1975, 54% of the U.S. independents' funds came from medium and short term bank debt. However, between 1975 and 1980, commercial banking's share shrank to 37%. The share contributed by drilling partnerships is expected to fall back to 26% or even less by 1990 due to changes in the tax laws which reduce their tax shelter aspects. Bank lending is expected to regain a 40% share. The share of long term debt is predicted to remain fairly stable at around 25% for the 1975-1990 period with the exception of a drop to 18% in 1980 due to high interest rates and a difficult bond market. Funds raised from new equity financing rose from 5% in 1970 to 9% in 1980 and are expected to remain stable at that level through 1990.

Commercial bank credit to U.S. firms undertaking domestic projects comes primarily from regional banks which have specialized staffs who understand both the world of finance and the nature of the oil exploration and production business. However, regional banks' expertise and experience is mainly in domestic lending, and interesting them in small foreign ventures may require considerable cultivation—particularly given the recent disorder created in the U.S. banking system as a consequence of imprudent energy-lending to domestic projects by some banks. Certain major money center banks have developed oil and gas lending departments which seek opportunities to lend to foreign projects. However, their experience is with very large projects, not with small independents undertaking development of small fields for local markets overseas. Some companies have secured funds from the Eurocurrency market even for projects in the United States.

Bank credit may take the form of loans, guarantee facilities, combined credit facilities, or leasing arrangements. Bank loans are made in one of three forms. A revolving credit allows the borrower to draw down and repay advances as needed up to an agreed aggregate amount outstanding. The borrower may be assessed a fee for any unused portion. A nonrevolving credit provides for drawdown and repayment according to a fixed schedule. Both revolving and nonrevolving credits are in the form of a promissory note. Alternatively, the bank may provide a note facility which enables the borrower to issue a series of term notes to lenders. The primary lender may then either retain or sell the notes.

Some banks have begun offering nonrecourse loans for oil and gas projects in which the bank assumes the risk for the project and participates in the rewards if the project is successful. That option might be available for independents seeking projects in developing countries, although probably rarely in the case of small shallow deposits. In such cases, it is important that the contractual agreement and the laws of the country permit transfer of an interest in the project to the financial institution.

Insurance companies and other nonbank institutions are also potential sources of funds. In cases where they are unwilling or legally constrained from undertaking overseas project risk, a bank can facilitate that borrowing by providing a "guarantee of payment" or a "standing letter of credit." Letters of credit, like loans, can be syndicated. This may consist of a single issuing bank which relies on the participant banks to protect it with counter guarantees, or alternatively, may receive guarantees from a number of banks with each being exclusively responsible for a portion of the loan. In the latter case, if one bank fails, the other banks are not obligated to cover the short payment. Beneficiaries usually prefer to deal with a single issuing bank, but lending limits may preclude this arrangement on large loans.

When a project will seek funding from a variety of sources, the sponsor may use a bank facility for setting up an overall financing plan through which it may obtain loans of various types, guarantees for private placements, export credits, or commercial paper. While this is much more complex than using a bank-funded facility, it allows more flexibility in seeking attractive funding opportunities, and simplifies matters with a single unified set of documentation and covenants.

In addition to bank loans, projects can obtain credit through government export credit agencies which support the export of the nation's industrial output. Insurance companies and other institutions have also been active suppliers of long maturity, fixed rate funds for oil and gas projects in the domestic market and may be possible sources for the type of small project being considered by this conference.

Drilling partnerships, which have enjoyed great popularity due to their tax advantages and have in the past contributed a large share of financing for the independents' domestic projects, have lost much of their attractiveness, even in the domestic market, as a result of recent changes in tax interpretations. It is therefore unlikely that U.S. drilling partnerships will serve as an effective vehicle for financing foreign projects which involve higher costs and greater risk. However, the drilling partnership may be a useful vehicle for the host countries' domestic companies to use for raising local capital.

In addition to direct borrowing from financial intermediaries, local partners may be able to secure financing through local and regional privately sponsored development banks and funds. For example, host governments may issue bonds to raise funds from either the international financial market or the domestic market. Financing secured from these sources may be on-lent to assist local private firms interested in becoming partners in local oil and gas exploration and development projects. Alternatively, the national government might establish funds to function as development banks, with the country providing retail distribution of finance for the small shallow oil and gas projects. Mexico and Brazil have used such funds in the mining sector. The U.S. Government has used a similar innovation to facilitate rural electrification and telephone projects. The Ivory Coast, Mexico, and Egypt have used such innovations to promote hydrocarbon development. Where the funds are financed through a specific tax on hydrocarbon products, the arrangement also functions as an effective energy conservation device. This provides a source of local currency. Alternatively the independent oil company may have to seek host country currency borrowings as well as dollars from the commercial banks.

Conclusion

Cash-strapped oil-importing developing countries have strong incentives to develop any indigenous energy resources, and shallow oil and gas reserves are a particularly tantalizing case. There may be many sound economic reasons why such reserves remain undeveloped after the strong oil market of the 1970s and, other things equal, today's declining oil prices make their development even less rational from an economic standpoint. However, where the technical aspects permit such projects to yield a positive return, a country may be able to structure the terms to make the development of such resources for the local market attractive to the right company. Independent oil companies actively seek such projects in the United States. However, their traditional ability to hold down costs relative to the majors will be eroded by the significant costs of going international, raising doubts about their ability to do the job for a cheaper price than the majors. A country can improve its chances of being able to negotiate an agreement for development of its resources by removing cost-creating institutional barriers, providing the required infrastructure, facilitating development of an indigenous oil industry and ancillary service and supply industries to work with the foreign oil company, and by facilitating the development of a healthy product market which will assure adequate return on the company's investment.

Financing will probably have to come from the traditional sources, largely commercial bank borrowing. U.S. regional banks are largely inexperienced in overseas projects, and the money center banks are accustomed to dealing with the majors on large export-oriented projects. Consequently, an outstanding case may be required to persuade either of these institutions of the merit of this new idea. However, if the economics are right, and the country and company have been able to agree to terms which both have strong incentives to honor, such a case should be possible. Insurance and private finance companies are also potential sources of funds. Export credit agencies extend credit for materials and equipment used in foreign projects. A country itself can assist this effort by establishing energy funds which on-lend funds to energy development projects and by facilitating the participation of local investment in the project.

References

Blitzer, Charles R., et al., 1984, An analysis of financial impediments to oil and gas exploration in developing countries: (prepared for delivery at the annual meeting of the International Association of Energy Economists), Massachusetts Institute of Technology Energy Laboratory, Cambridge, January, 37 pp.

Examines geographic patterns of private investment in oil and gas exploration and development. Investigates causes for skewing of private investment away from the oil importing developing countries. Concludes that distributional factors involving contractual relationships, taxation, and political risk are the major determinants of departures from true global economic efficiency in the flow of investment funds. Explains the use of financial analysis to help design contracts based on the comparative advantage of the country *vis-a-vis* the foreign suppliers of capital and technology in assuming risks and responsibilities. The authors argue that finding the correct balance can yield self-enforcing contracts and increase the economic gains of both parties.

Broadman, Harry G., 1984, Incentives and constraints on exploratory drilling for petroleum in developing countries: an econometric analysis: unpublished paper presented at the annual meetings of the American Economic Association, Dallas, Texas, December 28-30, 50 pp. Available from Mr. Broadman at Resources for the Future, 1755 Massachusetts Avenue, N. W., Washington, D.C. 20036.

Develops a multivariate pooled cross-section-time series regression model to analyze the determinants of oil exploration activity in a sample of 47 developing countries over the period 1970-82. The econometric results indicate that factors other than resource potential strongly influence the ability of developing countries to attract exploratory activity and suggest unduly restrictive contract and taxation schemes may be a major cause for low levels of exploration in geologically promising areas. Policy implications are drawn on the basis of the findings.

Broadman, Harry G., 1983, Petroleum prospects and investment in East Asian developing countries: unpublished paper presented at the conference on Energy Pricing, Supply and Demand in East Asian Developing Countries sponsored by the Woodrow Wilson International Center for Scholars, Washington, D.C., June, 19 pp. Available from Mr. Broadman at Resources for the Future, 1755 Massachusetts Avenue, N.W., Washington, D.C.. 20036.

Oil and gas revenue disclosures (summary of 300 public companies, 1980-1982) and Oil and gas reserve disclosures (summary of 375 public companies, 1980-1983): Arthur Anderson & Co., 711 Louisiana Street, Suite 1300, Houston, Texas, 222 pp. and 316 pp.

Data tables and analysis of oil and gas reserves, production, revenues, and costs by individual company and industry groups.

Energy finance, 1983, Gordon McKechnie (Ed.), Euromoney Publications Ltd., London, 275 pp.

A comprehensive, nontechnical descriptive handbook written by industry experts for the layman. Specific examples of products and companies are cited and case studies are employed in tracing common practices and general trends in energy finance and making projections for the near future. Six chapters are specifically dedicated to oil and gas exploration and development, four to infrastructure financing, and four to financing procedures and techniques.

National Petroleum Council, 1982, Third world petroleum development: a statement of principles: National Petroleum Council, Washington, D.C., 27 pp. plus appendices.

The National Petroleum Council is composed of a broad cross section of industry and consumer representatives who advise the U.S. Secretary of Energy. This study contains analysis and recommendations by the Committee on Third World Development of factors that affect the decisions of private petroleum companies to explore for and develop oil in the oil importing developing countries. It suggests measures the U.S. Government and the developing countries can take to encourage private companies in this effort. It does not include detailed resource assessments or critiques of the programs of specific developing countries.

U.S. Department of the Treasury, 1981, An examination of the world bank energy lending program: U.S. Department of the Treasury, Washington, D.C., July 28, 63 pp.

A U.S. Government review of the World Bank energy lending program in the context of overall investment in energy in the developing countries. Contains sections on investment flows to developing countries, impediments to private investment in developing country energy projects, and measures for removing obstacles which interfere with the flow of investment funds for energy projects.

Wollstadt, Roger D., 1983, Oil exploration in less developed countries: activities of private oil companies: Discussion Paper #028, American Petroleum Institute, Washington, D.C., August, 38 pp.

Surveys recent private oil company activity in 97 developing countries. Makes reference to private firms assisting in preparing petroleum exploration laws designed to attract foreign investment to specific countries. Discusses the nature of activity of the private companies in individual countries. Provides an index of the companies in an appendix. Data were compiled from trade publications.

40
The Independent Oil and Gas Company— An Essential for the Exploitation of Small, Shallow or Marginal Hydrocarbon Deposits in Developing Countries

G. C. L. Jones, J. H. Burney, and G. W. L. Cull

Introduction

Independent companies, large and small, have played, and continue to play, a significant role in the exploitation of small oil and gas accumulations, particularly in North America. Independents have also been active in many developing countries throughout the history of the petroleum industry. However, their role in developing countries diminished, faced with the realities of the 1950s.

Given the wider opportunities available to the major international companies and their ability and need to explore for and exploit the more substantial accumulations, even in hostile environments, the smaller deposits in developing countries will continue to be ignored by these companies.

The authors believe that given infrastructural support the small independent company can once more play an important role in ensuring the exploitation of these accumulations in developing countries.

The United Nations meeting on Petroleum Exploration Strategies in Developing Countries, organized by the Natural Resources and Energy Division of the United Nations Department of Technical Cooperation for Development, held in the Hague in March 1981, focused on the wide disparity in exploration effort between the developed and the developing countries. It was pointed out that a large proportion of the unexplored and underexplored basins around the world were situated in developing countries which lacked the resources, financial and technical, to undertake much or, in a number of cases, any of the exploration effort without assistance from the developed countries. A number of arguments were advanced for the exploration gap, including the need in certain countries for an improvement in political conditions, investment climate, and the availability of acreage on reasonable and stable terms. Certainly, new strategies needed to be devised to reduce this exploration gap.

One of the strategies proposed was for the promotion of onshore exploration for shallow hydrocarbon deposits, as this exploration effort could require more modest expenditures, could be more quickly initiated, and might be more attractive to a wider group of companies. If successful, these efforts will not only make a positive contribution to the economies of these countries by reducing their energy import bills, but will provide a base for further exploration efforts.

Because of the usually marginal economics, and their limited contribution to reserves, even in the U.S., the development of small, shallow hydrocarbon deposits has been left largely to the small independent companies. This is to be expected, given the limited resources of such companies, and the proportionately reduced financial exposure associated with shallow drilling. Often, the small independent is only able to finance limited projects, so that any discovery is vigorously pursued with active management attention.

It is estimated that onshore sedimentary basins in the developing countries of Latin America, Africa, and Asia occupy an area of over 23 million square kilometers. Some of the countries in these regions are energy exporters, but the majority are energy importers, and for these the development of any indigenous energy resources could be particularly important economically.

Trinidad and Tobago: A Case History

However, a number of factors will have to be faced if the development of shallow hydrocarbon deposits is to be encouraged. It is illustrative to this discussion to examine the petroleum history of one developing country which, over a period of close to one hundred years, has seen the full spectrum of companies, from majors and large independents to small independents and a national oil company.

Although its prospective land area is less than 3,000 square kilometers, more than 100 companies have sought to explore for, or produce oil and gas in Trinidad. Typical Trinidad hydrocarbon reservoirs are small, with lenticular sands and complex faulting. Most of the accumulations onshore are shallow, from near surface to 1,700 m in depth.

Its relatively hospitable climate and the availability of small leases no doubt attracted the large number of players. By the beginning of World War II, there were still 12-14 oil companies actively engaged in exploring for or producing oil in the country, among them only one major company. However, by the early 1950s, the writing was on the wall for many of the small independents.

The small independent producer, without refining capacity and markets, had to compete in price with low-cost Middle East crude. By 1960, only four companies were engaged in production onshore Trinidad. Three of these companies were majors, with international operations, cross-processing agreements, and markets. The fourth company was relatively insignificant in terms of production and potential acreage to be worth acquiring.

The first half of the 1960s was a period of relatively high activity in the country, as the majors sought plays at greater depths than before and ventured into the offshore areas. With only limited success on land, onshore drilling went into decline and two of the three majors had ceased drilling entirely by 1967. These two companies then embarked on a major retrenchment of their work forces and in order to reactivate the industry and maintain a level of employment for onshore operations, the newly independent country formed a partnership in 1969 with a small independent U.S. company to acquire the interests of one of the majors. The Trinidad government created a national oil company and purchased the assets of another major five years later. It is currently engaged in negotiations with the remaining major to acquire part or all of its assets. The Trinidad government has from time to time stated that it proposed to acquire the assets of the remaining small independent producer but had not done so in mid-1984.

About 30,000 wells have been drilled onshore in Trinidad, or 100 wells per square kilometer, and more than three-quarters of these wells were drilled by independent companies. Until the 1970s, except for one near-shore field in very shallow water, all the offshore exploration and development had been undertaken by the major companies. By analogy, more than half of the 25,000 wells drilled in the U.S. annually during the last decade have been drilled by independent companies.

This brief history illustrates at least one of the factors which affect exploration for and production of small hydrocarbon accumulations in a developing country albeit, one which is an exporter of energy. The independent, while it might have the technical resources to develop and produce, needs to market production to allow a reasonable return on investment. The quantity of reserves associated with shallow oil and gas accumulations discovered in recent years are insignificant to a major oil company's global reserve situation, and their exploitation is perceived to be a distraction from the company's more substantial interests. The motivation of the major oil company is to secure supplies of oil to feed its refineries and markets. Clearly, if there is no perceived hope of a surplus of oil for export, then the major oil company will not likely be attracted.

Large though they are, the financial and other resources of the majors are still finite. It is a fact of commercial life that in seeking the greatest return on investment, the majors' interests are not always in harmony with the national interests of the host country, be it a developing or developed country. The small independent company, therefore, is more likely to be willing to develop the small accumulation, where the investment and financial exposure are considerably reduced. However, even the smallest investor seeks to be rewarded for a successful venture.

It should be clearly understood, nevertheless, that the smaller the company, the larger his corporate risk in an exploration project, and the small independent will only be attracted to a country in which he can operate on reasonable commercial terms.

Some of the factors which need to be considered in order to attract the smaller independent companies to carry out exploration in developing countries follow.

Exploration Promotion

The first consideration in the exploration for hydrocarbon deposits is the expectation of finding oil or gas. How much is known of the geology of the area? In the early years of oil and gas exploration, the existence of hydrocarbon seepages or outcropping oil-bearing formations played an important role in selecting drilling sites. However, the absence of seepages or outcropping sands will not write off an area as non-hydrocarbon-bearing. Certainly, the regional geological setting must be defined and at least some seismic data made available to determine depth of sediments and structural and stratigraphic trends. These data must be supplemented by drilling data from previous wells or stratigraphic tests.

The quantity and quality of available geological data will vary from country to country and the acquisition cost of additional data and their interpretation will have to be considered in an exploration venture. If these costs are borne by the operator, their later treatment for taxation purposes will be relevant to the profitability of the venture if hydrocarbons are discovered. In other words, the operator should be aware of the treatment of all exploration costs at the outset.

Another factor with regard to geological information is its availability. The host country should determine the advisability of making all data, including drilling data, available to any party interested in exploration in the country. The availability in public domain of geological data has contributed significantly to hydrocarbon development in the U.S., U.K., and Canada.

Given the requirement of attracting companies to undertake exploration, the country must promote its hydrocarbon potential and provide commercial incentives for a discovery. An important element of an exploration promotion program is the preparation of a data package collating all existing data to be made accessible to companies with even moderate resources. The data package will be enhanced by an independent technical study assessing the country's petroleum potential, based on existing data.

Leases

More than likely, the mineral rights in most countries will be vested with the government rather than private owners. This is by no means a deterrent to exploration effort. A system offering rational lease sizes, appropriate to the country, in fact, can be more easily developed and controlled than if small parcels of land were held by individual owners. Several systems of awarding leases exist in various countries, and one suited to the needs of a particular country can be adopted or modified quite readily. Similarly, in areas where surface rights are held privately, arrangements for right of entry and compensation for entry are well known.

Leases should provide for relinquishment of part of the block after a given number of years. This feature, the duration of the leases, and their size should be embodied in all contracts. Licenses should be small enough to be managed by independent companies, much as was the case in Trinidad, and still is the case in the United States and, in part, in the United Kingdom. Lease terms, including work commitments, should also be carefully considered. These need to be sufficiently flexible to allow alternative development paths as more data become available during the exploration phase.

A commitment to a work program, and the inclusion of a penalty for failure to carry out the program, even if earlier work indicates the futility of further work, will only serve as a deterrent to an operator. Allowing farming-in and farming-out of interests, common in exploration ventures in developed countries, not only spreads the risk of exploration but stimulates a climate for wider participation, even by local companies or institutions, in exploration and development ventures.

Drilling and Support Services

Except in the countries where there is already some exploration or production, exploration efforts will be hindered by the absence of drilling rigs, wireline logging equipment, and cementing and other important support services. The cost of mobilizing these services and getting them to a drilling site will probably be the largest single cost in the initial exploration effort. These costs and the associated costs for maintenance of the equipment will be most important in determining the overall cost of drilling. Thus, means of sharing these costs should be explored, if at all possible.

Political Risk

Political risk should not be overlooked in discussing the requirements for promoting petroleum exploration ventures in a country. Clearly, no company, particularly a small one with limited resources, is willing to see an investment evaporate as a result of political whim or a change of government. It is no coincidence that petroleum exploration effort has been greatest in developing countries with the most stable systems of government. Changes in the system of government have often been associated with renegotiation of contracts. The award of leases, therefore, should have the approval of as many sectors in the country as feasible, and awards should be on a competitive basis so that there is no suggestion of favoritism when a lease is awarded.

Financial Aspects

Given the vast disparity in exploration activity between developed and undeveloped countries, there is clearly a need for a developing country to attract multinational or independent companies to undertake exploration. The prospect of discovery of hydrocarbons will attract this investment, but the company must be assured as well that it may recover its investment plus a reasonable profit without undue restriction. Many times a company will have debt obligations to meet in order to stay in business.

The tax arrangements between the exploration company's home country and the host country must be resolved before activity can occur. How will taxes paid in the host country by the exploration company be treated in its home country?

Whatever the terms of exploration—concession, production-sharing agreement or service contract—the operator putting up risk money must receive a commensurate return on investment as quickly as possible.

The presence of experienced financial institutions common to developed countries, while not essential, might stimulate the exploration effort. These might include institutions for trading in shares in the company and raising venture capital, which would permit the participation of host-country citizens in exploration ventures and the spreading of the associated risk.

Production After Discovery

Even after a discovery is made, there are factors to be considered before the discovery is exploited. These involve the problems of marketing the hydrocarbons to be produced.

For liquid hydrocarbons, the considerations will be the cost of production and transportation to a refinery. An existing refinery in the country will naturally be an advantage. Factors such as the quality and quantity of the crude will need to be considered, together with the characteristics of the refinery.

The host government might consider the guarantee of purchase of a certain level of production and assume responsibility for the crude at the wellhead or alternative arrangements for disposal.

If gas is discovered, its exploitation poses different problems from an oil discovery. Oil is easily transported, for example, by road tanker to several points.

First of all, the size of the discovery has to justify its development. Projects involving the supply of 10 million cubic feet of gas, with no previously existing infrastructure, have proceeded in Australia, for example. However, it will be the small independent who is likely to develop this size of accumulation rather than a major international company. If the country is not already a gas producer, obviously there will be no existing infrastructure for using the gas. Gas will need to be competitively priced with existing fuel for potential users to convert their plants. Alternatively, new markets for gas will need to be developed.

As with oil, for which free market prices should be obtained, most companies would like to see the price for gas at least at 80% of crude parity.

Nevertheless, there will be opportunities, not previously available, to be considered even if gas is discovered.

Conclusions

Independent companies will explore in the same areas onshore as the majors. However, the spread of acreage held by the independent will be smaller and they will be more likely to follow-up a small discovery, however marginal, more aggressively. The majors, on the other hand, have more opportunities for exploration and development, and consequently provide wider competition for allocation of funds.

However, the exploitation of shallow hydrocarbon deposits will only accelerate if the proper infrastructure support, favorable business environment, and legislative stability is provided by the developing countries.

41
The Role of a Regional Bank in Financing Oil and Gas Exploration in Developing Countries

J. Terry Aimone

Introduction

The Liberty National Bank is pleased to participate in this conference, for the subject of exploration and development of oil and gas reserves anywhere in the world has an impact on all of us. My comments focus on Regional Banks: where we fit in the Banking Industry; what comprises our market area; and how we view our client base. Liberty Bank's policy for oil and gas lending will also be discussed. Although this policy is primarily domestic in its scope, I will also address the role of the regional bank in financing oil and gas exploration and development in foreign countries.

Definition of a Regional Bank

Let us examine for a moment a Regional Bank. To do so requires a frame of reference and, if you will, a categorization (and ranking) of various sized banks.

Liberty Bank is a typical Regional Bank. Its charter (articles of incorporation) states specific purposes for which it conducts business and thereby provides financial services to the community. Remember, however, that the bank is a business, responsible to shareholders, and with a profit motive. The bank is not an extension of the government; it is a business which is regulated by both Federal and State governments. State laws specify, for example, whether a bank may have a branch office, as well as the types of business that branch may conduct. State law also defines the role of holding companies and the number of individual banks each may control. Each state has different laws governing its banks. Within the last year, the Oklahoma Legislature passed a multi-bank holding company law which allows a single holding company to acquire banks in the state, within certain deposit limits, but, under Federal law, not outside the state.

Regional banks typically have assets of $1 to $15 billion, and are located in cities such as Dallas, Denver, Detroit, Oklahoma City, Charlotte, and Atlanta. These banks tend to operate within one state and within certain geographic regions of the United States, offering cash management, trust, investment, and commercial loan services. They may have international departments but generally the smaller the asset size of the bank, the less likely it will have an actual international department or a desire to do business abroad.

Regional banks have a customer base referred to as the "middle market." The middle market is comprised of individuals and small corporations whose borrowing needs typically range from $1–$30 million.

Money Center Banks

Money center banks are those located in New York, Chicago, San Francisco, and Pittsburgh, with asset sizes ranging from $15 to more than 100 billion. All have extensive international, as well as domestic operations. Their client bases include not only the middle market (domestic operations) but large corporations, such as the Fortune 500 companies and multinational corporations domiciled both in the U.S. and overseas. Money center banks represent the top 15 banks in the United States.

International Banks

International banks are located in such cities as London, Zurich, New York, and Hong Kong. They have very large asset bases, probably greater than $50 billion. They deal mainly offshore and in the money centers of the United States, with operations worldwide. Their clients are located in the countries in which the bank does business. Thus, there are vast differences between the banking factions. Regional bankers have a defined niche within this service industry. Because of their asset size, regional banks are better suited to handle commercial business in their own geographic area.

Market Area

Generally, regional banks concentrate their marketing efforts within a defined region, specifically, their home state and the states contiguous to it. Some regional banks venture beyond their defined areas to pursue new markets in such fields as real estate, telecommunications, shipping and transportation, and energy lending. Most regional banks, however, look within their immediate market areas, where they can provide services to financially sound companies, while maintaining good yields and, therefore, profits.

Client Base

As in any service industry, the regional banker considers his relationship with a client to be of paramount importance. Two types of clients exist in energy lending: the individual client and the corporate client.

Individual

The individual client group may include engineers, geologists, landmen, and private investors, such as medical doctors, attorneys, and other professionals. Most individual borrowers have spent several years working for a major oil company, acquiring first-hand knowledge of the industry as well as local contacts which provide a base upon which to build their business. Some individuals will form a small corporation. They deal mainly in the local region and with people they know and trust. Investing in an international venture has not attracted this group of clients because the risks appear to outweigh the benefits.

Corporate

Corporate clients can be divided into two groups: private and public. Private companies can be comprised of one or two individuals, or more.

They capitalize their company by pooling assets, comprised of oil and gas properties. Their net revenue and overriding interest provide both the asset base for bank leverage as well as current cash flow to pursue additional exploration and development. These companies usually have low overhead and thin corporate structures. They will operate in localized areas, often within a few counties of a state.

Public, or junior companies, are set up similarly to the private companies but have chosen to "go public" in an attempt to raise outside capital. They operate in the same manner as private companies but have a better-defined corporate structure necessitated by Securities and Exchange Commission financial reporting requirements. These companies invest in domestic ventures, for investor confidence stems from corporate success, which in turn stems from known local or domestic ability to find hydrocarbons.

The large independent oil companies comprise a second group of public companies. Most oil-producing regions domicile at least one; in Oklahoma, Champlin Petroleum, Kerr-McGee, Mapco, Reading and Bates, and Noble Affiliates are examples. These companies have large banking needs which are provided by money-center banks. Regional banks often participate in credit syndications structured by the money center banks, who serve as agents.

Where They Do Business

Most of the money borrowed by the client base just described goes for domestic activity. Investor money raised through various tax shelter funds also goes to domestic exploration and development.

The international oil and gas business attracts only select operators. The major companies, and large independents, dominate the field. Traditionally, the majors do not borrow funds for their activities; they are, in fact, net depositors seeking the best rates for excess funds. The large independents borrow funds principally to meet needs for working capital. Loans to them are extended on the basis of strong balance sheets and the value (if secured) of domestic assets. These companies have complex infrastructures, great expertise, and the financial strength to deal abroad and can cope with the risks involved.

Regardless of the type of client, individual or corporate, public or private, their tendency is still to deal domestically rather than internationally.

Lending Policy

Liberty Bank's Energy Lending Policy, as an example of a regional bank, is one that has been established by the bank's executive management. It sets forth principles under which each potential credit situation is judged. To better understand how this policy is actually implemented, let me explain the basic tenets of lending to the oil and gas industry.

We lend money on the value of proved producing reserves. We do not lend money solely on exploration prospects, leasehold acquisitions, or development drilling. This is a dramatic statement, but certainly not without justification. Proved producing reserves generate revenue which can service interest charges and retire principal. Exploration prospects, leasehold acquisitions, and development drilling prospects do not generate revenue but consume cash. A client can use proved producing reserves to obtain a loan which can fund leasing, exploration, and development activities. Bank engineers will verify the ownership of the reserves by examining division-order title opinions, which reflect all the participants' percentage of ownership in each well. Subsequently, an economic model is produced demonstrating a future net revenue stream. This enables the bank lending officer to make a determination of loan value. To arrive at a final credit decision, the bank lending officer will also consider the concentration of wells in a single field, the concentration of the revenue stream, whether oil or gas, and the concentration of revenues from a small number of wells. Repayment terms must be set within the estimated half life of the revenue stream to further adjust the risk of dealing with natural phenomena.

This type of loan is made on a secured basis. Collateral includes a first mortgage on all leasehold interests as well as any surface equipment used in producing the wells. Documents are also filed which direct production revenues to the bank to be applied against the outstanding debt. The basic tenets, then, of oil and gas lending include lending against proved producing reserves, collateralizing the loan and controlling the revenues. As you realize, these factors govern domestic lending, but from a business standpoint, we would expect these same factors to govern international lending. The realization of that fact, however, is far different.

Lending to Developing Countries

In most developing countries, natural resources and their revenues are the sovereign right of the government, a position similar to that governing the natural resources in Canada, making it impossible to follow

United States domestic lending policy. The governments of many developing countries seek the hard currency which can be generated from the successful development of such natural resources as oil and gas. This currency helps to achieve a favorable balance of payments. These are the same dollars, however, that our lending institution seeks to control for the purposes of debt repayment.

Since the reserves are the sovereign right of the government, as are the revenues, the bank considers itself in a secondary position with respect to both collateralization and revenue controls. This is tantamount to being unsecured—a clear compromise to our stated lending policy.

It may be recalled that a bank is chartered to serve the client base in its community. When a new service is demanded, the Liberty Bank is prepared to investigate the feasibility of providing the new product or service to meet a particular need. As of mid-1984 there has been no demand from our clients to finance projects involving oil and gas exploration in developing countries.

Summary

It is hoped this will help you to better understand the position of a regional bank when it comes to financing oil and gas exploration in developing countries.

Regional banks are the third tier of a worldwide banking/financial institutional "network" comprised of international banks, such as Barclays Bank in England and Credit Suisse in Switzerland, and money center banks, such as Bank of America in San Francisco and CitiCorp in New York. Clearly, these international banks and money center banks have the resources, the expertise, and the experience to provide financing for oil and gas exploration in developing nations.

The client base of a regional bank typically transacts business in its own state or region. The opportunities for a regional bank to enter into any kind of international lending transaction, including those for oil and gas, are remote.

42
An Emerging Alternative— A Trading Company Providing Energy Development Capability

Douglas R. Courville

Over the last decade, significant changes have occurred in the international petroleum industry. These structural shifts within the industry and in international trade practices have created the need for a new approach to properly respond to the changing marketplace. The developments in the international petroleum industry over the last decade have been complex and far-reaching; however, there are three major categories of change on which we wish to focus and which create the need for the trading company approach. These three major developments are:

1. A tremendous increase in the scope and capability of the national oil companies.
2. The tenfold increase in the price of crude oil from 1973 to the present, accompanied by significant economic shifts which created both major problems and offered significant opportunities.
3. Increasing reliance on nonmonetary transactions in international trade, that is, barter and countertrade.

In the early 1970s, well over 80% of the movement of oil in the non-Communist world was controlled by multinational oil corporations based in Europe or the United States. Last year, less than 40% of non-Communist oil movements were so controlled. The factor accounting for this change is the rise of the National Oil Companies and increasing influence of the national oil ministries in a multitude of countries around the world, both inside and apart from OPEC. Today, National Oil Companies can range in scope from relatively small crude oil purchasing operations to fully-integrated petroleum producers, refiners, and marketing organizations.

The mammoth increase in the price of petroleum and the associated economic dislocations constitute the second key factor which we are considering. Considered from a strictly academic, historical perspective, it is almost absurd to talk about a "weakening" or "softening" of oil prices to $27 or $28 per barrel. It was only a decade ago that crude oil was selling for under $3 per barrel and only about 5 years ago that crude oil could still be bought for under $10 per barrel. We are all aware of many of the problems which were created by the major price increase which happened over a very short time frame.

However, with the problems came opportunities. Specifically, there are now a number of oil fields around the world—many of them shallow—which today can be commercially developed even though only five years ago they may have been considered uneconomic or marginally economic at best. The development of such petroleum reserves can help create jobs, bolster engineering and technical capabilities, and through import substitution, help manage the balance of trade of the country which is developing those petroleum reserves.

Another major economic change beginning in the early 1970s was the disintegration of the "fixed" foreign exchange rate system. This added greater complexity to international trade transactions. Also, by the mid-1970s major western banks had instituted the process of taking deposits placed with them by petroleum exporting countries and loaning the funds to other countries in the developing world. Only a few years ago, we benignly referred to this process as "recycling petro dollars"—but today we call it the "world debt crisis." Called by any name, world debt coupled with a volatile foreign exchange rate combined with still other influences have lead to a massive growth in the use of nonmonetary methods of international trade—barter and countertrade.

These, then, are the three key points of the rapid and severe shifts in the petroleum industry: the rise of the national oil companies, the rise in the price of petroleum, and the rise in the use of barter and countertrade for international trade transactions.

Equipment Suppliers

Despite these changes in the marketplace, the infrastructure for supporting petroleum exploration and development remains remarkably similar to what it was 10, 20, or even 30 years ago. Manufacturers of equipment for petroleum exploration and development are still very much oriented toward supplying the needs of the highly industrialized countries or at least the needs of companies which are home-based in the highly industrialized countries. This is really not surprising since it must be remembered that even with the "depressed" drilling activity in the United States today, it is still true that two out of every three active rigs in the non-Communist world are in the United States. More than three out of four active rigs are in North America or Europe.

However, most equipment manufacturers clearly recognize that the market outside North America and Europe clearly has the highest growth potential for business in the future. Unfortunately, most petroleum equipment manufacturers are not well equipped to respond to the specialized requirements and individualized approaches necessary in dealing with the multiple country markets which will be their key markets for the future. It was only a few years ago that equipment manufacturers had only a few dozen key customers (mostly in the United States). Today, there are at least two or three dozen key countries which the equipment manufacturers need to support, and many countries have multiple petroleum-related organizations.

Complicating the matter is the fragmentation within the petroleum equipment manufacturing and service industry itself. Given the overall size of the petroleum industry, the equipment and service vendors are actually fairly small. Even though there are some large and internationally well-known organizations providing equipment or services for petroleum exploration and development, almost all of them are associated with a particular specialty or fairly narrow range of products. There simply is no one company which even comes close to being able to provide a full range of equipment and services necessary for exploration and drilling operations.

Thus, we are faced with a marketplace which contains a number of customers (the National Oil Companies), each of whom has its specific organizational, political, cultural, and other special considerations. On the other side of the market, we have an even larger number of vendor companies, mostly operating out of the industrialized countries, mostly focused on one or a few specialty areas, and, with only a few exceptions, not able to service the global market in a very cost-effective manner.

494 Shallow Oil and Gas Resources

Fortunately, we have observed that the petroleum services and equipment vendors are becoming increasingly aware of the need to find new approaches for the global marketing and support of their products.

National Oil Companies

The continuing growth and evolution of National Oil Companies and the beginning of indigenous service company operations are also creating the need for new marketing approaches.

Obviously, practical, dependable, cost-effective hardware or equipment is necessary for exploration and production operations undertaken by the national oil companies. One must also note, though, that the technical support requirements of one national oil company may be different than those of another national oil company which, in turn, can be different than the technical support requirements of a large multinational oil corporation. Certainly, the capital requirements and funding mechanisms for national oil companies are quite different from those of the multinational major oil corporations. And, of course, there are other infrastructure needs and requirements which can differ markedly from country to country, and from company to company.

It is quite difficult for individual manufacturers and service vendors to respond adequately to these diverse needs.

Energy World Trade

As an answer to all of the criteria which have been outlined, a new trading company has been created: this company is Energy World Trade. Although we have been developing the conceptual framework and operational policies of Energy World Trade for over three years, it was formally launched only one month ago. We believe that Energy World Trade can act as a coordinating or "umbrella" type of organization which can combine the four key elements necessary to successfully respond to the changes taking place in the international petroleum marketplace:

1. Energy World Trade will offer a full range of "upstream" petroleum-related equipment and services. By the term "upstream" we are referring to those activities associated with exploration and production operations. This will be the Energy World Trade focus initially, although other aspects of petroleum operations will become increasingly a part of Energy World Trade's scope as the company

develops. Energy World Trade will combine the products and services of a multitude of vendors (both from the United States and from other countries) who are capable of providing superior quality products and services.
2. Energy World Trade will, in all cases, operate in a given country in a "partnership" fashion with a strong, well-established premier local organization which can provide the kind of effective, continuous local presence and participation which we feel is crucial to the success of Energy World Trade. We are not talking about the typical "agent" relationship which is common to the oil industry today; rather, we are referring to the development of a full-fledged "marketing associate" relationship with our in-country counterpart organization.
3. Energy World Trade will, both by expanding the size and scope of its trading activities and by maintaining a staff of employees and advisors skilled in such matters, make possible the delivery of products and services using multifaceted financial capabilities. Project finance, countertrade, barter, and other transactions all can supplement the traditional means of financing petroleum equipment and services purchases.
4. Energy World Trade brings together in its founding group a number of organizations with appropriate expertise and the kind of long-term and global orientation which is critical to an organization such as Energy World Trade.

- SUPRA Corporation is the prime mover behind Energy World Trade. SUPRA is a consulting and venture-development firm which specializes in energy development projects around the world. SUPRA's clients include national oil companies, U.S. oil corporations, and other firms and individuals associated with specific technical and entrepreneurial areas of energy development.
- Diamond Shamrock International Petroleum Company, the international subsidiary of Diamond Shamrock Corporation, an integrated oil and gas company, is another founding member of the Energy World Trade organization.
- Dravo Corporation is a worldwide engineering company with transportation and manufacturing operations and is the third initial participant in Energy World Trade.
- Energy World Trade also has a verbal commitment to be joined by a well-known international money-center bank. We hope to announce this additional participant within the next few months.
- A selected group of other firms are now considering participation in Energy World Trade.

Advantages of Energy World Trade

We feel that Energy World Trade can uniquely fill some of the functions of both buyer and seller in international trade transactions. That is, buyers (national oil companies or private enterprises) in the developing world can look to Energy World Trade to perform many of the clerical, administrative, and nonexecutive tasks of a purchasing department. The managerial and executive tasks of a purchasing operation will, of course, always remain with the buying organization. However, there may be a number of instances in which Energy World Trade can more cost-effectively handle many of the mundane, routine operations which are associated with a purchasing operation.

Similarly, we feel that equipment manufacturers and service vendors can rely on Energy World Trade to perform many of the functions of an international sales department. Although Energy World Trade can never approach the level of technical expertise that a given manufacturer may have relative to its specific product line, the Energy World Trade organization can clearly do a superior job of providing "on-site" and continuing presence to provide routine service functions and to maintain good communications if nonroutine services or support are required.

Energy World Trade is truly, then, a coordinating entity: it combines the products, skills, and technical expertise of a number of organizations so that the combined capability of the Energy World Trade organization is greater than the sum of its many individual parts.

The specific advantages of Energy World Trade for buyers of oilfield services and equipment will be numerous.

1. Energy World Trade will provide easy access to a full line of petroleum-related equipment, services, and technical support. Note, however, that the key word is "access"—Energy World Trade will not be determining or dictating the specifications or terms of petroleum development within a country. Rather, Energy World Trade will be working with indigenous entities to develop and implement such petroleum projects.
2. Energy World Trade, through its local marketing associates, will provide full-time, fully accessible contacts for account servicing.
3. Buyers have the opportunity to use Energy World Trade's resources to more cost-effectively perform many of the detail and administrative functions associated with purchasing.
4. Most importantly, international buyers will have the ability to deal with a premier trading group which can respond with dependability, flexibility, and a depth of expertise.

Examples

In order to help convey the ways in which Energy World Trade can assist in international energy development, I would like to list a few actual projects by way of example.

1. Energy World Trade has structured an arrangement whereby a country which has fairly large, but declining reserves can launch a major program of enhanced oil recovery. The Energy World Trade proposal includes the engineering work (to be done by a well-known international reservoir engineering company, assisted extensively by indigenous personnel) and all equipment and services necessary for oil production enhancement (including imports, domestic manufacture, and domestic assembly). The complete program is to be financed almost entirely by the sale of specific petroleum products which are excess to the country's current internal needs.
2. An affiliate of Energy World Trade has structured a proposal which will allow the use of excess refining capacity and an abundance of natural gas reserves in one country to be used as an upgrading medium for heavy oil reserves which are available in a neighboring country. The feasibility studies on this program have been negotiated at an extremely low cost. Negotiations are underway to determine ways in which project financing can be done.
3. Energy World Trade and its affiliates are currently working in three countries to assist in the establishment of oilfield services companies. Energy World Trade will provide technical assistance, required equipment and financial support for these developing companies.
4. Energy World Trade also plans to launch a number of programs which will allow for the indigenous manufacturing and/or assembly of components or sub-components of products which Energy World Trade supplies.

Conclusion

In short, we believe that Energy World Trade offers a new and responsive vehicle for providing petroleum development services, products, and technology to the rapidly changing and growing international marketplace.

43
Risk: The Wildcatter's Partner
Charles Robbins

A healthy petroleum industry has evolved in the United States built with private capital and initiative, with no government involvement through its growing years. Thousands of uses for oil and gas have been found, and the economy of this country and the world has been enriched beyond the wildest dreams of any prognosticator of a century ago.

The vast amount of expertise, based on experience gained in finding thousands of oil and gas deposits in the United States and other countries, and the sophisticated equipment that has been developed, and the experience gained in drawing up and implementing financial arrangements required for oil and gas projects, is now available for the development of the oil potential in other countries. This expertise is available from other countries too, including some of the national oil companies.

Many if not most geologists and others engaged in finding oil believe that oil and gas exist in shallow deposits in many if not almost all areas of the world, and that such deposits may contribute a very large proportion of oil and gas yet to be found.

But something is wrong. These deposits in developing countries are not being identified and brought into production as expeditiously as they might. Petroleum geologists-economists estimate that two-thirds of the oil and gas discoveries yet to be made will be in the developing non-OPEC countries. And economists predict that energy consumption will go up much faster in developing countries.

And so will the investment needed to expand energy programs in the oil-importing developing countries. The World Bank estimates that investment in the energy sector will have to increase from about 2 or 3% of gross domestic product in the late 1970s to about 4% of gross product over the next decade, to around $130 billion average a year. The financing of such a huge program will make heavy demands on all sources of capital, domestic and foreign.

Oil and gas development in the Third World was essentially static during the 1970s, despite the shocking increases in world prices. The world economy was damaged, and a recession ensued, which reduced capital flows and thereby any increase in oil exploration and development. The situation worsened into this present decade: oil exploration in 1983 in developing nations dropped 15%; 40% in Africa.

More than three-fourths of all drilling in the world has been done in the United States. Yet this country occupies only six percent of the world's land space and has five percent of the world's proven oil and gas reserves. Doesn't that say something to other countries where so little has been done to probe their energy potential?

Nature is wonderful, but when it comes to finding oil it is most baffling, enigmatic, inscrutable, mysterious, puzzling, frustrating, and, sadly, expensive. Searching for it requires careful preparations and the employment of all methods that can be afforded in that search. As we have learned these past few days, when satellite photographs are interpreted, seepages examined and their implications evaluated, geochemical and other geoscientific and engineering evidences weighed, the results can be used to predict the location of oil. Then someone must swallow hard and make a decision, and risk success or failure.

The oil-history books are full of fabulously successful finds by wildcatters, but most wildcatters tried their luck and failed, usually many times, before they succeeded. History gives short shrift to those who tried and tried, and tried again, and failed—with perhaps enough successes to keep them trying until their luck ran out completely and they disappeared from the scene, and the history books.

Lone individual or corporate entrepreneurs capable psychologically or financially to undertake risks that only sometimes pay off, modestly or even handsomely, are about nonexistent in developing countries. A few of the 80-plus national oil companies have tried to emulate the wildcatter, and have had modest success. Many are not audacious enough to take the risks. One can only wonder if they, or the government to which they report, can stand the criticism and the pressures that failures might bring. And, for that matter, will they be able to stand the pressure that success might impose—of changing the terms of agreement so as to increase their percentage of revenues that might be forthcoming? That has happened, which accounts for the reluctance oil entrepreneurs have in trying to work out deals with some developing countries.

Today the principle of national sovereignty over subsurface areas is well established throughout the world, with the notable exception of the United States. The United States government owns almost half of the total U.S. land areas, and it of course exercises sovereignty over all above and below surface properties. It leases subsurface rights to a small portion of its lands, and is under pressure to open up more lands for private exploitation.

Why developing countries have been so protective of their sovereignty is understandable, if one looks back at the exploitation of backward areas by rich and powerful nations down through time. It has always been thus, and may always be, and it can only be stopped by weaker countries learning how the developed world works, and how to deal with those countries that have the technology and the money. Education and training of those who represent their governments is absolutely essential.

Developing and developed countries and their nationals realize that lasting relations must be based on arrangements that are mutually beneficial—and as equally beneficial as it possible to devise—between contracting parties. It takes good sense, knowledgeable negotiators, and capital as well as confidence of the negotiators in themselves and their counterparts on the other side when considering the finding and development of petroleum reserves.

What difference would it make if nationals of developing countries had rights of ownership and development of their subsurface wealth? Would the world be much better off? Would developing countries—or at least those having petroleum resources—be able to improve their economies, and reduce their outflow of capital for foreign oil? The record of the United States is pretty conclusive.

It has been the wildcatter, the descendents of Colonel Edwin Drake, who made it all happen. It is he—whether individual or company—who has drilled most of the wildcat wells and found most of the oil. It is he who has provided those essentials of intelligence, experience, derring-do and risk capital that is today's oil industry and the base of the automobile, petrochemical, and hundreds of other industries essential to our well-being, and to the well-being of the world.

It might be appropriate to declare the wildcatter "The Man of this Century."

Insurance companies will protect individuals against almost all risks, but they will not accept the risk of finding oil. The wildcatter must take all that risk, or he might share that risk with the organization or government for whom he works. What share of the risk, and the revenues to be shared, if any, are matters for negotiation. All the risks must be evaluated, as well as possible, and then the risk taken. There is no getting around it.

Developing countries are not developing their energy resources as quickly as they might. Oil-importing countries are hard pressed, prices at which commodities can be sold on the world markets are down, and trade balances are in most instances negative, decidedly negative. Internal debt is growing at a frightening rate, as in the United States. American banks have loaned billions of dollars more than they should have, and so have to cut back sharply. And that hurts everyone.

But some of the countries have large undiscovered and undeveloped resources which must be found and produced. To do that will require taking large risks. These may be shared with oil companies; for that it is necessary to share the profits that successful production will provide.

We know everything about the past, but we don't understand all the implications. We know nothing about the future. We do know that one must take bold and imaginative steps if the future is to be fruitful.

44
OPIC Programs for the Energy Industry

R. Douglas Greco

The Overseas Private Investment Corporation ("OPIC") is a United States Government agency designed to promote the economic development of less-developed nations by assisting eligible U.S. investors in efforts to invest in or lend to projects, or develop contractual relationships, in the Third World. As a complement to the U.S. Agency for International Development, which finances projects on a government-to-government basis, OPIC's purpose is to facilitate the transfer of U.S. capital, technology, and management skills by helping American companies establish a broad variety of projects in Latin America, Africa, the Middle East, the Indian Subcontinent, Asia, and parts of Europe (Portugal, Greece, Yugoslavia and Romania).

OPIC assists U.S. investors in this effort through two principal programs: (1) the insurance of investments against certain political risks, and (2) the financing of such enterprises through direct loans and/or loan guaranties. These programs operate either independently or in conjunction with the programs of the Export-Import Bank of the United States, the World Bank (including the International Finance Corporation), the regional development banks, and foreign government supplier credit and political-risk insurance programs throughout the world.

While Congress views OPIC as a development agency, corporate America views OPIC as an insurance company and a source of recourse and nonrecourse financing. With its only office located in Washington, D.C., OPIC is managed by a 15-member board of directors, the majority of whom come from the private sector. All of OPIC's insurance and guaranty obligations are backed by the full faith and credit of the United States of America as well as OPIC's own substantial financial reserves.

Companies buy OPIC political-risk insurance for a number of reasons. While obviously most take coverage in order to quantify risks and receive compensation for losses, others are equally interested in the insurance for nonfinancial reasons. Because of the agency's vast experience in resolving investment disputes, some corporate executives place great importance on the fact that OPIC clients can turn to the U.S. Government to help them resolve major investment issues with host governments. OPIC's involvement brings with it the direct financial interest by a U.S. Government Agency in the outcome of any dispute; hence, OPIC's participation increases the likelihood that disputes will be settled amicably and in a professional manner. In addition, smaller companies which are deeply involved in Third-World projects will find that their bankers and major suppliers feel more secure when those exposures are insured against country risk by an agency of the U.S. Government.

OPIC provides insurance for most types of investments in energy exploration, development, and production, including those made pursuant to traditional concession agreements, production-sharing agreements, service contracts, risk contracts, and other agreements with host country governments. Coverage is available for up to 90% of the insured's investment, generally not to exceed $100 million per project.

OPIC's insurance provides coverage against the following risks:

1. *Currency inconvertibility* (Coverage A)—insurance against a deterioration in exchange control laws, regulations, and procedures (not including devaluation). There are two types of inconvertibility—active blockage and passive blockage. Active blockage entails an actual change in currency conversion laws, regulations, or procedures, causing a conversion delay of 30 days or more. Much more common is passive blockage, wherein the government simply fails to act on an application for foreign exchange which has been pending for 60 days or longer. For passive blockage, no change in the currency laws or regulations is required. In both active and passive blockage, the foreign currency, which is the subject of the claim, is delivered to OPIC at the U.S. Embassy in the host country and OPIC provides 99% of the U.S. dollar equivalent in the United States at the then prevailing rate of exchange.
2. *Expropriation* (Coverage B)—insurance against nationalization, confiscation, and creeping expropriation, including material changes in project-agreements contracts unilaterally imposed by the host government. In exchange for a transfer of the insured securi-

ties (e.g., shares, loan agreements, concession agreements or production-sharing agreements), OPIC pays to the insured the amount of its investment less cost recovery, if any, received pursuant to the project agreement.
3. *War, revolution, insurrection with/without civil strife* (Coverage C)—insurance against damage to physical assets caused by war, revolution, insurrection, and, if elected, civil strife. OPIC pays the least of the undepreciated original installed cost of the damaged item, the cost of repairing the damage, or the replacement cost.
4. *Interference with Operations* (Coverage D)—insurance against a cessation of operations forced by war, revolution, insurrection, or civil strife. Recovery for "interference with operations coverage" is measured in the same manner as for "expropriation coverage;" however, if at any time during the five-year period following a claim, the conditions causing the cessation subside, either the insured or OPIC may effect a reconveyance of project interests and claims payments (without interest) in order that project operations may resume.

The insured may elect to take one or more of the preceding coverages, but such election must be made when the insurance contract is originally issued. The term of the contract is twelve years, after which it may be extended by OPIC, at the request of the insured, for an additional eight years. While the contract may be terminated at will by the insured at any time, it may be terminated by OPIC only for a very limited number of reasons specified in the contract, such as nonpayment of premiums.

Eligible investors must be U.S. citizens, U.S. corporations, partnerships, or other entities more than 50% beneficially owned by U.S. citizens, or foreign entities at least 95% owned by these eligible investors.

An investment does not have to take any fixed form to be eligible for OPIC insurance. Coverage is available not only for conventional debt and equity investments but also for contributions of goods and services under a variety of contractual arrangements, including licensing and technical assistance agreements and construction, service, and export contracts. OPIC, however, is an incentive agency; therefore, it can insure investments in new projects, or new investments in existing projects, but it cannot insure existing investments in existing projects.

OPIC insurance is not available for investments in oil and gas exploration, development, and production projects in OPEC member countries. However, projects involving investments in other petroleum service operations and downstream petrochemical projects, as well as investments in other energy and mineral sources in OPEC countries, may be eligible.

Premium rates are determined by the risk profile for a particular project (Table 1).

A potential OPIC political risk insurance client must complete two principal forms. The first is the Request for Registration (OPIC Form No. 50), which usually must be submitted to OPIC before an investment has been made or legally committed to be made. This is necessary in order to evidence that OPIC programs, in fact, encouraged the investor to commit resources to the project. The second is the Application for Political Risk Investment Insurance (OPIC Form No. 52) which supplies OPIC with the information necessary both to complete the political-risk insurance contract and to judge the impact of the project on the economies of the host country and the United States.

The Request for Registration is a simple, one-page form designed to elicit the basic general information upon which a threshold determination of investor and project eligibility can be made. It asks for the name and address of the applicant and any insurance broker or authorized representative who may be designated by the investor; covers the eligibility of the applicant, the host country, and the foreign enterprise; requests preliminary information on the form and amount of investment and a one sentence description of the project; and asks several questions regarding levels of host government ownership, procurement, effect on the U.S. economy, and the investor's interest in other OPIC services.

OPIC recognizes that an investment project may change substantially from the time of conception to the date of implementation and allows the

Table 1
Base Rates for Oil and Gas Projects

Coverage	Exploration Period (%)	Development Production Period (%)
Inconvertibility	0.1	0.3
Expropriation	0.4	1.5
War/revolution/insurrection	0.6	0.6
with civil strife	0.75	0.75
Interference with operations	0.4	0.4
with civil strife	0.55	0.55
Primary standby (per coverage)	.075/.09	.25/.30
Secondary standby (per coverage)	.0075	.0075

investor to vary his actual amount and form of investment considerably without prejudicing its eligibility. The determination of eligibility is, however, always made on the basis of the investment as it is actually made rather than as it may have been described in the Request for Registration.

If the project is eligible in accordance with the limited information provided in the Request for Registration, and OPIC programs are available in the country, then OPIC will respond to the Request for Registration with a Registration Letter. Where time is of the essence, an investor may register his project by telephone and follow up with a written Request for Registration. In all cases, the date of registration will be the earlier of the date of the telephone conversation or the date upon which OPIC receives the Request for Registration.

It should be emphasized that the registration is not a binder and does not commit OPIC to issue insurance. Investors who wish to be absolutely assured of coverage will want to obtain an insurance contract from OPIC before the investment is made.

Once investment plans have been established, the investor should promptly complete and return to OPIC three copies of the Application for Political Risk Investment Insurance (OPIC Form 77). The purpose of this form is to elicit from the investor the definitive information upon which OPIC will determine investor, investment, and project eligibility; details of the investment to be insured and the insurance cover desired; the developmental effects of the project; and the effect of the project on the U.S. economy.

Where it is clear that before OPIC can issue an insurance contract it must have detailed information on the investor, the project, the investment, and the insurance coverage desired, investors may wonder why detailed information must be provided on the developmental impact of the project on the host country and its effects on the U.S. economy. As indicated above, OPIC's prime statutory purpose is to assist in the development process, and thus it needs to receive information regarding host government revenues from the project and the effect of the project on host country trade and balance of payments, including capital mobilization, downstream economic effects, environmental considerations, and other significant effects on local economic development and social wellbeing. With this information, OPIC can satisfy itself that support of the investment is in accord with its statutory purpose.

Concomitantly, as a U.S. Government agency, OPIC must be concerned with the effects of its actions on the U.S. economy. Thus, OPIC will inquire as to the effects of the investment on existing facilities in the United States, particularly regarding employment, as well as the effects on the U.S. balance of payments and U.S. exports.

In addition to political risk insurance, OPIC offers financial support to U.S. investors for the exploitation of energy resources once a commercially feasible project has been established. Support is provided through the extension of loan guaranties to U.S. institutional lenders financing the development and production phases of energy projects.

OPIC loan guaranties of up to $50 million are available for U.S. dollar loans made by U.S. financial institutions on terms approved by OPIC. Lenders are provided with an irrevocable guaranty that installments of principal and interest will be paid by OPIC within 10 days of notification that a borrower has failed, for any reason (commercial or political), to pay in accordance with the loan agreement.

Loan maturities typically range from 5 to 12 years, with appropriate grace periods before principal repayment begins. Guaranties may be applied to either fixed or floating rate borrowings, are normally issued on a nonrecourse basis, and can cover up to 50% of the cost of a new project or up to 75% of the cost of expanding an existing project.

The borrower pays OPIC an annual guaranty fee of between $1/2$ and $2^{1}/2$% of the outstanding principal. OPIC will typically seek to secure loan repayments through a mortgage of fixed assets, concession rights, and other project assets.

To be eligible, the borrowing company or U.S. project sponsor must be an established organization with a proven track record in the industry and must have the technical, financial, and managerial capability to carry out successfully the proposed project. The U.S. sponsor must own a significant equity interest in the project (generally 25% or more) and must be willing to establish sound debt/equity relationships that will not endanger the project through excessive leverage.

The host country should provide reasonable assurances of foreign exchange availability for repayment of the obligation. As with its insurance programs, OPIC financial services are not available for oil and gas projects in OPEC member countries, but may be available for other energy and minerals development projects in those nations.

45
A New Strategy for Improving Exploration Incentives in Developing Countries

Arthur E. Owen

Abstract

Technical and non-technical factors impacting upon petroleum exploration companies are examined, and a strategy is proposed whereby oil importing developing countries might achieve greater returns from exploration efforts via more effective management of technology and information presently in place, or through liaison with high-technology independents. Offsetting oil import costs and rising oil-finding costs, in fact, may require it. Four critical elements are recognized through which this can be achieved:

1. There is a wide variance between available technology and how effectively it is managed and applied to exploration problems in all ranks of companies, which is related to qualitative, functional, and prejudicial attributes in management and technical staffs. These conditions can be offset by technical upgrading, increasing mutual access, autonomy, and entrepreneurship in exploration departments, or through contractual assistance.

2. Present surpluses of technology and cumulative information can afford great tactical leverage to small, low-overhead companies or national agencies, in developing, packaging, and marketing complex exploration programs and attracting capital investors, a procedure well-known and practiced by most independent oil companies.
3. A more equitable balance between application of exploration technology from industrialized countries and exploration incentives among developing countries could accelerate activity in many sectors, mutually improving net gains against imported oil.
4. Incorporating into exploration programs those qualities which can enhance experience, opportunity, success, and moderate risk through:

- Seeking operator status in exploration/drilling programs.
- Strategic program mixes, including in-house wildcat and development plays and industry joint-venture prospects or plays.
- Improved quality and scope of exploration programs so as to achieve the widest tolerance of geological and economic conditions.

Introduction

Geological conferences most often stress solutions to technical and scientific problems. This paper, however, departs from this general theme to discuss some major factors impacting upon science and technology, at the same time formulating elements of a strategy by which national agencies within oil-importing developing countries (OIDCs), or corporations within industrialized countries, might more effectively manage petroleum exploration technology and information presently in place. Some problems are posed for which solutions may be difficult, as the most troublesome elements impacting upon our technology lie within the power and organizational structure of both governments and corporations. For example, how well does technology serve us when corporations find it necessary to cannibalize each other in their quest for a cheaper barrel of oil? We find an inequal global distribution of technology and information; not enough in OIDCs, and more than enough in industrialized countries. Explorers find decreasing economic incentives abroad due to large government takes among developing countries. Across the board, technology and information seem to be outstripping the industry's ability to apply it intelligently and/or economically to our common goal of finding new oil and gas reserves.

In view of the some 670 billion barrels of oil and 2,911 trillion cubic feet of gas which have been developed within the last 100 years, the industry may congratulate itself on a job well done. But finding the nearly equal amount of undiscovered, conventional reserves estimated to exist in competition with the OPEC countries who control 65% (or 436.5 billion barrels) of the world's known reserves, and in the face of high finding costs and price disincentives, will be much more difficult. It should be recognized that the industry is now on the threshold of a new era, wherein the indifference, the extravagance, and the "good-old-boy" mentality must be replaced with more intelligent management and effective modes of operation and oil-finding, and a spirit of international cooperation. If anything good has come of the energy depression, it is that it has hastened this long-needed process.

The energy outlook for OIDCs, with the possible exception of China, is presently poor, by virtue of the present world economy. Most OIDCs are faced with oil importing costs of $30/bbl, which may displace up to one-half or more of their foreign-exchange earnings from exports. High interest rates on foreign debts stifle economic growth and, so long as oil markets remain soft, increased financial aid to OIDCs from OPEC countries or funds from international banks seems unlikely. There has, however, been talk of a plan by the International Finance Corp. to share exploration risks and the financing of development projects with its affiliate, the World Bank, as coincentives to both developing countries and independent oil companies. Within the last few years, some $3.5 billion per year have reportedly been spent in developing countries outside of OPEC and Mexico, expenditures borne primarily by the international majors, European and Japanese corporations, Canadian and U.S. independents, and the OIDCs themselves.

Aside from the present economy, some key geo-political events also help explain the present energy situation. The industry has always been vulnerable to the influence of international politics. It has come full circle from the breakup of the colonial empires and American dominance of the oil industry to balkanization of the world and the emergence of OPEC, which ushered in a shift of energy power from the industrialized to developing countries. It seems ironic, therefore, that America now shares a problem with the OIDCs in its needs for oil and gas reserves beyond the immediate surplus. But, as Frank Ikard, a former President of the American Petroleum Institute, noted: "We are an industry of problem-solvers; we have a considerable accumulation of brain power and expertise in problem solving and, no matter who caused the problem, we will all suffer if it is not solved."

How, then, in the current world economy and soft oil and gas market, can those OIDCs and industrialized countries with petroleum potential more effectively cope with today's energy problems? Four critical elements are recognized through which this can be achieved:

1. More effective management of petroleum technology—or taking the slack out of the business.
2. Increased access to the present surplus of technology and abundance of information in industrialized countries by those in need, and their willingness to use it to improve and render themselves more competitive.
3. An improved spirit of international industry cooperation through exchange of technology and information from industrialized countries, and equitable contractual exploration incentives from OIDCs.
4. Improved quality and expanded scope of our exploration programs.

Management of Exploration Technology

Problems in petroleum exploration management may be divided into two categories:

1. Mismanagement or under-utilization of highly-developed in-house technology within oil corporations in general, including majors, independents, state-owned companies, etc.
2. Failure of certain corporations to acquire and apply state-of-the-art technology to the solutions of their problems. This could include many large oil companies or independents, OIDCs, governments, and large investors, such as banks and brokerage houses.

It really doesn't matter who you are. Organizations are composed of people, and all suffer various degrees of the same organizational and management problems. To cite some examples:

There existed a yet-undiscovered ten-billion-barrel oil field. Major Company A was the first to recognize its seismic manifestations. Major Companies B and C shortly afterward also interpreted the huge structure and invited Company A to participate in their bid for leases covering the seismic prospect. Company A had, however, drilled a number of unsuccessful wells in the area on lesser structures, and declined. Companies B and C won the lease, drilled, and discovered a ten billion barrel oil field. Later, Company B, perhaps sensing a weakness in Company A, took it over. Why did Company A stop short of participating in the huge discovery? Was it a lack of cash, lack of a well-balanced exploration program, or lack of a sense of strategy and consequent vulnerability to takeover? Was it, then, simply poor management?

Oil companies are increasingly being liquidated or merged into larger companies. While it may be difficult to defend against unwelcome takeovers, the target company is often said to lack adaptive, innovative, and introspective leadership. Takeovers in industrialized countries such as the United States are detrimental to the national interests. Net reserve gains to the country are zero; they monopolize credit lines for those who need or prefer to explore for oil and gas, such as independents who are responsible for significant domestic exploration and development. Mergers halt immediate incentives in the search for oil and gas for the acquiring company and terminally for the target company. Is the emerging monolith rendered more or less economically capable of maintaining its internal reserve requirements in the long run? Certainly a large chunk of the competition is eliminated—competition which really finds oil. One may have the finest exploration machine on earth, but if top management loses the company, what good is it?

An OIDC depending upon international contractors to develop one of its remote oil and gas provinces has what appears to be a major oil strike. A difficult and costly pipeline is laid into the area from the coast in time to see the play fizzle out. Did the OIDC depend too heavily upon its contractors? Was the large pipeline a miscalculation? Could the OIDC, with its own internal technology, breathe new life into the play? Perhaps it is a lesson to be learned in assessing the petroleum potential of remote landlocked basins.

Some U.S. banks have recently failed, or otherwise made bad energy loans, due to unrealistic estimates of industry activity. Predicting market trends is one thing; but why are poor technical assessments of petroleum reserves and oil and gas activity made? Is it, for example, in how bankers perceive the business, looking at track records and balance sheets on paper, measured quantities, but, on occasions, missing qualitative nuances of meaning or assessment of geological risk? Independent Company X, which is coasting on past exploration successes and now investing in real estate or some other unfamiliar field, might look good to the banks, but closer inspection of its plant, programs and people tells you it could be headed for trouble in its exploration programs.

One factor in the decline of some American technical corporations is the tendency only to be dominated by nontechnical management: lawyers, financiers, and accountants, who lack the professional sense of good technical strategy and management, especially among those who tend to be egocentric or unilateral in their decision-making. Similarly, when exploration departments are dominated by other professional disciplines or departments, as is sometimes experienced in the industry, an inferior product results.

Wide variances are thus observed between the technology which is available and the effectiveness with which it is managed or applied to exploration problems. In general, these conditions can be offset by technical upgrading, increasing mutual access, autonomy, and entrepreneurship in exploration and other departments. In this connection, there are four keys to management of technology:

- Maintain simplicity and adaptability in organizational structure, with high vertical and lateral communication and access.
- Among executives, divisions, managers, and staff, instill innovative, independent thinking, decision-making, and risk management.
- Recognize the power of human resources and attain maximum productivity from people through their management as people and not machines.
- Any corporation, government, or institution, either in the business of investing in or lending substantial sums of money to the petroleum industry, should incorporate into its organization competent technical staff, sufficient to intelligently direct its petroleum affairs. Even where expert outside consultation is used, effective in-house assimilation, adaptation, and use of this service and data source is needed.

Departmental autonomy and priorities are crucial. If exploration and production are the basis for the business, then a clear priority is established for an exploration department, for this department initiates the chain of events from inception of prospect that leads to its definition, land acquisition, drilling, and development stages. Because of the lead time, or turnaround time, needed from prospect to production to cash flow, advanced planning and implementation of exploration programs is essential. The exploration department, therefore, has primacy, and should be autonomous, managed by hard-core explorationists who answer only to (or are included in) top executive management, at the same time maintaining good rapport and communication with other departments. Among better-run exploration companies, including those from developing countries, exploration is the key to success.

Certain attributes in exploration managers need to be recognized in order to optimize effective management of exploration programs. For example:

- Earth scientists are either generalists or specialists. A good explorationist is a generalist, a carbonate petrologist or well-site geologist, a specialist. Exploration managers are ideally generalists, geologists or geophysicists who may have incorporated several specialties or many years of exploration experience into his or her career. They are thus

well-suited to strategic planning, developing an overview, or unifying interfacing disciplines.
- It is important to determine how the explorationists think—pragmatically or in abstract. Do they think primarily in practical or theoretical terms? There are critical differences in how they perceive and solve problems, and managers should be aware of this in considering an individual's observations and job assignments. Applied thinkers might excel in solving operational problems whereas the abstract thinker might excel in understanding and conceiving abstract geologic ideas and concepts and recognizing fundamental principles which tie things or events into a structural whole.
- Geologists who are managers should be geologists first, and "Harvard Business School graduates" second. A good manager is a state-of-the-art, hands-on manager who is a problem solver with highly-tuned technical insights and feeling for his exploration staff and programs.
- Creative oil finders are in the minority. They should be assigned as much geotechnical support personnel as they can manage. This not only extends their talents but is very cost-effective.
- Explorationists should have strong secondary skills in geophysics and petroleum economics, for these disciplines are increasingly interdependent with geology.
- Technical advisory committees, composed of top consultants and/or academicians in the field, or regular in-house brainstorming sessions, further amplify brain power.
- While it is preferred to operate with relatively lean geological staffs, it seems that fewer problems are encountered with huge capital expenditures for drilling projects than for staff overhead. Geologists generate ideas resulting in prospects and reserves. We purport to be in the business of oil finding, but often we see a disproportionate number of people assigned to the task of generating the "raw material" which an exploratory company runs on and, when the economy is down, earth scientists are the most vulnerable to economic cutbacks, a time in which advances in exploration can be most easily attained.
- Within many American petroleum corporations, geologists should be more highly regarded, valued, and upgraded from the enigmatic corporate subculture to which they seem to be relegated. There is an overriding tendency for many geologists to "escape" from the "think tank" into management before they have attained their full potential as technical oil finders. This is because they perceive management to be a more satisfactory means of improvement, advancement, and personal recognition. This factor alone leads to high rates of job turnover, especially among American companies.

Strategic Application of Exploration Technology and Information

Petroleum technology and information in industrialized countries can be used most advantageously by small, low-overhead companies or national oil companies in developing, packaging, and marketing complex exploration programs and attracting capital investors, a well-known procedure practiced by most independent oil companies.

While the industry is capital-intensive, it is also information-intensive, requiring the continual acquisition of new geological and geophysical information and the re-evaluation and reprocessing of older data. Scientific information proliferates exponentially in our companies, state and national agencies, universities, professional societies, and publications, doubling every 5.5 years.

The industry has known and practiced for a long time what John Naisbitt points out in his recent best-seller, *Megatrends,* that "information is an economic reality, the more and better information we have, the greater our chances of success will be. One of the most strategic resources of the future will be information because, with information, access to the economic system will be greatly facilitated." Information is thus viewed as "a new source of power in the hands of many, as opposed to money in the hands of a few." Correlated with the exploration business, this means what the independent oil company already knows—that high technology and information can render small cadres of explorationists technically competitive with much larger, less adaptive corporate entities, simultaneously facilitating their access to capital investors.

Growth within OIDCs must, therefore, come with technical upgrading, increased participation, and competition within the international oil and gas arena, increasing their access to information, and following the examples of great exploration and production companies which have emerged from developing countries, such as Pemex, Petrobras, Sonatrach, Petromin, Ecopetrol, Enap, and Etap.

The main problem facing international oil companies is the rapidly diminishing area in which they can most effectively explore for large oil deposits. High government takes in the old producing regions of the world tend to be emulated by other countries, to the point that countries without large reserves from existing production find it difficult to raise the capital necessary to establish new production. Falling oil prices and glutted world crude markets have forced many U.S. independents to abandon the international scene due to cash flow problems. In gingerly transcending the "black hole" of government spending around the world, developing countries somehow must use less of their foreign exchange to service debt, and more to foster economic growth. A lowering of the enormous U.S. deficit would reduce interest rates; increased international trade is needed. Stimulation of international oil and gas exploration among the OIDCs would help achieve net offsets to imported crude but, in many cases, either the technical and operating abilities of OIDCs need to be improved or better incentives must be offered to those willing to risk exploration capital.

If an OIDC such as China, for example, with the majors knocking at its door, is willing to furnish technology and to risk capital dollars to share in finding new oil reserves, this will facilitate increased production. While much potential is seen in China's offshore, and in a few of its less developed interior basins, such as the Tarim, Tsaidam, and Dzungaria, fortunes can change quickly. Too much dependence upon foreign technology, while beneficial to a point, could erode China's—or any OIDC's—internal incentives, and the benefits gained from deploying the technology itself, an essential factor in developing an effective, competitive industry.

On the other hand, if an OIDC is not China, and no one is knocking at the door, then some other kind of initiatives needs to be taken. If an OIDC with limited capital resources requires an exploration vehicle, its needs could be adequately met with a small, low-overhead organization modeled after an independent oil and gas company initially infused, if need be by experts from the outside. The primary purpose of the exploration company would be to evaluate the OIDC's petroleum potential, develop prospects to the fullest extent possible and, if merited, open the prospects for bidding on the international market. If the prospect had demonstrable potential and the terms were reasonable, heavy capital investors could be attracted to complete the exploration and/or drilling.

Those OIDCs wanting to develop their own oil and gas potential thus have two choices: either to operate alone with the aid of internal funds, or international loans and grants, in cooperative efforts together with adjoining countries, or to operate through contractual arrangements with international exploration companies.

Exploration Programs

Explorers must move aggressively to incorporate elements in exploration programs which can:

- Increase experience and expertise.
- Increase opportunity, but moderate risk.
- Increase success and lower finding costs.

If an exploration company is state-of-the-art, it should seek operatorship of exploration drilling programs. Being an operator means that a company is a leader instead of a follower; the sloppy, costly operating practices of others can be eliminated. Being an operator, in spite of its headaches, means taking the initiative in an area and, in performing, learning more from the experience than the joint-venture participants. The operator develops a better feel for the geology and technology of the area essential to developing a competitive edge. If an operator achieves profits for its investors or partners, and respects and keeps them informed, this will pay great dividends in terms of its reputation and its ability to raise joint-venture capital.

If a company operates in an area where a variety of oil and gas plays may be developed, then the program mix can be effective in maximizing exposure and opportunity and spreading capital risk. The mix may consist of varying percentages of the following play types:

- Those prospects or plays generated in-house, including:
  1. Shallow, low-risk, low-yield, developmental (so-called "bread & butter") type plays, which may tap extensive oil- or gas-saturated blanket sandstones and carbonates which, in a small company, can pay the overhead plus a modest profit.
  2. Deep or shallow high-risk, high-yield wildcat prospects which afford the opportunity for high profits and exponential growth.

Both types of plays may be drilled by the operator alone or variously promoted, often in joint ventures.

- Joint-venture prospects generated by competitors offer a wide variety of international and domestic opportunities. Nine out of ten are not worth participating in, and must be carefully screened relative to terms, economics, geology, operator or promoter, track records, and their intent to participate as a measure of confidence in their own prospects.

With respect to the so-called shallow or small oil and gas accumulations, the theme of this conference, these do not necessarily constitute easy and inexpensive exploration and development objectives. In the pursuit of oil and gas reserves, foreign or domestic, ideal targets may include 50–100 million barrel oil or gas-equivalent fields, but explorers will end up taking what is actually discovered and can be profitably developed. In international situations, assuming seismically-oriented exploration and moderate government takes, marginal reserves may easily approximate eight to ten million barrels. Small hydrocarbons might, therefore, be defined as the smallest fields which can be economically found and produced. Small conventional oil and gas deposits fall into two general categories:

- Shallow, stratigraphic, low-reserve gas- or oil-saturated, vertically and laterally extensive sandstones, limestones, and shales occurring in petroliferous sedimentary basins such as the Gulf Coast, Mid-Continent, and Appalachian areas of the United States. Economics of these deposits are often marginal and usually would not warrant developing in international situations by outsiders.
- Widely-disseminated structural or stratigraphic deposits whose exploration cost/logistics in sparsely-explored basins would roughly equal that of exploring for deep occurrences, save for the differences in drilling costs. In highly-drilled areas, the cost of finding small hydrocarbons may be less by virtue of more abundant subsurface control, but the frequency of occurrence of economically viable prospects will also be less.
- In sparsely explored areas it therefore would not be cost-effective to confine the search to small hydrocarbons or shallow horizons, but rather, the exploration effort should include all areas within limits of geophysical, surface/subsurface geological investigation and economic drilling tolerances.

Exploration programs should have a broad regional base so as to investigate broad geological and economic conditions, serving both as a basis for generating local subsurface or seismic prospects, and evaluating the activity of concessionaires, farmees, and competitors in a game of offensive and defensive strategy. The program should incorporate key methodologies, including basin classification and analysis; tectomorphic analysis using Landsat and/or radar imagery; source rock, thermal maturity, and hydrodynamic-susceptibility studies; subsurface mapping; aeromagnetic and gravity surveys; and Landsat reconnaissance studies required to more effectively orient seismic surveys.

Geological insights and ideas, and luck, are the most powerful tools in exploration.

Conclusions

Several types of internal and external problems are perceived to impact upon petroleum exploration technology. Although many problems cited may be peculiar to industrialized countries, they are all human problems based upon principles which can affect organizations and managements anywhere. Some solutions to these problems are:

- A need to be introspective about our organizations, to see that they are structured and professionally managed in a way which will optimize development and utilization of in-house technology. This may be essentially achieved by maintaining a state-of-the-art exploration apparatus and encouraging autonomy and entrepreneurship among our exploration divisions and managers under a system of tight corporate control.
- If capital and material resources are limited, compensating strategically and intellectually through better technology and information, ideas, and prospects will improve access to capital.
- Exploration programs should be of sufficient scope to exploit the widest range of geological and economic conditions, incorporating within them those qualities which will enhance experience, opportunity, success, and moderate risk.
- A better atmosphere of international understanding and cooperation can be created through a common goal of finding new oil and gas reserves, striving to achieve a better balance of exploration technology between industrialized countries and the OIDCs in exchange for equitable exploration incentives where outside exploration capital is needed.

Above all, OIDCs need to become more self-sufficient and competitive in the international market place for oil-company capital, promoting themselves, and accelerating their domestic oil and gas exploration and production.

Section VI

Legal and Institutional Considerations

46

Shallow Oil and Gas Development: Proposals for its Encouragement

Maxwell Bruce

The Problem

The development of shallow petroleum deposits must involve governments offering terms competitive with those offered by other countries and which enable oil companies to earn a profit commensurate with their risks and successes. A government can enhance a venture by reducing the risks in its control. However, and notwithstanding the pressures of higher-priced oil, the search for petroleum in oil-importing developing countries has been stagnant for the last decade (Palmer, 1983). It has been said in relation to such development, "(t)he key to increased exploration in third world countries does not of course lie with any one mechanism or type of institution. In the end, all alternatives will have to play their part—foreign oil companies, national oil companies, development assistance funds, new mechanisms for financing—none is exclusive of any other, and the extent of the effort required is large enough for there to be more than enough room for all" (Parra, 1982, p. 191).

Geology

Prospects for increased exploration in the lightly explored basins, mainly in developing countries, would appear to be good. However, in currently producing basins, the largest and most attractive structures were explored first and there is a strong incentive to re-evaluate older prospects. As the industry turns to second-tier prospects, prospective areas in developed countries have the advantage that the essential infrastructure is in place and that they are close to the markets. These circumstances explain much of the disparity in drilling intensity between developed and developing countries.

Developments based on shallow reservoirs, however, have been shown in the United States to have a high exploratory success rate. In the Appalachian Basin, for example, there was an increase in the number of wells drilled from 3,000 in 1971 to more than 10,000 in 1981. This represents more than 15 percent of all wells drilled in the United States. The reasons for the increase in drilling are the 89% success ratio and favorable price incentives (Brown, 1982, p. 48).

Data

Fundamental to the assessment of geological prospects is the accessibility of data. Before conducting its own examination of a prospective area, an oil company will want to review all available and relevant previously obtained data. For depths below about 2,500 ft, seismic prospecting, after geology, is the best method for gaining more knowledge of new basins and for reducing risks related to subsequent wildcat drilling. For depths above 2,500 ft seismic prospecting does not seem to work as well.

Only by the drilling of wildcat wells is it possible to evaluate the actual hydrocarbon potential of a zone. In principle, copies of the results of seismic work and of drilling carried out should be lodged with the appropriate agency of government and, within certain limits, publicly available.

Governments should insist upon the removal, within reasonable limits or after a reasonable period of time, of restrictions upon the dissemination of information. Copies of all data, including seismic work and drill logs, should be mandatorily lodged with host governments or their appropriate agencies, together with copies of all interpretations generated by operators. One important reason for retaining old data is that advances in technology, changing economics, or new interpretation methods may mean that areas previously abandoned may become attractive in the future. Only if a government retains all data is it in a position to generate the maximum amount of interest among oil companies.

There is a practice of informal trading or selling of data between oil companies which is apparently not prevented by nondisclosure provisions of legislation or contracts. Oil companies therefore tend to resist efforts directed towards freeing data, particularly towards freeing it quickly. They generally argue that it should remain confidential for extended periods, perhaps of the order of five to ten years, which of course makes it an inaccessible commodity tradeable for other comparable information.

Drilling Costs

One of the major advantages of the development of shallow occurrences of petroleum is the associated low cost of drilling. The cost of drilling does not rise proportionately to the depth reached but at a much higher rate. In the United States, using standard drilling equipment, drilling costs for shallow wells vary between $10 and $20 per foot, and, if smaller rigs are used, costs are as low as $8 to $10 per foot. This compares with average per foot costs of onshore drilling between 10,000 and 12,500 feet of $86, between 15,000 and 17,500 feet of $226 and below 20,000 feet of $379 (Belden, 1981). In addition, fewer appraisal wells are likely to be required for shallow prospects than in the case of a deeper field. These factors, taken with the high success rate already mentioned, should tend to encourage the exploitation of shallow petroleum resources.

Political Risk

A stable political and economic system in a host country is important to a potential entrant into a petroleum exploration and development arrangement therein. Oil companies may assess political risk economically, and impose on that account a financial ceiling upon their participation in a particular country. Another type of political risk is uncertainty caused by bureaucratic indecision and delay, although this is not confined to developing countries, nor is it typical of all of them. Such uncertainty is eventually expressed in terms of extra costs. The greatest concerns about political risk are expressed by the smaller, primarily North American, independent companies who have less scope to diversify across countries and less experience in dealing with governments in the developing world (Palmer, 1983).

Energy Policy

The purpose of developing shallow petroleum resources is to supplement or replace existing energy sources where it makes economic sense to do so. A country's energy policy and planning should address substitution of shallow petroleum resources at the local community level for existing sources of whatever sort, and not only of imported oil. The availability of locally produced oil or gas at an attractive price could foster the establishment of new rural industries, incuding petrochemical industries.

Fiscal Systems

The financial base upon which it rests is of fundamental importance to any arrangement for the exploitation of oil and gas, whatever the size of the resource. Some fiscal and taxation systems are extremely complex. From the point of view of shallow petroleum development, for local use, the most attractive are the simplest and the most flexible. Payments to the investor in the early years of production should be accelerated, thus reducing the period during which the oil company is at risk, and limiting the government profit share in the early years but increasing that share progressively thereafter as field profitability increases.

A successful project must generate sufficient funds for the oil company to earn an overall profit after taking account of losses from unsuccessful projects wherever they may occur and thus be able to sustain itself in business. Every oil company has many investment opportunities that compete for its capital, technology, and personnel. The company compares the expected economic benefits and the perceived risks for all such competing alternative investment opportunities, and ultimately chooses those it believes to be most attractive.

Exploration and development of petroleum is undoubtedly encouraged by reciprocal tax treaties, and their absence amounts to a tax bias against overseas exploration by U.S. oil companies. The ability to write off exploration and development costs is of course dependent upon a particular oil company's possession of offsetting income in the host country or in a tax-treaty country that does not have a "ring fence" system. Majors are more likely than independents to have such income, giving them an advantage over independents, who may have to wait many years for tax relief. As a consequence, developing countries are perceived as having a higher "income tax cost" than, for example, OECD countries, which reinforces the attraction to oil companies generally of operations in the OECD area.

Financing

The major oil companies raise capital through internally generated funds from operations, or by the sale of various debt instruments or equity securities. The capital required by independents, on the other hand, often comes from individual investors through the mechanism of public and private drilling funds and the sale of limited-partnership units. The raising of capital from local sources, if such exist, should be particularly appropriate in relation to the exploitation of shallow petroleum resources intended for community use. It has been pointed out that even the most generous fiscal terms used in the petroleum industry today will leave a sizeable gap between foreign company and host country economics. This gap may justify host country investment where private investment would not make sense.

Once a discovery has been made, and the existence of recoverable petroleum established, there ought not to be difficulty in obtaining financing, for a bankable security is at hand. It has been suggested that by borrowing some of the creative financing techniques used by the independent petroleum industry in the United States, some countries may be able to exploit their petroleum resources with a minimum of outside help. The analysis and implementation of these techniques under actual conditions will not be easy, but they do appear to offer viable alternatives (Batt, 1983). Foreign oil companies do not especially like to become involved in deals in which lending facilities are made available by public agencies. They look upon such involvement as eroding their own bargaining position with the host government, and would generally prefer to do their own financing.

Legislation

In order that any natural resource be exploited there must exist in the country concerned an adequate legal infrastructure so that a workable agreement can be reached between the interested parties. The laws can be broad or detailed. If the former, a burden is cast upon the government officials to fill in the details, and an undue strain may be imposed where such officials are inadequate in number or experience, and uncertainty may be created. On the other hand, detailed legislation may lack the elegance of the more succinct. In either case, it is unrealistic to expect a solution for every eventuality, some of which must perforce be unseen, so a degree of ministerial discretion is desirable.

For shallow resources of oil and gas, where speed of development and of recoupment of investment is all important, simplicity of approach is especially desirable. Not only should the legal norms be plain and straightforward, which is not always easy to achieve, but the administration of the legislation relevant to such resources should be expeditious and uncomplicated.

A draftsman of relevant legal provisions needs to define the concept of shallow. The depth criterion may vary from place to place and all scientists might not agree on a precise figure. There may be general consensus on a maximum of 5,000 ft onshore and a water depth of 50 ft offshore, beyond which an inexpensive jack-up rig could not be used. One might add to the depth criterion that of the budgetary cost for the project.

There appears to be no reason why title should not be broken down between the shallow and the deep subsurface, and a shallow subsurface title, extending to the 5,000-ft depth horizon granted to one explorationist and the rights below such horizon to another. This would allow simultaneous exploitation of both shallow and deep petroleum reservoirs in overlapping locations and for different purposes by different operators. It should impart a flexibility to the industry which under present legal concepts cannot engage in such multiple activities, and perhaps would establish a new dimension to the way petroleum resources can be exploited.

Permits

Exploration and development permits, upon simple and straightforward application, should be speedily available. Delays of many months or even a year, not unknown under contemporary practice, are incompatible with the concept of shallow petroleum development as envisaged by this paper. A "fast track" framework for shallow petroleum development is suggested. An application that a shallow petroleum development be designated as such would be the subject of a hearing. If the application were successful, an order designating the project as a shallow petroleum development would be issued, and would require officials to abridge the normal waiting periods and issue licences and permits expeditiously. An appendix to an extended version of this paper contains a brief enactment for insertion in national petroleum legislation as a basis for such a "fast track" system.

Such a system could be universally adopted. It might prove to be the most important single step towards the removal of legal constraints to shallow oil and gas development. Few obstacles are more intractable or more costly than delay.

Contracts

A wide variety of contractual forms covers the spectrum from the original oil concession to state ownership as found today. These include more sophisticated types of concession agreement as well as joint ventures and production-sharing agreements, and although the trend is towards increasing state ownership and control, foreign oil companies continue in most cases to be involved. It may be misleading to associate a particular type of contract or legislative regime with a certain level of financial return and control, for the same result may be obtained by a state by varying the different elements of the contract and/or the legislative regime. Over-rigid agreements should be avoided. Conditions geared to profitability and a flexible tax or royalty system which adjusts itself automatically to profitability are not only in the interests of all parties, but will also help to ensure the stability of investments.

In the development of shallow sources of oil and gas, the principal benefit consists of the product itself, the value of which can be measured by what would have been spent to import it. The costs consist largely of payments to foreigners for capital, technology, and management, including in some cases a management or service fee. Foreign oil companies will in all cases regard as essential the freedom to repatriate funds, unless a form of barter can be arranged.

Where an independent oil company finances its own exploration and/or development activities, it is important that it has the right to assign to its financing partners a portion of its interest in an agreement in return for the share of capital furnished by them. Whilst there is invariably a need for a mechanism to resolve disputes arising out of the interpretation of a petroleum agreement, in the case of shallow petroleum developments, it is important that such mechanism be rapid and simple.

Gas

Natural gas reserves may well be economic if a domestic market can be found, but private oil companies have shown little interest in exploring for gas fields where the output would be sold within a developing country. Natural gas, particularly associated natural gas, is one of the cheapest energy options in developing countries, especially in those with a large fuel-oil market. It is almost ideal as a fuel, having a high calorific value and being clean and efficient in use. Its difficulties lie in transportation and distribution. The relatively low level of natural gas use in the developing countries is the result of historical developments and of the limited interest of foreign oil companies in developing a source of energy that is not easily exportable.

A major problem in the development of natural gas lies in the absence of a worldwide market. Thus, whilst the commerciality of an oil field can be established in relation to expected future oil prices, the commerciality of a particular gas field cannot be determined without first identifying the specific market for that field's production and the price appropriate thereto. In the case of shallow gas resources for local consumption, and to replace imported fuels, the solution may lie in a price guaranteed by the government. Such price could most readily and logically be related to the cost of alternative fuels and be designed solely to serve the social purposes of the development.

Conclusions

Simply stated, the purpose of exploiting shallow resources of oil and gas is to make available for local and community use a cheap and handy energy source. It is not to enrich the national treasury. Nor is it to attract oil companies in search of additional supplies for the world market although it can, in a minor way, lessen the strain in the world market. Any draft legislation or agreement should, of course, be cast in the framework of the local legal system and of local public concerns. It is only possible in this paper to outline in broad terms those constraints that appear to be important. What can be achieved will ultimately depend almost exclusively upon fundamental nonlegal questions, those of economics and politics.

In summary, these are the main points, some of which are elaborated in the extended version of this paper:

1. *Access to exploration data.* Although not unique to shallow petroleum developments, it is of the greatest importance to them. Copies of all exploration data (including, where possible, interpretation) generated by whatever source and of every kind should be required to be filed with and retained by the relevant government department or authority and, within reasonable limits, to be freely available.
2. *Fiscal and tax systems.* No foreign oil company is likely to be prepared to enter into any arrangement for the development of shallow petroleum resources unless the potential for profit is at least as high, and possibly higher, than it would be in the case of any alter-

native use of its resources. The financial aspect of any arrangements will, of course, depend upon the relevant terms of the applicable legislation, including tax legislation. There should be no royalties or taxes payable until all outlays, including operating costs, have been recovered by the oil company. The simplest fiscal regime thereafter, involving perhaps income tax alone (with a holiday for an initial period) would be the ideal.
3. *"Shallow" defined.* A suggested definition is based upon depth coupled with the economic size of the project.
4. *Area description.* Local survey systems, where available, rather than references to latitude and longitude, are considered to be more appropriate for the description of shallow petroleum development areas.
5. *Plural uses of subsurface.* The same geographical area might be potentially exploitable both on account of its shallow and its deep reservoirs. In order to facilitate simultaneous (or successive) exploitation of reservoirs of varying depths by different operators, the novel idea is advanced of separate legal titles to the shallow (less than 5,000 ft in depth) and deep (below 5,000 ft in depth) subsurface.
6. *Bureaucratic delays.* Delays imposed by legislation and regulation are apt to be the most costly and frustrating aspects of shallow petroleum development. A "fast track" approach to the granting of permits and approvals, based upon the designation of a petroleum development as shallow, is proposed.
7. *Form of agreement.* The form of petroleum development agreement has evolved with the industry over the years. None is specifically concerned with exploitation of shallow occurrences. Of the usual forms, those of service contract and of direct management appear the most adaptable. Which was chosen would depend, among other things, upon whether the foreign oil company assumed the financial risks or whether the community did so.
8. *Right to assign and dispute settlement.* The assignability of an interest in the rights under a contract under which a foreign oil company puts up money and is exposed to risk may, in certain circumstances, be essential to the company's ability to finance. In agreements involving the development of shallow petroleum occurrences, quick and simple dispute settlement provisions are important.
9. *Price for natural gas.* In order to encourage the development of shallow natural gas by a foreign oil company, a sale price for the gas should be agreed, possibly related to the cost of alternative fuel, and guaranteed by the government.

Acknowledgment

Financial support for this research was provided by the Canadian International Development Agency. The views expressed are those of the author alone.

References

Batt, Robert, 1983, Energy in the eighties: New York, United Nations, pp. 45-51.
Belden, H. S., IV., 1981, An independent's assessment of small, shallow oil and gas development in the United States: paper delivered at UNITAR Conference on Small Energy Resources, Los Angeles, 9-18 September 1981.
Brown, Porter J., 1982, The future potential of the Appalachian Basin, with special reference to the Eastern Thrust Belt: Exploration and Economics of the Petroleum Industry, New York, Matthew Bender, Vol. 20, pp. 47-56.
Palmer, Keith, 1983, Private sector petroleum exploration in developing countries: Finance and Development, May, pp. 36-38.
Parra, Francisco, R., 1982, Petroleum exploration strategies in developing countries: London, Graham and Trotman, pp. 177-191.

47

A Producer's Perspective on Selected Issues in the Establishment of a Legal Regime to Encourage Marginal Petroleum and Natural Gas Development

Sheila S. Hollis

Introduction

The purpose of this article is to set forth in a brief form an overview of issues that confront a nation desiring to encourage development of smaller reserves of petroleum and natural gas. Many countries may have legal systems which contain mineral codes or precedential contracts which do not serve to remove barriers to expeditious development. Of course, there is no single right or wrong way for a nation to set up a framework for such development. However, it is important to be cognizant of the needs of the producer as well as the country and to recognize that the state of local technical expertise may require development arrangements of a complex nature.

This article sets out the basic formats for world petroleum development and production arrangements. There are huge variations on these simple formats and the creativity of both the host countries and producers, as well as input from international institutions, including the United Nations and the World Bank, have led to refinements of these arrangements.

But, for the country considering whether to embark upon a reserve development program designed to elicit production for local use, several factors must be weighed. The experiences of nations in setting up legal and policy regimes to encourage large reserve development provide useful tableaux. To that end, this article discusses each major format utilized in the world today for petroleum development—the concession, the production-sharing arrangement, and the service contract. In real life, arrangements at different stages of development may incorporate the best of all three basic types. Salient features of the three formats are given in Table 1.

Concession

The petroleum concession represents the oldest legal form under which contractors (concessionaires), both public and private, operate in host countries. Various types of concession agreements are still in use in a number of countries, including several that are industrialized. Characteristically, this type of arrangement puts control of the operations under management of the producer through the grant of a concession by the host government. The producer receives the exclusive right to explore and exploit a designated area for a specified term upon payment of bonuses and royalties, either in cash or in kind. The host country may obtain revenues through bonuses, royalties, surface taxes, compensation works (i.e, schools, public works, etc.), scholarships for studying abroad, and income and other taxes. In some cases the agreement may provide for participation by the host government in profits or specific activities covered by the agreements.

Petroleum concession agreements contain clauses that vary from country to country as well as from transaction to transaction. Further, the terms of various concession agreements have changed dramatically over the last several years and have been influenced significantly by the other types of contractual arrangements. The following are the key provisions found in concession arrangements:

1. Host government grants producer (concessionaire) exclusive right to explore and exploit designated areas for specified terms. Producer provides all capital funding, technical expertise, and other items needed to carry on the exploration and exploitation operation throughout duration of Agreement.
2. Total management of exploration and exploitation operations is in producer.

Table 1
World Natural Gas Reserves, Production, and Consumption, January 1983
(billion cubic meters)

Feature	Concession Arrangement	Production-Sharing Arrangement	Service Arrangement
Management	Total management of exploration and exploitation operations in producer/contractor.	Joint management by producer/contractor and host government with day-to-day operational control in producer/contractor.	Host government retains management control with day-to-day operational control in contractor initially (subject to shift on attainment of commercial production).
Revenues	To host government: generally royalty, bonuses, compensation, income and other taxes. To producer/contractor: production.	To host government: a share of production (in value or in kind), and may include royalty, bonuses, income and other taxes. To producer/contractor: his share of production with recovery of costs before any production split.	To host government: production, income and other taxes and may include certain bonuses. To producer/contractor: payment in value or kind (may be keyed to expenditure or production).
Disposition of production	Producer generally retains all with host government having the option to take royalty in kind; usually host government is given the option to satisfy certain domestic consumption needs; the host government may also be given participation rights.	After producer/contractor recovers costs, the remainder of the production is split between the host government and producer/contractor on the basis of profit and/or production.	Production to host government with the right to pay contractor in kind; contractor may possibly be given the right to purchase an additional portion of production.

Production Sharing

A production-sharing arrangement is essentially an agreement between the host government and producer/contractor that provides for joint control of operations. The host government designates certain areas for petroleum exploration and exploitation and the producer/contractor provides technical expertise and other services to conduct joint exploration and exploitation operations in the designated areas for specified terms. The producer provides all capital funding, at least prior to commercial production (exploitation). Thereafter, capital funding may be provided jointly by the producer and host government (whose portion may be funded out of its share of production). The producer first recovers its exploration and development costs from production. The remainder of the production is allocated between the parties.

The following are the key elements of most production-sharing arrangements:

1. Host government provides designated areas and producer provides technical expertise and services to conduct joint exploration and exploitation operations. Producer makes all capital expenditures, at least prior to commercial production. Thereafter, capital funding may be provided jointly.
2. Joint management by host government and contractor.

Service Arrangement

In a service contract (a variation of which is the so-called "risk service contract") the host government enters into an agreement, typically of short duration, under which contractor is to perform certain services specified therein. Payment to contractor for its services may be in cash, or in kind, and may be keyed to contractor's expenditures and possibly to production. Contractor may possibly be given the right to purchase a portion of production. These contracts are used in countries with both established petroleum infrastructures and established reserves.

The following are the key provisions in a typical service contract:

1. Host government enters into an agreement, typically of short duration, under which contractor is to perform certain services specified therein.

2. Host government retains management control throughout duration of contract. Day-to-day operational control initially in contractor may shift to host government upon attaining commercial production levels.
3. Payment to contractor for its services may be in cash, or in kind, and may be keyed to contractor's expenditures and possibly to production. Contractor may possibly be given right to purchase a portion of the production.
4. Risk on contractor may vary in degree on the basis of its financial commitment.

Regardless of the type of petroleum agreement or arrangement ultimately adopted by a host country, there are certain economic and non-economic considerations to be addressed by the host country. This portion of the article sets forth in summary form a discussion of the major considerations requiring policy decisions in developing a comprehensive system of law, regulation, and contract to expedite the development of accessible reserves.

Economic Considerations

Economic considerations are a primary concern of the host country and the producer/contractor. Petroleum laws of various countries have many diverse provisions for allocating costs and revenue between the government and the petroleum company or producer/contractor. Prior to actual production the government has several ways to generate revenue:

- Significant bonuses.
- Surface fees.
- Discovery bonuses.
- Production bonuses.

These payments increase the amount of capital required for petroleum operations, but are generally an insignificant source of government revenue and may not be desirable in smaller or marginal development. Once production has commenced, revenue can be generated by the government in a number of ways, including, use of:

- A royalty.
- An income tax.
- A production-sharing agreement.
- An excess profits tax or a sharing of net revenues.

These mechanisms are frequently used in combination. Care should be exercised to ensure that incentives for oil companies or producers/contractors exist at the inception of the petroleum program and that these incentives remain during production in order to provide maximum benefit to the country. If the government share is disproportionate and makes investment in the country less favorable than other investment opportunities available to the oil companies or producers/contractors, the development of petroleum resources in the country will be stymied. Additionally, caution is necessary in utilizing production sharing or excess profits tax mechanisms to avoid suffocating the exploration and development of relatively small and marginal prospects.

After production commences, payments to the government under any one or more of the preceding methods are generally based on levels of production or net revenue. If payments based on production are involved, there are at least two key factors to consider. First, an accurate system to measure production must be developed to ensure that the government obtains its correct share. Second, payments based on production may discourage the complete development of a field or may discourage the development of a marginal field since payments based on production generally increase marginal costs. However, it should be noted that various methods have been devised to address the marginal field problem and, although payments are still based on production, they can be based on a sliding scale of production, or a sliding scale based on some other factor such as price or location, which will avoid the marginal disincentive.

Payments based on net revenue (profit) have come into frequent use in modern petroleum arrangements. Under these types of arrangements the government will generally receive revenue later than if the payment was based on production. However, from the viewpoint of an oil company or contractor, this system is generally more acceptable than one based solely on production. Nevertheless, as with payments based on production, there are problems associated with collection of revenue based on net revenue or profit. In order to assess the amount due the government, as is the case with taxes, a mechanism to arrive at the amount of net revenue must be devised. Specifically, gross revenue must be measured, and provision for capital costs and operating costs must be made. In addition, a comprehensive system to monitor these costs must be developed for the payment program to function fairly and effectively from the standpoint of both parties.

Most countries use some combination of the production and net revenue-based methods for generating most funds. In any event, it is imperative that the government structure its petroleum legislation and agreements to permit the oil company or producer/contractor to obtain the maximum tax credit in his home country, for this will directly affect the amount of revenue the host government will be able to generate.

Finally, it should be noted that such matters as taxation, including the creditability of various payments, are possible subjects for treaty negotiations between the host country and the home countries of the potential contractors or oil companies. Although handling various tax matters through the use of a treaty provides an additional consideration, it should be recognized that negotiating a bilateral treaty can be a long and time-consuming process.

Noneconomic Considerations

In addition to economic matters, there are important noneconomic considerations that must be addressed by the host country. While some of these matters are already included in the legislation or regulation of certain countries, they generally may not be designed to encourage smaller reserve development. Thus, they are noted here for "fine-tuning" to take account of the special considerations for smaller resource development.

Exploration, Development and Production Periods and Commitments

The rights and obligations of the oil company or contractor to explore for, produce and develop petroleum from the contract area must be clear. During the exploration period the following matters should be addressed:

- Minimum work programs, both in terms of monetary expenditures and work obligations (i.e., drilling or seismic surveying).
- Duration of the exploration period, including provisions for successive relinquishment of the contract area.
- Specific obligations of the contractor to supply all drilling and other geological or geophysical records and information to the government.

During the development and production period, the following matters should be addressed:

- The definition and determination of commerciality, including the possibility of establishing objective criteria to make such determination (i.e., production at a predetermined rate or attainment of a specified rate of return). This should include an analysis of the expenditures necessary to utilize the petroleum or gas in a local market, including the costs of distribution and transportation facilities if necessary.

- The establishment of a set production rate and the delineation and establishment of a field development plan, including provision for the day-to-day operational decisions and approval of the overall operating plan.
- The establishment of a time period for production and development activities.

Training and Technology

In order to achieve the ultimate goal of governing petroleum operations, the petroleum agreements must provide for the training of nationals in all aspects of petroleum operations including the transportation and distribution of the product. Further, the use of domestic goods and services should be encouraged subject to availability and price competitiveness.

Domestic Consumption and Marketing

Many countries currently provide that the national demand for oil and gas must be met from local production. In these instances a provision designating a price and a production allocation for sales to the local market and transfer of the funds so generated must be provided. Generally, oil companies or producers/contractors shall be obligated to satisfy domestic consumption based on the ratio of their production to the total national production. The right to export and sell production remaining and the right to retain sales proceeds abroad must also be addressed.

Natural Gas

The petroleum agreement should provide for handling and disposition of natural gas produced by itself or in association with oil. Included should be discussion of pricing of the natural gas and the necessity for development of a transportation and distribution system.

Legal Provisions

Various legal issues should be addressed in the context of any type of petroleum regime or arrangement. These are concerned primarily with assuring performance and settling disputes. These should include provisions dealing with the following:

1. Ownership of assets.
2. Guarantees of performance.
3. Default and terminations.
4. Governing law.
5. Settlement of disputes.

Summary

Many of the issues relevant to large-scale development of petroleum and natural gas resources apply with equal force to the development of smaller reserves. The analysis must be made by the host country of what format of legal and policy systems it wishes to provide as a framework to encourage such development. The country must consider its desire to control the management and operations of a producer/contractor; the pricing of any reserves that may be developed; the ultimate disposition of these reserves, including an analysis of whether a transportation or distribution system must be constructed; the level of payments which should flow to the producer in view of the reduced reward potential for smaller reserves; and the level of flexibility the country wishes to allow producer/contractors and subcontractors in such matters as taxation, currency exchange, transfer of technology, and use of local personnel, goods, and services. With such a variety of issues, the country must carefully identify its goals and evaluate whether the creation of revenue flow or the provision of local energy from oil and gas to local markets is the keystone of its small oil and gas reserve development. In constructing its legal system and the regulations and contracts that flow therefrom, the country must thoroughly analyze and evaluate not only the economic but the social and political implications of its decision. It is hoped this article has set forth a basic exposition of the legal frameworks available to accomplish the country's desired ends.

48
A Natural Gas Energy System for Rural Communities in Developing Countries

Christopher F. Blazek

Introduction

Shallow small pools of natural gas exist in many sedimentary areas throughout the world. Although economically unattractive for development by large energy companies, these small pockets of gas may be an attractive energy source for rural villages where extensive transmission networks do not exist. For these villages the availability of natural gas would pave the way for an improved standard of living in the areas of cooking, heating, electricity, and water pumping. The objective of this paper is to outline a natural gas energy system from production to utilization for implementation in these rural communities.

This natural gas energy system starts with a portable cable tool drilling rig. This type of rig represents an inexpensive and practical drilling method for remote applications where skilled labor and machining equipment are at a premium. Presumably small wells would be discovered during water well drilling or as a result of large company exploration efforts. After well completion and testing, onsite equipment such as a dehydration unit, regulator, and heater may be required. The gas is then sent by pipeline to the rural community for distribution. For this paper it was arbitrarily assumed that the transmission distance would be 40.2 km and the rural community would consist of roughly 100 homes. At the city gate the transmission pressure, which would be roughly 515 kPa, is reduced for local distribution. A feeder line operating at higher pressures from the transmission line would supply the local gas-fired electrical generator.

Community Parameters

In order to determine system component sizing an estimate of the community energy needs must be made. These energy requirements should preferably be based on peak projected energy needs over the life of the project. In addition, factors such as population increases, per capita growth in energy demand, and increased water pumping requirements should all be factored into sizing calculations. Often such projections are difficult to make because historic energy and water demand patterns are based on different energy types which may be less convenient to use or less plentiful.

For this paper a number of general assumptions were made with regard to a hypothetical community size and energy and water needs. The community size was set at 100 homes with a total population of roughly 500 people. Single-family structures were assumed, spaced roughy 18.3 meters apart, with homes facing each other on two streets. Peak household electrical consumption was assumed to be $1/2$ kW with a growth rate reaching a peak of 1 kW over the life of the project. The total housing stock and population was assumed to double over the life of the project due to factors such as new births and migrations from the surrounding countryside. Total project life was not set, due to its dependence on reservoir life and the availability of other gas deposits in the immediate vicinity.

Residential natural gas energy demand was assumed to come from cooking, hot water, and space heating requirements. Space heating requirements may be negligible in equatorial regions but were included here for completeness. Typical peak cooking demand would occur for a conventional stove when all four of the top burners and oven were on. This peak cooking demand would be roughly 8,440 kJ/h for each burner and 12,660 kJ/h for the oven, or approximately 46,420 kJ/h total. Peak hot water heating demand would occur when the hot water burner comes on. This burner would be typically sized at 31,650 kJ/h. Space heating requirements were assumed to be met by a ventless space heater. This unit would be centrally located to meet as much of the household space heating requirement as possible. Space heating energy requirements are highly dependent on climatic conditions; for this paper it was assumed that peak demand would be roughly 31,650 kJ/h. These numbers represent peak appliance inputs per residence and as such would not increase over time. Therefore, a combined peak natural gas energy demand for each residence would be the equivalent of roughly 109,720 kJ/h.

Additional energy demand will come from local water pumping requirements. Table 1 presents representative water requirements for human, animal, and agricultural use. World Bank values were used to cal-

Table 1
Daily Water Requirement per Stated Category of Use

	Human		Animal			Agriculture
	Drinking Only (m³/person/day)	Drinking Plus Other (m³/person/day)	Large* (m³/head/day)	Moderate** (m³/head/day)	Small*** (m³/head/day)	
Source of Estimate						
IBRD[a]	--	.02[b]	.03	--	--	--
World Bank[c]	.005[d]	.03[b]	.04	.004	.001	55 m³/hectare/day[e]
United Nations[f]	.003[d]	.08[g]	--	--	--	0.8–1.23 m³/capita/year[h]

* Cattle, horses, asses, camels.
** Sheep, pigs, goats.
*** Poultry.
-- Indicates information not available.

Notes:
a. Source: International Bank for Reconstruction and Development: Village Water.
b. Defined as quantity of water for drinking and domestic use.
c. Source: Personal communication Mr. R. Steeds, World Bank. Data assumes tropical climate.
d. Minimum survival.
e. Assuming typical land, vegetable, and grain crops.
f. Sources: Water Development and Management: Proceedings of the U.N. Water Conference. Edited by Asit K. Biswas. Pergamon Press, 1978. Ambroggi, Robert P., "Water," Scientific American, 243: 101–116 (1980) September.
g. Defined as quantity sustaining an "acceptable" quality of life.
h. Range of values assuming the maintenance of a diet of 2,500 calories/day.
i. Animals categorized by author, based on livestock in Senegal.

culate water use, assuming a peak population of 1,000 people, 200 cattle, 400 goats, and 1,000 chickens. Irrigation requirements for farming were calculated on the basis of a four-hectare parcel. The resulting daily water requirements are roughly 260 m^3/day. Water for irrigation may be diverted for human consumption where an influx of users from a wider area may occur due to the availability of water. However, this probably would be relatively small compared to the quantity of water used for agricultural purposes.

Natural Gas Utilization Equipment

The availability of natural gas in the rural village opens up the possibility for a host of natural gas-fired equipment. Chief among these applications is electrical generation equipment and home appliances. For this paper it was assumed that all water-pumping requirements would be met by less expensive and easier to maintain downhole electric pumping equipment. Electricity would come from the natural gas-fired generator set.

As mentioned earlier, the water requirements were estimated to be roughly 260 m^3/day of water for our hypothetical rural village. Ideally, this water would be co-produced with the natural gas. However, the distance to the natural gas well and the likelihood that the water may be unpotable may make it uneconomical or unattractive to transport the water to the village. Therefore, two water well pumping systems may be required; one near the village and one at the wellhead to stimulate gas production. Table 2 presents the installed downhole pump costs for a range of pump capacities. For our hypothetical village, pumping requirements are roughly 10.8 m^3/h with a power requirement of approximately 5 kW, assuming a 91.4-meter head. This power requirement must then be added to the village power requirements when sizing the natural gas-fired electrical generator set.

Table 2
Estimated Installed Downhole Pump Costs (1984 U.S. $)

m^3/h	Power, kW	Installed Cost
1.0	.37	$1,500
4.0	1.86	$3,000
16.0	7.5	$5,000

Table 3
Engine-Generator Set Costs*

kW Rating	Estimated Cost of Set	Estimated Installation Cost	Total Installed Cost†
100	$ 61,000	$ 36,500	$ 97,500
200	96,000	58,000	154,000
300	161,000	97,000	258,000
400	187,000	112,000	299,000
500	240,000	145,000	385,000

* *From AGA Marketing Manual, 1981.*
† *Exclusive of transportation costs.*
Maintenance costs estimated to be 1¢/kWh.

The natural gas-fired electrical generator set can be easily sized by adding together the water pumping and projected residential power requirements. The initial electrical demand is estimated to be very small, since most homes presumably have very few electrical appliances in the beginning. Assuming ½ kW peak demand per home, initial peak residential demand is roughly 50 kW. This would eventually grow over time to roughly 200 kW. With the addition of the water pumping power requirements it would appear that a 100 kW generator would suffice at the start of the project, with a second one added at a later date as demand grows. Costs for these natural gas-fired internal combustion engine-generator sets are presented in Table 3 as a function of size. For our hypothetical village, a 100 kW engine-generator set would cost roughly $100,000 installed (AGA, 1981). Table 4 presents a list of natural gas engine manufacturers in the U.S. Based on a 12,660 kJ/kWh efficiency for the engine-generator set, peak gas consumption of a 200 kW machine would be roughly 75.5 NM^3/h assuming 33,527 kJ/NM^3 gas.

The other major application for natural gas in the community is for residential heating and cooking. The basic appliance package for each home would consist of a cooking stove, instantaneous hot water heater, and small unvented space heater. A wide variety of basic cooking appliances exist ranging from one-burner range tops for under $50 to stoves with four burners on top and an oven for over $200. For this analysis, peak cooking-gas consumption assumes four top burners, each with an 8,440 kJ/h capacity, and an oven burner sized at roughly 12,660 kJ/h. Therefore, peak cooking gas demand would be roughly 46,420 kJ/h.

The second gas appliance for the home would be an instantaneous water heater. Unlike conventional storage tank water heaters commonly

Table 4
Natural Gas Engine Manufacturers—U.S.*

Manufacturer	Available Range of Power (kW)	(Hp)	Speed Range (rpm)
Caterpillar Tractor Co., Engine Division	45–694	63–930	1,200–2,800
Colt Industries, Fairbanks Morse Engine Division	477–6,711	650–9,000	514–900
Cooper Energy Services, Cooper-Bessemer	671–19,067	900–13,500	275–600
Cooper Energy Services, Superior Products	134–1,975	180–2,650	450–1,000
Cummins Rio Grande Sales & Service, Albuquerque, NM[†]	73–400	110–600	900–1,800
Dresser Clark Division, Dresser Industries, Inc.	1,492–8,877	2,000–11,900	300–360
Ingersoll-Rand Company	410–4,476	550–6,000	330–1,200
International Harvester	57–169	76–227	1,500–3,600
Stewart & Stevenson Services, Inc.	22–1,119	30–1,500	270–2,900
Waukesha Engine Division Dresser Industries, Inc.	22–2,122	30–2,845	720–2,600
White Engines, Inc.	40.3–106.6	54–143	2,400–3,600

* *From AGA Marketing Manual, 1981.*
† *Cummins Rio Grande is exclusive worldwide distributor of Cummins Natural Gas Engines.*

used in the United States, an instantaneous water heater is a small wall-mounted unit weighing as little as 6.8 kilograms that heats water upon demand. Residential units are typically sized from 31,650 kJ/h to 211,000 kJ/h, with capacities of 1.9 liters to 11.3 liters/minute at a 55.5°C temperature rise. For our hypothetical community it was assumed that a small 31,650 kJ/h unit with a capacity of 3.785 liters/minute would be sufficient to meet the hot water needs of the household. This unit would be mounted on the outside kitchen or bathroom wall for direct venting to the atmosphere, would cost roughly $125, and would have a steady-state efficiency of 80%.

In addition to cooking and hot water demand, each residence may also require some space heating, depending on location. For this paper, a small unvented space heater for room heating was chosen to minimize initial cost and to avoid venting problems that would otherwise require additional house modifications and cost. Unvented space heaters are available in sizes of 7,385 kJ/h to 42,200 kJ/h. For this paper, it was assumed that the typical residence would have a 31,650 kJ/h space heater, which would cost approximately $150. Figure 1 illustrates the three types of appliances suggested in this paper. Total peak gas consumption for these three appliances is roughly 109,720 kJ/h or, assuming 33,527 kJ/NM^3, 3.26 NM^3/h. Although this peak gas demand may seldom if ever be reached, the gas distribution and transmission equipment discussed in the following section must be sized to meet this potential demand.

Local Gas Distribution and Transmission System

Based on the assumptions made above, peak gas demand for our hypothetical community could range from a high of 365.0 NM^3/h initially, assuming 100 homes and a 100 kW electric generator, to nearly 786.7 NM^3/h in later years, assuming community expansion to 200 homes and a 200 kW electrical generator. Gas from the wellhead would be transported at relatively high pressure (308.2 kPa to 790.8 kPa) to the community. At this point, the gas pressure would be reduced before injection into the local distribution system. Gas requirements for the gas-fired internal combustion electrical generator would be taken directly from the transmission line to the generator set with one pressure regulator reducing the pressure to the inlet pressure requirements of the generator set. In

Unvented Space Heater   Gas Range   Instantaneous Water Heater

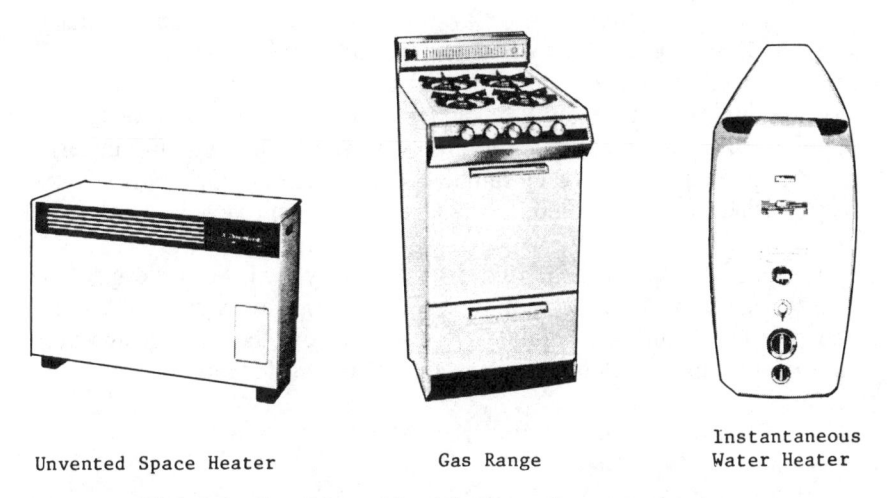

Figure 1. Possible residential gas appliance package.

this manner the generator would not interfere with the operation of the distribution system and the pipe feeding the generator would be kept down to a minimum diameter to avoid additional costs.

The local distribution system could operate in the range of 515 kPa to 102 kPa. At some point in the system the pressure must be reduced below 104.8 kPa for proper appliance performance. This can be performed going into the distribution system or just prior to the gas meter at the house.

Materials of construction for the distribution and transmission system could include steel, copper, or plastic pipe. With few exceptions the natural gas distribution companies in the U.S. have turned to plastic pipe for most of their distribution needs. Factors which have made this practical include lower cost, easier installation, and lower maintenance costs. Plastic materials which have been used in the U.S. include ABS, CAB, PVC Type I and II, and PE Types 2 and 3. Based on field experience and recent industry trends, PE (polyethylene) has emerged as the most suitable material for natural gas use. For this reason, PE was chosen for the distribution and transmission system of our hypothetical community.

PE Types 2 and 3 are more commonly referred to as medium and high density, respectively. Medium density PE pipe is manufactured by a number of U.S. suppliers as PE 2306, Aldyl "A," TR 418, and Gulf 9300. High density polyethylene is manufactured as PE 3306, PE 3406, PE 3408, and Driscopipe 7000. Manufacturers of PE pipe include Phillips Driscopipe, Plexco, Nipak, DuPont, and Gulf, to name a few. High density PE differs from medium density PE primarily in its ability to withstand higher internal operating pressures and better weatherability.

A number of different criteria can be used in sizing the distribution and transmission systems. Depending primarily on the projected gas demand growth rate, the systems can be sized to meet the current peak demand, with additional loops added when demand increases, or the system can be initially sized to meet gas demand over the life of the project. Transmission and distribution system life can be greater than 50 years for PE, depending on local climatic conditions. The first general sizing criterion has the advantage of reducing initial costs. Since this is probably of paramount concern for developing countries, this approach is emphasized here.

In sizing the transmission and distribution system, both the gas flow rate and the standard dimension ratio (SDR = pipe average outside diameter/pipe minimum wall thickness) must be calculated. The design pressure of the pipe is calculated using the following formula:

$$P = [2St/(D - t)] \times F = [2S/(SDR - 1)] \times F$$

where   P = Design pressure in kPa
S = The hydrostatic design basis (kPa)
   For PE 2306, PE 3306, and PE 3406, S = 8,618 kPa @ 23.5°C
   For PE 3408, S = 11,031 kPa @ 23.5°C
t = Specified minimum wall thickness (mm)
D = Specified outside diameter (mm)
F = Design factor for plastic pipe.*

SDR = Standard dimension ratio = $\dfrac{\text{pipe average outside diameter}}{\text{pipe minimum wall thickness}}$

The gas flow in the pipeline can be calculated using the flow formula for a smooth pipe operating at pressures above 108.3 kPa:

$$Q_h = \frac{57.33 \times 10^6}{G^{0.425}} \left[\frac{P_1^2 - P_2^2}{1} \right]^{0.575} \times d^{2.725}$$

where Q_h = Gas flow rate (NM³/h)
G = Specific gravity of gas (air = 1.0, natural gas = 0.65)
P_1 = Pipe inlet pressure (kPa)
P_2 = Pipe outlet pressure (kPa)
l = Length of pipe (m)
d = Pipe internal diameter (mm)

Using these formulas it was calculated that an 88.9 mm OD PE 3408 pipe with an SDR of 11 would have sufficient capacity for the transmission system. This line would be capable of carrying nearly 363.2 NM³/h of gas, assuming a maximum allowable operating pressure (MAOP) of 790.8 kPa and a pressure drop to roughly 206 kPa over the 40.2 km transmission distance. This capacity is slightly larger than the presumed 362.5 NM³/h peak gas demand. A second "loop" or parallel 88.9 mm OD pipe could be installed at a later date, when gas demand increased. Otherwise, a 114.3 mm OD PE 3408 line with an SDR of 11 would have sufficient capacity for the projected growth rate of the community. This pipeline

---

* Design factor, as specified by the Department of Transportation, is 0.32 for plastic pipe used in gas distribution services; design pressure may not exceed 790.8 kPa.

could carry roughly 720.7 NM^3/h of gas, assuming an MAOP of 790.8 kPa and a pressure drop of roughly 206 kPa over the 40.2 km transmission distance. This capacity is almost the same as the 730.1 NM^3/d gas demand projected over the life of the system for our hypothetical community.

The 88.9 mm OD and the 114.3 mm OD PE 3408 pipelines differ not only in material cost but also in potential installation costs. The 88.9 mm OD PE 3408 is available in coiled pipe lengths of 152.4 meters per coil whereas the 114.3 mm OD PE 3408 is only available in 12.2 meter straight pipe sections. Therefore, for a 40.2 km section, the 88.9 mm OD pipe would require 254 connections versus 3,300 connections for the 114.3 mm OD pipe. All connections would be made using the butt heat-fusion technique. Although this technique avoids the additional cost of mechanical or socket fusion fittings, the 114.3 mm OD is clearly more labor intensive. Additionally, the 88.9 mm OD pipe can be "plowed-in," whereas the 114.3 mm OD pipe must be placed in a cleared trench. The "plowed-in" method is less expensive and quicker than trenching.

The cost of PE 3408 piping as quoted by Plexco (1984) is $12,230 (f.o.b. U.S. port) for a 3,048 meter truckload of 88.9 mm OD SDR 11 coiled piping. Each truckload carries 20 coils arranged in stacks of four coils. The coils weigh approximately 299.4 kg each and have a diameter of 2.44 meters and a height of 0.63 meters. The 114.3 mm OD SDR 11 pipe, which costs $27,850 (f.o.b. U.S. port) per truckload, is stacked in 43 pipe packs that are 12.2 meters long.[2] Eight packs make one truckload for a total length of 4,194 meters. Each pipe length weighs approximately 39.75 kg. Judging from the material and labor cost differences of the 88.9 mm OD and 114.3 mm OD pipes, it would appear potentially more cost effective to go with the 88.9 mm OD pipe initially, with a second loop added later when demand required the added capacity.

Installation costs were not estimated in this paper due to the variability in such factors as transportation costs, terrain, soil conditions, and mechanical or labor-intensive installation methods. Although PE 3408 has excellent weatherability properties and has been installed on the ground in some situations, it is recommended that the pipe be buried and properly marked to avoid damage from such outside forces as falling trees, animals, or farmer's plows. Burial to a depth of 0.6 meters should be sufficient in tropical or subtropical climates. Deeper burial may be required in colder regions to avoid ice formation in the pipeline. A metal wire must be buried on top of the pipe to permit locating the pipe.

The pipeline entering the village would feed both the local distribution system and the electric generator set. The generator set was assumed to be placed approximately 152.4 meters from the pipeline and village for safety and noise factors. One coil of PE 3408 26.7 mm OD SDR 11 pipe would be used to connect the electric generator to the pipeline. This coil costs approximately $90 (Plexco, 1984). A butt fusion tapping tee (88.9 mm OD26.7 mm OD) costing roughly $8 would be used to connect the feeder line to the transmission line (Plexco, 1983). A pressure regulator with a price of approximately $123 would reduce the transmission pressure below 170.3 kPa just before the electric generator set. If desired, a gas meter can be inserted in the line to measure gas consumption. This meter costs approximately $1,245, f.o.b., and is rated for outdoor installation, is temperature compensating, has a maximum inlet pressure rating of 790.8 kPa and a capacity to 85 m^3/h.[4] An $8.20 transition coupling would be used to connect the plastic pipe to the steel fittings of the regulator.

The outlet of the transmission line feeds into the distribution system. The distribution system was assumed to be 975.4 meters long with 100 service lines of approximately 12.2 meters each. The village distribution system configuration is presented in Figure 2. Pressure would be reduced from the transmission pressure of 206 kPa to 790.8 kPa to the distribution pressure of roughly 115.1 kPa to 135.8 kPa. The pressure regulator to reduce the transmission pressure to the distribution pressure would cost approximately $132, f.o.b. (Walter Norris Corp., 1984). The 88.9 mm OD butt-tee fitting to connect the two distribution legs to the transmission line costs $15. Gas valves for each distribution leg are roughly $50 each.

Each leg of the distribution system would consist of 88.9 mm OD PE 3408 pipe. PE 3408 was used for the distribution mains since the same 88.9 mm OD pipe was also used for the transmission line and the bulk price of $400.98/100 meters was applicable. Otherwise, PE 2306 88.9 mm OD IPS SDR 11.5 pipe would have been used, which lists for

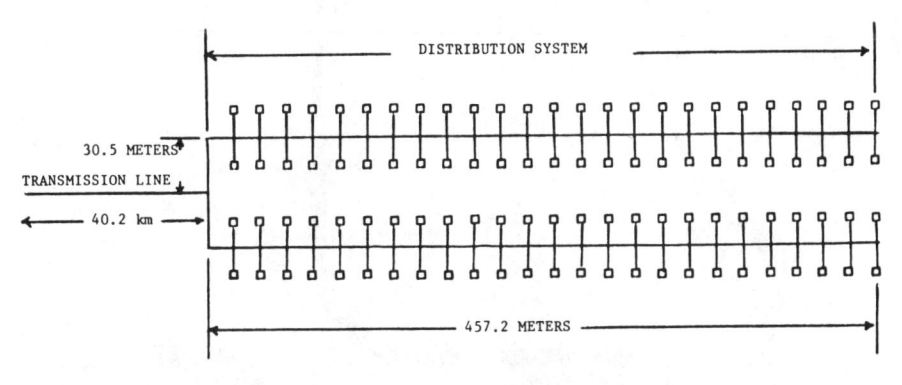

Figure 2. Initial distribution system layout.

554 Shallow Oil and Gas Resources

$469.51/100 m in less than truckload quantities. For these lower pressures the Weymouth approximation was used to calculate pipe diameter. This equation is:

$$Q = 3.35 \times 10^6 \times d^{2.667} \times \sqrt{P_1^2 - P_2^2} \times 1/\sqrt{L}$$

where
- Q = m³/day
- P_1, P_2 = Pipe inlet, pipe outlet pressure (kPa)
- L = Kilometers
- d = Inside diameter, millimeters

Using the Weymouth approximation the service lines were sized to 21.3 mm OD PE 3408. Approximately 121.9 meters of 21.3 mm OD PE 3408 with a SDR of 9.3 is required. In less than truckload quantities, this pipe costs $43.77/100 meters.[2] The service line is connected to the 88.9 mm OD main with an 88.9 mm OD × 21.3 mm OD socket-outlet fusion tapping-tee costing $7.25. The 12.2 m long service line to each house would be connected to the service regulator by a service riser pipe with transition fittings from 21.3 mm OD PE 3408 to the 26.7 mm OD steel fittings of the regulator as shown in Figure 3. The service riser costs ap-

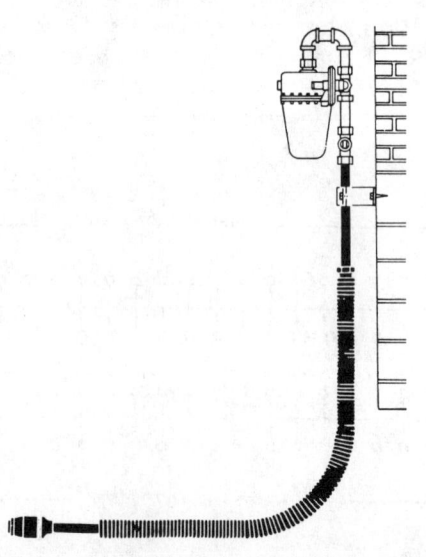

Figure 3. Meter and regulator mounted on service pipe riser [5].

proximately $16.50 when purchased in lots of 100. All plastic-to-plastic connections with the 21.3 mm OD pipe are socket fused whereas the main 88.9 mm OD connections are butt fused. The riser pipe is fixed to the structure, and the regulator/meter set is supported by this connection. The service regulator reduces the distribution pressure of approximately 108 kPa to 136 kPa psig to the 102 kPa required for proper appliance performance. Each regulator, when purchased in bulk quantities of 100 units, costs approximately $22.50 (Walter Norris Corp., 1984). The gas meter, if desired, costs $68.70 when purchased in similar quantities (Walter Norris Corp., 1984). A short length of steel pipe would extend into the house from the gas meter to complete the service connection.

All joints in the transmission and distribution system are butt, socket, or saddle fusion joints with the exception of mechanical fittings for all transition fittings of plastic to metal pipe. Polyethylene pipe, as a general rule, should only be joined by the heat fusion method. Threaded fittings or solvent cement connections are not possible with PE pipe. Fusion joints are relatively simple to make by trained personnel. Training is also an easy procedure. The fundamental procedure requires simultaneously heating the pipe and/or fitting surfaces to be joined with fusion tools at the prescribed temperature. The heat is then removed and the two surfaces are mated and held until the joint solidifies. In general, butt fusion joints are used on all 88.9 mm OD pipe sizes and larger, but can also be made on small diameter pipes. Larger socket fusion joints, which require a coupling fitting, are usually reserved for 60.3 mm OD sizes or smaller to provide support around the joint area. In either case, properly made fusion joints are as strong as the pipe itself. Saddle fusion joints are made when connecting the service tapping tees to the main and service lines. Figure 4 shows the fusion equipment and examples of field joints. Fusion equipment requirements for our hypothetical community distribution and transmission systems include the butt fusion machine, and socket and saddle fusion tools. The butt fusion cart assembly, with all support tools and heaters for a 33.4 mm OD to 114.3 mm OD PE pipe, costs roughly $2,700 f.o.b., as quoted by the Rigid Tool Co. (1984). Socket and saddle fusion tools will cost approximately $1,000 f.o.b. in addition, (Rigid Tool Co., 1984).

Shallow Well Development

Shallow well drilling can be performed using either the rotary or cable-tool methods. The most prevalent method today is the rotary, in which a bit is attached to a revolving steel pipe to bore a hole. Drilling

Figure 4. Fusion joining equipment and typical field joints.

fluid, or "mud," is pumped down the pipe to remove debris and cool the bit. The mud is pushed up the annulus of the pipe and removed. The older cable-tool method uses a heavy bit and stem attached to a cable. The bit is used to chip into the formation by alternately raising and dropping it. Water is flushed into the hole and bailed out to remove the debris. Portable cable-tool rigs are available for drilling relatively shallow wells, can be operated by one man, and are relatively simple to move and operate.

For our hypothetical community it was assumed that a truck-mounted cable-tool rig capable of drilling to about 450 meters would be sufficient for shallow well development. Such a rig would cost approximately $75,000 equipped with the tools required to drill 450 meters. Actual drill rig parts come from a variety of suppliers. A breakdown of the costs includes:

Truck chassis	$10,000
Rig equipment mounted on truck	$50,000
Cable and drilling tools	$15,000

Additional drilling costs include labor, piping, and cable and tool wear. Typical drilling rates would be approximately 15.2 meters in 8 hours in soft material and 4.5 meters in 8 hours for hard rock. Fuel and wear on the tools and cables are roughly $3.25/meter of hole. A Schedule 40 steel pipe casing would be used in the upper hole to prevent loose soil from falling into the hole. Once into hard rock, casing would not be required. Casing pipe costs are roughly $19.50 to $26.25/m for a 152.4 mm OD Schedule 40 steel pipe.

Figure 5 presents the downhole layout for a shallow well completed in limestone. Multiple wells of this configuration would probably be neces-

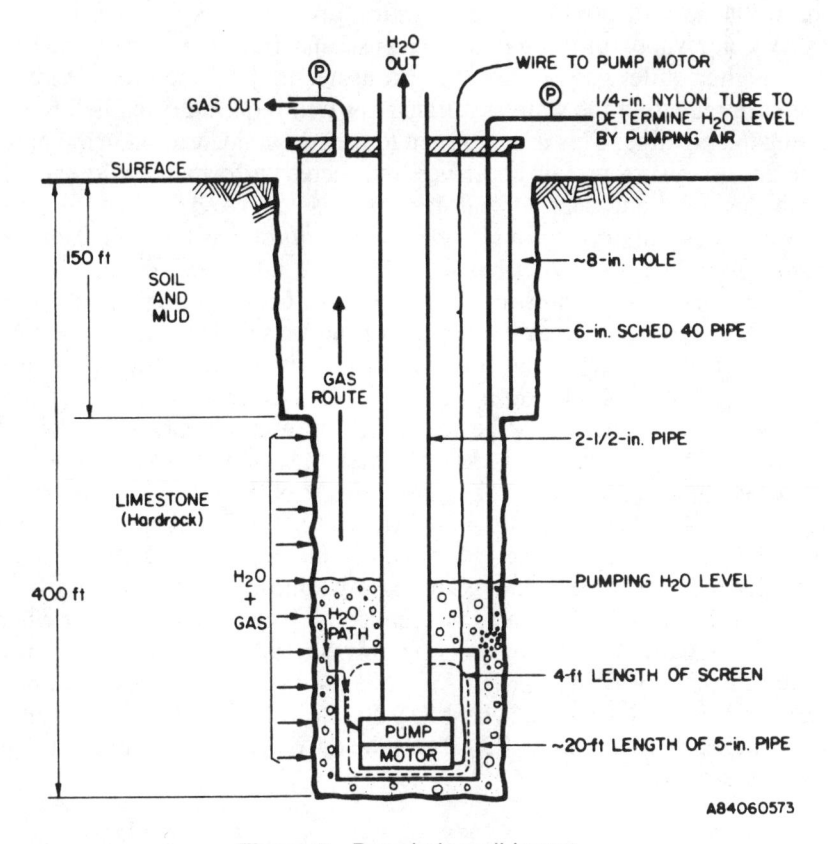

Figure 5. Downhole well layout.

sary due to their relatively low production capacity. This well has a 152.4 mm OD Schedule 40 steel pipe driven to 45.7 m through soft material. An open hole extends through the limestone to a depth of 121.9 m. Natural gas enters the wellbore as a by-product of water pumping. An electric downhole water pump is located near the bottom of the hole and is connected to the surface by a 63.5 mm OD pipe. Pump costs are the same as those presented earlier. Water and entrained gas are drawn to the hole where the gas separates and flows up the annulus. To facilitate gas separation, a 6.1-m length of 127 mm OD pipe is placed around the downhole pump which is surrounded by a screen with 0.15 to 0.25 mm slots. Pressure gauges at the wellhead monitor gas pressure and downhole water levels.

The gas exiting the wellhead will vary in pressure depending primarily on the depth of the hole. For our hypothetical case, the gas pressure would be reduced by a pressure regulator costing about $150, depending on well pressure, to roughly 790.8 kPa. The expansion of the gas from the wellhead will produce a temperature drop that may cause the saturated water vapor in the gas to condense and freeze. For deeper wells with higher outlet gas pressures a gas heater may be required. Prior to pressure reduction the water-saturated gas will require drying before entering the pipeline. This is performed to prevent condensation in the pipeline and to ensure proper appliance and meter performance. To accomplish this a deliquescent natural gas dryer is suggested.

A deliquescent dryer provides clean, dry natural gas through chemical absorption. The dryer is filled with desiccant tablets which slowly dissolve as they absorb moisture from the gas. The resulting liquid flows through the packed bed to the bottom of the vessel where it is automatically drained. Desiccant consumption varies with gas moisture, pressure, and temperature. Assuming a saturated gas inlet pressure of 1,825 kPa the moisture content would be reduced from approximately 801 kgs of water/MM NM^3 of gas to 314 kgs of water/MM NM^3 of gas. At this rate desiccant consumption would be 208 kgs/MM NM^3. Deliquescent dryers for natural gas service are available from Van Air Systems Inc. For our hypothetical community energy needs, one model DS12-7.2 dryer with a capacity of 8297.8 NM^3/day assuming a wellhead gas pressure of 1825 kPa would be required initially. A second unit would be added later when demand increased. This unit costs roughly $3,190, f.o.b., and has a desiccant capacity of 54.4 kgs (Van Air Systems, Inc., 1984). Desiccant purchased in bulk 22.7 kgs bags cost $48.50, f.o.b., from Van Air Systems. After pressure reduction to 790.8 kPa the dryed gas would be sent to the transmission system.

Summary

The hypothetical community discussed in this paper requires a shallow-well cable-tool drilling rig, gas conditioning equipment, transmission line, distribution system, generator set, water pumps, and residential gas appliances. Table 5 presents a summary of the projected cost for each system component. All costs are in 1984 dollars and do not include transportation, installation, or labor. Depending on the planning philosophy, the system can be sized to meet present peak demands with capacity added when required or the system can be sized to meet all peak project gas demand over the life of the system. The first approach, which is less expensive in the short run, would cost roughly $333,600 for the materials and equipment alone. The most expensive part of the system is the 40.2 km transmission line. Considerable saving would occur if the natural gas well were closer to the village. Additional savings are also possible in the drilling operation; the cable-tool drilling rig may be rented or the drilling operation contracted, lowering the cost of the drilling.

Table 5
Natural Gas Energy System Equipment Costs*
(Sized for Initial Peak Demands—100 Homes)

Shallow well equipment		
Cable-tool drill rig		$ 75,200
Material costs (122 meters)		7,200
Dryer costs		3,190
Regulator cost		150
	Subtotal	$ 85,740
Transmission and distribution system		
Transmission line		$161,500
Distribution system		
Mains		4,160
Services		2,910
Regulators and meters		9,120
Generator service		1,470
Fusion equipment		3,700
	Subtotal	$182,860
Gas-fired generator set		
100 kW Unit		$ 61,000
Water pump		4,000
	Subtotal	$ 65,000
	Total	$333,600

* Exclusive of transportation and installation costs.

References

AGA Marketing Manual—Natural Gas Prime Movers, 1981.
Perfection Corp. Brochure, Permasert™ System Price List, PSPL-1.
Plexco-Amsted Industries Catalog, 1983, PE 2306 (Orange) and PE 3408 (Black) Fittings General Information—Pricing and Packaging Data, Revised January 15, 1983.
Plexco-Amsted Industries, 1984, private communications (Mr. Jerry Wielgat), May.
Rigid Tool/P+S—Plastic Pipe Tools and Equipment Brochure and private communications, May 1984.
Van Air Systems, Inc., 1984, private communications (Mr. George Currie), June.
Walter Norris Corp., 1984, Rockwell Industrial Gas Equipment, private communications (Mr. D. Brand), May.

49
Utilization of Small Shallow Heavy Crude and Tar Sands Resources

Joseph Barnea and Ramon Omana

Introduction

Since its creation, the UNITAR/UNDP Information Centre for Heavy Crude and Tar Sands has been interested in the utilization of heavy crude oil and tar sand (bitumen) resources by developing countries. This "utilization" poses the following questions:

1. Does the developing country have any of these resources at all?
2. Does the developing country know that it has this type of resource?
3. Has the developing country evaluated and quantified its resources?
4. Has the developing country taken steps for the exploration and development of those resources?
5. Under the present situation is it economically and technically feasible to try to exploit the resource?

It is not uncommon to learn that some developing countries do not know that they have resources—or have chosen not to know—as indicated by some answers to the Centre's First International Survey of Heavy Crude and Tar Sands, conducted in 1983. Often, information is kept confidential for a variety of reasons outside the scope of this paper. In most cases, when resources are reported, the information is incomplete. Despite the lack of information, it may be possible for a developing country to utilize its heavy oil and tar sands resources to lessen the impact of oil imports in its economy. It is the purpose of this paper to explore the latter idea, to give a general description of the resource, and to present general guidelines to the utilization of heavy crude resources in rural areas.

Definition of the Resource

Heavy oil and bitumen are naturally occurring, high molecular weight hydrocarbons which may be found in the solid or liquid state. The definitions of the UNITAR Working Group on Definitions are stated as follows:

- Crude oils and (natural) bitumens may be characterized first by viscosity and then by density.
- In determining the international resource base, viscosity may be used to differentiate between crude oils on the one hand and bitumen on the other. Subsequently, density should be used to differentiate between heavy crude oil and other crude oils.
- Bitumens have viscosities greater than 10,000 mPa.s (centipoises). Crude oils have viscosities less than or equal to 10,000 mPa.s. These viscosities are gas-free measured and referenced to original reservoir temperature.
- Extra-heavy crude oils have densities greater than 1,000 kg/m^3 (API gravities less than 10 degrees). Heavy crude oils have densities from 934 to 1,000 kg/m^3 (API gravities from 20 to 10 degrees inclusive). These densities are referenced to 15.6°C (60°F) and atmospheric pressure.

Heavy Crude

Of the two parameters used to define the resources, viscosity and density, viscosity is of great practical significance because it determines to a large extent the manner of production of the resource. It can be said that the more viscous the material the higher the production and transportation costs. It may be noted that since viscosity varies inversely with temperature, most production processes for heavy crude use either steam or other heat sources to reduce the viscosity of the material and make it pumpable.

Apart from the higher density and viscosity, heavy crude usually contains a metal and sulphur content higher than conventional oils. Because of these characteristics, the cost of refining is higher for heavy crude oil than for light. Differentials on the order of $3 to $6 per barrel are not uncommon.

In spite of these handicaps, heavy crude has two significant advantages. It is widespread and could be available in many areas and regions where light crude has not been discovered or does not exist. The second advantage is that in most countries the exploration costs for heavy crude and tar sands are very low because heavy crude was encountered while exploring for light crude. This includes the finding of surface shows, seepages, etc. For example, in Surinam, oil companies explored for light crude, found none, and disregarded indications for heavy crude. Recently, the newly formed Surinam National Oil Company, with the help of a loan from local banks, has begun to develop the heavy crude. The company will produce, by the end of this year, about 1,000 barrels per day, which will cover about 25% of the local oil demand. The heavy crude being developed in Surinam is relatively low in sulphur and lends itself for direct burning in power stations.

Most of the heavy crude which has been developed around the world is in shallow formations not deeper than 3,000 feet. However, there are heavy crude reservoirs deeper than 10,000 feet. Deep deposits have one advantage, namely, the reservoir temperature is higher and as a result, the viscosity of such heavy crude is lower. Conventional steaming of heavy crude becomes costly and inefficient if the steam has to be injected deeper than 3,000 to 4,000 feet, due to heat losses.

Shallow heavy crude fields are in operation in many countries, including Argentina, Brazil, Colombia, Venezuela, Syria, and Thailand. Among the industrialized countries, there are hundreds of shallow heavy crude fields in California and Texas in the United States and in Alberta and Saskatchewan, Canada.

Tar Sands

Tar sands contain bitumen with a viscosity exceeding 10,000 mPa.s. Tar sand is often found as a surface-mineable resource, which offers two advantages: the mining cost can be very low ($.10–.50/metric ton) if the overburden is small; and more than 90% of the bitumen contained in the material can be extracted. The bitumen can be separated from the sand by a hot water or other process and upgraded in an appropriate refinery.

Criteria for the Utilization of the Resources

The utilization of the resource requires that any associated project must be technically and economically feasible. The basic requirements are that (1) they should not be too far from a rural area, to minimize the cost of pipelines and/or transport and (2) the reservoirs or deposits should be able to stand production for at least 15 to 20 years. Under these conditions, where the infrastructure already exists, a local refinery could be upgraded to handle the heavy crude and bitumen.

The Range of Cost

As indicated earlier, the main element in the production of heavy crude is the production cost, not the exploration cost. The cost is determined by many factors, including the already mentioned density and viscosity of heavy crude, its depth, and the location where it is found. In addition, the infrastructure available in the region (such as roads, electricity, supplies, availability of refineries, etc.) has a bearing on the cost, the value or price of the commodity produced, and the market. The cost may range between $4 to $12 per barrel. The value, or price paid for heavy crude before refining, ranged between $20 to $25 per barrel in 1984.

Local Conditions

In rural areas, if kerosene and fuel wood are expensive it may be advisable to develop shallow heavy crude if a sufficient market exists for gasoline, diesel oil, fuel oil, and asphalt.

Alternatively, if good transportation exists in the area where the heavy crude is found, it could be transported to urban centers, refineries, or ports for further processing and/or export.

The production and processing of heavy crude is somewhat more complex than that of conventional oil but, as Surinam demonstrates, the proper technology can quickly be obtained if trained engineers are available. A paper available at this conference, on the production of heavy crude with 95% water cut, demonstrates how a careful selection of pumping equipment makes the extraction of heavy crude, under the conditions discussed in that report, cost effective.

In rural areas, the development of shallow heavy crude, assuming that it is a fairly small reservoir, say less than 50 million barrels of oil-in-place, greatly depends on the availability of population and infrastructure. There are numerous small deposits located in inaccessible jungle areas, which probably cannot be developed economically until an infrastructure exists and part of the jungle has been developed. The latter applies to desert areas if there is no population within 300 miles, although as a rule, desert development costs are somewhat lower than those costs in mountain or jungle areas.

Where to Look for Heavy Crude and Tar Sands

In rural areas where oil or gas are not found, it may make sense to investigate the possibility of finding heavy crude or tar sands. The first step should involve a review of all drilling records in sedimentary areas to check for hydrocarbon indications, whether drilling for oil, gas, water, or minerals. If such indications exist, further evaluation should be carried out by an exploration geologist.

A second step involves the review of literature on seepages or tar sand outcrops in the region. If such seepages or outcrops could be located again, a specialist should study them to advise on further action.

If old drilling records confirm the existence of heavy crude and/or tar sands, and the geological evaluation is positive, then funds will have to be raised for the exploration, which will hopefully provide information on the size of the deposit, the quality of the material, and its viscosity and density. When that information is available, a decision has to be made on the extent to which the discovery is to be developed, based on an estimate of local market demand in the rural or other areas to which the products could be transported or exported under attractive economic conditions. The transport of the heavy crude could be facilitated either by the addition of biosurfactants or by mixing the heavy crude with light crude. The mixing of heavy crude with light, in many countries, avoids the necessity of installing an upgrader near the wellhead and also avoids the construction of heated pipelines. The mixing with light crude has a further advantage, namely, the new mixture will have lower density, higher quality, and practically any refinery will be able to process it. However, this method can become very costly to the operator, depending on how and where the mixing is to be done and what volumes of light crude are involved.

In the case of tar sands, the bitumen must be separated from the sand near the mine and that may require, even for small projects, an investment running into several millions of dollars. But the bitumen could then be mixed with light crude and would also become transportable to refineries.

In exceptional cases, the bitumen might be of sufficient quality to burn directly in local power stations; work is in progress to develop a process to make this possible. This will require a fairly high grade deposit with at least 10% oil by weight per ton of material. If in interior rural areas, the best solution is to set up a small refinery (1,000 b/d) for the processing of heavy crude, then such refinery might provide gasoline for transport, diesel oil for equipment, and fuel oil for burning. It would also have, as a by-product, asphalt which could be used for road construction, for roofs, and other purposes.

Conclusion

A developing country with small, shallow heavy crude and tar sands resources can utilize them in most cases to lessen the impact of oil imports on its economy. The UNITAR/UNDP Information Centre for Heavy Crude and Tar Sands in New York is willing to help, by providing pertinent information to those developing countries in the initial information-gathering stage.

50
Gas Development Strategy and the Commercialization Process in International Exploration and Production Contracts

Mohsen Shirazi

Introduction

Natural gas is among the largest of the world's primary energy resources; its share in 1982 total primary energy consumption was about 18%. In developing countries, however, this share was only 7%. The relative importance of gas varies greatly across countries, from 43% in Pakistan to 5% in Brazil and even zero in many other developing countries.

At the end of 1982, world proven gas reserves amounted to about 87 trillion (10^{12}) cubic meters, or 79 billion tons of oil equivalent (t.o.e.). This is equivalent to roughly 86% of known world oil reserves. The ratio of gas reserves to oil reserves has increased from 39% in 1960 to 75% in 1980 and to the current level of 86% in 1983. Based on this trend and various geological studies, the ratio of new gas discoveries to new oil discoveries may continue to increase until, on an energy equivalent basis, gas reserves exceed oil reserves.

Gas reserves and consumption are unevenly distributed across regions and countries (Figure 1). Within the developing countries, the non-oil-exporting developing countries accounted for 1.7% of total world gas reserves and 2.5% of world consumption. The oil-exporting developing countries, on the other hand, hold more than 40% of world proven gas reserves and only a modest 9% share of world consumption. The United States and the Soviet Union remain the two largest producers and con-

568 Shallow Oil and Gas Resources

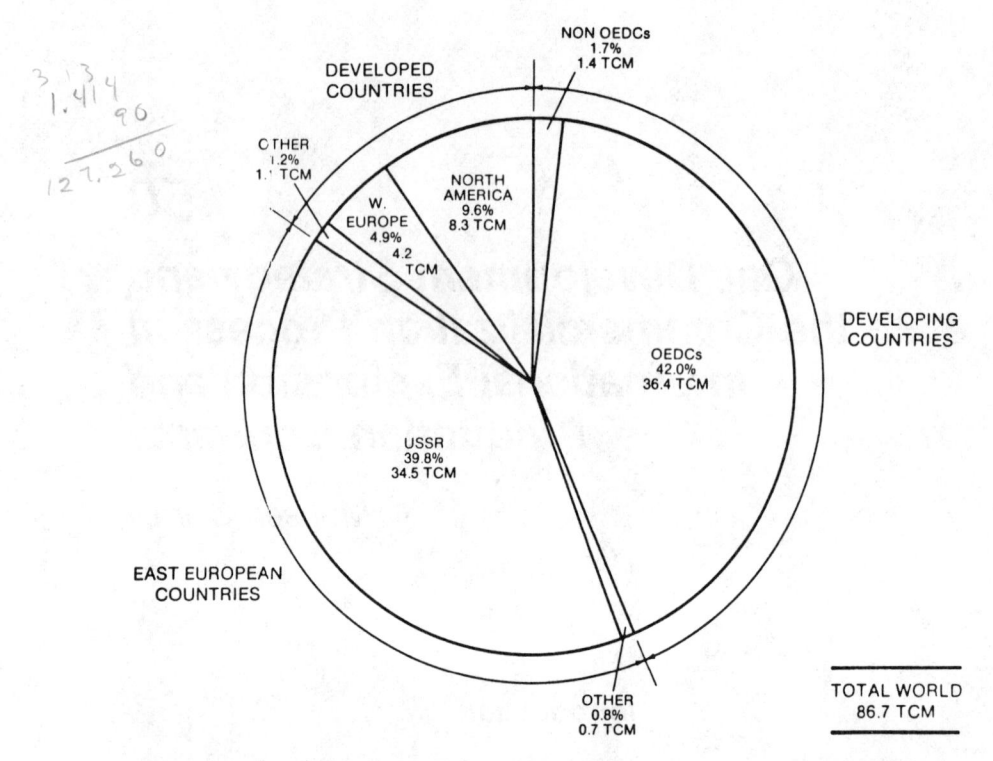

Figure 1. Developing countries' natural gas reserves in a world context, as of January 1983 (trillion cubic meters).

sumers of natural gas, with the Soviet Union well in the lead in terms of proven gas reserves. Western Europe is the third major consumer of natural gas, accounting for 13.5% of 1982 total world consumption (Figure 2). Japan's consumption of natural gas is relatively small (1.7% of total world consumption), but it is noteworthy to mention that Japan's production satisfies only 8% of its consumption; the balance is imported from the Far East and Middle East in the form of Liquefied Natural Gas (LNG). Eastern Europe accounts for about 41% of the world proven gas reserves, although 98% of this is located in the Soviet Union.

As Table 1 indicates, the developing countries' marketed production represented only 51% of their gross production, compared to 87% in the developed countries and 98% in Eastern Europe. In the oil-exporting developing countries, more than half the produced quantities of natural gas were flared. Although gas producers have recently put more emphasis on

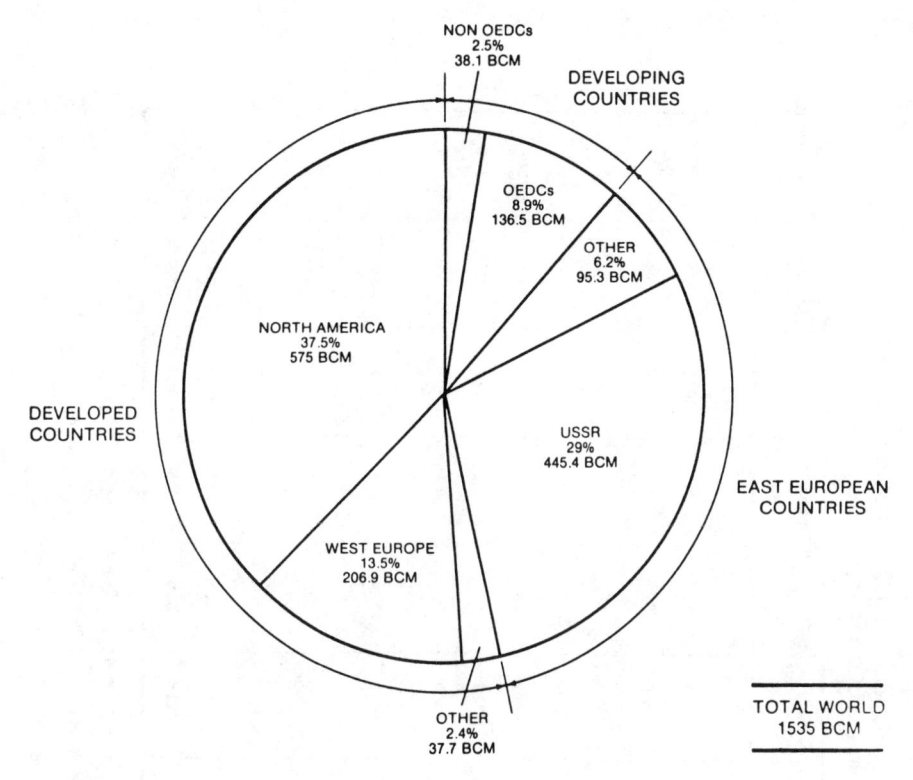

Figure 2. Developing countries' natural gas consumption in a world context, January 1983 (billion cubic meters).

gas reinjection as a form of natural gas conservation, flaring is still common. In 1982, Iran flared 60% of its gross production, Saudi Arabia 71%, Nigeria 85%, Iraq 84%, and Gabon 92%. The domestic gas consumption of developing countries in 1982 was only 0.5% of their 38 trillion cubic meters of gas reserves. Prospects for natural gas development and local utilization in the developing countries are therefore considerable.

Gas Commercialization Process

The primary objective of this paper is to evaluate the need for an efficient gas commercialization process from the exploration and production phase to the market place and end use. A second and equally important

Table 1
World Natural Gas Reserves, Production, and Consumption, January 1983
(billion cubic meters)

	PROVEN RESERVES (BCM)	% OF WORLD	GROSS PRODUCTION (BCM)	MARKETED[3] PRODUCTION (BCM)	M. PRODUCTION AS % OF GROSS PRODUCTION	M. PRODUCTION AS % OF WORLD	R/P[4]	CONSUMPTION (BCM)	% OF WORLD
DEVELOPING COUNTRIES									
O.E.D.Cs[1]	36408.9	42.0	363.14	175.4	48	11.4	100.3	136.5	8.9
NON O.E.D.Cs	1430.1	1.7	47.96	36.2	75	2.4	29.8	38.1	2.5
SUBTOTAL	37839.0	43.7	411.1	211.6	51%	13.8	92.0	174.6	11.4
DEVELOPED COUNTRIES									
NORTH AMERICA	8357.2	9.6	663.5	571.7	86	37.3	12.8	575.0	37.5
WEST EUROPE[2]	4242.3	4.9	189.2	172.6	91	11.2	22.4	206.9	13.5
JAPAN	20.0	—	2.0	2.0	100	0.1	10.0	26.0	1.7
AUSTRALIA & NEW ZEALAND	1064.0	1.2	14.9	13.9	93	0.9	71.4	11.7	0.7
SUBTOTAL	13683.5	15.7	869.6	760.2	87%	49.5	15.7	819.6	53.4
EASTERN EUROPE									
USSR	34505.0	39.8	513.7	500.9	98	32.6	67.2	445.4	29.0
OTHER EAST EUROPEAN COUNTRIES	654.2	0.8	63.2	62.2	98	4.1	10.4	95.3	6.2
SUBTOTAL	35159.2	40.6	576.9	563.1	98%	36.7	60.9	540.7	35.2
TOTAL WORLD	86681.7	100%	1857.6	1534.9	83%	100%	46.7	1534.9	100%

Sources: The World Bank, Cedigaz, The Economist Intelligence Unit,¹ Petroleum Economist.
1) OEDCs = Oil Exporting Developing Countries.
2) Six West European Countries (Finland, Spain, Luxembourg, Switzerland, Belgium, and Denmark) are considered gas consumers only.
3) Marketed Production = Gross Production — Reinjection — Flared — Other Losses.
4) Reserves/Gross Production Ratio.

objective is to identify the issues and obstacles which may deter or delay an efficient commercialization process.

Gas commercialization is often treated in the same way as oil development despite clear differences at various phases of development, such as exploration and production (E&P) contracts, project development, financing, marketing, and institution building. As a result, gas has been either flared or has remained undeveloped despite marketing opportunities. Therefore, two procedures regarding oil and gas development will be analyzed and compared, and possible gas development strategies and policies which would support their implementation will be identified.

A Sequential Review of Gas Development From Exploration to Market Place and Comparison with an Oil Development

Any petroleum exploration activity may result in the discovery of oil and/or gas. In the case of an oil discovery, the approach toward development and commercialization is normally well established and the steps to be taken are defined in great detail in the agreement signed by the company and the state. However, the same expertise does not exist in gas, partly because the domestic gas industry in developing countries is still young or nonexistent. Therefore, in addition to all of the usual technical factors which influence the development of oil resources, gas development strategies have to be tailored to the social, political, and economic environments of each situation. By this statement, we do not intend to overlook the fact that similar factors exist in the commercialization process of oil; however, we believe that the nature, intensity, and magnitude of the problems are somewhat different for the gas commercialization process.

In Figure 3, the basic steps from exploration and production (E&P) to the marketing phase of oil and gas are compared. Phases I and II on the chart, which deal with exploration activities, are practically the same for oil and gas. Phase III is related to the evaluation of an oil or gas discovery. In the case of an oil discovery, the commercial evaluation can be made reasonably quickly once the main physical parameters of the discovery have been assessed. In the case of international license agreements, the development phase can begin promptly after agreement on commerciality has been reached between the operating group and the host country. Upon completion of this phase, oil will be sold in the best available domestic and/or international oil markets.

In the case of a gas discovery, commerciality cannot be established until the conditions for gas utilization and marketing have been defined and

572 Shallow Oil and Gas Resources

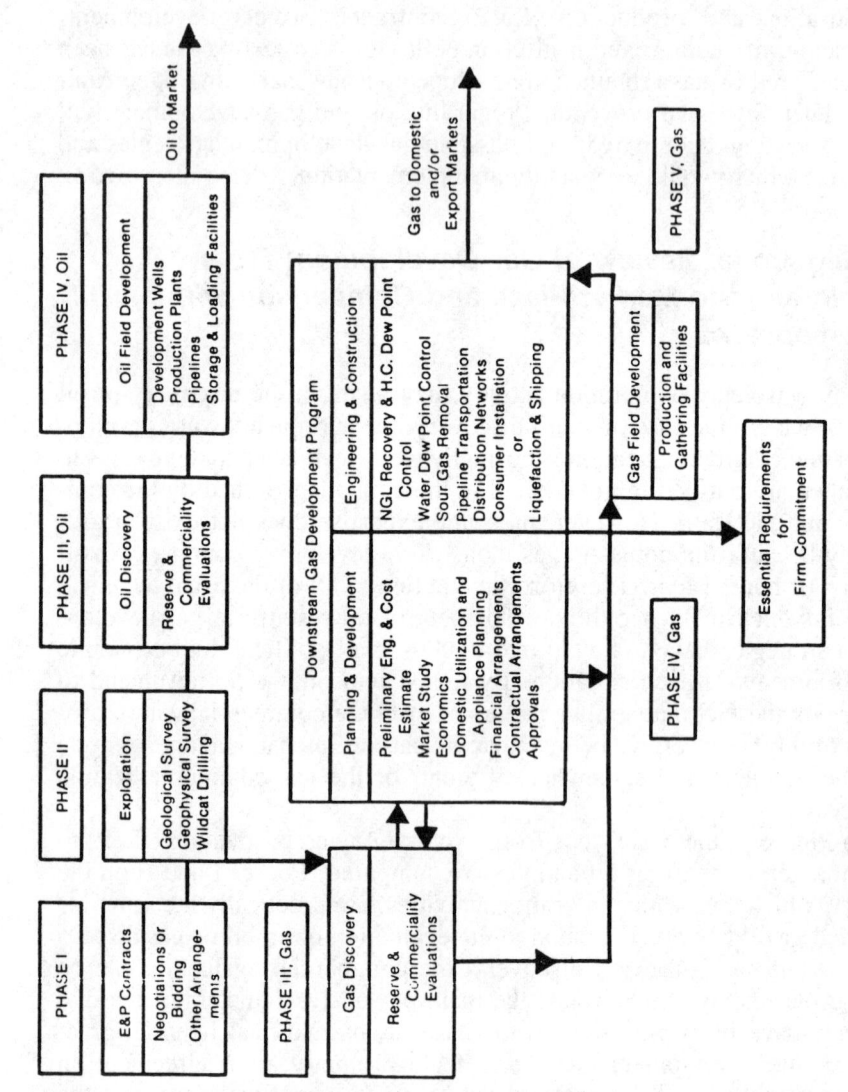

Figure 3. A sequential comparison of gas and oil developments.

the development of the field has been closely coordinated with a "Downstream Gas Development Program."

A Downstream Gas Development Program is defined as an integrated chain of activities downstream of the oil and/or gas field development operations for the commercial sales and delivery of natural gas and/or LNG to the buyers in the domestic market or for export. It also includes the design, construction, operation and maintenance of the facilities necessary to receive, measure, treat, process, and transport natural gas from the field development operations to the domestic consumers or export buyers. Further, it includes all economic and technical studies as well as contractual arrangements.

In view of the preceding definition and the sequential review of the gas development, we can conclude that in the absence of a fully identified Downstream Gas Development Program, the commerciality of a gas field cannot be established and development of a gas field would not be advisable. The evaluation of a gas field and design of a Downstream Gas Development Program should proceed in parallel until the parties of the E&P contract, who are responsible for the development of the gas field, and the gas authorities, who are responsible for the Downstream Gas Development Program, are able to reach agreement on the physical, financial, economic and contractual parameters. These are indicated in Figure 3 as essential requirements for a firm commitment from each of the parties involved.

Design of a Downstream Gas Development Program

Successful implementation of a gas development program depends on the coordinated development of an integrated chain of activities which include studies and investments on both the supply and demand sides. The full development of a large gas field and Downstream Gas Development Program, including approximately two years of initial studies, may take up to five years (Figure 4). This may be somewhat less for smaller fields although the process remains the same regardless of size. Because of the close link between development of the gas field, the transmission and distribution network, and the market, there are a number of decision points at which studies and investments must be completed *simultaneously* before the next step can be selected. These points include:

1. Assessment of the availability of and markets for natural gas under various price/cost assumptions, for 10–20 year projections.
2. Comparison of various gas development schemes in terms of their technical, economic, and financial feasibility.

574 *Shallow Oil and Gas Resources*

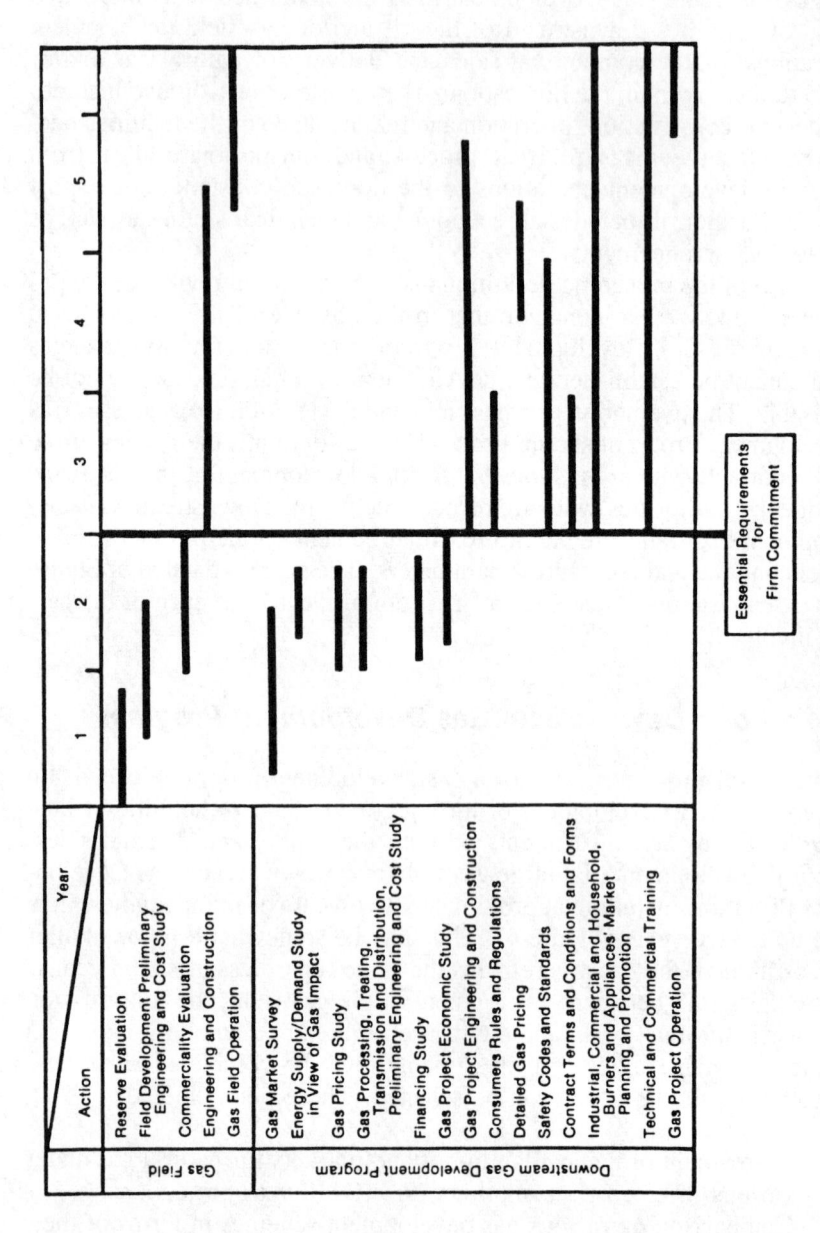

Figure 4. A coordinated gas development plan.

3. Determination of a master development plan within which individual projects could be formulated.
4. Formulation of economic, financial, and manpower policies which will ensure the efficient development of gas and the technical and financial viability of the producer, transmission, and distribution companies and users.
5. Formulation of individual projects, financing, commitment, and efficient implementation in accordance with a coordinated timetable.

Assessment of the Gas Availability and Markets

The first discovery well(s) provides only a preliminary indication of the commercial potential of a discovery. Normally, more wells need to be drilled and studies undertaken before reliable evaluation of the recoverable reserves can be carried out. Before such a program is undertaken, a preliminary assessment of the size of the market as well as regional and sectoral demand information will be required. In order to assess the size of the market, however, an estimate of the cost of gas development is also needed.

At this stage already a potential conflict exists between the producer—public or private—and the institutions in the host country that are concerned with gas development and utilization. Producers are reluctant to provide information which could influence their ability to negotiate a fair gas price and consumers are unable to commit themselves to gas development because of the uncertainty of supply. The consumers are also reluctant to undertake a comprehensive market survey before they are sure gas is available in commercial quantities and at reasonable cost. Although this conflict has delayed gas development in several countries, it can be avoided if a government is willing to:

- Systematically review the market for gas as part of its periodic energy review and assess the range of costs which would be acceptable to major groups of consumers.
- Provide a clear policy toward producer and consumer prices, linking them to the value of the fuel to be displaced.

Such information is not costly to generate and would provide a framework, at the time of discovery, within which producers could assess their future profitability and make the corresponding investment decision more rapidly. In the case of foreign oil companies, the host-government national oil company should include clauses in the exploration-production agreement which would indicate clearly how the price for gas is determined. If the agreements in force do not have such a clause, a procedure should be established and made available to the companies for this purpose.

Comparison of Various Gas Development Schemes

Once a set of projections has been agreed upon for both demand and supply, alternative development schemes can be designed. At this stage, only conceptual engineering will be required to assess the comparative technical and economic feasibility of alternative schemes under various demand assumptions. This analysis should also include a comparison of alternative technologies and approaches to field development and infrastructure (larger diameter pipes initially versus subsequent compression), as well as a comparison of the various codes of practice used in the industry world-wide (maximum operating pressure, safety regulation, unit of measurement, demand risk), which would affect the design and therefore the cost of the facilities to be built.

It might be important at this stage to secure the assistanc of an organization with operating experience in order to avoid "wrong" technical decisions, which might prove costly in the future. Judgement would also have to be passed on long-term policy issues which will affect the economic viability of alternative schemes, such as reserve depletion and dedication policies, interruptible versus continuous supply of large consumers, duration of supply contracts, and pricing policies, both in economic (level, structure) and administrative terms (measurement, bill collecting, tariffs). The analysis should clearly identify the impact of alternative choices on the feasibility of the whole scheme over the *long term* and enable decision makers to evaluate the "global cost" for the *country* of alternative policies. Too often, decisions which affect the future are made on the basis of short-term considerations and cannot be modified at a later stage. In this early stage of gas development most developing countries require considerable assistance to avoid such mistakes.

Selection of Market Development Plan

The previously described analysis should result in the selection of a market development plan in which discrete projects could be formulated. This market plan would in many cases be a series of compromises aimed at minimizing risks over time. It should provide a flexible framework which could be amended as new information on gas availability becomes available or as new markets are developed. Typically the market plan selected would cover a period of 10 to 15 years and would have to be reviewed frequently, particularly in the early years.

Policy Formulation

As indicated earlier, the successful development of gas resources will require the formulation of adequate pricing policies, regulations, and incentive schemes which will ensure that gas resources will be developed in a timely fashion and that consumers will undertake the necessary investments to use the gas when it becomes available.

Experience has shown in developed and developing countries that to be successful such an exercise must be carefully planned and a single institution must be in charge of natural gas development so as to facilitate the evaluation of alternative schemes. This is particularly important in developing countries, which are at the beginning of natural gas development, when several options can be considered for the export as well as the domestic use of gas resources. While each development scheme is country-specific, a large body of expertise has emerged in those countries that have introduced natural gas into their energy balance over the past three decades. This expertise can benefit and facilitate the development of gas resources in developing countries.

Individual Projects

Once the Downstream Gas Development Program has been agreed upon, specific projects can be identified, prepared, and implemented. Numerous agreements, arrangements, and contracts must be negotiated and entered into before implementation plans for these projects can be drawn up. For example, in the case of export projects agreements, governmental approvals must be obtained prior to a final decision in order to proceed with the projects:

- Shareholder agreement between the host country and the international companies.
- Formation of joint companies, if necessary, for each of several phases of the program.
- Long-term gas-sales agreements between the producers and the transmission company and between the transmission company and the LNG company.
- Negotiations between the LNG company and its overseas customers for sales of LNG on a f.o.b. or c.i.f. basis.

- Arrangements for the financing and construction of all phases of the project; approvals from government authorities in the host country, and those in the importing countries in the case of LNG.

The coordination of these steps can be complex, particularly when the implementation periods of the discrete investments differ. Implementation of the component with the longest implementation period cannot normally commence until financing has been secured for all investments. In addition, the various contracts must assign the costs associated with any failure to complete all investments according to schedule.

Institutions

In view of the complexity of the issues involved in gas development, both in the resource and market areas, a strong institution with expertise in natural gas has to be developed in order to negotiate supply contracts with producers, develop, transport, distribute, and market gas, and implement the chosen policies at the government level.

Difference between Oil and Gas Commercialization

The main points that differentiate between *oil* and *gas* exploration, production and commercialization can be summarized as follows:

1. Commercial evaluation and development work for production and marketing of oil takes place within the framework of the E&P Contract whereas a Downstream Gas Development Program, as defined earlier, is usually an operation independent of the field-development activity. Thus, it requires a different structure with more support from the host country's government.
2. Gas development generally requires a larger lead time than oil development, is more capital intensive and requires larger up-front investments, particularly for the transmission and distribution of gas.
3. The sale of oil is made in the best available domestic or international oil market, whereas marketing of gas requires special long-term assessment, with the producer and the customer tied together for the life of the project. Furthermore, the transmission and distribution of gas is sometimes done by public entities and on a country-wide basis, or if done by the private sector, it is often regulated by the government.

4. Gas requires a different type of expertise and specialized institution to carry out the activities defined under the Downstream Gas Development Program.

Gas Commercialization Process in International Exploration and Production (E&P) Contracts:

An E&P Contract is the first vehicle for developing a gas field and a Downstream Gas Development Program. Although international E&P contracts are well designed for oil activities, they often do not provide the gas provisions necessary for pragmatic commercialization of a discovered gas field.

Most of the production-sharing contracts either do not deal with gas at all or if they do, they are inappropriate from the standpoints of approach, economics, and governing regimes. Producer transfer prices and/or market prices are only rarely mentioned. Under a typical production-sharing contract regime, the government receives percentages of production (about 50%-70%), royalty or national needs (10%-15%), and the contractor's net profit as tax (about 50%). This regime results in a division of the net income of 70%-90% to the host country and 10%-30% to the contractor. In the case of an oil discovery or a sizable gas discovery, especially in the vicinity of industrial markets, the development of oil and gas fields under such a regime might be economically attractive. A good example is the North Sea gas projects. However, in most of the developing countries, when a gas discovery is small, or the market is remote, development of the gas field under the production-sharing regime may not prove to be commercially viable. Under an appropriate regime, however, the same development might become attractive, both to the host country and the contractor. Finally, in a typical production-sharing contract, the operational procedure and production-sharing arrangements are not practical for gas operation nor are they in line with the long-term nature of the gas program.

In view of this, we will review a few examples from a typical E&P Contract.

Under the definition article, petroleum is defined as oil and natural gas and all other substances covered by the contract. Therefore, the contract addresses all activities related to oil as well as natural gas. The contract under one of the items specifies that each party shall separately own and shall have the right to take in kind, and dispose of its participating interest share of that portion of the petroleum produced and saved from the contract. Petroleum shall be delivered to each party or its nominee at the outlet flange of the field *storage facility*. . . .

It is obvious that natural gas is not stored like oil in the field and this is neither a standard nor an economical field practice.

Another item indicates that following a commercial discovery and prior to commencement of production from any discovered field or fields, the parties participating *in the development of such field(s)* shall, by mutual agreement, promulgate a set of rules governing the scheduling and other necessary details for the offtakes of available petroleum by such parties. . . . Notice how smoothly and quickly the text has shifted from the commercial discovery phase to the development phase, which is quite true of an oil discovery. However, in the case of a gas discovery, both the commercial evaluation and development of a gas field is absolutely contingent upon the development of a Downstream Gas Development Program. In the case of some domestic and/or export projects, this could be a bigger problem than the whole E&P effort.

Gas Commercialization Process in Recent E&P Contracts

A new trend has been observed in recent E&P Contracts. In the past, the tendency was to design a petroleum arrangement based mainly on oil operation with some provisions for natural gas, namely: the flaring of associated gas, the right for the government to take at no cost the associated gas not used in petroleum operation, and the principle that all the provisions of the petroleum arrangement apply to gas operations (the Mutates Mutandes principle).

As mentioned earlier, due to the specific problems raised by gas commercialization, new provisions have been included in some of the recent E&P contracts. Representative examples of recent improvements in contractual terms are:

1. *Commerciality and new agreement*
   When a gas discovery is made, the parties negotiate to reach a new agreement providing for:

   - Cost recovery.
   - Production sharing rates and procedure.
   - Contract terms which might be different from the contract terms for oil.
   - A model contract specifying that if the contractor declares the discovered field noncommercial, the national company has the right to develop the field. The contractor is allowed to participate in the development project at any time until the end of the development

upon payment of the development cost plus x% of the cost incurred by the national company. The advantage of this provision is in allowing the government to develop a gas discovery under a sole risk clause while giving the oil company a back-in right, or in forcing the oil companies to make an early decision to avoid the x% penalty.

2. *Taxation and profit sharing regimes*
A few contracts provide for lower taxation or better terms in the case of gas discovery. For example:

- The production-sharing rates are subject to negotiation to allow the contractor to be able to obtain a reasonable economic benefit, i.e, one contract provides 70-30 (Government-Contractor) profit sharing for gas instead of 85-15 for oil.
- A reduced royalty rate of 5% or 10% for gas instead of 12.5% for oil.

3. *Gas Price Guideline*
Some contracts provide pricing guidelines with reference to the price of natural gas from other sources and the market price of competing sources of energy or alternative feedstocks.

The preceding improvements are in the right direction, but they are limited to only a few E&P contracts. Many contracts still lack the necessary gas provisions and a sound gas commercialization process. Therefore, these improvements should be extended and integrated so that new contracts fully address the complexities of gas discovery and development programs.

Conclusion

Up to this point we have been talking about the complexity of gas development. Over the past 30 years natural gas has furnished an increasingly important share of the world energy supply. In 1982, about 18% of the world's energy was supplied by gas. On the other hand, only about 10% of world gas is consumed in the developing countries, which hold more than 40% of the world reserves. This indicates that there is a large gas-demand potential in developing countries.

The World Bank projections indicate a potential four-fold increase in gas production in developing countries, provided the constraints on rapid gas development are removed. Domestic gas supply provides the most reliable, clean, and convenient source of energy; gas is ready to be used by consumers essentially right out of the gas network. As discussed earlier, the E&P contract is the first vehicle in gas development. A sound gas commercialization process in a gas oriented E&P contract is an essential first step and a key factor in the development of a gas industry in developing countries. This must be complemented by a gas development strategy to ensure an efficient development. Principal elements of these two key factors were reviewed in this paper.

In addition, consideration and support from the industry and the international community are essential to further gas development in developing countries. They could be obtained by addressing related issues and prospects at international conferences such as this, and in other forums on gas, energy and development problems in the developing countries. More appropriate would be a specific international panel on natural gas development and utilization in the developing countries, with participation of all partners involved; oil companies, gas utilities, engineering, equipment and construction firms, aid and financial institutions, and developing countries with a potential for gas development.

51
Recovery: The Saskatchewan Way
Paul Schoenhals

Introduction

Saskatchewan shares with the developing countries of the world the fact that it is primarily, at least from a historical perspective, an agriculturally-based economy within Canada's national economy. However, it is fortunate in being richly endowed with a wide variety of petroleum and mineral resources. The basic policy of the present government with respect to the petroleum and mineral sectors is to encourage the development of these resources in a manner which helps to diversify the provincial economy; to provide an overall stimulus to the economic development of the province; and to provide a reasonable return to both the province and the oil and mineral industries. Ministers of Energy and Mines throughout the developing world are faced with a similar challenge. Therefore it is highly likely that the experience of Saskatchewan may be a useful case study for governments from the developing world to consider.

Two of the specific issues to be addressed by this First International Conference on the Development of Shallow Oil and Gas Resources, are particularly relevant to the recent policy initiatives of the Government of Saskatchewan. They are:

- To promote cooperation between oil and gas companies and governments that produce or have potential to produce oil and natural gas.
- To involve independent oil and gas companies, in addition to the major companies in the exploration and development from developing countries.

Saskatchewan shares the concern of the governments of developing countries that small and independent local companies must be fully involved in the exploration and production of local resources. Moreover, we firmly believe that, in order for governments to design policies which are effective in guiding the development of the oil industry, they must establish and maintain a cooperative relationship with industry.

A further example of how the experience of Saskatchewan is of relevance to the efforts of developing country governments is that it provides a good "middle of the road" model of how resources can be developed effectively through the combined efforts of government and industry. In the developing world, petroleum and mineral resources tend to belong solely to the state. This has restricted exploration and development of oil and gas resources to either large multinational or government-owned corporations. On the other hand, in countries such as the United States, resource ownership resides primarily with private land owners. Such a system permits fuller participation by small or medium size companies. In Saskatchewan, while the state owns the majority of the petroleum resources, small local firms as well as large multinationals are active in the development of these resources.

There are four significant issues which must be addressed by any jurisdiction before it can effectively develop its resoures. They are:

- Do adequate resources exist for economic development or exploitation?
- Does the technology exist to exploit these resources at economic costs?
- Do adequate and suitable markets exist for these resources?
- Is there an appropriate fiscal regime in place to allow economic development of these resources?

While Saskatchewan has considerable reserves of light and medium oil which are recoverable through conventional techniques, the future of our oil industry lies in exploiting our vast reserves of heavy oil through various enhanced oil recovery techniques. The current government policy is targeted at developing both these resources with a short-run focus on light and medium oils combined with a concentration on our heavy reserves over the longer term.

The long-term strategy for heavy oil will require a government policy which addresses each of the four issues previously listed. This policy will be outlined in a later section of the paper. As regards our policy for light and medium oils, the only operative constraint to the full development of these resources is the establishment of an appropriate fiscal regime. The Government has been very successful in implementing such a fiscal system and its experience is of relevance to governments of developing countries who have little or no constraints as regards the first three issues but who are grappling with the issue of how to set up an appropriate system of taxes and incentives which will ensure the full development of their resources.

Therefore, while this paper may be of some interest to the representatives of the oil industry, it is primarily addressed to the planners and decision-makers from developing country governments. The experience of the Saskatchewan government is instructive in indicating which pitfalls to avoid and how best to proceed in developing an effective system of taxes and incentives for the oil industry. Furthermore, our experience over the past few years has given us an appreciation about what constitutes good, effective, and lasting policy towards the oil and gas industry. Specifically, we have found five key attributes to good policy.

First, it is important to realize that good policy is better served by a system which provides consistency of rules rather than consistency of results. Too often in the past, governments have attempted to base their policies on predictable results. Unfortunately, the world is unpredictable and changing circumstances soon undermine the assumptions upon which policies had been based. As a result, governments are forced to revise their policy. Thus it is better to choose a policy which is defined by a set of rules which have sufficient built-in flexibility to remain operative and relevant under a variety of situations.

Second, good policy must instill confidence in industry in order to remain effective. As industry cannot plan and make investment decisions in a world of changing policy direction, they have little confidence in policies which are predicated upon predictable results and dogmatic views of the world.

Third, it is important to realize that the benefits of a strong and vibrant oil and gas industry go beyond the level of royalty and taxes which government collects.

Fourth, for policy to remain effective, it is important that industry be consulted throughout the policy formulation process. This is not to say that industry sets the policy but that governments must admit that no one knows industry and its problems and possibilities better than the companies themselves.

Finally, good developmental policies must be relevant to the smaller, local firms connected with the oil industry and not simply be geared to the larger multinational firms.

It was with these guidelines in mind that my government sought to create a new system of incentives for our oil and natural gas industry. Before I discuss the attributes of our new incentive program, I would like to briefly summarize the geological features of the oil industry in Saskatchewan as well as the economic and institutional context in which these incentives were formulated.

Saskatchewan's Oil Industry: A Brief Overview

The Province of Saskatchewan has benefitted from oil and gas development activity for approximately 50 years. Although the search for oil began in 1906 with the drilling of our first well, it was not until 1934 that the first commercial gas well was discovered in the province and it was 1944 before our first commercial oil discovery was made.

The majority of Saskatchewan's hydrocarbon potential discovered to date has been in the form of crude oil. While greater emphasis is currently being placed on natural gas exploration and development in the province, the remarks contained within this paper have been restricted to the crude oil situation.

Interest in oil exploration and development in Saskatchewan did not gain any real momentum until after 1947 when the Leduc field was discovered in the neighboring province of Alberta. During the 1950s and early 1960s a concentrated exploration effort resulted in the discovery of the majority of today's producing pools.

Production grew from one million barrels in 1950 to 52 million barrels by 1960. Due to the rather rapid production declines experienced under primary recovery methods, the early- to mid-sixties were devoted in large part to enhancing recovery by waterflooding. Those efforts were very successful and production rose to a level of 93 million barrels in 1966, our highest production year. Saskatchewan's crude oil productive capacity has been in a general state of decline since that time with the exception of the last year or so which will be discussed later.

Currently, Saskatchewan is the second largest oil producer in Canada, with 10% to 15% of our country's total, or approximately 60 million barrels per year. Of this amount, approximately 75% is produced from lands owned by the provincial government, 1% from lands owned by the Government of Canada and the remaining 24% from lands held by the private sector.

Much of the early oil exploration and development in the province was carried out by major oil companies. While there are several hundred companies and individuals with a working interest in Saskatchewan's oil production today, a the majority of our crude is owned and controlled by less than 50 companies.

Most of Saskatchewan's oil is contained in relatively shallow reservoirs, located between 2,000 and 5,000 feet. The average well currently produces approximately 20 barrels per day and thus most of our production would be considered marginal by international standards. However, Saskatchewan has been able to develop this resource because the producing areas were highly accessible and drilling costs were relatively low. Today, the average well in the province costs between U.S. $150,000–$190,000 to drill and complete.

This advantage has been offset somewhat by the higher operating costs associated with the handling of water which is produced with the oil. Today approximately four barrels of water are produced with every barrel of oil recovered. Because of the marginal nature of our oil industry, it has been particularly sensitive to changing economic conditions over the years.

Saskatchewan produces a wide range of crudes, from light sweet varieties to high sulphur, bitumen-type heavy crudes. Our current slate is made up of approximately equal increments of heavy, medium, and light crudes. While most of the light and medium crudes are sold to refineries in Eastern Canada, much of the heavy and high-sulphur crude has been exported to the U.S. where refineries capable of handling these crudes exist. Because we rely heavily on exports to the U.S. and because those exports are regulated by the federal government rather than the province, Saskatchewan has suffered from some market instability in the past.

Saskatchewan's oil industry enjoyed a period of relative stability from the early 1950s to the early 1970s. Crude prices and drilling and operating costs increased moderately and oil producers enjoyed limited intervention from the provincial and federal governments.

Responses by Government and Industry to the OPEC Price Shocks

The OPEC price shock of 1973 abruptly ended this stability. The Government of Saskatchewan at the time reacted to the situation with new taxation legislation which sought to prevent the realization of large windfall profits by a predominantly foreign-controlled industry. Attempts were also made to gain greater control of the industry by expropriating the mineral rights of large freehold interests and by establishing a publicly owned oil and gas corporation.

The Canadian Government also reacted with major changes in its taxation system. Crude oil prices were regulated in an attempt to protect Canadian consumers from highly volatile energy prices and provide the Canadian manufacturing and processing industries with a possible competitive advantage in international markets.

In retrospect, the policy responses by both governments were over-reactions. The prevailing governments lacked a proper understanding of industry and generally were suspicious that increased profits would simply be removed from the province and the country without any reinvestment. Their reactions, therefore, were hastily formulated. Thus, consultation and even communication between governments and industry were sorely lacking at the very time when they were most needed.

Industry responded to the interventionist actions of the two levels of government in a predictable fashion. Drilling activity, interest in Crown minerals, and crude oil production levels all dropped dramatically.

Although the Government of Saskatchewan retained its high royalty and tax levels for the oil industry, it realized the need for continued oil development in the province. In order to encourage this development it introduced an incentive program which effectively reduced royalties and taxes for those continuing to invest in the province.

The program involved a complicated system of credits and grants that gave preferential treatment to large producing companies and encouraged less than efficient spending by the industry. Though there were problems with the system it did encourage industry to increase spending in this province.

By the late 1970s, industry had regained some of its lost confidence and drilling in the province increased. The second OPEC price shock, in 1979, served to fuel the expectations of industry that better days were just ahead. This speculation of better times saw a new drilling record of 1,498 wells established in the province during 1980.

This renewed optimism was, however, short-lived. In October 1980, the federal government introduced its infamous National Energy Program. Although the Program had several laudable objectives, it was premised entirely on an underlying assumption of ever-increasing world oil prices. In essence, the program's major results were increased taxes and considerably more regulation and control of the oil industry.

The timing of the program could not have been worse. International oil prices began to show signs of decreasing rather than increasing. Canada, as well as other parts of the world, began to feel the effects of the growing recession. As a result, oil demand fell and interest rates began to rise. The downturn in industry activity that followed was even more dramatic than that experienced in 1974 or 1975. The industry in Saskatchewan appeared to be in dire straits.

Saskatchewan's Energy Recovery Program: The Short Term

The situation just described is the one in which the present government of Saskatchewan found itself following the election in early 1982. The foregone economic opportunities associated with a stagnating industry were painfully obvious. Investment in the province was reduced, jobs were being lost, production was falling and, with it, government revenues.

Clearly, one of the primary objectives of the new government was to increase economic activity in the resource sector. Furthermore, the increased activity had to translate into increased economic benefits for the province as a whole. Given the economic recession at the time, these objectives appeared to ignore reality. As a first step in the process of attempting to achieve those objectives, a novel approach was tried—the industry was consulted. The advice they gave was incorporated within an oil recovery program implemented in July, 1982. It included the following features:

- More reasonable royalty and tax levels.
- A success-oriented incentive program.
- Simplified administrative procedures.
- A stable yet flexible fiscal framework.

The success of the program hinged on the effectiveness of an incentive program which could generate an immediate increase in conventional drilling activity in the light and medium oil-producing areas. These incentives had to take a form which recognized the geological and institutional characteristics of our oil industry. They had to be cost effective in eliciting a quick and significant response by industry. Institutionally, the most important feature of Saskatchewan's oil industry is the predominance of a large number of small firms. Thus, the new incentives had to be designed with these firms in mind.

The vehicle selected to achieve the short term objectives was a system of royalty holidays. The holidays ranged from 1 year of royalty-free production for shallow develoment wells to 5 years of royalty-free production for deep exploratory wells. The first attribute of such a program is that it rewards success and not activity for activity's sake. That is, production must be gained from the well before any benefit is realized. This encourages efficient drilling and is effective from an industry viewpoint because recovery of investment is accelerated. The other aspect of such a program is that it does not discriminate between large and small firms and thus provides smaller firms with equal opportunities to increase their activity, or even to establish themselves within the province.

The response to the royalty holidays by industry was overwhelming. In 1983 a new drilling record of 1,843 wells was established. Oil production rebounded from 51 million barrels in 1982 to 60 milllion barrels in 1983. For the first time in 15 years the reserves additions from new drilling surpassed the amount of oil produced, for a net increase in remaining recoverable oil reserves. Perhaps the most dramatic result, however, was the increased interest in the leasing of provincial oil and gas rights. Revenues for the province from the disposition of these rights amounted to $108 million in 1983. The comparable figure for 1982 was $34 million.

The success of the program is continuing into 1984. During the first six months of this year 1,100 wells were drilled as compared to 650 during the same period last year. Revenues from the disposition of oil and gas rights this year are also considerably ahead of those received during the first half of last year.

This immediate, and what now appears to be lasting, response was achieved at a relatively low cost in foregone royalties. The estimated cost in foregone royalties for 1983 and 1984 is approximately $125 million. That amount will be more than offset by the increased land-sale receipts alone. Moreover, a revenue base is being established which will generate returns to the province for many years to come.

The royalty holidays have been extremely successful in terms of generating increased conventional drilling activity. As Saskatchewan's sedimentary basins have been extensively explored, with estimates of 80% of our oil potential having already been discovered, the royalty holiday program represents but a short-term means of achieving the province's oil development objectives.

Saskatchewan's Energy Recovery Program: The Longer Term

However, the future of Saskatchewan's oil industry lies in the enhancement of recovery from existing oil pools, particularly the heavy oil deposits of the west-central and northwest areas of the province. These heavy oil deposits are estimated to contain 25 billion barrels of oil in place. Using primary and secondary recovery methods, less than 10% of this resource can be recovered. However, thermal recovery techniques hold the potential to recover at least 35% of these huge heavy oil reserves. Without doubt, heavy oil development will attract more and more attention as the light oil reserves in Saskatchewan and in other parts of the world continue to decline.

To reiterate, the four basic issues which must be addressed in formulating an oil development policy include adequacy of the resource, the existence of suitable recovery technologies, the availability of suitable and adequate markets, and an appropriate fiscal regime. The adequacy of Saskatchewan's heavy oil resource base has been confirmed. The heavy oil development strategy being pursued by the province therefore addresses the remaining issues.

Saskatchewan will be relying largely on the private sector to develop its heavy oil potential. Because of the capital requirements and associated technological and financial risks, the collective efforts of industry will be required. While the smaller independent oil companies dominate the conventional oil development scene in the province today, it is expected that the larger oil companies will play the major role in heavy oil development.

To attract the risk capital necessary for heavy oil development, appropriate rewards must be offered. The first part of our heavy oil strategy, therefore, deals with fiscal policy. Saskatchewan has very recently developed a separate royalty and tax structure for enhanced oil recovery (EOR), which addresses the high capital and operating costs and technological risks associated with this type of development. Under this new structure, the province will impose very low royalties and taxes until the EOR investments have been recovered. Following project payout, a reasonable sharing of the net operating revenues from the project will occur.

The second part of Saskatchewan's heavy oil strategy is to provide support and encouragement for the development of the new technology required, to allow more heavy oil to be recovered and utilized. A $30 million funding program, jointly supported by the Governments of Canada and Saskatchewan, was recently established to encourage heavy oil and other fossil fuel research, development and demonstration projects within the province.

Even if producible, Saskatchewan's heavy oils have faced some marketing problems in the past. Demand for this low gravity, high-sulphur crude has traditionally been tied to asphalt demand. Given the somewhat limited and seasonal asphalt market in Canada, this heavy oil has had to find a home in the northern tier of the United States. Although this area would appear to offer a stable market for this crude in the short- to medium-term, a shortage of diluent for blending purposes could limit future heavy oil development plans. These heavy oils must be blended with 20–25% condensate, by volume, to enable the crude to be moved by pipeline. There is also concern that producing heavy oil for export out of Canada may not be an effective way of addressing national energy-security objectives.

Not surprisingly, the third part of the province's heavy oil development strategy addresses marketing. A key element in our marketing strategy has been to encourage upgrading of the heavy oil to a more desirable form within the province. Upgrading, by removal of the sulphur and either the removal of carbon or the addition of hydrogen, not only improves the marketability of the crude, but also overcomes the pipeline transportation problem.

Significant progress in this respect has been made during the past year. In August 1983, an announcement was made regarding the addition of upgrading facilities to the Consumers' Co-operative Refinery complex in Regina. With these new facilities, 50,000 barrels per day of light-oil feedstock from a neighboring province will be displaced by Saskatchewan heavy oil and made available to Eastern Canadian refineries. Earlier this summer (1984) an announcement of a second upgrader in the Lloydminster area of Saskatchewan was made. The proposed Husky Oil upgrader will require 54,000 barrels per day of heavy oil of which half will be supplied from Saskatchewan. If all goes well, both projects will be operational by the end of the decade.

Although heavy oil development offers many economic benefits to the province, it is an intensive process with major land use and water implications. Because of this, the government has addressed an additional concern about the future development of our heavy oil reserves: its environmental implications. These environmental concerns are being addressed in the fourth part of our strategy, which deals with an appropriate regulatory framework for heavy oil development. It is a consideration which countries attempting to promote oil and gas development should not overlook. With limited additional effort, developments such as this can proceed in harmony with the surrounding environment.

Conclusions

A strong and profitable oil and gas sector can be a key instrument to further economic development. The economic benefits to both the owners and the developers of petroleum resources can be very significant. In order to maximize these benefits, governments must develop and implement sound resource management policies. Clearly, any policy for such a key developmental sector must be carefully targeted and cost effective. However, of equal importance, the policy must result from a relationship of cooperation and mutual respect between government and the entire industry. This is based on neither casual empiricism nor textbook theories but on our experience over the last two years. This approach has been tried in Saskatchewan and it works.

52
The Management and Development of Oil and Gas Mineral Resources on the Osage Indian Reservation

George E. Revard and Newell Barker

Introduction

The Indian Territory Reserve, now Osage County, Oklahoma, embracing 1,470,559 acres, was purchased from the Cherokee Nation by the Osage Indians on June 14, 1883. Since the Reserve was purchased pursuant to treaty, the Osages were excluded from the General Allotment Act of 1887, and their lands were not allotted until they agreed thereto in 1906. The Act of 1906 allotted the lands of the Reservation to the members of the tribe in severalty, with each member receiving an average of 658 acres, and reserved all minerals to the use and benefit of the tribe, with some small exceptions. The mineral reservation comprises 1,469,079.63 acres, from which over 1.1 billion barrels of oil and 1.7 trillion cubic feet of gas have been produced.

The 1906 act is the organic law governing administration of current Osage affairs, and marks the beginning of special legislation enacted by Congress for the exclusive benefit of the Osage Tribe and its individual members. The act gives the Tribal Council authority to manage the mineral estate, but with approval of its actions by the Secretary of the Interior. This approval has been delegated to the agency superintendent, the United States Geological Survey and other federal and state agencies being excluded. During the early 1930s the USGS began to supervise oil and gas operations on Indian lands. By that time the Osage Agency, responding to the activity on the Osage Reservation, already had in place an active Branch of Minerals to oversee tribal interests and perform the trust responsibility of the Bureau of Indian Affairs.

The tribe has been vocal in stating that it does not want anything to jeopardize the trust status of its mineral estate. Because of the varied surface ownership in Osage County the tribe finds itself in Federal Court an average of three times a year. Most cases involve ingress and egress to the mineral estate. In all cases the tribe is represented by the U.S. Attorney's Office. Three Department of the Interior field solicitors are stationed at the Osage Agency to handle litigation and other legal matters pertaining to the mineral estate. The tribe is well aware of the advantage afforded to it by this federal support. Less than 200,000 surface acres in Osage County are still owned by restricted Osage Indians and the leasing of this land and other operations are managed by the Osage Agency. The remaining 1,270,000 surface acres are held by individuals; this land is not under the control of the Federal Government, but the mineral estate is intact. The largest portion of the Bureau of Indian Affair's area budget, which is supported by the tribe, goes to the Minerals Branch. The tribe, through P.L. 93-638 contracts (contracts to the tribe funded by the Bureau of Indian Affairs), has computerized the entire operation under the supervision, and with the technical assistance of the Bureau of Indian Affairs.

General Operations

The Osage Agency superintendent has the delegated responsibility of managing all oil and gas operations on the Osage Reservation. This includes approval of oil and/or gas mining leases and approval of drilling, workover, and plugging applications. The superintendent also maintains accurate records of all production, and income received; appraises damage and collects compensation on restricted Indian lands; reviews all incoming well records for conformity with Agency standards; monitors lease operations to prevent pollution (surface and subsurface); and conducts lease operations prudently. The Minerals Branch has a staff of 27 employees (13 Bureau of Indian Affiars employees and 14 tribal employees) to supervise engineering, lease compliance, office, computer, and accounting operations.

The superintendent of the Osage Agency is also responsible for managing all other mineral operations, including mining leases for sandstone, gravel, sand, clay, and limestone.

Existing oil and gas leasing operations in Osage County are conducted as follows. A specific tract of land for an oil, gas, or combination oil and gas mining lease is nominated for auction at an upcoming Osage Oil and Gas Lease Sale by a prospective lessee. A nominating bid must accompany the nomination in order for the tract to be considered for listing at the sale. The amount of the nominating bid is reviewed by the Osage Agency minerals staff, and if it is considered to be equitable for an opening bid, the tract is advertised for sale at public auction. The lease sales are held three times per year on the third Wednesday of February, June, and October. An oral auction is conducted at each lease sale and the highest bidder is awarded a lease on the tract in question after approval by the Osage Tribal Council and agency superintendent. The successful bidder on a specific tract must obtain a performance bond and file all necessary corporate papers before his lease is effective. There were approximately 2,100 oil leases, 200 gas leases, and 1,700 combination oil and gas leases in force in 1984.

The lessee may file an application to drill a well after he has obtained a valid oil mining lease. Prior to filing an application he must contact the landowner and discuss the route of ingress and egress to the well and the proposed location site. He must also pay the landowner a commencement fee of $300, which is a credit toward the total damage, before submitting his application to the Osage Agency for approval. As soon as the proper procedure has been completed he submits a drilling application to the Osage Agency and he is granted a drilling permit.

If a drilled well is nonproductive, it is plugged with cement in accordance with Osage Agency requirements to protect potential fresh water or other potentially productive zones, to prevent contamination. After this has been done, the landowner makes claim to the oil lessee and damage claims are settled. If the well is productive, all necessary equipment is installed, the surrounding area is returned as nearly as possible to its natural state, and the landowner makes claim to the lessee for alleged damages due to oil operations.

The well is produced until it is determined by the lessee to be uneconomical. At that time he must make application to the Osage Agency for approval to plug the well. The oil, gas, or combination oil and gas lease is terminated on a specific tract as soon as production ceases.

In the event of possible oil theft reported to the tribe or the Bureau of Indian Affairs (BIA), or discovered by Agency field inspectors, the lease is immediately sealed off by the superintendent. The Federal Bureau of Investigation (FBI) is notified and the Osage County Sheriff's office is called. The tribe, the BIA, FBI, and county law enforcement officials have procedures for dealing with theft. Over the past ten years, there have been numerous reports of theft on the Osage Reservation; in most cases, the alleged thefts were the result of poor communication between pumpers, truckers, BIA, and the tribe. One case of actual theft resulted in prosecution.

Osage Agency Engineering Section

The engineering section is under the supervision of the senior petroleum engineer and is responsible for:

1. Evaluation of area nominated for lease sale and preparation of comments for sale bulletin.
2. Issuance of drilling, workover, saltwater disposal, and plugging permits, if applicable.
3. Review of oil and gas leases for development requirements.
4. Processing of completion and workover reports for wells in Osage County.
5. Review of seismic applications for consideration for approval.
6. Review and processing of "Natural Gas Policy Act" gas-price determinations.
7. Coordination of saltwater injection and disposal activities with Environmental Protection Agency (EPA) representatives.
8. Resolution of questions concerning mineral operations in Osage County.
9. Resolution of technical problems pertaining to waterfloods and all other mineral operations.

Osage Agency Field Operations Section

The authority for lease compliance on the Osage Reservation is contained in 25 CFR Part 226. The field section consists of a section chief, 6 field inspectors, 2 gaugers, and a clerk-typist.

The Osage Reservation is divided into 6 working areas, each assigned to a field inspector. Radio-control pickup trucks provided by the BIA and the Osage Tribe are utilized by the inspectors during their daily patrols of the leases. In an effort to maintain lease compliance, the inspector varies his schedule from day to day and does not routinely check any particular lease. His main responsibilities are to:

1. Police field operations to see that lessees are in compliance with federal regulations and lease terms.
2. Monitor leases to assure that they remain on production or are canceled.

3. Inspect terminated leases prior to releasing bonds to be sure lessees have performed necessary cleanup operations.
4. Issue required plugging instructions and witness plugging operations on all wells.
5. Review leases for spill prevention, counter-measure, and control plans prior to approval.
6. Acts as a mediator in disputes between mineral owners and lessees.
7. Inspect and appraise oil-operation damages on restricted property and collect damage reimbursement.

If a lease is found to be in need of repair, is causing pollution, or is in non-compliance with regulations and lease terms, a "Notice of Lease Inspection" is sent to the lessee and he is notified to correct the deficient conditions by a specified date or he will be subject to lease shutdown, cancelation, or fine. If the lease problem must be corrected immediately, the field man has the authority to shut down any wells causing pollution; he then contacts the lessee to resolve the problem.

When a lessee fails to make timely corrections of violations, as provided by the "Notice of Lease Inspection" or by order of the superintendent (25 CFR 226.30), penalties may be imposed as authorized under Section 226.42 and 226.43.

The Osage Reservation is divided into 2 work areas, each of which is assigned to a specific gauger. The gauger is responsible for spot gauging oil contained within a tank battery before the oil is measured by a representative of the purchasing company. The gauger's analysis of the amount of oil sold is compared with the purchaser's statement of the amount purchased and the income received. If the amount determined by the purchaser does not agree with that determined by the agency gauger, then the purchaser is requested to explain the discrepancy. If the reason is not satisfactory, the lessee is obligated to pay royalty on the difference in the amount removed from the lease. Very few discrepancies have been found in the measurements. The gauger is required to check all tanks to see that they are properly locked and sealed; if they do not meet Osage Agency criteria, then they may be sealed by the gauger until the problem is rectified.

Osage Agency Lease and Accounting Section

This section is under the supervision of the realty specialist who must:

1. Receive and review oil and/or gas nominations from prospective lessees and prepare lease sale bulletin.

2. Issue oil, gas, and other mineral mining leases and permits.
3. Process lease assignments.
4. Maintain records on oil, gas, and other minerals production and income.
5. Review lease terms and initiate paperwork to cancel or terminate leases for failure to pay rental or maintain production.
6. Review gas contracts.
7. Maintain plat books indicating location and status of all wells and leases.
8. Maintain all lease records and files pertaining to mineral operations.

Osage Agency Automatic Data Processing (ADP) System

The Osage Agency currently uses an IBM System 34. It has 128K of memory and 128 megabytes (70% being utilized) of disk storage. The memory can be expanded to 256K and the disk can be expanded to 256 megabytes.

There are 10 workstations and 5 printers directly attached to the existing communications port; 64 more devices can be added. Current applications include:

1. Minerals accounting, consisting of:
   A. Production history by lease
   B. Rental contracts
   C. Production and royalty accounting
   D. Lease status and performance tracking
   E. Electronic fund transfers
2. General-ledgers-fund accounting
3. Payroll
4. Word processing
5. Individual Indian accounting
   A. Annuity ledgers
   B. Quarterly minerals payment
   C. Daily-transactions posting ledgers

Each of these systems is supported with a variety of screen inquiry and listing programs.

Osage Agency Electronic Funds Transfer

In June 1979 the Osage Agency initiated the Electronic Funds Transfer system to deposit royalty received from oil and gas purchasers. In mid-

1984, approximately 98% of royalty income is received through electronic funds transfer. A request has been made to the Office of Finance Management, Albuquerque, NM, for approval to use a local bank for minor deposits for electronic funds transfer processing. This would speed up deposit of funds received from small royalty oil purchasers, lease bonus deposits on sale dates, and oil operation damage income received for individual Indian accounts.

Projected Annual Oil and Gas Activities

1. *Leasing:* approximately 350 leases are sold at lease sales during the year. Each lease normally consists of 160 acres. Based on the present price of oil, it appears that leasing activity will continue at the present level or possibly increase in the future as the energy needs of our country increase.
2. *Geophysical:* approximately 50 geophysical exploration permits are issued each year. Normally, existing access is used but in some instances the equipment is moved over pasture or wooded land in fair weather to drill the required shot holes and conduct seismic surveys. Damages are paid to the landowner for all geophysical work conducted on his property. It is anticipated that this activity will continue at the same pace in the future.
3. *Geological:* the Osage Nation has entered into a contract (pending final approval) with Earth Satellite Corporation, Chevy Chase, Maryland, to prepare a geologic remote sensing study of the Osage Reservation to complement the subsurface geological studies currently available at the Osage Agency. This project has been designed so that the data produced will stimulate exploration interest among oil companies likely to be interested in exploring for oil and gas on tribal lands.
4. *Drilling:* approximately 700 drilling permits are issued by the Osage Agency on an annual basis. Of all wells drilled, approximately 71% are completed as oil wells, 7% as gas wells, 9% as service wells, and 13% as dry holes, which are plugged and abandoned. The drilling activity should continue at its present level (see Table 1).
5. *Production:* approximately 9,900,000 barrels of oil and 10,100 million cubic feet of gas are produced annually in Osage County. The oil production will decline at a rate of approximately 0 to 3% per year, based on the present drilling activity. The gas production has stabilized, and future activity will depend on gas prices and demand (see Table 2).

Table 1
Well Drilling and Completion Activity, Osage Indian Nation, 1973–83

Year	Well Locations	Oil	Gas	*Other	Dry
1973	296	117	33		62
1974	653	182	84		118
1975	482	309	64		127
1976	564	306	33		112
1977	632	431	33		145
1978	664	415	38		147
1979	903	397	36		158
1980	1,192	553	37		190
1981	992	584	45		168
1982	658	508	43	63	100
1983	713	463	47	58	84

* Salt Water Disposal, Water Supply, Gas Injection, Water Injection, Temp. Abandoned (T.A.).

Table 2
Oil and Gas Production, Osage Indian Nation, 1973–83

Year	Oil Production (bbl)	Gas Production (mcf)
1973	10,467,123	2,038,521
1974	10,066,196	2,171,007
1975	10,280,296	2,867,904
1976	10,955,531	5,006,719
1977	11,146,128	9,364,811
1978	10,966,516	8,124,228
1979	10,394,374	9,036,862
1980	9,895,217	10,597,031
1981	10,317,439	9,677,573
1982	9,986,684	9,901,406
1983	9,872,351	10,106,574

6. *Transportation:* approximately 3,500 miles of oil-company-maintained roads and approximately 2,000 miles of county-maintained roads are utilized by oil lessees for the day-to-day operations of their properties. Approximately 16,800 wells located in the county are checked on a daily basis by an oil-field pumper, who normally makes his rounds in a pickup truck. Workover units, drilling rigs, cementing trucks, acidizing trucks, and logging and service trucks utilize the roads to drill approximately 700 wells per year and to repair or work over approximately 3,000 previously-drilled wells

per year. The roads are utilized by oil-trucking firms that purchase and truck approximately 30% of all oil produced in Osage County. The trucks carry approximately 160–200 barrels of oil per load. Approximately 200 miles of new roads are constructed annually for new oil operations. Approximately 8,000 miles of oil, gas and salt-water lines traverse the county to deliver products for sale or disposal. Approximately 150 to 200 miles of lines are installed annually for new oil operations. It is anticipated that this level of activity will continue in the future.

7. *Abandonment:* approximately 150 wells are plugged and abandoned every year in Osage County. In accordance with agency regulations, the casing is cut off three feet below ground level and the site is returned as near to its natural state as possible on abandonment. It is anticipated that the same level of activity will continue in the future.

8. The Osage Tribe pays a gross production tax of 5% to the State of Oklahoma for royalty oil and gas. Part of this tax is returned to Osage County for schools and roads.

53
The Development of Small Fields: The Case of Guatemala

Douglas Rosales Juarez

Background

Guatemala is situated in the northern part of Central America, immediately to the south of Mexico, and has an area of 130,000 square kilometers (50,200 square miles) and approximately 7 million inhabitants (Figure 1).

Even though the country has great potential in other sectors, the basis of the economy is agriculture. At present, the most important export goods include coffee, sugar, cotton, and bananas. In 1983 oil occupied third place among exports.

There is a reliable network of roads and landing-strips and an oil pipeline 237 kilometers in length which has a capacity of up to 50,000 barrels of oil per day. This pipeline runs from the southern part of Petén to the Atlantic coast, where port facilities for the export of oil are available.

The country's consumption of hydrocarbons is barely 24,000 barrels of oil per day; in 1983 an average of 6,980 barrels per day was being produced, so that levels of production for export will be easily attainable.

The Ministry of Energy and Mines (MEM), which is responsible for the petroleum sector, signs petroleum-operations contracts on behalf of the government, since there is no state petroleum enterprise. The General Hydrocarbons Office, one of the Ministry's five General Offices, is responsible, among its other functions, for seeing that the contracts just mentioned are carried out.

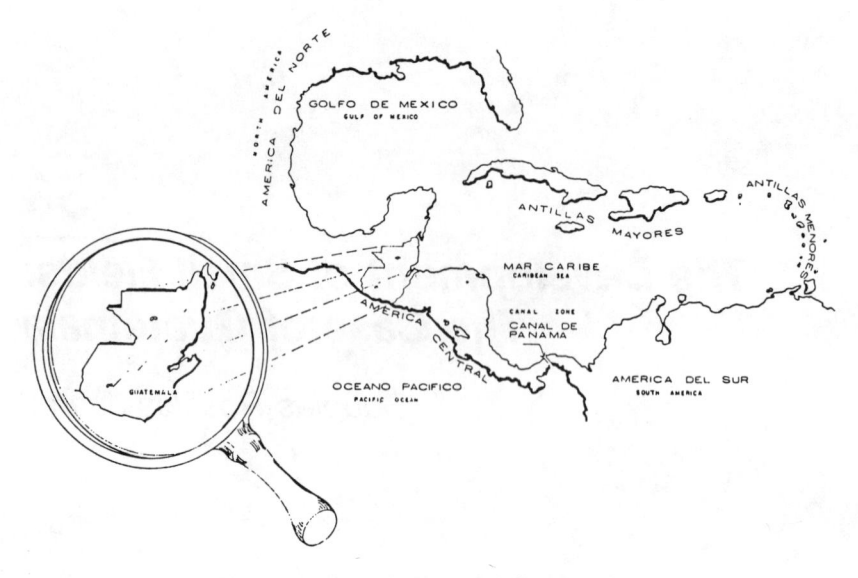

Figure 1. Location map of Guatemala.

Sedimentary Basins

There are four sedimentary basins in the country with hydrocarbon potential:

1. South Petén or Chapayal Basin
2. North Petén or Paso Caballos Basin
3. Amatique Basin
4. Pacific Basin

The Petén basins are extensions of the basins whose formations are being explored and exploited in that region of Mexico. The two basins have an area of approximately 40,000 square kilometers (15,440 square miles) and are made up of a thick sedimentary series. According to available information, these basins are among the deepest and least explored in the world.

Chapayal Basin

This basin's environment of sedimentation is transitional and ranges from continental (deltaic) facies to marine platform; the lithology varies from limestone, dolomites, and anhydrites to clastic sediments.

Since the geological conditions vary throughout the basin, it is not possible to establish a pattern of traps. To date, more than 50 structural prospects have been identified; the majority of the exploratory wells have been seeking some of these structural targets (anticlines, salt domes) at different levels of the Cobán Formation (member C). However, some geologists think that the true targets lie at greater depth (Todos Santos) in reef or stratigraphic traps, such as facies changes, which have hardly been studied or explored at all.

Oil has been discovered in this basin in the salt dome of Tortugas (1972) and in the anticlines of Rubelsanto (1974), Chinajá-Oeste (1977), Yalpemech (1980), Caribe (1981), San Diego (1981), and Tierra Blanca (1983). The deposits are located in lagoonal-facies limestone of the Early to Middle Cretaceous system.

Paso Caballos Basin

The environment of deposition of the Paso Caballos Basin is of the sabkha type, as is shown by the high proportion of evaporite rock. More than 6,000 meters (20,000 feet) of sediment fill the geosynclinal axis in the northern part of the Cordillera Misteriosa (Berner, 1975).

The dolomitization and fracturing of the limestone has generated secondary porosity. The drilling targets in this basin have been located within the Cobán Formation (member C) and in the subjacent clastic Todos Santos Formation. The structures are fewer in number and of lesser structural relief, although of greater area, than those of the Chapayal Basin. For the six wells drilled in this basin, one, the Texaco-Amoco Xan-1, produced oil; the other wells located only thick deposits of sedimentary rocks showing indications of hydrocarbons.

Amatique Basin

This might be considered as an eastern extension of the Chapayal Basin, but it differs therefrom in that its facies changes took place during Cretaceous time. On the other hand, it shows Tertiary clastic sediments which are not found in the Chapayal Basin.

The Tertiary stratigraphic sequence drilled in the Manglar-1 well (1,935 m) shows an alternation of limestone and sandstone, marly limestone, and, towards the base, gypsum-bearing limestone and dolomite.

A second well, Manabique 1-C, was drilled to a depth of 4,230 m; the Tertiary series included clastic sediments, limestone, dolomites, and anhydrites. The Cretaceous system is represented by dolomites, limestones, and siltstones. The oldest rocks drilled are Jurassic limestones and dolomites. This is not a salt basin and the evaporite sediments are mostly of anhydrites and gypsum-bearing limestones. However, geological information about the basin is limited, since only four wells have been drilled in it, two on the landward side (Castillo Armas and Livingstone) and the two on the seaward side.

Pacific Basin

The Pacific Basin is the least explored of all. It is a coastal strip approximately 56 kilometers (35 miles) wide, made up of Tertiary clastics and pyroclastics. Four wells have been drilled through a thick series of clays and sandstones. Texaco drilled one of the land wells and Esso drilled another offshore. Recently, a scientific expedition found signs of solidified methane.

Exploratory Work

As of December 31, 1983, 63 wells have been drilled (excluding five wells which were abandoned and replaced by others) in the country's sedimentary basins, of which nine have been classified as development wells (Table 1).

A total of 54 exploratory wells have been drilled in Guatemala (135,691 meters or 445,066 feet), and in the Chapayal Basin 36 exploratory wells have been drilled (85,847 meters). Of this total, 52 were originally classified as exploratory wells; however, only 15 of them actually reached the target formation. As a result, only these 15 wells may properly be called exploratory wells. Other wells reached the target formation but were inadequately tested, and others were abandoned following

Table 1
Wells Drilled According to Location
(1983)

Location	Number of Wells	Drilling Meters	Drilling Feet
Chapayal	45*	103,780	340,398
Paso Caballos	10†	31,434	103,104
Amatique	4‡	10,101	33,131
Pacifico	4	8,309	27,255
Total	63	153,624	503,888

Wells were not included which were drilled only to a certain depth in order to drill new ones:
* Three wells not included.
† One well not included.
‡ Two wells not included.

mechanical breakdowns. The development wells drilled in the Chapayal Basin total 17,933 meters (58,821 feet).

Under the National Petroleum Law, which predates the Hydrocarbons Law (September 1983), exploration and exploitation contracts were signed for six areas, of which four were returned to the State: the first in 1981 (Area BB, operator, Getty Oil) and the other three at the end of 1983 and beginning of 1984 (areas AA, E and D, Hispanoil operating the first two, and Texaco the last).

Present activity (Figure 2) is in Area L, where an exploratory well (operator, Texaco) is being drilled and two more are planned for 1984-85, and Area I (operator, Elf Aquitaine), site of the country's present crude-oil production, where an additional exploratory well will be drilled in 1984.

Most of the exploratory drilling in the country was completed under the National Petroleum Law (the first well was begun in 1979); on the last day of 1983, 23 exploratory wells, 3 development wells and 2 injection wells had been drilled on the basis of this legislation, making a total of 76,429 meters (250,687 feet) of exploratory drilling and 7,420 meters (24,339 feet) of development.

Discoveries and Commercial Fields

On the basis of the exploratory work completed to date in the Petén basins, the prospects of large discoveries in the Paso Caballos Basin are very promising; in the Chapayal Basin, at least in the members of the Cobán formation which have been tested to date, experience indicates that the fields might have reserves of up to 20 million barrels of recoverable oil per structure. The majority of the wells drilled in the Chapayal Basin which have reached the Cobán formation have shown indications of hydrocarbons; seven of the eight discovery wells in Guatemala are located in Chapayal, so that the probability of finding hydrocarbons in this basin is fairly high.

Taking into account the exploratory wells which did not reach their targets, were inadequately tested, or abandoned because of mechanical breakdowns, it may be said that the probability of discovering petroleum in the Chapayal Basin is approximately 20%. Any attempt to establish a percentage of probability for the other basins is much more uncertain. Only one exploratory well has found natural gas in Guatemala, the Chisec Este 1 (1982), which was drilled in the Chapayal Basin in Area I; during a drill-stem test (DST) it produced 30,000 m^3 of gas per day.

The majority of the discoveries in Chapayal have been of medium and light crudes, 22° API and higher. The first discovery well in the Paso Caballos Basin, the Xan 1 (1981), found heavy crude (16° API) esti-

608 Shallow Oil and Gas Resources

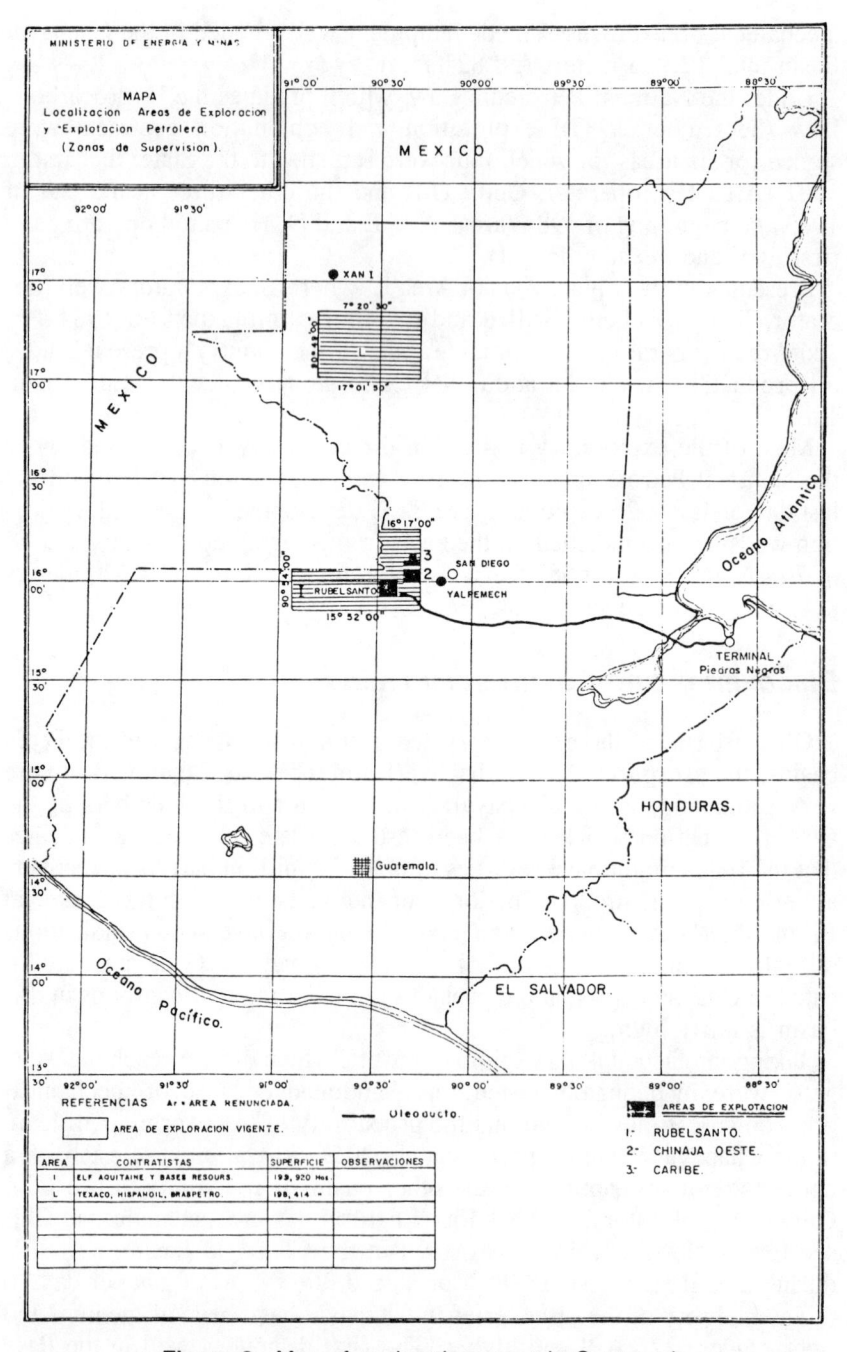

Figure 2. Map of exploration areas in Guatemala.

mated at 40 million barrels of oil in place. Thus, we can say that fields with widely varying characteristics may be discovered in Guatemala. It has been verified that there are small fields, and there is also evidence that there may be large fields; the crudes are of varying API gravity, and there remain target formations which have not been explored. Production from the smallest fields, although modest, could make a significant contribution to the country's self-sufficiency. Six fields, with a production of 4,000 barrels of crude oil per day each, would be enough to ensure self-sufficiency, even without taking present production into account.

To date there are three exploitation areas, Rubelsanto, Chinajá Oeste, and Caribe. The commercial exploitation of the first two began in March, 1980; all three are situated in Area I, where the operating company is Elf Aquitaine Guatemala.

Cumulative production to March 1984 was a little over 9 million barrels of crude oil with an average API gravity of 28° and the remaining recoverable reserves amounted to 10.2 million barrels; this does not include the Caribe discovery nor does it take into account other reestimates which have been made (Figure 3 and Table 2).

Table 2
National Production and Sales of Crude Oil in Guatemala, 1979–1983
(barrels)

Period	Domestic Sales	Exports	Total Production
Total 1978	215,200		215,200
Total 1979	568,700		571,400
4th quarter 1979	135,317	--	135,317
1st quarter 1980	165,017	--	360,615
2nd quarter 1980	133,881	396,275	462,529
3rd quarter 1980	124,919	243,042	368,699
4th quarter 1980	106,408	142,297	275,564
1st quarter 1981	166,461	263,736	451,736
2nd quarter 1981	201,696	123,763	399,298
3rd quarter 1981	222,641	140,583	334,885
4th quarter 1981	164,391	133,500	307,568
1st quarter 1982	167,500	396,200	564,400
2nd quarter 1982	161,400	281,600	508,500
3rd quarter 1982	187,600	435,500	579,400
4th quarter 1982	213,600	432,700	639,800
1st quarter 1983	130,800	429,600	620,600
2nd quarter 1983	117,800	610,600	703,800
3rd quarter 1983	57,000	560,000	662,000
4th quarter 1983	39,300	605,800	562,900
1st quarter 1984	163,000	362,900	515,500

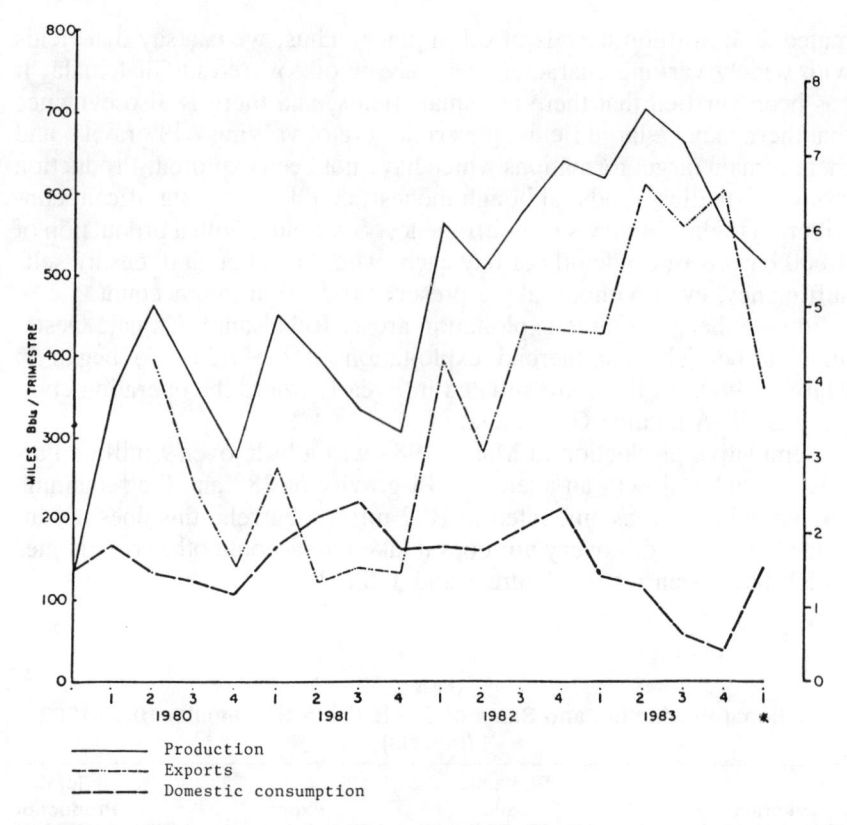

Figure 3. Cumulative production, export, and internal consumption of oil in Guatemala, 1980–84.

Hydrocarbons Law

With the developments in the world economy of recent years, which have been reflected in the deterioration of the regional and national economies, the importance of the petroleum factor has been obvious, quite apart from other considerations as to the value of this product, such as security of supply or energy independence. In the light of what has been said, there can be no question of the country's need to explore for petroleum reserves, all the more so since a hydrocarbon potential does exist.

The Ministry of Mines and Energy has identified the following goals to be attained through petroleum exploration and exploitation in Guatemala:

1. In the short term:
   - Signing of new exploration and exploitation contracts.
   - Promotion of investment in the country so as to produce new jobs and other benefits for the national economy.
2. In the medium term (up to three years):
   - Attainment of national self-sufficiency in petroleum, contributing to the country's energy independence and correction of the balance of payments.
   - Optimum use of State utilities, seeking the point of balance at which the State earns a profit but investment in petroleum exploration is still attractive.
3. In the long term:
   - Fostering the growth of the country's economy.
   - Guaranteeing the supply of crude oil and petroleum products.
   - Carrying out a complete exploration of the country's sedimentary basins in order to determine the resources available to it.
   - Maximizing the profits and participation of Guatemalan industries by offering materials, equipment, services, and personnel for petroleum development.

It is clear that, with the introduction of the National Petroleum Law, many of these goals will not be attainable within the time-framework, and studies have therefore been carried out as to reforms that might be needed. Such reforms would be designed to make possible the exploitation of all types of fields—those which today would be classified as marginal or small, as well as those with reserves of heavy crudes.

There is a wide variety of petroleum legislation in force in the world, and the large majority of developing countries exploit their petroleum resources by means of contracts. The system of concessions has become a thing of the past owing to the unhappy experience of some countries, when the rights of the State over its resources were seriously affected.

The new legislation, the Hydrocarbons Law, provides the legal basis by means of contracts, for the development of any kind of petroleum operation.

The introduction of the "Production-Participation Contract," in which the State assumes no risks, creates legal, technical, and economic conditions designed to make investment in petroleum exploration and exploitation attractive. The economic arrangements are flexible and based on the realities of the country's situation, taking into consideration the geological characteristics and prospects, local exploration costs, and the quality of our crudes. The economic and geological risks are reduced by making small fields and/or heavy crudes profitable.

Production-Participation Contract

The provisions of the production-participation contract which will be applied to the allocation of new exploration and exploitation blocks are set out in the Hydrocarbons Law and its Regulations. Chief among these are:

1. A contract may have a maximum duration of 25 years and cover a maximum of six blocks. Land blocks may each be up to 50,000 hectares (193 square miles) in area and offshore blocks up to 80,000 hectares (309 square miles). At the end of the fifth year of the contract, 50% of the original area must be given up. The whole of the area, with the exception of exploitation zones, must be returned to the State at the end of the sixth year of the contract. An exploitation zone may not exceed 10,000 hectares (39 square miles) unless the size of the field is greater.
2. By signing the contract the contractor enters into certain commitments, divided into phases—one mandatory, the other optional. The development of a field designated as commercial must be carried out with due diligence; the Regulations of the Hydrocarbons Law specify the modalities and scope of this principle.

    The contractor may dispose freely of his production, except that he must sell to the State a sufficient amount to satisfy the internal market or a maximum of 55% of the oil, whichever is greater.
3. During the exploitation of a field the contractor must pay the corresponding royalty to the State and he uses the remaining production to recover, as a first priority, all the exploration and exploitation costs relating to the contract area. The disposable hydrocarbons, i.e., the production remaining after payment of royalties and recovery of costs and expenses, is divided between the contractor and the State in accordance with a scale based on the level of production. The scale for distribution of the disposable hydrocarbons will be established in the bid, on the basis of a maximum of 70% for the contractor in the case of fields with low productivity.
4. In order to make the exploitation of heavy-crude fields economical, a new system will be established for determination of the royalty, which will vary according to the monthly average of API gravity, beginning with 5% for oil of 15°API and increasing by 1% for each additional API degree. The minimum of five percent

is the same as the royalty for natural gas and gas condensate. The average API gravity of the crude oil now produced in the country is 28°, which corresponds to a royalty of 18%.
5. There is an administrative tax of 100,000 quetzals payable on signature of the contract. This tax plus capitation dues, of which the contractor may keep 30% for capitation of his own workforce, are considered to be operating costs and are therefore 100% recoverable. The per-hectare taxes are 0.50 quetzals per year for the exploration area and 5.00 quetzals per year for the exploitation zone, which are also recoverable.
6. With regard to the payment of income tax, in any given period 100% of all exploration and exploitation costs and expenses incurred in the country may be deducted. The maximum rate of tax is 42%, and losses incurred in one fiscal year may be transferred to subsequent periods until they are completely recovered. Income tax paid in Guatemala may be credited against United States taxes.
7. The price of each type of crude oil produced in the country is determined on the basis of international market prices, account being taken of the official export prices of the exporter countries in the area and of prices in the spot market.
8. The contractor may import free of duty the materials and equipment which he requires for his operations.
9. Funds may be transferred freely within and outside the country. The Guatemalan quetzal maintains official parity with the United States dollar. The General Regulations of the Hydrocarbons Law cover any variation in the official rate of exchange in order to maintain the value of investments at a constant level.
10. The contractor has the right to build an oil pipeline for transporting his production when alternative transport facilities do not exist, and a contract is signed for this operation. The oil pipelines may be used by all producers without discrimination as to services and rates.
11. Provision is made for the assignment or transfer of rights to other companies.
12. All information and data obtained must be communicated to the Government; certain information may remain confidential for a period of two years.

Allocation of New Blocks

The Ministry of Energy and Mines plans to issue an invitation to tender for the signing of contracts in potential petroleum-bearing areas of the Chapayal and Paso Caballos basins and it will also entertain with great interest bids with respect to the other sedimentary basins.

Although the Hydrocarbons Law provides that contracts for petroleum operations may not be signed unless bidding has taken place, the legislation does allow the Ministry of Energy and Mines to reach agreements in principle with the companies concerned with regard to blocks, surface exploration work, drilling, the degree of participation in production of both parties, and other contractual details. Such agreements are used for the purpose of bids for the chosen blocks and the bidder can maintain or upgrade his initial offer with a view to signature of an exploration and exploitation contract. This flexible arrangement makes it possible for interested companies to work out these very important issues with the Ministry of Energy and Mines within the legal framework so that bids can be received for areas in which interest is shown.

Economic Analysis

The flexibility of the production-participation contract just described can be illustrated by applying it to three cases in which various figures are used for the number of exploratory wells drilled, the number of drilled dry wells, the volume of recoverable reserves discovered per well, and the API gravity of the oil. The contract is applied to three cases: in the first two, only one field is discovered; in the third there are two fields and several exploratory wells.

In the cases analyzed, the price structure is uniform and conservative; the price is adjusted for each type of crude according to its API gravity. The price remains fixed for the first two years, increases by five percent annually for the next two years and increases by ten percent until the field is exhausted. Prices, investments, and operating and transport costs were determined on the basis of 1984 and in the analysis, quantities have been discounted by seven percent.

For the purposes of the analysis it was assumed that the company would not have recourse to loans for its investment.

The following scale of participation in the production of disposable hydrocarbons was applied:

	Company	State
0– 5,000 bpd	70%	30%
5,000–10,000 bpd	60%	40%
10,000–20,000 bpd	50%	50%
20,000–50,000 bpd	40%	60%
over 50,000 bpd	30%	70%

Tables 3 and 4 summarize the data used as the basis for the economic analysis and some of the results.

Table 3
Summary of Results of the Economic Analysis, 1984
(million Quetzales)

	\multicolumn{6}{c	}{Cases}
	1A	1B	2A	2B	3A	3B
Total revenue	133.3	275.8	116.1	240.3	836.1	729.9
Transportation cost	23.1	44.9	23.1	44.9	127.2	127.2
Revenue	110.2	230.9	93.0	95.4	708.9	602.7
Investments	23.2	54.9	23.2	54.9	157.7	157.7
Operating expenses	14.0	26.3	14.0	26.3	75.9	75.9
Operating income	73.0	149.7	55.9	114.2	475.3	369.1
Payments to the state	54.3	111.5	39.7	81.4	351.9	260.7
Net cash flow	18.7	38.2	16.2	32.8	123.4	108.4
Years of repaid investment*	2	3	2	3	3	3
Net present value 25% of project	2.0	1.0	0	−1	14.0	7.9
Rate of return (%)	27.9	26.2	25.4	23.8	31.1	28.5
Exploration wells[†]	1	2	1	2	5	5
Discovery wells[†]	1	1	1	1	2	2

* Years since initial production
[†] Number of wells

FOB Price, crude oil 30° API = 27.79 Q/barrel (1984).
FOB Price, crude oil 22° API = 24.25 Q/barrel (1984).

Table 4
Summary of State Income
(million Quetzales)

	\multicolumn{6}{c	}{Cases}
	1A	1B	2A	2B	3A	3B
Production participation	16.2	32.9	14.3	29.2	110.0	98.6
Royalties	22.0	46.2	11.2	23.4	141.8	72.3
Real estate tax	16.1	32.4	14.2	28.8	100.1	89.9
Total	54.3	111.5	39.7	81.4	351.9	260.7

Case 1

It is assumed that one exploratory well is drilled in the first three years of the contract and that a field of five (case 1A) or ten (case 1B) million barrels of 30° API oil is discovered.

Case 1A

An analysis is made of the possible result of a production-participation contract under which an exploratory well is drilled in the third year of the contract, with the positive result of the discovery of 5 million barrels of recoverable 30° API oil.

The economic analysis indicates that the investment is recovered in two years from the start-up of production. The project has a rate of return on investment (ROR) of 27.9%.

Case 1B

In this case it is assumed that an exploratory well is drilled in the third and fourth years of the contract and that the latter well discovers a field of 10 million barrels of recoverable 30° API oil. The economic analysis shows that the investment, despite being more than double that of case 1, is recovered in three years from the start-up of production. The project has an ROR of 26.2%.

Case 2

This case shows the economic effect of the discovery of a field of crude oil of 22° API under the same conditions as in cases 1A and 1B; the alternatives are designated Case 2A and Case 2B respectively. In addition to the adjustment of the price of the crude, the royalty automatically becomes 12%.

Case 2A

Despite the reduction in API gravity from 30° to 22°, the investment is recovered in the same period of two years. The ROR decreases by 2.5% to 25.4%.

Case 2B

In this case, as in the previous one, the reduction in the gravity does not affect the investment-recovery time because of the compensatory effect of the reduced royalty, and the ROR decreases by 2.4% to 23.8%.

Case 3

In this case there are six years of exploration and five exploratory wells are drilled, as follows: one in the second year of the contract (dry); one in the third year (discovery of a field of 10 million barrels of recoverable oil); one in the fourth year (discovery of a field of 20 million barrels of recoverable oil); and one in each of the fifth and sixth years of the contract (dry).

Case 3A

The combined economic analysis of the discovery and exploitation of the two fields and of the three dry wells shows that the investment is recovered in three years from the start-up of production and that the ROR is 31.1%.

Case 3B

As in the previous cases, the reduction in API gravity does not have a great effect; there is no change in the investment-recovery time and the ROR decreases by 2.6% to 28.5%.

Conclusions

1. The exploration carried out in the country in recent years, despite the difficulties encountered, has verified that the sedimentary basins do have the capacity to generate and store hydrocarbons.
2. The sedimentary basins have been little explored and there are unexplored targets at various levels in the formation which, up to the present, has been considered the target, as well as an additional target in the recent verification of the presence of reefs (La Soledad well, 1982).
3. The option available to the contractor of choosing the location and size of the exploration area facilitates the participation of companies of any size, operating alone or in partnerships.
4. The option of handing back the exploration area if the results of drilling are unpromising eliminates the possibility of acquiring excessive commitments.
5. The economic exploitation of small fields helps to meet a country's demand for hydrocarbons, a fact which is of greater importance in countries whose consumption is relatively low.
6. The priority given to recovery of investments reduces the risks to the investor.

7. In addition to the priority given to recovery of investments, the requirement of less participation in the production of disposable hydrocarbons reduces the level of marginality of small fields.
8. The system of varying the royalty according to the API gravity of the crude oil compensates for the discovery of less valuable oil and prevents any great reduction in the rate of return on investment, since the investment-recovery time remains unchanged.
9. The production-participation contract provided for in the Hydrocarbons Law makes the exploitation of small fields profitable.

Section VII

Transfer of Technology

54
The Use of Simulators in Operational Skills Training for Developing Countries

L. Russell Records

Introduction

The lack of technical and skills training is one of the greatest impediments to energy self sufficiency for developing countries. For many years, most third world nations have relied exclusively on Western technology and personnel for the development of their petroleum reserves. Local personnel were used only in unskilled and manual labor jobs. Many nations are now working to reverse this trend; some have outlawed the use of expatriate workers in mid-level oil field jobs like driller and toolpusher. This shift in policy puts a strain on the ability of the operators to provide an adequate supply of trained workers, and inexperienced workers are the largest source of accidents and equipment damage.

The use of computerized simulators, sometimes costing millions of dollars, has become common in many industries that rely on highly skilled operators. Examples of such operators are airline pilots, power plant workers, and of course, the military. With the advent of powerful new microprocessors, the cost of simulation training equipment has become more affordable. Since 1979, Digitran has developed eight different models of engineering and training simulators, with prices ranging from U.S. $50,000 to U.S. $500,000. Digitran simulators have been instrumental in preventing the occurrence of blowouts, which are generally caused by human error and may cost millions of dollars. The initial cost of the simulator is small compared to the cost of having untrained personnel on the rig.

When effectively utilized, modern computerized training simulators can increase the skill level of workers, particularly in mid- to senior-level field personnel. They also provide a wider range of training experiences in a short time than might be seen in an entire career of daily field work, making these people more prepared to deal with new problems. They also make possible a degree of standardization of operating practices. By reducing the occurrence of problems, drilling time and time lost to accidents is reduced, resulting in a decrease in the cost of developing the energy resources, and the protection of the lives of valuable drilling and operations personnel.

Recent advances in computers, color graphics, and simulation technology have reduced the cost and increased the realism and reliability of industrial simulators, including those now available for use in the oil field. This paper describes several such simulators, and outlines how they are used to provide an increased level of skills training to operating personnel. It also describes alternative methods, and compares the economics and other considerations.

Description of Training Methods

Definition of Training

Training can be defined as the successful transfer of skills and knowledge from one person to another, in effect creating two capable persons where there was only one before. Training methods can be very simple—explaining the new skill, then asking questions—or very complex—like those used to train Space Shuttle crews. The essence of training remains the same; information is transferred from one person to another; then the information is verified by testing. If the testing process shows weaknesses in the transferred information, that portion of the information is reinforced, then re-tested.

How Is Training Conducted in Developing Countries?

Four main methods for petroleum training programs are used in developing countries. Digitran has been actively involved in all four methods.

1. *Expatriate instructors.* Many U.S. companies specialize in one to six week training programs in petroleum-related subjects, such as blowout prevention, drilling problems, and well planning, using expatriate instructors. These instructors usually have experience

with western drilling companies, and may have worked in the local area as a drilling consultant. In some areas, training is done by foreign drilling contractors as part of their contract.
2. *Local personnel sent abroad.* Many good training programs are held by major oil companies, service companies, training companies, and universities in the United States, Canada, and Europe. Personnel in developing countries spend several months in these programs, learning the latest in technology and field practices.
3. *Local courses.* Most national oil companies have formed training departments, using local instructors who have been trained in local universities or petroleum institutes. Many of these personnel have received some training in the West, are familiar with local drilling conditions, and will be fluent in the local languages. They may not be up to date on the latest procedures and equipment, however, and, in many cases, have difficulty in keeping current.
4. *On-site instructor training.* One of the most effective methods of providing up-to-date training programs is continual outside training of the local instructors. This method keeps them updated on the latest technologies, yet maintains their familiarity with local conditions and languages. If the training is conducted on-site, the local instructors will not need to be away for an extended time, and more instructors can be trained for the same expenditure.

All instructors should rotate between field jobs and instructional positions to maintain their credibility and professional competence. This period of rotation may range between one and three years.

Barriers to Effective Training

There are several problems unique to training in developing countries. These include:

1. *Language*—The instructor may not speak the local language well. Many dialects may exist within a country. Oil field terminology must be learned. Oil field mathematical units vary widely from country to country.
2. *Culture*—Students may not be able to assimilate technical information due to lack of technology-based culture. Most Western technical personnel have been around modern tools and machinery since childhood, and have no difficulty understanding the use and benefit of technology.
3. *Educational level*—Students will have widely varying educational levels due to lack of general educational programs. Some will have

received no formal education whatsoever, yet some may have advanced technical training.
4. *Work habits*—Students may not understand the need to work hard or study hard. Once educated, they may feel that manual work is degrading. A person's status many times depends on his educational level. The "Protestant work ethic" is not found worldwide.
5. *Technology*—Students may be totally unfamiliar with current oil field technology. As oil field equipment depends more and more on advanced technologies like computers, lasers, and digital electronics, the technology "gap" is widening.

Training Methods

Training can be performed either dynamically or nondynamically (passively). The word "dynamic" means a two-way exchange of ideas, with the students involved in the training exercise. Dynamic training uses actual equipment or simulators. Nondynamic methods use textbooks, equipment models, and pictures, and do not depend to a high degree on response from the students.

"Best" methods combine dynamic and nondynamic methods. Students must first have a "context;" this can be gained through experience and exposure to the working environment. Then nondynamic methods can be used to present material which may reinforce existing knowledge, or may give new knowledge. Having the "context" will enable them to assimilate the new information by allowing them to relate it to things that they already know. Nondynamic methods may include manuals, films, videos, models, and pictures. Good sources for these materials are PETEX and Gulf Publishing Company.

Once students have established a context and have received formal classroom training, they are prepared for "dynamic" training, using actual equipment or simulators.

In many situations, the most effective way of conducting this hands-on training is through the use of advanced training simulators. This topic will be explored in the next section.

The Role of Simulators in Dynamic Training

Advanced training simulators are used increasingly in dynamic training programs around the world, due to their low cost, flexibility, and safety.

Classrooms vs. on-the-Job Training

Much of the training must be performed on-the-job (OJT) in order to give the students a chance to function in the actual work environment. Skills that require the use of hand tools are best learned while doing. Classroom training works best as an adjunct to OJT, in which rules, procedures, and theoretical information can be presented to enhance the student's job knowledge and skills.

Simulator use is a form of OJT in which the job environment is simulated. The simulator environment must be realistic enough so that the students get the feeling that they are "on the rig."

Training Rigs vs. Simulators

Many vocational/technical training groups use training rigs for basic job training. The skills taught in this environment include:

1. Location and function of rig equipment.
2. Pipe handling.
3. Use of tongs, slips, spinning chain.
4. Basic rig safety.
5. Equipment maintenance and repair.

Advantages of Training Rigs

Training rigs are the most effective way of providing a realistic work environment for learning basic drilling-rig skills. The students should learn to safely conduct most basic drilling operations, and will learn to maintain and repair the equipment on the training rig. A well-maintained training rig can be a showpiece for the drilling training center, and may increase enrollment and industry support. One U.S. oil company has developed the concept of a training rig that is used to actually drill oil and gas wells under contract. Every person on the rig is a trainee, from the roustabouts to the rig supervisor.

Disadvantages of Training Rigs

Training rigs are normally surplus equipment that has been donated because it is no longer effective for field use. This usually implies that it has become outdated, or that maintenance costs are too high for efficient use.

In many cases, these rigs are run-down and do not have a complete set of equipment. Even if the rig is provided at no cost, the set-up and restoration cost can be very high. Equipment will have to be repaired and renovated.

Extra safety equipment will have to be installed. During use, maintenance and fuel costs can be quite high, and there are safety hazards that do not exist in classroom or simulator training.

Training rigs are very inflexible; in most cases the wells already exist and students just learn to trip pipe in and out. Some equipment operation with the pumps and rotary can be practiced. Training rigs can be quite useful for repair and maintenance training.

Training rigs cannot be used effectively to teach students how to handle most unexpected downhole problems, and certainly not well control problems. In some places, test wells have been constructed that allow students to pump out nitrogen or methane kicks, but again only a very narrow range of training is possible.

Using Simulators for Advanced Skills Training

Simulators are used for more advanced skills training than can be conducted on a training rig. A modern, complete rig-floor simulator can duplicate all main functions of the drilling rig, and can be constructed to exactly replicate any class of rig that is commonly used. A rig floor simulator can be used in lieu of a training rig for some familiarization training for beginning personnel.

In some cases, it is not necessary for the simulator to look exactly like the rig equipment. For advanced level personnel, a portable simulator that represents the equipment with graphics or miniature panels is quite adequate. In such training, the course is intended to teach operating procedures and theory rather than the details of equipment operation, and it tends to be more generic in content.

Specifically, five major considerations apply to the use of simulators in training.

1. *Simulator use is primarily an economic decision.* In most cases, using simulators for training is less expensive than using actual equipment for advanced training programs. Advanced skills training cannot be conducted with training rigs due to the hazard, lack of flexibility, and expense per training session. Rig floor simulators can teach some basic skills, though this is better done on a training rig or on the job.
2. *Simulators are safer than on-the-job training.* Inexperienced hands on the rig or in the plant can interfere with normal operations and cause safety problems. Mistakes made by inexperienced personnel

can be costly and possibly catastrophic. It is virtually impossible for personnel to be injured in simulator training.
3. *Simulators are more flexible than training rigs or attempts to perform training on actual equipment.* Simulators can be programmed to very closely duplicate a wide variety of possible situations, and are especially flexible in regard to well geometries, formation characteristics, and drilling fluid properties. The simulators are also very flexible in regard to the types of equipment and tools used, including bits, pumps, blowout preventers, casing, drillpipe, electric motors, and gearboxes. This flexibility makes them superior to training rigs for advanced training programs, where the ability to simulate a wide variety of situations is crucial.

 Using simulators, the student operators can perform any operation without damaging the equipment. Almost any well can be programmed, and many problem situations can be introduced by the operator. Exercises can be "frozen" while the operator explains the situation to the students, then can be continued at the same point. Exercises can be speeded up using time multiplication. This allows the exercises to be completed in a much shorter time than in real life. This allows more training to be done, and more students to be trained in a much wider variety of possible situations than with real equipment or training rigs. Simulators can record student performance data, and score students.
4. *Simulators can be almost as realistic as actual equipment, but realism is a non-linear function of cost.* The most cost-effective approach is to strive for realism in equipment appearance and function at the expense of totally realistic response and visual displays. Space flight simulators can replicate all aspects of the mission, including sound effects, movement, cockpit gauges, and out-the-window displays, at a cost of many millions of dollars. A simulator for drilling training with that degree of realism would also cost several million dollars. Simpler engineering-quality models can be used at no loss of training realism, and simplified color graphics systems can be used that provide the necessary information at a small fraction of the cost of three-dimensional full-scale visual systems. The latter type of graphics is available, however, if a customer wants to pay the price.
5. *Simulators are preferable to nondynamic forms of training under the following scenarios:*

- Operational rather than theoretical training. Although simulators can now provide some theoretical training, they are best suited to operational training.

- Sense of real-time operation needed. The primary use of simulators is to provide an environment in which real-time decision-making and equipment operation skills can be developed. If mistakes are made, the student will immediately get feedback indicating the results of his actions.
- Students need a feel for equipment responses. A properly-implemented simulator will have equipment responses that closely approximate actual equipment in either normal or degraded operation.

Simulator Types

Simulators may either be specific or general in nature:

1. General types
   Precise appearance and function of equipment is not essential to accomplish most training objectives. The use of a generic type is dictated by cost constraints and alternative systems. A simulator may have to be used by several groups who are using different equipment in the field. The simulator must be able to replicate the principal functions of the different types of equipment, but will look exactly like none of them. *Examples:* chemical plants, drilling rigs, introductory training.
2. Specific types
   These are built to look and function as closely as possible to equipment that will be used by the operator. These are used in life-critical operations, or when differences in target systems are limited. These are usually built to exactly replicate the appearance and function of the target system. *Examples:* Aircraft simulators, nuclear power plants, spacecraft training simulators.

Problems in Using Simulators in Developing Countries

Effective Instructor Training

In most developing countries, the instructors have had little or no exposure to the operation or use of training simulators. It is difficult to conduct an effective training program for the customer's instructors that will truly make them "masters" of the system. For the Rig Floor Simulator, the training process begins with a 1–2 week introductory course on operation and maintenance, then continues with a refresher course after the instructors have gotten some hands-on experience. After this training, the instructor will be capable of preparing training exercises for the students, as well as operating and maintaining the simulator system. We have prepared additional instructor training materials to assist the instructors in getting the maximum benefit from the system. However, we have found that it may still take up to a year of hands-on experience for an instructor to be able to accomplish all exercises that the simulator is capable of performing.

Difficulty in Obtaining Export Licenses

Modern simulators fall under export restrictions that apply to computers. Approval of licenses takes 3 to 6 months, and in some cases will not be approved. Licenses are subject to approval by the U.S. Departments of Commerce and Defense, and COCOM (NATO).

Electric Power Stability/Availability

Computer equipment needs stable power. This can be provided by regulated power supplies.

Environmental Stresses (Heat, Humidity, Dust, Sand)

The simulator must be provided with an indoor environment with temperature and humidity control, and reasonable protection from dust and sand. The simulator can operate to 90°F (32°C), but this will significantly shorten the life of its electronic components. Dust and sand particles will get into circuit card slots and prevent good contact. Equipment in dusty areas should be sealed and have dust filters on cooling fans. Dust and sand will also reduce the life of data stored on floppy and hard disks.

Diagnosing Hardware Problems

A computer-based simulator is a complex electro-mechanical system and is subject to component failures. When these problems occur, the instructors must be able to locate and replace failed components quickly, with a minimum of training and special equipment. Although service help is available from the U.S., it takes several days for a serviceman to arrive on site. This problem can be alleviated by design reliability and by designing the system for user maintainability.

Building exceptional reliability into the system. The system can be constructed to be exceptionally reliable, with little or no preventive-maintenance requirements. This is accomplished by using military standard or heavy-use industrial equipment, or by overdesign of power supplies, heat sinks, connectors, and all electronic components. It is also possible to build in "fault tolerance" with multiple, redundant components. These will add to the production costs of the system, and a system that is built to these standards will be more expensive to purchase, but will have less downtime and unscheduled maintenance. The replacement-parts cost also will be higher than for a less reliable system.

Designing for user maintainability. The simulator can specifically be designed for user maintainability by special design, including:

- *Minimum unique parts.* With the minimum number of unique parts, the user can maintain an adequate supply of spare parts, and find and replace faulty components. The fewer the number of unique parts, the fewer the unique problems that can develop. The electronic circuit cards in each panel are the same, so if the user has learned to maintain one panel, he can maintain any panel.
- *No special tools for maintenance.* The simulator electronics are designed to be maintained without special tools. The only tools required are simple, common hand tools like screwdrivers and nut drivers. All components can be easily removed for replacement. Checking the voltage levels inside the electronics is made easy by a set of lights to show if the power level is lower or higher than the correct value.
- *Provide user training.* The simulator manufacturer must provide maintenance training to the users. Easy-to-understand service manuals must also be provided.

Building-in automatic fault diagnostics. The simulator should be capable of finding faults inside the simulator hardware. This can be accomplished through several features:

- *Self-testing*—The instructor can push one button, and the system will automatically do a self-test.
- *Computer and systems diagnostics*—To find more subtle problems in the main computer, the instructor should be able to run a complete diagnostic on the computer and system.
- *Built-in failure reporting*—If hardware failures occur, the simulator should inform the instructor what has failed, and what to replace.

Repairs and Replacement Parts

The instructor should not be expected to repair electronic circuit boards that have been replaced. A 12-month spare parts kit should be provided with the simulator. When parts fail, the instructor should be able to remove the failed part, replace it with the spare part from the spares kit, then exchange the failed part for a new one to replenish his spare parts kit. The normal turnaround time for a replacement part can vary from four days to four weeks, depending on location of the simulator. Export licenses for spare parts replacement must be approved at the time the simulator is installed.

An Approach to a Petroleum Training Center

The most cost-effective approach to petroleum training in developing countries may be the establishment of a permanent, regionally located petroleum training center.

These are the general requirements for a small petroleum training center to be located in an overseas location.

Courses

The following courses would be offered:

1. *Introduction to drilling and completions.* These courses would be conducted on many different skill levels.
   - Rig equipment operation and maintenance
   - Drilling techniques and practices
   - Drilling problems
   - Basic reservoir characteristics
   - Completions
   - Well testing

2. Drilling well control (driller, toolpusher, drilling supervisor).
   - Blowout prevention equipment
   - Causes of blowouts
   - Detection of kicks
   - Shutin procedure
   - Kill methods
   - Kick circulation

3. Rig safety (all levels).

   - Hazardous areas and equipment
   - Safe use of tools
   - Personal injury prevention
   - Hydrogen sulfide problems

4. Basic job skills (support personnel).

   - Electric motor repair
   - Diesel engine repair/maintenance
   - Welding
   - Rig electrical systems

5. Advanced job skills.

   - Roughneck
   - Derrickman
   - Driller
   - Mud engineer
   - Other

Personnel

A small petroleum training center might have a total of five employees. These could include:

1. *School manager,* who also may be an instructor with the following duties:

   - Personnel management and scheduling
   - Industry interface
   - Course enhancement
   - Budget control

2. *Two instructors.* Two qualified instructors should be hired who can teach any of the courses offered and can handle the administrative details of the courses in their off-duty time, book courses, and also teach the simulator training sessions. They should be conversant in the native language, as well as familiar with oil field terminology and mathematical units. The instructors will be trained by the manufacturer in the operation of the simulator, but they need more detailed knowledge of the subject matter that they will be teaching.

Instructors should be selected based on the following general criteria:

- Knowledge of the skills to be taught.
- Field experience in using the skills.
- Desire to instruct.
- Previous teaching experience.

3. *Two administrative workers.* The administrative workers will handle typing, student files, invoicing, and routine matters.

Facilities

The school should be located in an area easily accessible to the students. It needs to have a reception and administration office, a manager's office, a break room, and at least two classrooms, one of which will contain the simulator. The other may contain other audio/visual equipment. A storage room for student manuals is also needed.

The instructors should conduct many of the training courses, particularly the courses stressing safety and equipment, in the field. This will ease the scheduling problem for rig personnel because they can be scheduled when the rig is not busy. The courses are effective since the training is conducted on equipment that the crews will use every day.

Equipment

Equipment will be needed to handle the presentation of audio-visual training material to the students. Since the school will need to schedule its courses in advance, it will need such communications equipment as telephones and a telex machine. A simulator and student manuals are also included in the equipment list.

1. Video tape machine with several training movies.
2. Slide projector and several equipment slide shows.
3. Equipment models.
4. Classroom furniture.
5. Simulator equipment.
6. Telex system.
7. Student manuals for all courses.
8. Copier machine.
9. Telephones.

Simulators Used for Oil/Gas Drilling Training

Digitran offers a wide range of simulator systems for use in petroleum training programs. These include:

Kick Control Simulator

The Kick Control Simulator is a portable well control simulator that consists of two student panels and a plug-in terminal. The drillers panel allows one student to control the pumps, drilling operation, blowout preventers, and manifold valves. The choke panel allows a second student to control the remote hydraulic choke and monitor the drillpipe and casing pressure. The plug-in miniterminal is used by the instructor to program different exercises, introduce problems, and control the simulation. The Kick Control Simulator is used for blowout prevention and kick circulation problems.

Well Systems Simulator

The Well Systems Simulator (WS1000) is a transportable drilling and well control simulator that consists of two student panels and an instructor's console. The WS1000 can fully simulate many drilling and well control problems. It consists of the rig control panel, which has a full complement of drilling controls, such as pumps, rotary, drawworks, brake, and pipe handling controls, as well as the blowout preventer panel and choke manifold panel. It also includes the choke panel that allows the operator to monitor and control drillpipe and casing pressure by means of the remote hydraulic choke.

Rig Floor Simulator

The Rig Floor Simulator is a full-size replica of the equipment normally found on a typical rig. It can be constructed to exactly duplicate an actual rig, though this is expensive, and normally it is constructed to replicate a specific class of rigs or generic equipment. The individual panels can be constructed to replicate a particular manufacturer's equipment.

The Rig Floor Simulator normally includes the following panels:

1. Driller's control panel
2. Driller's data display panel
3. Choke panel

4. BOP closing panel
5. Pit/flow monitor panel
6. Color graphics display
7. Standpipe manifold
8. Choke manifold
9. Instructor's console

There are many optional panels for the Rig Floor Simulator. These include:

10. Subsea BOP/diverter control panel
11. Motion compensator panel
12. Drilling fluids control panel
13. Directional drilling panel
14. Cementing head
15. Hardcopy printer

Directional Drilling Simulator

The Directional Drilling Simulator is an advanced engineering and training simulator that teaches drilling and supervisory personnel the most economic method of drilling directional wells. A directional well is one in which the target is located at some horizontal distance away from the surface location.

The Directional Drilling Simulator consists of up to five student color graphics consoles with keyboards, an XY Plotter, a hardcopy printer, and a computer with disk drives.

Production and Workover Simulator

In order to evaluate the productivity of a well, the well may be subjected to pressure and flow tests. These tests will allow estimates of the total oil and gas reserves. Students must learn to conduct these tests accurately.

Many accidents occur when wells which have been producing for a time, stop producing, and must be reconditioned or "worked over." The Production and Workover Simulator provides a means of teaching students the correct methods of performing pressure and flow testing, along with basic reservoir characteristics. It will also allow them to practice the best methods for conducting well "killing" operations and the ways to safely deal with problems that may arise.

The PAWS consists of a full-size wellhead, a pumping control unit, pipe manifold with valves and two chokes, one fixed and one variable. It also includes the instructor's console with the computer and disk drives.

How Are Simulators Actually Used?

In a typical petroleum training center, the simulator is a key part of the training program. Normally, students will spend about two hours in the classroom for each hour on the simulator. They are normally trained as rig crews of 3–5 men at each time, and they may be trained in several of the crew positions. The time spent in the classroom is used to introduce and discuss the theoretical material; the students are then put on the simulator to reinforce the information and be tested. The final certification depends on their performance on the simulator, as well as a written test.

Generally, the simulator is used for the following types of training:

Equipment Familiarization and Operation

A typical drilling rig consists of many control and display consoles that the rig crew must learn to operate. Each will have its own set of controls and gauges, and a unique operating procedure. The simulator is used to teach the students how to operate, calibrate, and adjust each of the control panels found on the rig.

Procedures Training

In carrying out the drilling plan, many sets of procedures will have to be followed by the rig crew. The simulator is an excellent medium for procedures training.

Testing and Certification

One of the principal uses of the simulator is for proficiency testing and certification of the operating personnel. The testing process involves giving the students an operational situation, giving them abnormal conditions, and measuring their responses. They must be able to safely and economically control the situations presented to them in order to pass the certification. In the United States, the United Kingdom, and Norway, this process is being used for certification of drillers, toolpushers, and company representatives in blowout control.

Engineering Uses

In addition to its training uses, the simulators have many engineering applications for which they can be utilized to assist in the solution of field problems. These applications include:

Verification of Well Plans

Well plans that have previously been prepared can be evaluated by programming on the simulator. The well plans can be segmented by drilling phase (spud, surface casing, intermediate casing, production casing) and the bit and hydraulics programs can be checked. The casing program can also be completely evaluated for kick tolerance considerations. The simulator will compute all fluid volumes, and the operator can then estimate the drilling fluid and cement costs.

Determination of "Best" Procedures

Within the scope of the simulation models, the simulator can be used to analyze the impact of alternate procedures in dealing with expected operational problems. As an example, the Digitran Simulator has been used to develop well control procedures for drilling in as much as 11,000 feet of water for the Ocean Margins Drilling Project now underway.

Well Planning

With the addition of well-planning software, the simulator computer system and instructor's console can be used to interactively plan straight hole or directional wells. Once the well is planned, it can be printed out or plotted for use in the field. The planning process allows for many checks of the proposed well plan against the occurrence of abnormal situations like kicks, extremely heavy mud weights, and underground blowouts. A drilling economics package could be developed to provide an estimate of the drilling time and costs.

Who Buys Simulators?

In general, the simulators are now being used for well control training and certification, though the use of the simulators in drilling training is increasing every year. The five main groups of users are (1) oil companies, both public and national; (2) drilling contractors; (3) industry-supported schools; (4) research institutes; and (5) training institutes and universities.

The prices of the Digitran simulators range from U.S. $50,000 to U.S. $360,000, though with all available options, the price may exceed U.S. $500,000.

Summary and Conclusion

The use of advanced training simulators is increasing as developing countries attempt to become self-sufficient in drilling and production technology. Simulators are very cost-effective when compared to other training methods. Most types of advanced training can be done most effectively using simulators.

Simulator purchases must be carefully throught-out to ensure that they will function properly in the available environment and that a training staff is present in support. The users will have to provide maintenance support, with assistance from the manufacturer.

Simulators are only a part of the overall effectiveness of a well-executed training program. The success of the effort depends on the orchestration of all elements of the program.

55
Geological and Geophysical Technology Transfer Between U.S. Universities and Developing Countries

John D. Pigott and John S. Wickham

Finding and utilizing natural resources is essential to the economic development of any country, and requires a group of well-trained people to accomplish those ends. Historically, those people have come from the developed countries who have, over decades and centuries, invested in education, research, and development. In order to control their own economic development, developing countries need to have an equally well-trained group of citizens.

In most developed countries, the investment in education, research, and development is institutionalized in the universities. Technology transfer from these universities to developing countries has been occurring for decades, and most developing countries have made their own investments in education, research, and development to meet the need for economic development.

Universities in the United States and elsewhere have been very successful in passing on new and established technology to their own citizens. People from developing countries have moved into these existing programs and generally have received little guidance on which universities or programs are best suited to their particular needs. For those countries and institutions which need additional transfer of technology from the developed countries in the form of well-trained citizens, programs can be organized in established universities to make that transfer more efficient and effective.

There are two elements essential to good education and training in technology, general principles and applications. Universities in the United States tend to concentrate on the general principles and leave the applications to the industrial sector. A good program for technology transfer to developing countries should address both general principles and applications. The applications should be designed to meet the needs of a particular country and are best accomplished in that country.

The University of Oklahoma as an Example

Geology and geophysics are essential to the development of energy, mineral, and water resources. The University of Oklahoma has the organizational structure to transfer both general principles and applications and can serve as an example.

The School of Geology and Geophysics at the University of Oklahoma concentrates on the education of students in general principles. It is typical of most departments in universities in that it has established degree programs aimed at providing students with a solid foundation in the principles of geology and geophysics. Like most academic departments, applications are not the main purpose of the degree programs although students are typically exposed to applied problems.

In addition to the School of Geology and Geophysics, the University of Oklahoma has established the Rock Systems Laboratory to give students experience in applications of geology and geophysics to applied problems. The Rock Systems Laboratory takes people already educated in the general principles of geology and geophysics and gives them experience working with applied problems under the direction of an experienced professional.

These two organizations within the University of Oklahoma can be effectively utilized to provide both aspects of technology transfer to developing nations: The School of Geology and Geophysics for education in general principles, and the Rock Systems Laboratory for experience in applied problems.

The School of Geology and Geophysics

Organized in 1901, the school has been an important factor in educating professionals for the petroleum industry in the United States. At present, it has over 3,000 graduates worldwide who are active in the industry. The school offers degree programs at the bachelor, master's, and doctoral levels in geology and geophysics. For most industrial positions in the

United States, the master's degree is considered essential for exploration and production technology.

The master's degree is designed to give students a solid background in the principles of geology and geophysics as well as a foundation in science, mathematics and engineering. To be admitted into the master's program, students must have had introductory courses in all the sub-disciplines of geology and geophysics. In addition, they are expected to have had introductory courses in physics, chemistry, calculus, biology, computer science, and some advanced work in one or more of these subjects.

Students in the master's program typically take nine advanced courses as part of their degree program. Three of them are designed to give breadth in geology and geophysics, two advanced training in science, mathematics, or engineering, and the remaining four, specialization in geology and geophysics. In addition, students must write a thesis on independent work they have performed in geology or geophysics.

Students typically take a minimum of two years, including summers, to complete this program.

Rock Systems Laboratory

The Rock Systems Laboratory does research on applied problems in the following areas of geology and geophysics:

- Ore deposits
- Basin analysis
- Fracture systems
- Sedimentary systems
- Exploration geophysics
- Deformation structures of the shallow crust
- Maturation, characterization, and migration of petroleum

Because of its broad scope, the Rock Systems Laboratory can take an integrated or systems approach to geological and geophysical problems of an applied nature.

The senior staff of the Rock Systems Laboratory consists of faculty members of the University of Oklahoma, most of whom are in the School of Geology and Geophysics. Students working on applied problems do so under the tutelage of these experienced professionals. Because most of the people involved in the Rock Systems Laboratory are also part of the School of Geology and Geophysics, an efficient program of technology transfer can be easily accomplished. The work done by students on applied problems with the Rock Systems Laboratory can be integrated into their degree program in the School of Geology and Geophysics.

Facilities and Equipment

The Energy Center, a new building for Geoscience, Petroleum, and Chemical Engineering, is presently under construction and will provide first-class space for work in advanced technology. The Geoscience Library has over 80,000 volumes and is one of the most complete geoscience libraries in the United States.

The School of Geology and Geophysics has recently purchased over $1.5 million worth of equipment. The latest purchase was for computer hardware and software for seismic data processing, graphics, and image processing, which provides the latest technology in processing geological and geophysical data.

Geophysical equipment includes both Sercel and DFSIII seismic recording systems, a cryogenic magnetometer to measure extremely weak magnetism, and portable gravimeters and magnetometers.

Geochemical equipment includes automated X-ray diffraction and fluorescence units, a stable isotope mass spectrometer, gas and liquid chromatographs, C-H-N analyzer, neutron activation spectrometers, atomic absorption, scanning electron microscopes, and an accelerator for proton induced X-ray emission (PIXE) and prompt nuclear reactions (PNR).

Other facilities include equipment for structural modeling, rock mechanics, fission track dating, fluid inclusion analysis, and hydrothermal mineral synthesis.

Cost

The investment in time and money is considerable, but the payoff over a person's lifetime gives the investment tremendous return. For students with the equivalent of a U.S. bachelor of science degree, the time needed is about three years. The first two years are involved in learning general principles of science, mathematics, geology, and geophysics, leading to a masters degree in geology or geophysics. The third year is devoted to the solution of applied problems in the student's own country.

Not counting travel to and from the U.S., food, lodging, tuition, and fees for a student at the University of Oklahoma would be approximately $10,000 per year. The cost of solving applied problems in the developing country would be considerably more, and depend on the nature of the problems, the number of people from the Rock Systems Laboratory required, and the availability of equipment in the developing country. $100,000 per year for five or six students would probably be a minimum.

Student Selection

Because of the large investment in time and money, students must be selected with care, using the following guidelines:

1. Excellent command of English (for U.S. Universities).
2. Background in science, mathematics, geology, and geophysics, equivalent to a U.S. bachelor of science degree.
3. Ability to learn quickly.
4. Ability to lead and communicate.

At least five or six students should enter the program together. They would be able to help each other with course work and with adjustments to a new culture. Because the last year of applied work has high fixed costs, the larger number of students would reduce the cost per student.

Summary

People who understand science and engineering technology are essential to developing countries if they are to control their own economic future. Universities in the developed countries are in the business of technology transfer to their own citizens, and have provided this service to people from developing countries on an individual basis for decades.

There are two elements to technology transfer: (1) education in general principles, and (2) applications. Universities have traditionally emphasized the general principles, and few have the internal organizations to work in applications.

At the University of Oklahoma, the traditional education in general principles resides in the academic departments and schools. In the School of Geology and Geophysics, we have recently organized the Rock Systems Laboratory to give students experience in applied problems.

If they so desire, developing nations can use this type of organizational structure to (1) educate their own citizens in the latest technological principles, (2) provide experience in the applications of that technology, and (3) find technological solutions to important problems within a particular country.

This can best be accomplished through organized cooperation between a university and officials in the developing nation to: (1) select the appropriate students, (2) identify the applied problems to be solved, and (3) work together with professionals in the developing nation to tutor the students in the process of solving those problems.

56
Fossil Energy Assistance Activities of the United States Agency for International Development

Charles Bliss and Pamela Baldwin

Introduction

Energy development and production have been recognized by the United States Congress as vital elements in the development process and reflected in the foreign assistance legislation. This recognition is reflected by the Congressionally-mandated projects of technical assistance and training administered by the Office of Energy in the Agency for International Development's Bureau for Science and Technology. These Washington-based projects provide assistance to developing countries for activities ranging from energy planning to support for development of fossil fuels, biomass, and renewable resources, and serve to initiate new activities and to support related activities undertaken by USAID Missions overseas. Where resource development requires large capital investment, AID's operations are designed to stimulate this investment by lending institutions such as the World Bank and regional development banks, as well as by private investors.

Two AID projects offering technical assistance and training, respectively, are relevant to the potential for the development of shallow oil and gas resources. AID's Office of Energy is pleased to be able, through these projects, to join the roster of sponsors for this Conference and to support the participation of a significant number of officials from developing countries.

The scope of the Conventional Energy Technical Assistance Project includes all fossil energy forms—oil, gas, coal, oil-shale, peat, tar sands, and, for convenience, geothermal—and efforts related to their exploration and production. The Conventional Energy Training Project offers fellowships to qualified professionals from developing countries to study in U.S. academic institutions, to pursue training in specialized short courses, or to undertake internships in U.S. industries. The fields of study or practice relate to geology, geophysics, petroleum engineering, and a number of other fields related to fossil fuels and electricity.

The overall goal for both projects is to enhance the ability of oil-importing developing countries to reduce their dependence on imported oil through the development of their indigenous energy resources. The technical assistance project assists in specific situations, while the training project enhances the capabilities of personnel and institutions.

Technical Assistance

Technical assistance provided under this project can include these typical activities:

- Reviewing, at the country level, what is known about the conventional energy resources of the country. This can involve collection, review, compilation and summarization of all geological and geophysical data related to prospects for finding and developing conventional energy resources.
- Filling gaps in the basic elements of geological information, particularly through the development of geological maps and satellite images that can be used to identify areas for energy resource exploration.
- Processing and interpreting existing raw geological and geophysical data that have either not been interpreted previously or have been interpreted under conditions that have failed to provide the country with the maximum amount and best quality of information. Recent technological breakthroughs in data processing have made it possible to learn much more about resource prospects from the reinterpretation of existing data. In addition, it may be in the interest of the countries involved to commission an independent interpretation of data previously evaluated by private oil companies, particularly when the companies involved have decided on the basis of the data to cease exploration efforts in the country. A second look, based on new techniques and up-to-date economic factors, could lead to renewed interest in exploratory drilling.

- Assisting the host country in developing a comprehensive energy resource exploration strategy. Using existing geological and geophysical data, the plan could identify promising areas for further geophysical work and for exploratory drilling and would provide timetables, budgets, and implementation strategies.
- Evaluating the country's existing manpower and institutional capabilities in the conventional energy field and making recommendations for strengthening those capabilities. Particular attention is paid to identifying training needs that can be met through the companion education and training project. In addition, opportunities for on-the-job training during activities under this project can be identified. The latter category of training may include the development of technicians who can interpret Landsat images or aerial photos, work in the field on geophysical studies, and assist in geophysical data interpretation.
- Providing assistance in the organization, management, technical-personnel training, and operation of government entities involved in energy resource exploration, development, and production. Such entities can include geological institutes; ministries of energy, mining, planning, economic development, finance, etc.; and government oil companies.
- Conducting new geophysical field work, such as remote sensing studies, magnetic or gravity studies, and seismic surveys. Since seismic surveys tend to be expensive they are not likely to be undertaken until extensive study reveals that no other private or public sources of funding are available to the requesting country, and that gathering and analyzing new seismic data could significantly advance the country's overall exploration effort and improve its chances of attracting private capital for exploratory drilling.

A typical example of support in these areas is the program of activities now being conducted in Morocco in support of the national oil and gas office (Office National de Recherches et d'Exploitations Petrolieres, ONAREP). Since 1982, ONAREP has been assisted in its exploration operations designed to develop better geological and geophysical knowledge of Morocco's hydrocarbon resource base and in its drilling operations to determine the extent of its natural gas finds. The services of the assisting personnel have been provided by Bechtel National, Inc., and its subcontractor Woodward-Clyde Consultants under contract with the Agency for International Development. These services are planned to extend for several years into the future. The goals of such activities are:

- Expanded quantity and quality of available information about the geological potential for indigenous fuel resources.

- More developed institutional mechanisms and the technical capability needed to identify and develop their energy resources.
- Direct applications of expertise and technology to help plan, organize and implement national programs of conventional energy exploration.

The nation should then be able to mobilize public and private capital and technology for conventional fuel exploration and development, and to plan and manage the process of fuel development in a manner that advances its overall economic development progress and improves the quality of life for its citizens.

Training

The Conventional Energy Training Project (CETP) offers:

- Up to two years of study in a U.S. university, leading to a master's degree in science, engineering, or other energy-related fields, such as natural resources law, economics, and management.
- Specialized nondegree courses offered by leading U.S. organizations in energy-related research, technology, consulting, or field operations and targeted at mid-career energy professionals.
- Internships with U.S. companies to provide practical experience and build skills needed to explore, develop, and produce fossil energy, and nonnuclear electricity.

Typical fields of study include:

- Geology and geophysics (coal, oil, gas, oil shale and geothermal resources).
- Petroleum engineering (reservoir management, refining, enhanced oil recovery).
- Coal mining engineering and coal use (including beneficiation, liquefication and gasification).
- Electrical engineering (power production, conservation, rural electrification, distribution, and load management).
- Energy resource planning and management (economics, law and policy analysis, demand forecasting, and conservation).
- Civil engineering (hydropower, geothermal, and other conventional resources).
- Mechanical engineering (boilers, cogeneration, and conservation).
- Chemistry and chemical engineering (petroleum refining, petrochemical production).

By participating in this project, capable professionals can enrich their personal knowledge and experience, thereby enhancing their ability to contribute to meeting the energy needs of their own countries. All CETP participants are nominated by their employing agencies, which must demonstrate their commitments to using the skills gained in the program by certifying their agreement with the terms of the proposed training, pledging to employ the participant in an appropriate position on his/her return home, and by paying for the participant's air fare and continuing salary during the training period.

In the CETP's first two years, nearly 240 persons received training under its auspices, including 31 participants who have been awarded master's degrees and 60 more who are now studying toward such degrees. Industry internships under the program have varied. They have included, for example, a year-long exposure to the operations of an electric utility company that produces power from coal for two utility engineers who will be involved in their country's design and construction of a coal-fired power plant. For another participant from an oil-producing country, an internship in enhanced oil recovery contributed toward sustaining the country's petroleum output. Others have gained from first-hand exposure the U.S. approaches to oil and gas well logging, cementation, seismic analysis, and laboratory testing of coal samples. Nondegree technical courses attended by nearly 100 persons have covered such topics as petroleum geology and engineering, energy conservation for utility and industry engineers, utility operations, petroleum management for senior oil company officials, and electric system planning for utility planners and managers.

Operations

Technical Assistance

Activities can generate in either of two modes. The initiative may come from the host country itself, by identifying a need which fits the scope of the project, and bringing this to the attention of the USAID Mission in that country. The Mission may have a project in place that can accommodate the need, or, if not, the Mission may refer the request to the Office of Energy for consideration. The initiative may come from the Office of Energy and be advertised to the Missions. USAID sponsorship of a number of participants for this conference is a case in point.

Projects are usually undertaken by a qualified firm or organization working with counterpart personnel in the host country.

Training

For a candidate to be selected, he or she must meet these qualifications:

- Be nominated by his or her government.
- Have a firm commitment for a job in his or her own country upon return, that will utilize the training received in this project.
- Have a first university degree in an appropriate field.
- Speak and read English well enough to succeed in graduate-level and professional training.
- For academic placement, Graduate Record Examination scores must be acceptable.

To apply for this project the candidate should contact the Energy Officer or the Training Officer at the USAID Mission in his or her country to obtain a training application and to learn which ministry in his or her government should receive the request to be nominated for the training. He or she should then collect all necessary materials and information to accompany the completed application form.

The ministry which nominates candidates should contact the USAID Mission. The Mission will provide further information about the project and will forward the applications. For a nominee to be selected for this project, he or she must have guaranteed future employment in an organization that will benefit from the training received.

AID funding provides tuition and fees or other educational costs; travel within the U.S.; an orientation program in Washington, D.C.; limited English language training, if necessary; a living allowance; books and supplies; a thesis allowance; and health insurance for the participant. International travel is generally arranged and paid for by the candidate's government.

Under contract to AID, the Institute for International Education administers the Conventional Energy Training Program and arranges placement at appropriate universities or other training centers.

57
The United Nations Assistance Programme in Petroleum
United Nations Department of
Technical Co-operation for Development

Introduction

The United Nations is not a newcomer to the field of petroleum. Early in its history, the Economic and Social Council considered the establishment of a United Nations Petroleum Commission. The Committee on Natural Resources, formed in 1970, has had energy, including petroleum, high on its agenda. Hardly a year has passed when the United Nations has not held a meeting or issued a report on some aspect of the world petroleum situation.

However, it is in the area of technical cooperation that the United Nations system has made its greatest contribution to the solution of the problems involving petroleum development, particularly in developing countries. At any given time, United Nations technical experts are providing assistance and advice to governments at their specific request on concrete aspects of petroleum exploration, development, and management in the context of the particular country's situation and requirements. They have helped with petroleum development in Bolivia, Chile, India, Trinidad and Tobago, Malta, Turkey, Syria, Indonesia, and Saudi Arabia, among others. The Indian Institute for Petroleum Exploration, which the United Nations helped to set up and finance, has already trained hundreds of university graduates in advanced petroleum work. Long before energy became the subject of widespread public attention, the United Nations was advising on oil, natural gas, and oil shale exploration in many developing countries.

The following pages provide some information on United Nations technical cooperation activities in petroleum and highlight in particular the work of the Natural Resources and Energy Division of the Department of Technical Co-operation for Development (DTCD), which has played a predominant role in the field of petroleum resources within the United Nations system, mainly in the area of technical cooperation.

Petroleum contributes about 80% of the total requirements of commercial energy in the developing world. If the developing countries are to achieve their economic and social development targets, their requirements for petroleum will increase for many years to come. While most developing countries remain largely unexplored or insufficiently explored areas, some estimates indicate that up to half of the world's undiscovered oil resources may be found within their boundaries. It is the primary objective of the United Nations technical cooperation program to assist developing countries in strengthening their technical, institutional, and manpower capacities to develop the indigenous petroleum potential they may possess.

Scope of UN Technical Cooperation Activities in Petroleum

In its broadest sense, technical cooperation includes:

- Executing projects requiring experts, training, and equipment.
- Providing advisory services.
- Carrying out short-term missions to identify exploration requirements.
- Providing fellowships.
- Organizing meetings.
- Undertaking studies.

As the main UN executing agency for projects in the petroleum field, the department maintains a staff of experts in petroleum economics, legislation, geology, geophysics, and engineering which is well-equipped to furnish an extensive range of technical support and advice to interested governments. The many examples of activities include assistance in the formulation of petroleum strategies, legislation, and policies, including appropriate terms and conditions for cooperation with transnational oil corporations; initiating petroleum exploration efforts; reevaluating petroleum exploration data; application of modern methods and techniques in petroleum reservoir engineering and deep drilling; and assessment of energy exploration data and estimates of financial requirements for further exploration.

Over the past two decades, the United Nations has assisted the petroleum exploration and development programs of most of the world's developing countries (Table 1). It is now active in most phases of petroleum resource development. Large-scale projects, involving expenditures of $400,000 or more, have contributed significantly to the efforts of some developing countries.

Although the Department of Technical Co-operation for Development is implementing the bulk of the United Nations assistance in the petroleum sector, several of the specialized agencies of the United Nations as well as the United Nations regional commissions have their own programs in petroleum. The World Bank provides loans for petroleum development and, more recently, exploration. The United Nations Industrial Development Organization (UNIDO) assists in development of petroleum-based processing operations, especially in the petrochemical industries. The objectives of the United Nations Centre on Transnational Corporations (UNCTC) include strengthening the negotiating capacity of developing countries in their dealings with transnational oil corporations. The United Nations Conference on Trade and Development (UNCTAD) covers many facets of the petroleum sector, such as international trade, security of supply, access to markets, and finance. The International Labor Organization (ILO) has been concerned with employment, working conditions, and training in the petroleum sector, and the United Nations Environment Program (UNEP) has been involved in studies of the environmental aspects of production, transportation, processing, and use of hydrocarbons.

The role of DTCD in international technical cooperation is primarily one of a technical rather than a financing organization, apart from the modest funds available to it from the portion of the regular budget of the United Nations specifically set aside for technical cooperation purposes.

The funding of the department's technical cooperation activities in petroleum has been provided mainly by the United Nations Development Programme (UNDP). Technical cooperation projects financed by UNDP are undertaken in response to specific requests to UNDP from the governments of member countries. During the last two decades more than 100 technical cooperation and preinvestment projects in petroleum involving close to $100 million of UNDP expenditures have been implemented by DTCD. A roughly equivalent amount was provided by the recipient governments in the form of national staff, facilities, and local operating costs.

Over the years, the department has witnessed a steady expansion of requests for technical cooperation and advisory services in the area of petroleum. However, it was not until the latter part of the 1970s that both the number and size of projects in petroleum registered a substantial increase. Thus, in 1975, eight projects were executed by the department in

Table 1
Countries That Have Received Assistance in the Field of Petroleum Under Technical Cooperation Projects Executed By UN/DTCD

Country	Number of Projects
Albania	1
Algeria	1
Bahamas	1
Bangladesh	2
Barbados	1
Belize	1
Bolivia	2
Brazil	1
Burma	1
Chad	1
Chile	2
China	4
Costa Rica	2
Cyprus	1
Djibouti	1
Ecuador	3
Equatorial Guinea	1
Ethiopia	1
Grenada	1
Guyana	1
India	10
Indonesia	2
Jamaica	1
Kenya	2
Lebanon	1
Lesotho	1
Liberia	2
Malaysia	1
Malta	3
Mauritania	1
Morocco	1
Mozambique	2
Niger	1
Pakistan	1
Panama	1
Peru	1
Saudi Arabia	1
Seychelles	1
Syrian Arab Republic	2
Trinidad and Tobago	2
Turkey	3
Yugoslavia	1
Zambia	1
Total (43 countries)	71

the field of petroleum (Table 2). At the present time, the department is executing 29 UNDP-financed petroleum projects in 18 developing countries (Table 3).

Catalytic Role of UN Projects

While the DTCD program in petroleum is the largest technical cooperation program in the petroleum field within the United Nations system, it is relatively small, especially in view of the needs of the developing countries.

In spite of the modest amount of financing offered by the UN system, UN technical cooperation projects have often played an important catalytic role in many developing countries, especially at an early stage of their petroleum endeavors. This applies in particular to strengthening of institutions concerned with the development of petroleum as well as to providing the critical impetus to attracting the interest of the international oil industry in a given prospective area. The same may also be said of upgrading national capabilities through training, provision of experts or acquisition of modern technology.

In many instances, the United Nations is the organization of last resort in petroleum assistance. This may apply to sensitive areas, such as assistance in the formulation of petroleum exploration and development agreements, where objectivity is essential, or in the transfer of the latest technology and know-how, where the United Nations provides an independent means of access.

Table 2
UN/DTCD-Executed Petroleum Projects in 1975

Barbados	Petroleum Engineer/Economist
Bolivia	Centre for Petroleum Development
Chile	Petroleum Exploration in Central and South Chile
Ecuador	Strengthening of the Petroleum Sector
El Salvador	Petroleum Exploration and Legislation
India	Intensification of Offshore Exploration for Crude Oil
Malta	Petroleum Exploration
Turkey	Establishment of a Petroleum Development Center
Total	8 projects

Table 3
UN/DTCD-Executed Petroleum Projects as of July 20, 1984

Albania	Fellowship in petroleum sector and other areas.
Bangladesh	Strengthening the Planning and Implementation Cell of the Ministry of Petroleum and Mineral Resources.
Cayman Islands	Petroleum development.
Chad	Fellowship in petroleum exploration.
China	Training center for oil exploration and development techniques.
China	Deep borehole surveying.
China	Oil well sand control techniques.
China	Geophysical prospecting for petroleum in Guizhou Province.
Guatemala	Assistance in petroleum.
India	Deep drilling technology.
India	General consultancy for oilfields.
India	Geophysical data processing.
India	Training in offshore production, transportation and terminal operations.
India	Three-dimensional geophysical (seismic) surveys.
India	Advanced techniques in reservoir engineering.
Indonesia	Upgrading of LEMIGAS Oil and Gas Training Centre.
Kenya	Advisory services in petroleum.
Lebanon	Training in the field of modern methods of petroleum analysis.
Malaysia	Assistance to PETRONAS.
Malta	Offshore seismic survey and drilling program.
Morocco	Domestic utilization of natural gas.
Mozambique	Strengthening of the state secretariat for coal and hydrocarbons.
Mozambique	Assistance in petroleum exploration.
Poland	Fellowships in energy and other areas.
Romania	Black Sea offshore drilling and production.
Romania	Increasing the national oil recovery factor.
Turkey	Establishment of a petroleum development center (Phase II).
Turkey	Group training in the petroleum industry.
Yugoslavia	Development and utilization of Aleksinac oil shales (Phase II).
Total	29 projects.

* *Three additional projects are expected shortly to become operational: China (enhanced oil recovery techniques), Ghana (technical assistance to Ghana National Petroleum Corporation) and Togo (assistance in petroleum exploration).*

Typical Project

It would be difficult to describe a typical UN-executed project in petroleum, for each one is different and tailored to the specific needs of the requesting country. Technical cooperation projects range in size from a single expert for several months to large-scale projects lasting a number of years, costing millions of dollars, and requiring the setting up of a team of experts. These projects normally include, in varying degree, expert personnel, equipment, contract services, training, and fellowships. A selection of completed projects in petroleum (Table 4) may be broadly divided into five categories, although this classification is to a large degree arbitrary:

- Assistance for the establishment of petroleum research and training institutes.
- Support for carrying out petroleum resources investigations and surveys.
- Strengthening of existing petroleum institutions, including national oil companies, in developing countries.
- Supply of high-level expertise and technology.
- Training through the grant of fellowships.

Several examples of such projects are given in detail in the appendix.

Table 4
A Selection of Past Petroleum Projects Undertaken by UN/DTCD

Country	Project Title
Bolivia	Center for petroleum development.
Chile	Offshore exploration for petroleum.
Chile	Petroleum exploration in Central and South Chile.
Ecuador	Strengthening the petroleum sector.
India	Institute for Petroleum Exploration, Dehra Dun (Phase I).
India	Institute for Petroleum Exploration, Dehra Dun (Phase II).
Malta	Petroleum exploration.
Panama	Petroleum legislation.
Syrian Arab Republic	Strengthening exploration drilling capacity of the Syrian Petroleum Company.
Trinidad and Tobago	Seismic survey in the Marine Area between Trinidad and Tobago.
Turkey	Establishment of a petroleum development Center.
Burma	Oil well drilling training.
Bangladesh	Petrobangla pipeline training.
India	Oil exploration.
Niger	Petroleum drilling expert.
Algeria	Industrial and marketing surveys on petroleum derivatives and natural gas.
India	Intensification of offshore exploration for crude oil.
Albania	Assistance to the Petroleum and Gas Institute.
Turkey	Assistance for Petroleum Prospection and Studies.
Zambia	Oil exploration.
Ecuador	Establishment of a management control system in the petroleum sector.

Latest Trends in Project Composition

Accompanying the overall growth in the number and size of the technical cooperation projects in petroleum, there have been some changes in the composition of these projects. Table 5 presents expenditures by major component of UNDP-financed, DTCD-executed petroleum projects over the three-year period between 1979 and 1981.

While the individual components have fluctuated from year to year, personnel (experts and consultants) accounted for 25% of the total; fellowship training, 16%; equipment, 56%; and subcontracts, 3%. Expenditures on miscellaneous items have been excluded from these calculations.

This breakdown is to some extent indicative of several recent trends in the nature of UN-supported technical cooperation projects. First, the large share of expenditures for equipment may not be typical for DTCD-executed projects. Nevertheless, it does underline a shift in priority of the recipient governments which became even more noticeable in several new petroleum projects submitted in the period 1983-1984, reflecting the need of these countries for sophisticated modern technology.

Second, there is a tendency to the use of short-term consultants instead of long-term experts on most projects. In fact, only one of the projects currently in operation has a chief technical adviser (a project manager) assigned to assist the national project director in the management of the project. This is also linked with the increasing involvement of governments in project formulation and implementation and the emphasis on the use of qualified national personnel as experts whenever possible. Table 6 gives an indication of the number and scope of expert and consultant services provided on petroleum projects from 1979 to 1981. For 1984, the department anticipates recruiting 24 short-term consultants for the operational projects (Table 7).

Table 5
Annual Expenditures on UN/DTCD/Executed
Petroleum Projects in 1979, 1980 and 1981

Year	Number of Projects	Personnel	Fellowships	Equipment	Subcontracts	Total
1979	16	39.5	10.7	49.8	0.0	100.0
1980	16	44.4	11.9	31.1	12.6	100.0
1981	21	12.7	18.4	68.4	0.5	100.0
Total		24.9	15.5	56.3	3.3	100.0

Project Components (%)

Table 6
Experts on Petroleum Projects in the Field in 1979, 1980, and 1981

Country of Assignment	Expert Title
1979	
Ecuador	System accountant
	Computer programmer
	Reproduction expert
Turkey	Macropaleontologist
	Petroleum reservoir engineer
	Drilling mud engineer
	Project manager (Petroleum Development Centre)
	Petroleum production consultant
	Sedimentologist
Yugoslavia	Consultant (oil shale)
Total	10
1980	
Bangladesh	Petroleum exploration geologist
	Economist planner
	Consultant/statistician
Cayman Islands	Oil refinery adviser
Ecuador	Computer programmer
	Reproduction experts (2)
	System accountant
India	Consultants/geophysical data processing (4)
Kenya	Petroleum adviser
Malta	Seismic expert (field data gathering)
	Exploration consultant
	Seismic expert (computer processing)
	Seismic expert (interpretation)
Turkey	Project manager (Petroleum Development Centre)
	Macropaleontologist
	Organic maturation expert
	Sedimentologist
	Petroleum production consultant
Yugoslavia	Consultants (oil shale in situ retorting) (2)
Zambia	Senior consultant (petroleum exploration)
Total	25
1981	
Cayman Islands	Petroleum legislation expert
China	Instructor in petroleum engineering
	Specialist in petroleum drilling hole deviation and deviation control

Table 6 continued

Country of Assignment	Expert Title
India	Consultant on geophysical data processing
Kenya	Petroleum adviser
Turkey	Project manager (Petroleum Development Centre)
	Petroleum geologist
	Macropaleontologist
	Clay mineralogist
	Organic maturation expert
	Petroleum production consultant
	Petroleum consultant
Yugoslavia	Consultant/environmental protection
	Consultant/mining engineer
	Consultant/oil shale retorting technology
Total	17

Table 7
Experts on Petroleum Projects to be Recruited in 1984

Country of Assignment	Expert Title
Belize	Chief technical adviser
	Seismic interpreter
	Micropaleontologist
	Carbonate petrographer
China	Petroleum production consultant
	Seismologist
	Petroleum geologist
	Vibroseis expert (data acquisition) (2)
	Vibroseis expert (data processing) (2)
	Sand control engineer (2)
	Sand control laboratory specialist
India	3-D seismic expert (2)
	Enhanced oil recovery experts (4) (polymers, micellar/surfactant and miscible gas floods)
Malaysia	Curriculum development experts for PETRONAS (2)
Morocco	Natural gas distribution expert
Turkey	Expert on seismic methods

Third, training continues to be a major component of virtually every project. In addition to formal fellowships outside their countries (Table 8), national counterpart personnel receive considerable on-the-job training from the international staff assigned to the project and consultants may be required to give lectures as part of their project assignments.

Finally, the proportion of subcontracts, while small in the past, is expected to increase substantially in the near future. The need for a timely delivery of some highly specialized and complex piece of equipment or service often demands the use of business and consulting firms contracted by the United Nations.

Petroleum Assessment Missions

In 1977, the General Assembly requested the Secretary-General by resolution 32/176 to prepare estimates of the financial requirements for the next 10 to 15 years for the exploration and location of natural resources, including petroleum, in interested developing countries. Following the recommendations of a group of experts, it was decided that these needs should be assessed in part on the basis of short-term visits of technical missions to assist in making on-site reviews and, if desired, in identifying specific projects for the improvement of the geoscientific base and for further exploration. The Department of Technical Co-operation for Development has been the principal entity responsible for carrying out the assessment missions mandated by the General Assembly resolution 34/201 and financed from resources within the existing regular budget of the United Nations as well as funds-in-trust established for this purpose by the Government of Norway.

A total of 33 assessment missions were undertaken between 1980 and 1983. Twenty-six reports have been completed estimating financial requirements and other needs for further petroleum exploration in these countries. The reports cover a wide variety of topics relating to petroleum resources. These include a summary of the geology of the country and of the geological surveys that have been conducted; a history of petroleum exploration and development activities; the country's infrastructure and technical capacity for exploration and development; governmental organizations for petroleum exploration and training; governmental policy and legislation with respect to foreign investment in the petroleum sector; and an appraisal of financial, personnel, and technical requirements. The reports vary substantially with respect to the emphasis given to each of these categories. The information in the reports on geological surveys and hydrocarbon occurrences, together with the capacity of gov-

Table 8
Fellowships in the Field of Petroleum in 1979, 1980, and 1981

Country	Field of Study	Number of Fellowships
1979		
Ecuador	Organization and methods	2
India	Deep drilling technology	2
	Mud engineering	2
Romania	Enhanced oil recovery methods	1
Turkey	Unspecified (Petroleum Development Centre)	9
	Numerical reservoir evaluation	1
Total		17
1980		
China	Drilling techniques/study tour	2
Romania	Offshore platform design	1
Turkey	Unspecified (establishment of a Petroleum Development Centre)	7
	Refinery operations	1
	Natural gas production	1
	Refinery processes	2
	Geoelectrical exploration for oil	1
	Oil and gas fields production	1
	Numerical reservoir evaluation	2
	Pipeline transportation	1
	Economic and legal aspects	1
	Distribution	1
	Drilling management	1
Yugoslavia	Oil shale/study tour	7
Total		29

ernmental institutions for carrying out geological investigations and petroleum exploration, should be of considerable value in identifying further technical assistance and financial needs. Moreover, the reports may also provide an important basis for attracting external investment.

The aggregated total financial requirements for the 26 countries for which reports are completed amount to between $23.437 and $23.716 million (Table 9).

In a related activity, UN/DTCD and UNCTAD have been cooperating in the preparation and distribution of a questionnaire on Technological

Table 8
continued

Country	Field of Study	Number of Fellowships
	1981	
China	Drilling techniques	5
India	Seismic software applications	3
	Computer hardware maintenance	2
	System programming	1
	Geophysical data processing/study tour	1
	Well testing and completion techniques	2
	Offshore platforms and pipelines	1
	Onshore terminal operations	4
	Management of offshore drilling operations	7
	Well workover (offshore)	2
	Oil and gas production technology	4
	Offshore operations of processing platforms	3
	Platform configuration	1
	Cementing offshore wells	1
Indonesia	Unspecified (upgrading LEMIGAS Oil and Gas Training Centre)	5
Romania	Offshore drilling operations and equipment	1
	Offshore drilling and production platforms	1
	Offshore production technology	1
	Improved oil recovery	1
Turkey	Unspecified (Petroleum Development Centre)	11
	Oil and gas field production	2
	Formation evaluation	1
	Unspecified (group training in petroleum industry)	1
Total		63

Capacity in Developing Countries for Petroleum Exploration. The survey represents another part of the effort to implement General Assembly resolution 34/201, specifically as it relates to the transfer of technology. The results of the questionnaires will be analyzed and reported to the relevant bodies of the United Nations. At the same time, the results of this survey will be utilized in evolving appropriate programs and projects of technical cooperation in developing countries. As of July 20, 1984, 43 questionnaires have been received and the data are being analyzed by UN/DTCD and the UNCTAD Secretariat (Table 10).

Table 9
Decade Financial Requirements for Petroleum Exploration
(US$ million)

	Infra-structure	Exploration	Total
Africa (18)			
Benin	24	246	270
Botswana	–	99–189	99–189
Comoros	0	0	0[i]
Ethiopia	(Report pending)		
Ghana	N/A	106–160	106–160
Madagascar	N/A	324	324
Malawi	–	–	0[ii]
Mali	N/A	315–320	315–320
Mauritania	N/A	69	69
Morocco	N/A	2,225	2,225
Mozambique	N/A	115–205	115–205
Seychelles	1.5	74.5	75[iii]
Sierra Leone	(Report pending)		
Somalia	3	172	175
Sudan	16	5,750	5,766
Tanzania	8	1,170	1,178
Togo	–	30–55	30–55
Tunisia	–	2,383	2,383[iv]
		Subtotal	13,131–13,395
Middle East (2)			
Syria	(Report pending)		
Turkey	(Report pending)		
Asia and the Pacific (6)			
Bangladesh	18	430	448
Fiji	3	360	363
India	N/A	3,000	3,000

Advisory Services

In addition to projects and assessment missions, staff of the Natural Resources and Energy Division, at times supplemented by outside consultants, has carried out short-term advisory missions to most developing countries. Financed from the United Nations Regular Budget, these advisory services respond to Government requests for assistance regarding the formulation of petroleum strategies; proposals for petroleum exploration programs; petroleum legislation and exploration and development agreements; evaluation of petroleum exploration data; advice on petro-

Table 9 continued

	Infra-structure	Exploration	Total
Nepal	3	240	243
Pakistan	–	4,640	4,640
Philippines	–	411	411
		Subtotal:	9,105
Latin America and the Caribbean (7)			
Dominican Republic	35–50	700	735–750[v]
Ecuador	(Report pending)		
Guatemala	N/A	286	286[vi]
Guyana	N/A	88	88
Jamaica	–	92	92
Peru	(Report pending)		
Uruguay	(Report pending)		
		Subtotal:	1,201–1,216
		Grand Total:	US $23.437–23.716 million
		(seven reports still pending)	

[i] *No estimate was given by consultant.*
[ii] *Less than US $500,000.*
[iii] *Estimate for 5 years only.*
[iv] *An additional requirement of US $1,273 million was estimated for the development of new discoveries.*
[v] *Development for commercial production was estimated to require another US $2,400 million.*
[vi] *Development cost was estimated to total US $390 million for the same period.*

leum refining and distribution of petroleum products, proposals for utilization of natural gas, and issues concerning the development of oil shale resources.

Particular mention should be made of the assistance provided to developing countries in petroleum development agreements, petroleum legislation, and preparation and evaluation of competitive bidding for licenses, permits, and contracts. Many developing countries have, to a large extent, relied on transnational corporations to finance and carry out petroleum exploration activities. However, such arrangements often require complex negotiations in areas where many developing countries do

Table 10
Joint UN/DTCD-UNCTAD Questionnaire Survey of Technological Capacity of Developing Countries for Petroleum Exploration
(Status as of July 1984)

Africa (14)	Americas (16)	Asia (13)
Central African Rep.	Barbados*	Afghanistan*
Gambia	Belize	Cyprus
Lesotho	Brazil*	Jordan
Liberia	Chile*	Korea, Rep. of
Malawi	Dominica	Malaysia*
Mauritius	Dominican Republic	Nepal
Morocco*	Ecuador*	Oman*
Niger	Grenada	Papua New Guinea
Seychelles	Guatemala*	Philippines*
Sierra Leone	Guyana	Qatar*
Tanzania	Jamaica	Sri Lanka
Tunisia*	Mexico*	Thailand*
Zambia	Panama	Yemen, Dem.
Zimbabwe	Suriname	
	Uruguay	
	Venezuela*	

Questionnaires completed: 43

* 15 petroleum producers, of which 7 are petroleum exporters.
Uncompleted questionnaires were returned from Bhutan, Cameroon, Laos, Singapore, and Upper Volta.

not have sufficient experience. This and the growing government role in the development of indigenous petroleum resources have led to an increasing demand for technical cooperation services in this area. In this respect, more than 60 countries have received assistance since the early 1970s (Table 11).

Information Functions

In addition to its technical cooperation programs undertaken at the country level, the Department of Technical Co-operation for Development organizes and convenes conferences, seminars, and *ad hoc* groups of experts on various technical and economic aspects of petroleum resources exploration, development, and utilization. Specific subjects range from state petroleum enterprises in developing countries and petro-

Table 11
Advisory Missions on Petroleum Economics and Legislation

Africa (14)	Americas (19)	Asia (10)	Europe (3)
Burundi	Argentina	Afghanistan	Greece
Ethiopia	Belize	Bahrain	Malta*
Gabon	Bolivia	Bangladesh	Portugal
Gambia	Brazil	China	
Ghana	Colombia*	India	
Kenya*	Costa Rica*	Lebanon	
Liberia*	Dominican Republic*	Malaysia*	
Madagascar	Ecuador	Pakistan	
Mauritania*	El Salvador	Philippines	
Mozambique	Guatemala	Yemen Arab Republic	
Niger*	Guyana*		
Senegal	Haiti		
Seychelles	Honduras		
Sierra Leone	Jamaica		
	Nicaragua		
	Peru*		
	St. Lucia		
	Suriname		
	Uruguay		

* More than one mission.

leum administration to petroleum resources classification and techniques of petroleum development. The more recently held meetings include the United Nations Meeting on Petroleum Geology, which was held in March 1980 in Beijing, China with participants from 20 developing countries, the United Nations Meeting on Petroleum Exploration Strategies in Developing Countries, which took place in The Hague, the Netherlands in March 1981 and which brought together some 150 participants from 50 countries, and two meetings in 1982. The United Nations Workshop on Geophysical Methods in Petroleum Exploration, organized in cooperation with the Government of Norway, held from August 16 to 20 in Norway, and the United Nations Meeting on Oilfield Development Techniques, organized in cooperation with the China National Oil and Gas Exploration and Development Corporation, held from September 7 to 14 in Daqing, China. A complete list of meetings on petroleum is presented in Table 12.

Table 12
Meetings on Petroleum

United Nations Interregional Seminar on Techniques of Petroleum Development, New York, January 23–February 21, 1962.

United Nations Interregional Seminar on Petroleum Administration, Port-of-Spain, April 16–27, 1968.

United Nations Symposium on the Development and Utilization of Oil Shale Resources, Tallinn, August 26–September 4, 1968.

Ad Hoc Panel of Experts on Projections of Demand and Supply of Crude Petroleum and Products, New York, March 9–18, 1971.

United Nations Interregional Seminar on the Development and Utilization of Natural Gas, Moscow, October 12–28, 1971.

United Nations Interregional Seminar on Petroleum Refining in Developing Countries, New Delhi, January 22–February 3, 1973.

United Nations Meeting on Co-operation Among Developing Countries in Petroleum, Geneva, November 10–20, 1975.

Ad Hoc Group on the Classification and Measurement of Oil and Gas Resources, New York, March 28–April 4, 1977.

United Nations Symposium on State Petroleum Enterprises in Developing Countries, Vienna, March 7–15, 1978.

Intergovernmental Group of Experts on Mineral and Energy Exploration, New York, April 17–21, and July 10–21, 1978.

United Nations International Meeting on Petroleum Geology, Beijing, March 18–25, 1980.

United Nations Meeting on Petroleum Exploration Strategies in Developing Countries, The Hague, March 16–20, 1981.

United Nations Workshop on Geophysical Methods in Petroleum Exploration, Gaustablikk, Norway, August 16–20, 1982.

United Nations Meeting on Oilfield Development Techniques, Daqing, September 7–14, 1982.

In addition to meetings, the department's information functions are carried out through the preparation and publication of studies and reports on specific issues in the petroleum sector of particular concern to the developing countries. Table 13 lists most of the publications issued by the department in the field of oil, natural gas, and oil shale. Of particular interest may be the recently published proceedings of the United Nations Meeting on Petroleum Exploration Strategies in Developing Countries. The book provides an analysis of the petroleum potential of the developing countries, examines the impediments to the full exploitation of this potential and presents the views of some of the world's foremost authorities on the subject of how petroleum exploration and production can be accelerated in areas where it is now lagging.

Finally, two periodic publications prepared by UN/DTCD regularly carry articles, reports, notes, and other items concerning petroleum re-

Table 13
Publications on Petroleum

Petroleum Exploration: Capital Requirements and Methods of Financing. (United Nations Publication, Sales No. 62.II.B.3.)

Techniques of Petroleum Development. Proceedings of the United Nations Interregional Seminar on Techniques of Petroleum Development, New York, 23 January–21 February 1962. (United Nations Publication, Sales No. 64.II.B.2.)

Utilization of Oil Shale: Progress and Prospects. (United Nations Publication, Sales No. 67.II.B.20.)

Petroleum in the 1970s. Report of the *Ad Hoc* Panel of Experts on Projections of Demand and Supply of Crude Petroleum and Products, New York, 9–18 March 1971. (United Nations Publication, Sales No. E.74.II.A.1.)

Proceedings of the United Nations Interregional Seminar on Petroleum Refining in Developing Countries, New Delhi, 22 January–3 February 1973. Volume I: Report of the Seminar and Economic Aspects. Volume II: Technical Papers. (Published by the Indian Oil Corporation of the Government of India with the co-operation of the United Nations.)

Petroleum Co-operation Among Developing Countries. Proceedings of the United Nations Meeting on Co-operation Among Developing Countries in Petroleum, Geneva, 10–20 September 1975. (United Nations Publication, Sales No. E.77.II.A.3.)

State Petroleum Enterprises in Developing Countries. Papers presented at the United Nations Symposium on State Petroleum Enterprises in Developing Countries, Vienna, 7–15 March 1978. (Published for the United Nations by Pergamon Press, New York, 1980.)

Petroleum Geology in China. Principal lectures presented to the United Nations International Meeting on Petroleum Geology, Beijing, 18–25 March 1980. (Published for the United Nations by Penn Well Publishing Company, Tulsa, Oklahoma, 1981.)

Petroleum Exploration Strategies in Developing Countries. Proceedings of a United Nations Meeting on Petroleum Exploration Strategies in Developing Countries, The Hague, 16–20 March 1981. (Published in co-operation with the United Nations by Graham and Trotman Limited, London, 1982.)

sources. *Natural Resources Forum,* a quarterly journal published on behalf of the United Nations by D. Reidel Publishing Company in the Netherlands, and *Natural Resources and Energy Newsletter,* a bi-monthly published by the United Nations, are distributed free of charge.

Conclusions

The technical cooperation activities of UN/DTCD in the field of petroleum reviewed here reflect the diverse needs of a wide range of developing countries at various stages of development, possessed of differing resource endowments, pursuing a spectrum of development policies. As a result, the United Nations technical cooperation program has been called upon to deal with practically every area and issue of the petroleum sector in their technological, economic, legal, and administrative aspects in responding to the requests for assistance from the developing countries.

Appendix: Project Examples

The following examples of projects illustrate the kind of assistance in petroleum provided by the United Nations over the last two decades.

Research and Training Centers

Institute for Petroleum Exploration, Dehra Dun, India

A request for assistance from the United Nations in setting up a research and training wing within the Oil and Natural Gas Commission (ONGC) was submitted in 1960 and the project became operational in 1962. Its purpose was the establishment of an institute to undertake applied research in the exploration and production phases of the oil industry; conduct refresher training courses for technical personnel already serving with the ONGC; and train new recruits required for the planned expansion of the ONGC's operations.

The project, consisting of two phases, was completed in 1968 with a total expenditure of $1,869,400 from UNDP and the equivalent of $2,525,350 contributed by the government. In 1967, the Institute moved to new buildings constructed for the purpose at Dehra Dun. A total of 38 reports were produced by the staff of the Institute, both United Nations experts and counterpart personnel, dealing with various technical aspects of the activities of the Oil and Natural Gas Commission. The United Nations provided a total of eleven categories of experts, including the project manager, and a number of consultants visited the project. During Phase I of the project, ONGC provided a total of 132 counterpart staff, of whom eight were senior technical personnel. The Phase II plan of operations called for a total of 300 counterpart staff, a figure which indicates some measure of the rate of expansion of the Institute during the life of the project. Thirty-one fellowship training courses for ONGC personnel were provided outside India by the United Nations.

The Institute was organized into the following six divisions: Geology, Geophysics, Chemistry, Drilling, Production, and Training and Documentation. During the course of the project the Institute took an active part in all the operations of the Commission, that is, in research and exploration for oil and gas fields, their development and exploitation, as well as in the transportation of petroleum. After the Institute started providing assistance to ONGC, 16 fields were discovered in the western region of India and 3 fields in Assam.

At the time of the termination of the project, the ONGC had been provided with a first-class technological facility equipped with the most up-to-date instruments, capable of investigating almost any type of problem arising from the operations of the Commission. The Indian staff of the Institute had been trained to use the facilities provided and was fully capable of carrying on the work of the Institute. During subsequent years, the Institute played an important role in the work of the Oil and Natural Gas Commission of India.

Training Center for Drilling Techniques in China

The United Nations received a request for assistance in the establishment of a training center to provide a continuing education program in modern drilling technology. The objectives of the project comprise setting up the Training Center for Drilling Techniques, with up-to-date teaching facilities as well as an advanced laboratory; upgrading the level of drilling-technology training through the adoption of advanced training concepts, methods, and materials; upgrading the level of technical drilling personnel through a specialized training program; and development of qualified trainers in drilling technology to undertake technical training in various oilfields in China. The United Nations inputs consist of several experts in petroleum drilling, fellowship training in petroleum engineering and petroleum drilling practices, and equipment, including a simulator. The budget includes the Government's contribution of ¥4,136,800 and a UNDP input of $500,000.

This project is presently (mid-1984) nearing completion. The first course, a refresher course on drilling fluids, was offered in 1981 and attended by some 50 senior drilling engineers from all over China. This was the first of three courses to be assisted under this project. In 1982, the Center gave 12 courses, mainly for senior engineers, held by members of its own faculty who have received training under the project. At present, the Training Center uses classroom facilities of the East China Institute of Petroleum and the new classroom building completed in mid-1982, which was part of the Government contribution-in-kind to this project. The training program has already been established and all the training facilities installed. The Center will eventually train 200 technicians from China and other Asian countries annually through courses in abnormal-pressure direction, well control, well planning, computer-aided drilling optimization, and directional drilling.

Since 1983 the project has expanded the scope of the existing Training Center for Drilling Techniques into a Training Center for Oil Exploration and Development Techniques.

Petroleum Development Centre in Turkey

The Petroleum Development Centre of the State Oil Company (TPAO) in Ankara, Turkey was established with the assistance of the United Nations in 1976 under the first phase of a project to extend over a 10-year period. The UNDP contribution was approximately $3 million. Matching resources were provided by the Government of Turkey, in addition to a large cost-sharing component.

The Centre is an integral and important part of the TPAO organization and is designed for research and development of geological and geophysical exploration and drilling and production techniques. The various laboratories now in operation are equipped with the most modern equipment for sedimentological, paleontological, and palynological investigations; reservoir engineering (including PVT); core analysis, chromatography for analysis of drilling muds, cement, and water; and geophysical data processing.

A large number of fellowships providing training abroad in different fields were included in the project, in order to assist in staffing the Centre. TPAO is planning to expand the Centre during the next five years to include research sections for refining and petrochemical operations and possibly also for oilfield equipment and materials. High on the priority list of work are investigations into the enhanced recovery of heavy oil from the Bati Raman oilfield.

Surveys

Seismic Survey of the Marine Area between Trinidad and Tobago

This project had its origin in a request in 1968 from the Government of Trinidad and Tobago for United Nations assistance in investigating the possibility of sedimentary deposits, potentially attractive for exploration for hydrocarbon accumulations, being present beneath the territorial waters north of Trinidad and west of Tobago, and between the two islands. This part of Trinidad and Tobago Continental Shelf had not been considered attractive for such exploration by private industry. A previous United Nations-sponsored airborne magnetometer survey of the area had, however, yielded information suggesting the possible presence of a substantial column of bedded deposits overlying a magnetically-responsive "basement" terrain.

The Seismic Survey of the Marine Area between Trinidad and Tobago was established as a one-year project costing $690,171, with the United Nations as Executing Agency. Owing to the highly specialized nature of marine seismic techniques, it had been decided from the start to subcontract the survey operation to a geophysical contractor.

The project was carried out to the full satisfaction of the Government and the United Nations. The results aroused a considerable interest within the oil industry and, subsequently, the Government granted licenses for further exploration in the project area.

Chile

A United Nations technical cooperation project in Chile, which included the appraisal of the oil and gas potential of the offshore Pacific area and Straits of Magellan in Chile, was completed in 1976. The project was financed by a contribution of $2.3 million from the United Nations Development Program and Government counterpart funds. Preparatory activities began in 1971 with an aeromagnetic survey, followed by a feasibility study. Two marine seismic surveys were carried out in the offshore Pacific area and one in the Straits of Magellan. These surveys led to an oil-drilling exploration program; natural gas was discovered in one of the wells. In addition, a digital recording facility was set up in one of the Chilean land seismic surveys, and a sophisticated minicomputer for processing seismic data was installed in Santiago. A fellowships program led to the training of Chilean engineers in data processing and geophysical interpretation.

Support to Government Institutions

Geophysical Data Processing in the Oil and Natural Gas Commission of India

The Oil and Natural Gas Commission of India needed a high-speed seismic computer system, including hardware and software, in order to clear the backlog of seismic data processing work and to keep pace with the data generated in the field. The Government of India has therefore requested United Nations assistance in acquiring the most modern seismic computer system, through international tender, and in training the staff of the Commission in its efficient operation and maintenance.

A total sum of $1.7 million from UNDP has been budgeted for this purpose. The project provides for the establishment of a fully equipped and manned seismic data processing center at Dehra Dun, India. In addition to supplying the computer system with hardware and matching software, it also provides training of the Commission staff, and installation and commissioning of the computer system at Dehra Dun. Furthermore, a study tour of similar operational seismic data processing centers elsewhere in the world by the Commission staff and fellowships has been budgeted to provide additional in-depth training in the operation and maintenance of the system.

As the executing agency for the project, the Department of Technical Co-operation for Development, through its Natural Resources and Energy Division, has provided technical advice in the selection and evaluation of a suitable system and contract services for its procurement and installation. Proposals for the subcontracts have been submitted and evaluated and the contract was signed last year. The Department also arranged the training program in the United States of America for the staff of the Oil and Natural Gas Commission concerned with operating and maintaining the system.

Other

In Ecuador, a United Nations technical cooperation project to strengthen the Petroleum State Company and the Hydrocarbon Division of the Ministry of Natural Resources in various aspects of petroleum exploration and development was completed at the end of 1977. A contribution of $1 million from the United Nations Development Programme and equivalent resources from the Ecuadorian Government have enabled a team of six international experts to provide technical assistance in the fields of petroleum legislation, accounting, marketing, oil production, geology, geophysical interpretation, and petroleum economics.

A project in Kenya provided for a petroleum expert to serve as an overall adviser to the Government on petroleum strategy and policy. He also reviewed existing agreements, arrangements and policies governing the procurement, refining, and processing of crude oil, and examined the current refinery with a view toward improving recovery and reducing residual heavy fuel, which was being reexported.

The objective of a project in Mozambique is to provide assistance to the Government in formulating a petroleum exploration strategy involving appropriate terms and conditions for cooperation with transnational oil corporations. This consists of (a) carrying out speculative offshore seismic surveys by specialized geophysical contractors at their own cost; (b) carrying out onshore seismic surveys by contractors with costs to be borne by the Government; (c) formulating appropriate petroleum legislation, a model contract for exploration and development and a fiscal code; and (d) preparing seismic data packages for sale to transnational oil corporations and other entities such as state petroleum enterprises interested in undertaking petroleum exploration activities in the country.

High-Level Expertise and Technology

Oil Shale In Situ Reporting in Yugoslavia

The Government of Yugoslavia decided to investigate the possibility of producing oil from oil shales in the Aleksinac Basin and requested assistance from the United Nations in the implementation of this project. A project, approved in June 1979, called for a UNDP contribution of $US 112,400 and a government contribution of Din 6 million, over a period of two and a half years. The UNDP input consisted of international experts, fellowship training abroad, and some equipment.

The project consisted of three phases. Phase I comprised geological exploration to determine such characteristics as the permeability, depth and thickness of oil shale deposits, distribution and variation of coal content, etc. Phase II of the project consisted of consultations with foreign experts, training of personnel, visits to oil shale installations abroad, and acquisition of some laboratory equipment and documentation. Phase III involved the preparation of a preliminary feasibility analysis on a combined surface and modified in situ development project. The study, prepared by the Mining Institute at Belgrade, answered questions regarding the technological and technical aspects of the process, capital investment requirements, processing costs, etc.

The results obtained from the first three phases of this project provided answers regarding the possibility of applying in situ technology to Aleksinac Basin oil shales. The fundamental output was a prefeasibility report based on the latest techniques applied to Aleksinac conditions and their economics in Yugoslavia. The results represent a starting point for phases IV, V, and VI of the project, which will consist of the development of a process flow-sheet for pilot-plant facilities, the construction of a pilot plant, and running of the pilot-scale process to investigate and determine the technological parameters of the retorting process. This stage of the project is now being initiated with an estimated UNDP component of $704,200, including cost sharing by the Government.

Geophysical Prospecting for Petroleum in Area of Carbonate Deposition of Southern Guizhou Province (Southern China)

This project is providing advanced equipment and expertise for a geophysical survey in the Southern Guizhou province, using seismic reflection to be followed by thorough geological interpretation to detect potential subsurface traps. The United Nations assistance will enable Chinese experts to become familiar with advanced geophysical prospection techniques in marine carbonate basins. A vibroseis crew and a portable conventional seismic crew, using air drilling equipment, will be operating in the project area and more than 1,000 kilometers of vibroseis regional seismic lines will be completed and processed. Following the completion of the vibroseis survey a comprehensive report will be prepared on the subsurface regional geology of the Southern Guizhou Province, including recommendations for further seismic surveys and, if possible, exploration drilling for petroleum. The project includes consultants, fellowship training, study tours and various equipment, including installation, maintenance, and training supplied under subcontracts. The cost of UNDP inputs for the project is approximately $4 million while the Government contributes $850,000 as cost sharing.

Three-Dimensional Geophysical (Seismic) Surveys in India

The Oil and Natural Gas Commission of India has recently decided to obtain a 3-D seismic facility and approached the United Nations for assistance in acquiring through international tenders the most modern 3-D seismic unit with telemetric equipment, complete with a set of sophisticated energy sources, and including the training of ONGC staff in its operation and maintenance, both at the supplier's home base and on the job in India. The United Nations assistance will take the form of a consultant's services, a subcontract for the supply of the complete 3-D seismic system, including installation and commissioning at the site in India, individual training, and a study tour by ONGC staff to field sites of similar operational 3-D seismic systems in the world. The financing for the project will come from the Government (Rs. 3,814,000 in kind and $1,617,000 as cost sharing) and from UNDP ($1,204,779).

Training

Upgrading of LEMIGAS Oil and Gas Training Centre in Indonesia

The United Nations Department of Technical Co-operation for Development is currently providing assistance to the LEMIGAS Oil and Gas Training Centre. Through a UNDP contribution of $170,000 and the Government's equivalent input in kind, the senior teaching staff of the Centre will receive specialized practical fellowship training in countries with advanced technology in the fields of oil and gas and related areas. Short term (2-3 months) fellowships for up to 20 members of the faculty are envisaged under this project in a variety of fields, including gas technology, corrosion engineering, offshore drilling, petroleum geology, training management, petroleum administration and accounting, etc.

Other

The expertise of the staff of the Turkish Petroleum Company (TPAO) is currently being further enhanced through a DTCD-executed project which provides for in-house training as well as training abroad. The United Nations project provides for the services of a number of consultants to review and advise the various departments of TPAO on a variety of operational issues. A team of experts is to perform an in-house training exercise for the exploration department in the preparation of a basin study. A number of fellowships abroad and some equipment are also included. The total UNDP input is about $750,000, to which the Government contributes about a third as cost sharing, in addition to an equivalent of $749,764 in kind.

Index

Abiogenic theory of petroleum origin, 12-13
Aeromagnetic survey, 195
 in Zambia, 198-202
Africa, wildcat drilling statistics, 427-432. *See also individual countries.*
Airborne magnetics, 117
Airborne landform analysis, 60-62
Airborne photography
 landform analysis, 60-61
 microtechniques site selection, 117
Airborne radiometrics, 117
Alabama, 215-225
Alaska, 7, 21-24
Amazon Basin, 16
American Gas Association (A.G.A.), 297
Anadarko Basin, 352-356
Anticlines, 6
 and landform analysis, 60
Apical theory and microtechniques, 120
Appalachia, 337, 367-371, 524
Argentina, 9, 346
Arrehnius-derived plot, 40
Artificial lift, 227. *See also* Pumps.
Asphalt, 4
Australia,
 photogeologic mapping, 61
 source rock studies, 25

Back-arc basins and maturation studies, 42
Bahrain, 7
Baku, eternal fires, 4-5, 46, 48
Bangladesh, 389-403
Banks. *See also* Financing.
 regional, 485-490
 types of, 485-487
Barefoot wells, 367-371
Basin maturation, 35-42
Basins. *See individual countries.*
Belden and Blake Corp., 213
Bengal Basin, 389-403

Bitumen, 4
 definition, 135
 deposits, 155-158
 utilization, 561-566
Blow-down, 330-331
Bolivia, 9
Brazil, 16-17, 346
 photogeologic mapping, 61
Bureau de Recherches et de Participations Minieres (BRPM), 173-176
Burma, landform analysis, 60

California, 253-254, 259
Cameroon, source-rock studies, 17-21
Canada
 Athabasca, 48
 bitumen, 155
 drilling depth statistics, 432-433
 energy policies, 583-592
 first wells, 5
 heavy oil, 158
 Saskatchewan development, 583-592
 seeps, 48
 wildcat wells, 425-426
Central America, wildcat wells, 427-432. *See also individual countries.*
Chile, United Nations oil surveys, 673
China
 natural gas availability, 346
 source-rock studies, 31-32
 Training Center for Drilling Techniques, 671
 United Nations, projects, 675-676
Club of Rome, 347
Coal energy, 363
Colorado, 74, 181-192
Commercialization of gas, 567-582. *See also* Natural gas.
 oil vs. gas, 578-579
Compaction. *See* Subsidence.
Completion units, 208
Compressibility, 256-258
Compression cost, shallow gas field development, 317

679

680 Shallow Oil and Gas Resources

Compressor capacity, equations for, 328–329
Computer
  databases, 99–112
  models, 40–42
  simulators, 621–638
Conventional Energy Training Project (CETP), 648–650
Concession agreements, 534. See also Contracts.
Contracts
  concession agreements, 534
  economic considerations, 537–539
  energy development, 442–454, 481–483, 527–529, 573–578, 579–582
  Guatemala, case study, 610
  independent oil companies, 462–470, 481–484
  natural gas development, 540, 573–578
  noneconomic considerations, 539
  production-participation, 612–618
  production sharing, 536
  resource commercialization, 579–582
  service arrangements, 536–537
Costs
  comparisons by target depth, 425–439
  of compression, 317
  drilling, 432–437, 525
  exploration, 64, 279–281
  heavy crude utilization, 564
  insurance, 266–267
  landform analysis, 64
  natural gas energy system, 543–559
  production, 281–285
  shallow gas development, 329
  shallow oil drilling, 525
  technology transfer, 642
Costa Rica, seeps, 55
Cratonic crustal basins, 38–42

Data, availability of, 99–112, 279–281, 480–481, 516–517, 524
Databases, 99–112
De Golyer, Everett Lee, 7
Denver Basin, 181–192
Developing countries. See also individual countries.
  energy policies, 441–454
  exploration
    financing, 441 ff.
    incentives, 509–520
    independent oil companies, 459–476

  independent oil companies, 459–476
    case history, 478–480
    contracts, 462–470, 481–484
    and financing, 470–474
  natural gas energy systems, 543–559
  Overseas Private Investment Corporation, 505–508
  training, 621–638
  United States universities, 639–644
Development. See Natural gas; Shallow oil and gas.
Domes, 60
Dominican Republic, seeps, 55
Downstream Gas Development Program, 573–576
Drilling
  costs, 432–437, 525
  depths, 3, 432–433, 525
  equipment, 208
  low-cost, 281–283
  methods, 210–211
  microdrilling, 245–252
  schedules, 317–322
  statistics, 425–437
  technology, 158–162
  in United States, 164–169
  wildcat wells, 425–432
Dry holes, 10–12
Dynamic programming, 316

Economic considerations. See also Costs; Financing.
  in contracts, 537–539
Ecuador, United Nations projects in, 674
Energy. See also Hydrocarbons; Natural gas; Oil; Shallow oil and gas.
  coal, 363
  cost-effective, 345–365
  nuclear, 361–363
  policies
    Canada, 583–592
    developing countries, 441–454
    exploration, 526
    Guatemala, 610–618
  synthetic, 364
  systems, 543–559
  technology substitute analysis, 356–360
Energy Resources Institute, 111
Energy World Trade, 494–497
Enhanced oil recovery, 235–241
Environmentalists, 8
Equations-of-state, for gas, 298, 300–312
Equatorial Guinea, seeps, 55

Index   681

Equipment
  sizing equations, 326-329
  suppliers, 493
Essaouira Basin, 174-178
Eternal fires, 4-5, 46, 48
Exploration
  data sources, 99-112, 279-281,
    480-481, 516-517, 524
  developing countries, 455-458,
    509-520
  economics, 125-129
  financing, 441-458
  geochemical techniques, 115-116
  history, 3-10
  incentives, 509-520
  landform analysis, 59-65
  microtechniques method, 113-123
  policies, 526
  satellite remote sensing, 67-75,
    77-94
  seepsand, 45-55
  source-rock studies, 12-32
  seismic reservoir analysis, 95-97
  sparsely tested areas, 11-34
  technology management, 512-514
  temperature-history studies, 11-32,
    35-43
Explosives, 235-236
Exxon, 8
Europe, 435-437. See also individual
    countries.

Fang Basin, 287-293
Field development optimization,
    316-331
Financing
  in developing countries, 455 ff.
  exploration, 441 ff.
  by independents, 456, 459-474
  sources, 456-458, 472-474
    regional banks, 485-490
    trading company, 491-497
  shallow gas exploration, 527
Flaring, of natural gas, 568-569
Flow equation for gas wells, 327-328
Fracturing (hydrofracturing), 368-371
Fuel Oil Age, 6

Gamma-ray surveying, 116
Gas. See Natural gas.
Gas cycling, 330-331
Gas flow production optimization,
    317-331
Gas Research Institute, 297
Geochemical techniques, 115-116
Geology, 3, 11-12. See also individual
    countries.

Geomorphologic evaluation. See
    Landform analysis.
Geophysics, 3, 115-116
Geothermal gradients, 38-40
Ghana, seeps, 55
Golden Lane, 6
Guatemala
  development, 603-618
  exploration, 604-609
  legislation, 610
Guyana,
  seeps, 55
  source-rock studies, 25-30

Halliburton Company, The, 368
Hazards. See Safety.
Heading, in naturally flowing wells,
    227, 232
Heavy crudes
  definition, 135
  deposits, 155-158
  with high water content (case
    studies), 215-225
  recovery and subsidence, 258-260
  utilization, 561-566
High energy gas fracturing, 235-241
Hydrocarbons. See also Natural gas;
    Oil; Petroleum; Shallow oil and
    gas.
  bitumen, 4, 135, 155-158, 561-566
  heavy crudes, 135, 155-158,
    215-225, 258-260, 561-566
Hydrofracturing, 368-371

Illinois Basin, 90
Incentives
  exploration, 480-483, 509-520
  investment, 505-508
  resource development, 442-454, 523,
    532
Independent oil companies. See also
    Wildcat wells.
  developing countries and, 459-478
    case histories, 478-480
  contracts, 462-470, 481-484
  financing, 470-474
  role in oil discovery, 4
India
  Indian Institute for Petroleum
    Exploration, 651, 670-671
  natural gas availability, 346
  United Nations programs, 673-674,
    676
Indonesia
  first wells in, 5
  landform analysis, 60

682   Shallow Oil and Gas Resources

maturation studies, 42
production-sharing contracts, 9
satellite remote sensing, 69
shallow gas reserves, 405-422
small communities and shallow gas reserves, 405-422
United Nations projects, 676
Inorganic theory of petroleum origin, 12-13
Insurance
  costs, 266-267
  investments in developing countries and, 505-508
International Finance Corp., 511
International Institute of Applied Systems Analysis (IIASA), 356-360
International Monetary Fund, 457-458
International investment
  case histories, 478-480
  contracts, 481-484
  by independent oil companies, 462-467, 477-478
Investment, in developing countries, 455-458, 503, 508
Iran
  first wells in, 6-7
  landform analysis, 60
Iraq, 7
Italy, 254

Jamaica, 42
Japan
  natural gas availability, 346
  subsidence in, 254
Joiner, Dad, 7
Jordan, seeps in, 55

Kathmandu Valley, 373-386
Kenya, 55
  United Nations projects, 674
Kerogen, 23
  development into hydrocarbons, 35-37
  types of, 18
Kerosene Age, 6
Kinetics, and oil and gas development, 35-40
Kuwait, 7

Landform analysis, 59-65
Landsat Multispectral Scanner (MSS), 69-70, 77-78
  imagery, 67, 69, 84

Leases. *See also* Contracts.
  case history: Osage Indian Reservation, 397-399
  developing countries, 481-482
Legal considerations, 530-531, 533-541
Lineaments, 85
Logistic Substitution Analysis, 357-360
Logs
  seismic, 96-97
  sonic, 95-96
LOM (level of organic metamorphism) technique, 40
Lopatin method, Waples-calibrated, 40
Loss control. *See* Safety.

Magnetoelectrics, 121-122
Malagasy Republic, 55
Malaysia,
  maturation studies, 42
  natural gas availability, 42
Mapping, landform analysis, 63-64
Marchetti Market Penetration Analysis, 360
Marine landform analysis, 63
Mathematical models,
  production performance analysis, 315-337
  stability predictions, 277-284
  subsidence, 260-261
Maturity studies, 12-21, 35-40
  case studies, 21-32
Methane. *See* Natural gas.
Mexico
  oil development, 9
  subsidence, 254
Microdrilling
  costs, 245-246
  environmental impact, 251
  logging, 251
  production, 252
  safety, 250
  services, 250
  time, 246, 248, 250
Micromagnetics, 120
Microseepages, 85
Microtechniques method of shallow oil and gas exploration, 113-123
  vs. conventional methods, 115
  costs, 116
  effectiveness, 116
Middle East, 6-7. *See also* individual countries.
Morocco, 173-178, 647
  ONAREP, 176-177, 647
  seepages, 173

Mozambique, 674
Multiple regression method of Yu, 40

National oil companies, 492-494. *See also individual countries.*
Natural gas, 12
  advantages of, 359
  availability, 346-356
  co-generation topping plant, 337-343
  commercialization, 567-582
  contract provisions, 540, 573-578
  developing countries, 373-386, 543-559
  downstream gas development program, 573-576
  energy system, 543-559
  field development, 329
    compression cost, 317
  flaring, 568-569
  history of production, 8
  least-cost energy source, 345-365, 543-559
  production optimization, 317-331
  well, under partial water drive, 315-331
Natural-gas-fired co-generation topping plant, 337-343
Naturally flowing oil wells, stability predictions, 227-284
Nebraska, 181-192
Nepal
  natural gas development, 373-386
  seeps, 55
New Zealand, 346
Norway, 661, 667
Nuclear energy, 337-338, 361-363

Occupational Safety and Health Act (OSHA), 267
Oceanic crust, 40
Offshore shallow oil and gas, 149-152
Oil
  ceiling, 32-38
  companies, 11. *See also* Independent oil companies; National oil companies.
  drilling activity, 164-165
  crisis, 8, 61-62
  floor, 38
  potential, 4
  seeps, 5
  wells, 281
    safety, 265-277
    stability predictions, 277-284
Oil-importing developing countries (OPIC). *See* Developing countries.

Oklahoma, 6, 593-601
Oklahoma Geological Survey, 6
Onshore shallow oil and gas, 149-152
OPEC. *See* Organization of Petroleum Exporting Countries.
Optimization of field development, 316-331
Organic theory of petroleum origin, 12-13
Organization of Petroleum Exporting Countries (OPEC), 347, 492, 505, 511, 587-588
Orinoco oil belt, 149
Osage Indian Reservation, 593-601
Overseas Private Investment Corp. (OPIC), 593-601

Pakistan, 346
Paleostructure, 127-128
Panama, 55
Passive-decay model of basin maturation, 42
PDS Database, 108-110
Pennsylvania, 5
Permian Basin, 7
Peru, 5
Petroleum. *See also* Oil; Shallow oil and gas.
  exploration
    history, 3-12
    source-rock studies, 12-32
    origin, 12-23
Petroleum Administration for Defense (PAD), 145, 149
Petroleum Data Systems (PDS), 135, 149
  databases, 99-100
  example applications, 101-108
Petroleum training center, 631-633
Photogeologic mapping, 61
Pitch lakes, 48
Poland, 5
Portugal
  landform analysis, 63
  seeps, 55
Possible oil-bearing sediments, 4
Powder River Basin, 892
Pratt, Wallace, 8
Pre-Rif nappe, 173
Price controls, United States federal, 8-9
Production
  developing countries, 455-458
  equipment, 208-209
  field development, 207-208
  financing, 455-458
  low-cost, 281-285

684 Shallow Oil and Gas Resources

optimization, 317-331
safety, 265-277
system model, 317, 326-329
Production-sharing agreements, 9, 536
Progressive cavity pumps, 243-244, 284
Propellants, for well stimulation, 235-241
Proppants, 368
Prospects, rating, 125-129
Prudhoe Bay, 7
Pumps, 227
  low cost systems, 283-285
  progressive cavity, 243-244, 284
Pyrolysis, 14, 27

Radialfrac, 235-241
Radiometrics, 116, 121
Regional banks and developing countries, 485-490
Resource occurrences, 133-172
Risk
  reducing through OPIC, 503-508
  shallow oil exploration, 525
  wildcatters and, 494-502
Risk-service contract, 536
Rockefeller, John D., 5
Royal Dutch Shell, 5
Royalties, 444, 537
Rumania, 5
Rural communities, energy systems for, 543-560

Safety, 265-277
  costs, 266-267
  for loss control, 268-275
Salt domes, 60
Saudi Arabia, 7, 61
Saskatchewan. *See* Canada.
Satellite remote sensing, 67-75, 77-94
  case studies
    Indonesia, 69
    United States, 69-74, 78-82, 86-94
  cost, 68
  methodology, 84
Satellite landform analysis, 63
Secondary recovery, with Radialfrac, 240
Sedimentary basins. *See individual countries.*
Seislogs, 96-97
Seismic reservoir analysis, 95-97
Seismography, and microtechniques, 115-116

Senegal, 203-204
Service arrangements, 536-537
Seeps, oil, 5
  analysis of, 53-54
  clues to, 47
  Canada, 48-52
  content of, 45-46
  dead, 46, 48, 52
  definition, 45
  direct surface exploitation, 46
  economic value of, 46, 48, 52
  eternal fires, 46, 48
  as guides to deposits, 52-55
  landform analysis, 60
  live, 46, 52
  microseepage, 85
  microtechniques and, 120
  occurrences of, 46, 49-51, 53, 287
  Thailand, 287
  Trinidad, 48
  United States, 53
  Venezuela, 48
Shallow oil and gas
  definition, 3, 133-134
  development incentives, 523-532
  geology, 3
  exploration
    computers and, 99-112
    databases, 99-112
    economics, 125-129
    history, 3-10
    landform analysis, 59-65
    microtechniques method, 113-123
    satellite remote sensing, 67-76
    seeps and, 45-55
    seismic reservoir analysis, 95-97
    source-rock studies, 11-32
    temperature history studies, 11-32, 35-43
  natural gas
    availability, 346-365
    Bangladesh, 398-403
    barefoot wells, 367-371
    co-generation topping plant, 337-343
    commercialization, 567-582
    contracts, 540, 573-578
    development, 329
    drilling, 525
    energy system, 543-559
    equation-of-state calculation, 298-312
    field development, 317
    field operational limits, 316
    fluid property prediction, 297-313
    hydrofracturing, 367-371
    Indonesia, 405-422

Nepal, 373-386
  reservoirs under partial water
    drive, 315-331
  well development in rural
    communities, 555-559
occurrences
  by depth, 139-145
  by geography, 135-138
offshore, 149-152
oil
  case study of field development,
    209-212
  enhanced oil recovery, 235-241
  equipment, 243-244
  field development, 209-212
  heavy crude, 213-225, 561-566
  high energy gas fracturing,
    235-241
  low-cost production, 279-285
  microdrilling, 245-252
  Morocco, 173-178
  production safety, 265-278
  Senegal, 203-204
  subsidence, 253-264
  surface indications, 3, 68, 85-86
  survey of resource occurrences,
    158-162
  stability predictions, 227-234
  Thailand, 287-293
  United States, 162-164, 181-192
  wells, categories of, 207
    naturally flowing, 227-234
    stimulation, 235-241
  Zambia, 198-202
onshore, 149-152
potential, 12-43
Shuttle imagery radar (SIR), 69
Simulators, training, 621-638
  vs. training rigs, 625-626
  types, 634-636
  uses, 636-638
Soil-air analysis, 122
Somali Republic, 55
Sonic logs, 95-96
Source-rock studies, 12-21
  case studies, 21-32
South America, wildcat wells in,
  427-432
Soviet Union. *See* U.S.S.R.
Sparsely tested exploration areas,
  11-34
Spindletop, 5-6
Stability prediction, 227-234
Standard Oil Company, 5, 8
Stimulation, 235-241
Stress and subsidence, 255-258
Stripper oil, 207

Subsidence,
  classification, 254-256
  heavy oil, 260
  mathematical evaluation, 260-261
  shallow oil production and, 253-264
Surface indications of shallow oil, 3,
  68, 85-86
Suriname, 55, 244
SUPRA Corp., 494-497
Synthetic fuels, 364
System analysis model, 326-329
Sweden, 245

Taiwan, 346
Tanzania, 55
Tar sands
  definition, 135
  utilization, 561-566
Taxes
  contracts, 537-539
  development, 445-454
  exploration, 526
Technology transfer to developing
    countries
  simulator training, 621-638
  United States Agency for
    International Development,
    645-650
  United States Universities, 639-644
Technique of operation (TOR) method
    of loss control, 271-275
Tectonics, 35-43
Temperature-history studies, 12-21,
    40-42
  case studies, 21-32
Texas, 5, 7, 78-82, 209-212
Thailand, 287-293
Thematic mapper, 73-74
Thermal Alteration Index, 23
Thermal-maturation studies, 12-21
  case studies, 21-32
Thermodynamic equation of state for
    gas, 298-312
Three Mile Island, 337-338
Time-temperature interaction, 16-17,
    35-40
Tobago
  independent oil companies and,
    478-480
  United Nations Surveys, 672-673
Trading companies, as energy
    development financiers, 491-497
Training
  Conventional Energy Training
    Project, 648-650
  petroleum training center, 631

686 Shallow Oil and Gas Resources

simulators and, 621–631
training rigs, 625–626
United Nations projects, 670
Trinidad
  independent oil companies, 478–480
  seeps, 48
  United Nations projects, 672–673
Turkey, 672

Uganda, 55
Uinta Basin, 14
UNITAR, 561
United Kingdom, 347
United Nations
  agencies for assistance, 651–677
  UNITAR, 561
  United Nations Assistance
    Programme in Petroleum,
    651–677
United States
  Agency for Petroleum Development,
    645–650
  Alabama
    Gilbertown field, 215–221
    South Carlton field, 221–225
  Alaska
    Prudhoe Bay, 7
    source-rock studies, 21–24
  Appalachia
    barefoot wells, 367–371
    shallow oil, 524
    shut-in gas, 337
  bitumen, 155
  California, 253–254, 259
  Colorado, 74, 181–192
  Denver Basin, 181–192
  drilling
    activity, 164–169
    depth statistics, 432–433
  exploration, 162–164
  federal lands, 8
  landform analysis, 60, 63
  natural gas,
    consumption, 567–568
    distribution, 145, 149, 152
    policies, 347–348, 361–365
  Nebraska, 181–192
  Oklahoma
    anticlines, 6
    Osage Indian Reservation, 593–601
  Pennsylvania, 5
  price controls, 8–9
  satellite remote sensing, 69–72
  shallow wells, 425
  Texas
    first wells in, 5
    Navarro formation, 209–212

Permian Basin, 7
satellite remote sensing, 80–82
Spreen Ranch Prospect, 78–79
subsidence, 253–254
universities, and technology transfer,
  639–644
Utah, 73
wildcatters, 500
United States Geological Survey, 8, 117
United States Agency for International
  Development, 645–650
University of Oklahoma, 298
  Petroleum Data System, 99–100,
    135, 149
  technology transfer, 640–642
U.S.S.R.
  bitumen, 155
  first oil wells, 5
  natural gas consumption, 567–568
Utah, 73

Van Krevelen diagram, 18, 23–29
Venezuela
  first wells in, 5
  landform analysis, 60
  seeps, 48
  subsidence, 253–254, 259
Vitrinite reflectance, 21

Water disposal, case studies, 215–225
Water-flooding, 367
Water influx, mathematical models,
  316, 322–326
Weeks, Lewis, 9
Wells. See also Shallow oil and gas.
  naturally flowing, 227–234
  production rates, 322–326 for gas
    equation
  stimulation, 235–241, 368–371
West Germany, 346
Wildcat wells
  exploration, 62–64
  risk, 499–502
  United States, 425–426
Williston Basin, 86–88
World Bank, 461, 499–502
  exploration, 457–458, 511
  loans, 176, 196, 204
  OPIC, 503
World wars, 6–7, 62

Yemen, 55
Yugoslavia, 55

Zambia, 675

A000013110178